Audio Source Separation and Speech Enhancement

Audio Source Separation and Speech Enhancement

Edited by

Emmanuel Vincent
Inria
France

Tuomas Virtanen
Tampere University of Technology
Finland

Sharon Gannot
Bar-Ilan University
Israel

Registered Offices
John Wiley & Sons, Inc., 111 River Street, Hoboken, NJ 07030, USA
John Wiley & Sons Ltd, The Atrium, Southern Gate, Chichester, West Sussex, PO19 8SQ, UK

Editorial Office
The Atrium, Southern Gate, Chichester, West Sussex, PO19 8SQ, UK

For details of our global editorial offices, customer services, and more information about Wiley products visit us at www.wiley.com.

Wiley also publishes its books in a variety of electronic formats and by print-on-demand. Some content that appears in standard print versions of this book may not be available in other formats.

Library of Congress Cataloging-in-Publication Data

Names: Vincent, Emmanuel (Research scientist), editor. | Virtanen, Tuomas, editor. | Gannot, Sharon, editor.
Title: Audio source separation and speech enhancement / edited by Emmanuel Vincent, Tuomas Virtanen, Sharon Gannot.
Description: Hoboken, NJ : John Wiley & Sons, 2018. | Includes bibliographical references and index. |
Identifiers: LCCN 2018013163 (print) | LCCN 2018021195 (ebook) | ISBN 9781119279884 (pdf) | ISBN 9781119279914 (epub) | ISBN 9781119279891 (cloth)
Subjects: LCSH: Speech processing systems. | Automatic speech recognition.
Classification: LCC TK7882.S65 (ebook) | LCC TK7882.S65 .A945 2018 (print) | DDC 006.4/54–dc23
LC record available at https://lccn.loc.gov/2018013163

Cover Design: Wiley
Cover Images: © 45RPM/iStockphoto;
© franckreporter/iStockphoto

Set in 10/12pt WarnockPro by SPi Global, Chennai, India
Printed in Singapore by C.O.S. Printers Pte Ltd

10 9 8 7 6 5 4 3 2 1

Contents

List of Authors

Shoko Araki
NTT Communication Science
Laboratories
Japan

Roland Badeau
Institut Mines-Télécom
France

Alessio Brutti
Fondazione Bruno Kessler
Italy

Israel Cohen
Technion
Israel

Simon Doclo
Carl von Ossietzky-Universität
Oldenburg
Germany

Jun Du
University of Science and Technology
of China
China

Zhiyao Duan
University of Rochester
NY
USA

Cédric Févotte
CNRS
France

Sharon Gannot
Bar-Ilan University
Israel

Tian Gao
University of Science and Technology
of China
China

Timo Gerkmann
Universität Hamburg
Germany

Emanuël A.P. Habets
International Audio Laboratories
Erlangen
Germany

Elior Hadad
Bar-Ilan University
Israel

Hirokazu Kameoka
The University of Tokyo
Japan

Walter Kellermann
Friedrich-Alexander Universität
Erlangen-Nürnberg
Germany

Zbyněk Koldovský
Technical University of Liberec
Czech Republic

Dorothea Kolossa
Ruhr-Universität Bochum
Germany

Antoine Liutkus
Inria
France

Michael I. Mandel
City University of New York
NY
USA

Erik Marchi
Technische Universität München
Germany

Shmulik Markovich-Golan
Bar-Ilan University
Israel

Daniel Marquardt
Carl von Ossietzky-Universität
Oldenburg
Germany

Rainer Martin
Ruhr-Universität Bochum
Germany

Nasser Mohammadiha
Chalmers University of Technology
Sweden

Gautham J. Mysore
Adobe Research
CA
USA

Tomohiro Nakatani
NTT Communication Science
Laboratories
Japan

Patrick A. Naylor
Imperial College London
UK

Maurizio Omologo
Fondazione Bruno Kessler
Italy

Alexey Ozerov
Technicolor
France

Bryan Pardo
Northwestern University
IL
USA

Pasi Pertilä
Tampere University of Technology
Finland

Gaël Richard
Institut Mines-Télécom
France

Hiroshi Sawada
NTT Communication Science
Laboratories
Japan

Paris Smaragdis
University of Illinois at
Urbana-Champaign
IL
USA

Piergiorgio Svaizer
Fondazione Bruno Kessler
Italy

Emmanuel Vincent
Inria
France

Tuomas Virtanen
Tampere University of Technology
Finland

Shinji Watanabe
Johns Hopkins University
MD
USA

Felix Weninger
Nuance Communications
Germany

Preface

Source separation and speech enhancement are some of the most studied technologies in audio signal processing. Their goal is to extract one or more source signals of interest from an audio recording involving several sound sources. This problem arises in many everyday situations. For instance, spoken communication is often obscured by concurrent speakers or by background noise, outdoor recordings feature a variety of environmental sounds, and most music recordings involve a group of instruments. When facing such scenes, humans are able to perceive and listen to individual sources so as to communicate with other speakers, navigate in a crowded street or memorize the melody of a song. Source separation and speech enhancement technologies aim to empower machines with similar abilities.

These technologies are already present in our lives today. Beyond "clean" single-source signals recorded with close microphones, they allow the industry to extend the applicability of speech and audio processing systems to multi-source, reverberant, noisy signals recorded with distant microphones. Some of the most striking examples include hearing aids, speech enhancement for smartphones, and distant-microphone voice command systems. Current technologies are expected to keep improving and spread to many other scenarios in the next few years.

Traditionally, *speech enhancement* has referred to the problem of segregating speech and background noise, while *source separation* has referred to the segregation of multiple speech or audio sources. Most textbooks focus on one of these problems and on one of three historical approaches, namely sensor array processing, computational auditory scene analysis, or independent component analysis. These communities now routinely borrow ideas from each other and other approaches have emerged, most notably based on deep learning.

This textbook is the first to provide a comprehensive overview of these problems and approaches by presenting their shared foundations and their differences using common language and notations. Starting with prerequisites (Part I), it proceeds with single-channel separation and enhancement (Part II), multichannel separation and enhancement (Part III), and applications and perspectives (Part IV). Each chapter provides both introductory and advanced material.

We designed this textbook for people in academia and industry with basic knowledge of signal processing and machine learning. Thanks to its comprehensiveness, we hope it will help students select a promising research track, researchers leverage the acquired cross-domain knowledge to design improved techniques, and engineers and developers choose the right technology for their application scenario. We also hope that it will be useful for practitioners from other fields (e.g., acoustics, multimedia, phonetics, musicology) willing to exploit audio source separation or speech enhancement as a pre-processing tool for their own needs.

May 2017   *Emmanuel Vincent, Tuomas Virtanen, and Sharon Gannot*

Acknowledgment

We would like to thank all the chapter authors, as well as the following people who helped with proofreading: Sebastian Braun, Yaakov Buchris, Emre Cakir, Aleksandr Diment, Dylan Fagot, Nico Gößling, Tomoki Hayashi, Jakub Janský, Ante Jukić, Václav Kautský, Martin Krawczyk-Becker, Simon Leglaive, Bochen Li, Min Ma, Paul Magron, Zhong Meng, Gaurav Naithani, Zhaoheng Ni, Aditya Arie Nugraha, Sanjeel Parekh, Robert Rehr, Lea Schönherr, Georgina Tryfou, Ziteng Wang, and Mehdi Zohourian

May 2017   *Emmanuel Vincent, Tuomas Virtanen, and Sharon Gannot*

Notations

Linear algebra

x	scalar
$\mathbf{x}$	vector
$[x_i]_i$	vector with entries x_i
$(\mathbf{x})_i$	ith entry of vector $\mathbf{x}$
$\mathbf{0}_I$	$I \times 1$ vector of zeros
$\mathbf{1}_I$	$I \times 1$ vector of ones
$\mathbf{X}$	matrix
$[x_{ij}]_{ij}$	matrix with entries x_{ij}
$(\mathbf{X})_{ij}$	(i, j)th entry of matrix $\mathbf{X}$
$\mathbf{I}_I$	$I \times I$ identity matrix
$\mathcal{X}$	tensor/array (with three or more dimensions) or set
$\{x_{ijk}\}_{ijk}$	tensor with entries x_{ijk}
$\mathrm{Diag}(\mathbf{x})$	diagonal matrix whose entries are those of vector $\mathbf{x}$
$\mathbf{X} \circ \mathbf{Y}$	entrywise product of matrices $\mathbf{X}$ and $\mathbf{Y}$
$\mathrm{tr}(\mathbf{X})$	trace of matrix $\mathbf{X}$
$\det(\mathbf{X})$	determinant of matrix $\mathbf{X}$
$\mathbf{x}^T$	transpose of vector $\mathbf{x}$
$\mathbf{x}^H$	conjugate-transpose of vector $\mathbf{x}$
x^*	conjugate of scalar x
$\Re(x)$	real part of scalar x
J	imaginary unit

Statistics

$p(x)$	probability distribution of continuous random variable x
$p(x \mid y)$	conditional probability distribution of x given y
$P(x)$	probability value of discrete random variable x
$P(x \mid y)$	conditional probability value of x given y
$\mathbb{E}\{x\}$	expectation of random variable x
$\mathbb{E}\{x \mid y\}$	conditional expectation of x
$\mathbb{H}\{x\}$	entropy of random variable x
$\mathcal{N}(\mathbf{x} \mid \boldsymbol{\mu}, \boldsymbol{\Sigma})$	real Gaussian distribution with mean $\boldsymbol{\mu}$ and covariance $\boldsymbol{\Sigma}$
$\mathcal{N}_c(\mathbf{x} \mid \boldsymbol{\mu}, \boldsymbol{\Sigma})$	complex Gaussian distribution with mean $\boldsymbol{\mu}$ and covariance $\boldsymbol{\Sigma}$
$\hat{x}$	estimated value of random variable x (e.g., first-order statistics)

σ_x^2	variance of random variable x
$\hat{\sigma}_x^2$	estimated second-order statistics of random variable x
$\mathbf{\Sigma_x}$	autocovariance of random vector $\mathbf{x}$
$\hat{\mathbf{\Sigma}}_{\mathbf{x}}$	estimated second-order statistics of random vector $\mathbf{x}$
$\mathbf{\Sigma_{xy}}$	covariance of random vectors $\mathbf{x}$ and $\mathbf{y}$
$\hat{\mathbf{\Sigma}}_{\mathbf{xy}}$	estimated second-order statistics of random vectors $\mathbf{x}$ and $\mathbf{y}$
$\mathcal{C}^{\text{cost}}(\boldsymbol{\theta})$	cost function to be minimized w.r.t. the vector of parameters $\boldsymbol{\theta}$
$\mathcal{M}^{\text{objective}}(\boldsymbol{\theta})$	objective function to be maximized w.r.t. the vector of parameters $\boldsymbol{\theta}$
$\mathcal{Q}(\boldsymbol{\theta}, \cdot)$	auxiliary function to be minimized or maximized, depending on the context

Common indexes

I	number of microphones or channels
i	microphone or channel index in $\{1, \dots, I\}$
J	number of sources
j	source index in $\{1, \dots, J\}$
T	number of time-domain samples
t	sample index in $\{0, \dots, T - 1\}$
L	time-domain filter length
τ	tap index in $\{0, \dots, L - 1\}$
N	number of time frames
n	time frame index in $\{0, \dots, N - 1\}$
F	number of frequency bins
f	frequency bin index in $\{0, \dots, F - 1\}$
ν_f	frequency in Hz corresponding to frequency bin f
$x(t)$	time-domain signal x
$x(n,f)$	complex-valued STFT coefficient of signal x

Signals

x_i	input signal recorded at microphone i
$\mathbf{x}$	$I \times 1$ multichannel input signal, e.g. $\mathbf{x}(t) = [x_1(t), \dots, x_I(t)]^T$
$\mathbf{X}$	matrix of input signals, e.g. $\mathbf{X} = [x_i(t)]_{it}$ or $\mathbf{X} = [x(n,f)]_{fn}$
$\|\mathbf{X}\|$	input magnitude spectrogram, i.e. $\|\mathbf{X}\| = [\|x(n,f)\|]_{fn}$
$\mathcal{X}$	tensor/array/set of input signals, e.g. $\mathcal{X} = [x_i(n,f)]_{ifn}$
s_j	point source signal
$\mathbf{s}$	$J \times 1$ vector of source signals, e.g. $\mathbf{s}(t) = [s_1(t), \dots, s_J(t)]^T$
$\mathbf{S}$	matrix of source signals, e.g. $\mathbf{S} = [s_j(t)]_{jt}$
c_{ij}	spatial image of source j as recorded on microphone i
$\mathbf{c}_j$	$I \times 1$ spatial image of source j on all microphones
$\mathcal{C}$	tensor/array/set of spatial source image signals, e.g. $\mathcal{C} = [c_{ij}(n,f)]_{ijfn}$
a_{ij}	acoustic impulse response (or transfer function) from source j to microphone i
$\mathbf{a}_j$	$I \times 1$ vector of acoustic impulse responses (or transfer functions) from source j, mixing vector

A	$I \times J$ matrix of acoustic impulse responses (or transfer functions), mixing matrix
u	$I \times 1$ noise signal

Filters

$\star$	convolution operator
w	single-output single-channel filter (mask), e.g. $\widehat{s} = w^*x$
w	single-output multichannel filter (beamformer), e.g. $\widehat{s} = \mathbf{w}^H\mathbf{x}$
W	multiple-output multichannel filter, e.g. $\widehat{\mathbf{s}} = \mathbf{W}^H\mathbf{x}$

Nonnegative matrix factorization

$\mathbf{b}_k$	kth nonnegative basis spectrum
B	matrix of nonnegative basis spectra
$h_k(n)$	kth activation coefficient in time frame n
$\mathbf{h}(n)$	vector of activation coefficients in time frame n
H	matrix of activation coefficients

Deep learning

H	number of layers
h	layer index in $\{1, \ldots, H\}$
K_h	number of neurons in layer h
k	neuron index in $\{1, \ldots, K_h\}$
$\mathbf{Z}_h$	matrix of weights and biases in layer h
g_h	activation function in layer h
$g_{\mathcal{Z}}$	multivariate nonlinear function encoded by the full DNN

Geometry

$\mathbf{m}_i$	3D location of microphone i with respect to the array origin
$\ell_{ii'}$	distance between microphones i and i'
$\mathbf{p}_j$	3D location of source j with respect to the array origin
r_{ij}	distance between source j and microphone i
θ_j	azimuth of source j
φ_j	elevation of source j
c	speed of sound in air
$\Delta_{ii'j}$	time difference of arrival of source j between microphones i and i'

Acronyms

AR	autoregressive
ASR	automatic speech recognition
BSS	blind source separation
CASA	computational auditory scene analysis
DDR	direct-to-diffuse ratio
DFT	discrete Fourier transform
DNN	deep neural network
DOA	direction of arrival
DRNN	deep recurrent neural network
DRR	direct-to-reverberant ratio
DS	delay-and-sum
ERB	equivalent rectangular bandwidth
EM	expectation-maximization
EUC	Euclidean
FD-ICA	frequency-domain independent component analysis
FIR	finite impulse response
GCC	generalized cross-correlation
GCC-PHAT	generalized cross-correlation with phase transform
GMM	Gaussian mixture model
GSC	generalized sidelobe canceler
HMM	hidden Markov model
IC	interchannel (or interaural) coherence
ICA	independent component analysis
ILD	interchannel (or interaural) level difference
IPD	interchannel (or interaural) phase difference
ITD	interchannel (or interaural) time difference
IVA	independent vector analysis
IS	Itakura–Saito
KL	Kullback–Leibler
LCMV	linearly constrained minimum variance
LSTM	long short-term memory
MAP	maximum a posteriori
MFCC	Mel-frequency cepstral coefficient
ML	maximum likelihood
MM	majorization-minimization

MMSE	minimum mean square error
MSC	magnitude squared coherence
MSE	mean square error
MVDR	minimum variance distortionless response
MWF	multichannel Wiener filter
NMF	nonnegative matrix factorization
PLCA	probabilistic latent component analysis
RNN	recurrent neural network
RT60	reverberation time
RTF	relative transfer function
SAR	signal-to-artifacts ratio
SDR	signal-to-distortion ratio
SINR	signal-to-interference-plus-noise ratio
SIR	signal-to-interference ratio
SNR	signal-to-noise ratio
SPP	speech presence probability
SRP	steered response power
SRP-PHAT	steered response power with phase transform
SRR	signal-to-reverberation ratio
STFT	short-time Fourier transform
TDOA	time difference of arrival
VAD	voice activity detection
VB	variational Bayesian

About the Companion Website

This book is accompanied by a companion website:

https://project.inria.fr/ssse/

The website includes:

- Implementations of algorithms
- Audio samples

Part I

Prerequisites

1

Introduction

Emmanuel Vincent, Sharon Gannot, and Tuomas Virtanen

Source separation and speech enhancement are core problems in the field of audio signal processing, with applications to speech, music, and environmental audio. Research in this field has accompanied technological trends, such as the move from landline to mobile or hands-free phones, the gradual replacement of stereo by 3D audio, and the emergence of connected devices equipped with one or more microphones that can execute audio processing tasks which were previously regarded as impossible. In this short introductory chapter, after a brief discussion of the application needs in Section 1.1, we define the problems of source separation and speech enhancement and introduce relevant terminology regarding the scenarios and the desired outcome in Section 1.2. We then present the general processing scheme followed by most source separation and speech enhancement approaches and categorize these approaches in Section 1.3. Finally, we provide an outline of the book in Section 1.4.

1.1 Why are Source Separation and Speech Enhancement Needed?

The problems of source separation and speech enhancement arise from several application needs in the context of speech, music, and environmental audio processing.

Real-world speech signals are often contaminated by interfering speakers, environmental noise, and/or reverberation. These phenomena deteriorate speech quality and, in adverse scenarios, speech intelligibility and automatic speech recognition (ASR) performance. Source separation and speech enhancement are therefore required in such scenarios. For instance, spoken communication over mobile phones or hands-free systems requires the separation or enhancement of the near-end speaker's voice with respect to interfering speakers and environmental noises before it is transmitted to the far-end listener. Conference call systems or hearing aids face the same problem, except that several speakers may be considered as targets. Source separation and speech enhancement are also crucial preprocessing steps for robust distant-microphone ASR, as available in today's personal assistants, car navigation systems, televisions, video game consoles, medical dictation devices, and meeting transcription systems.

Audio Source Separation and Speech Enhancement, First Edition.
Edited by Emmanuel Vincent, Tuomas Virtanen and Sharon Gannot.
© 2018 John Wiley & Sons Ltd. Published 2018 by John Wiley & Sons Ltd.
Companion Website: https://project.inria.fr/ssse/

Finally, they are necessary components in providing humanoid robots, assistive listening devices, and surveillance systems with "super-hearing" capabilities, which may exceed the hearing capabilities of humans.

Besides speech, music and movie soundtracks are another important application area for source separation. Indeed, music recordings typically involve several instruments playing together live or mixed together in a studio, while movie soundtracks involve speech overlapped with music and sound effects. Source separation has been successfully used to upmix mono or stereo recordings to 3D sound formats and/or to remix them. It lies at the core of object-based audio coders, which encode a given recording as the sum of several sound objects that can then easily be rendered and manipulated. It is also useful for music information retrieval purposes, e.g. to transcribe the melody or the lyrics of a song from the separated singing voice.

This is an emerging research field with many real-life applications concerning the analysis of general sound scenes, involving the detection of sound events, their localization and tracking, and the inference of the acoustic environment properties.

1.2 What are the Goals of Source Separation and Speech Enhancement?

The goal of source separation and speech enhancement can be defined in layman's terms as that of recovering the signal of one or more sound sources from an observed signal involving other sound sources and/or reverberation. This definition turns out to be ambiguous. In order to address the ambiguity, the notion of source and the process leading to the observed signal must be characterized more precisely. In this section and in the rest of this book we adopt the general notations defined on p. xxv–xxvii.

1.2.1 Single-Channel vs. Multichannel

Let us assume that the observed signal has I *channels* indexed by $i \in \{1, \dots, I\}$. By channel, we mean the output of one microphone in the case when the observed signal has been recorded by one or more microphones, or the input of one loudspeaker in the case when it is destined to be played back on one or more loudspeakers.[1] A signal with $I = 1$ channels is called *single-channel* and is represented by a scalar $x(t)$, while a signal with $I > 1$ channels is called *multichannel* and is represented by an $I \times 1$ vector $\mathbf{x}(t)$. The explanation below employs multichannel notation, but is also valid in the single-channel case.

1.2.2 Point vs. Diffuse Sources

Furthermore, let us assume that there are J sound *sources* indexed by $j \in \{1, \dots, J\}$. The word "source" can refer to two different concepts. A *point source* such as a human

---

1 This is the usual meaning of "channel" in the field of professional and consumer audio. In the field of telecommunications and, by extension, in some speech enhancement papers, "channel" refers to the distortions (e.g., noise and reverberation) occurring when transmitting a signal instead. The latter meaning will not be employed hereafter.

speaker, a bird, or a loudspeaker is considered to emit sound from a single point in space. It can be represented as a single-channel signal. A *diffuse source* such as a car, a piano, or rain simultaneously emits sound from a whole region in space. The sounds emitted from different points of that region are different but not always independent of each other. Therefore, a diffuse source can be thought of as an infinite collection of point sources. The estimation of the individual point sources in this collection can be important for the study of vibrating bodies, but it is considered irrelevant for source separation or speech enhancement. A diffuse source is therefore typically represented by the corresponding signal recorded at the microphone(s) and it is processed as a whole.

1.2.3 Mixing Process

The mixing process leading to the observed signal can generally be expressed in two steps. First, each single-channel point source signal $s_j(t)$ is transformed into an $I \times 1$ source *spatial image* signal (Vincent *et al.*, 2012) $\mathbf{c}_j(t) = [c_{1j}(t), \ldots, c_{Ij}(t)]^T$ by means of a possibly nonlinear spatialization operation. This operation can describe the acoustic propagation from the point source to the microphone(s), including reverberation, or some artificial mixing effects. Diffuse sources are directly represented by their $I \times 1$ spatial images $\mathbf{c}_j(t)$ instead. Second, the spatial images of all sources are summed to yield the observed $I \times 1$ signal $\mathbf{x}(t) = [x_1(t), \ldots, x_I(t)]^T$ called the *mixture*:

$$\mathbf{x}(t) = \sum_{j=1}^{J} \mathbf{c}_j(t). \tag{1.1}$$

This summation is due to the superposition of the sources in the case of microphone recording or to explicit summation in the case of artificial mixing. This implies that the spatial image of each source represents the contribution of the source to the mixture signal. A schematic overview of the mixing process is depicted in Figure 1.1. More specific details are given in Chapter 3.

Note that *target* sources, *interfering* sources, and *noise* are treated in the same way in this formulation. All these signals can be either point or diffuse sources. The choice of target sources depends on the use case. Also, the distinction between interfering sources and noise may or may not be relevant depending on the use case. In the context of speech processing, these terms typically refer to undesired speech vs. nonspeech sources, respectively. In the context of music or environmental sound processing, this distinction is most often irrelevant and the former term is preferred to the latter.

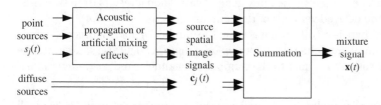

Figure 1.1 General mixing process, illustrated in the case of $J = 4$ sources, including three point sources and one diffuse source, and $I = 2$ channels.

In the following, we assume that all signals are digital, meaning that the time variable t is discrete. We also assume that quantization effects are negligible, so that we can operate on continuous amplitudes. Regarding the conversion of acoustic signals to analog audio signals and analog signals to digital, see, for example, Havelock *et al.* (2008, Part XII) and Pohlmann (1995, pp. 22–49).

1.2.4 Separation vs. Enhancement

The above mixing process implies one or more distortions of the target signals: interfering sources, noise, reverberation, and echo emitted by the loudspeakers (if any). In this context, *source separation* refers to the problem of extracting one or more target sources while suppressing interfering sources and noise. It explicitly excludes dereverberation and echo cancellation. *Enhancement* is more general, in that it refers to the problem of extracting one or more target sources while suppressing all types of distortion, including reverberation and echo. In practice, though, this term is mostly used in the case when the target sources are speech. In the audio processing literature, these two terms are often interchanged, especially when referring to the problem of suppressing both interfering speakers and noise from a speech signal. Note that, for either source separation or enhancement tasks, the extracted source(s) can be either the spatial image of the source or its direct path component, namely the delayed and attenuated version of the original source signal (Vincent *et al.*, 2012; Gannot *et al.*, 2001).

The problem of echo cancellation is out of the scope of this book. Please refer to Hänsler and Schmidt (2004) for a comprehensive overview of this topic. The problem of source localization and tracking cannot be viewed as a separation or enhancement task, but it is sometimes used as a preprocessing step prior to separation or enhancement, hence it is discussed in Chapter 4. Dereverberation is explored in Chapter 15. The remaining chapters focus on separation and enhancement.

1.2.5 Typology of Scenarios

The general source separation literature has come up with a terminology to characterize the mixing process (Hyvärinen *et al.*, 2001; O'Grady *et al.*, 2005; Comon and Jutten, 2010). A given mixture signal is said to be

- *linear* if the mixing process is linear, and *nonlinear* otherwise;
- *time-invariant* if the mixing process is fixed over time, and *time-varying* otherwise;
- *instantaneous* if the mixing process simply scales each source signal by a different factor on each channel, *anechoic* if it also applies a different delay to each source on each channel, and *convolutive* in the more general case when it results from summing multiple scaled and delayed versions of the sources;
- *overdetermined* if there is no diffuse source and the number of point sources is strictly smaller than the number of channels, *determined* if there is no diffuse source and the number of point sources is equal to the number of channels, and *underdetermined* otherwise.

This categorization is relevant but has limited usefulness in the case of audio. As we shall see in Chapter 3, virtually all audio mixtures are linear (or can be considered so) and convolutive. The over- vs. underdetermined distinction was motivated by the fact that a determined or overdetermined linear time-invariant mixture can be perfectly separated by inverting the mixing system using a linear time-invariant inverse (see Chapter 13). In practice, however, the majority of audio mixtures involve at least one diffuse source (e.g., background noise) or more point sources than channels. Audio source separation and speech enhancement systems are therefore generally faced with underdetermined linear (time-invariant or time-varying) convolutive mixtures.[2]

Recently, an alternative categorization has been proposed based on the amount of prior information available about the mixture signal to be processed (Vincent *et al.*, 2014). The separation problem is said to be

- *blind* when absolutely no information is given about the source signals, the mixing process or the intended application;
- *weakly guided* or *semi-blind* when general information is available about the context of use, e.g. the nature of the sources (speech, music, environmental sounds), the microphone positions, the recording scenario (domestic, outdoor, professional music), and the intended application (hearing aid, speech recognition);
- *strongly guided* when specific information is available about the signal to be processed, e.g. the spatial location of the sources, their activity pattern, the identity of the speakers, or a musical score;
- *informed* when highly precise information about the sources and the mixing process is encoded and transmitted along with the audio.

Although the term "blind" has been extensively used in source separation (see Chapters 4, 10, 11, and 13), strictly blind separation is inapplicable in the context of audio. As we shall see in Chapter 13, certain assumptions about the probability distribution of the sources and/or the mixing process must always be made in practice. Strictly speaking, the term "weakly guided" would therefore be more appropriate. Informed separation is closer to audio coding than to separation and will be briefly covered in Chapter 16. All other source separation and speech enhancement methods reviewed in this book are therefore either weakly or strongly guided.

Finally, the separation or enhancement problem can be categorized depending on the order in which the samples of the mixture signal are processed. It is called *online* when the mixture signal is captured in real time by small blocks of a few tens or hundred samples and each block must be processed given past blocks only, or few future blocks introducing tolerated latency. On the contrary, it is called *offline* or *batch* when the recording has been completed and it is processed as a whole, using both past and future samples to estimate a given sample of the sources.

---

2 Certain authors call mixtures for which the number of point sources is equal to (resp. strictly smaller than) the number of channels as determined (resp. overdetermined) even when there is a diffuse noise source. Perfect separation of such mixtures cannot be achieved using time-invariant filtering anymore: it requires a time-varying separation filter, similarly to underdetermined mixtures. Indeed, a time-invariant filter can cancel the interfering sources and reduce the noise, but it cannot cancel the noise perfectly. We prefer the above definition of "determined" and "overdetermined", which matches the mathematical definition of these concepts for systems of linear equations and has a more direct implication on the separation performance achievable by linear time-invariant filtering.

1.2.6   Evaluation

Using current technology, source separation and dereverberation are rarely perfect in real-life scenarios. For each source, the estimated source or source spatial image signal can differ from the true target signal in several ways, including (Vincent *et al.*, 2006; Loizou, 2007)

- *distortion* of the target signal, e.g. lowpass filtering, fluctuating intensity over time;
- residual interference or noise from the other sources;
- *"musical noise" artifacts*, i.e. isolated sounds in both frequency and time similar to those generated by a lossy audio codec at a very low bitrate.

The assessment of these distortions is essential to compare the merits of different algorithms and understand how to improve their performance.

Ideally, this assessment should be based on the performance of the tested source separation or speech enhancement method for the desired application. Indeed, the importance of various types of distortion depends on the specific application. For instance, some amount of distortion of the target signal which is deemed acceptable when listening to the separated signals can lead to a major drop in the speech recognition performance. Artifacts are often greatly reduced when the separated signals are remixed together in a different way, while they must be avoided at all costs in hearing aids. Standard performance metrics are typically available for each task, some of which will be mentioned later in this book.

When the desired application involves listening to the separated or enhanced signals or to a remix, sound quality and, whenever relevant, speech intelligibility should ideally be assessed by means of a subjective listening test (ITU-T, 2003; Emiya *et al.*, 2011; ITU-T, 2016). Contrary to a widespread belief, a number of subjects as low as ten can sometimes suffice to obtain statistically significant results. However, data selection and subject screening are time-consuming. Recent attempts with crowdsourcing are a promising way of making subjective testing more convenient in the near future (Cartwright *et al.*, 2016). An alternative approach is to use objective separation or dereverberation metrics. Table 1.1 provides an overview of some commonly used metrics. The so-called *PESQ* metric, the segmental signal-to-noise ratio (SNR), and the *signal-to-distortion ratio* (SDR) measure the overall estimation error, including the three types of distortion listed above. The so-called *STOI* index is more related to speech intelligibility by humans, and the log-likelihood ratio and cepstrum distance to ASR by machines. The *signal-to-interference ratio* (SIR) and the *signal-to-artifacts ratio* (SAR) aim to assess separately the latter two types of distortion listed above. The segmental SNR, SDR, SIR, and SAR are expressed in decibels (dB), while PESQ and STOI are expressed on a perceptual scale. More specific metrics will be reviewed later in the book.

A natural question that arises once the metrics have been defined is: what is the best performance possibly achievable for a given mixture signal? This can be used to assess the difficulty of solving the source separation or speech enhancement problem in a given scenario and the room left for performance improvement as compared to current systems. This question can be answered using *oracle* or *ideal* estimators based on the knowledge of the true source or source spatial image signals (Vincent *et al.*, 2007).

Table 1.1 Evaluation software and metrics.

Software	Implemented metrics
ITU-T (2001)	PESQ
Taal *et al.* (2011)[a]	STOI
Loizou (2007)[b]	Segmental SNR
	Log-likelihood ratio
	Cepstrum distance
BSS Eval (Vincent *et al.*, 2006)[c]	SDR
	SIR
	SAR
Falk *et al.* (2010)	Speech to reverberation modulation energy ratio

a) http://amtoolbox.sourceforge.net/doc/speech/taal2011.php.
b) http://www.crcpress.com/product/isbn/9781466504219.
c) http://bass-db.gforge.inria.fr/bss_eval/.

1.3 How can Source Separation and Speech Enhancement be Addressed?

Now that we have defined the goals of source separation and speech enhancement, let us turn to how they can be addressed.

1.3.1 General Processing Scheme

Many different approaches to source separation and speech enhancement have been proposed in the literature. The vast majority of approaches follow the general processing scheme depicted in Figure 1.2, which applies to both single-channel and multichannel scenarios. The time-domain mixture signal $\mathbf{x}(t)$ is represented in the time-frequency domain (see Chapter 2). A model of the complex-valued time-frequency coefficients of the mixture $\mathbf{x}(n,f)$ and the sources $s_j(n,f)$ (resp. the source spatial images $\mathbf{c}_j(n,f)$) is built. The choice of model is motivated by the general prior information about

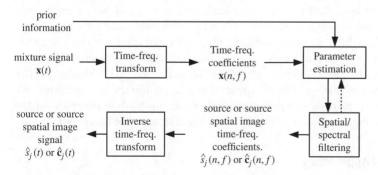

Figure 1.2 General processing scheme for single-channel and multichannel source separation and speech enhancement.

the scenario (see Section 1.2.5). The model parameters are estimated from $\mathbf{x}(n,f)$ or from separate training data according to a certain criterion. Additional specific prior information can be used to help parameter estimation whenever available. Given these parameters, a time-varying single-output (resp. multiple-output) complex-valued filter is derived and applied to the mixture $\mathbf{x}(n,f)$ in order to obtain an estimate of the complex-valued time-frequency coefficients of the sources $\hat{s}_j(n,f)$ (resp. the source spatial images $\hat{\mathbf{c}}_j(n,f)$). Finally, the time-frequency transform is inverted, yielding time-domain source estimates $\hat{s}_j(t)$ (resp. source spatial image estimates $\hat{\mathbf{c}}_j(t)$).

1.3.2 Converging Historical Trends

The various approaches proposed in the literature differ by the choice of model, the parameter estimation algorithm, and the derivation of the separation or enhancement filter. Research has followed three historical paths. First, microphone array processing emerged from the theory of sensor array processing for telecommunications and focused mostly on the localization and enhancement of speech in noisy or reverberant environments. Second, the concepts of independent component analysis (ICA) and nonnegative matrix factorization (NMF) gave birth to a stream of blind source separation (BSS) methods aiming to address "cocktail party" scenarios (as coined by Cherry (1953)) involving several sound sources mixed together. Third, attempts to implement the sound segregation properties of the human ear (Bregman, 1994) in a computer gave rise to computational auditory scene analysis (CASA) methods. These paths have converged in the last decade and they are hardly distinguishable anymore. As a matter of fact, virtually all source separation and speech enhancement methods rely on modeling the *spectral* properties of the sources, i.e. their distribution of energy over time and frequency, and/or their *spatial* properties, i.e. the relations between channels over time.

Most books and surveys about audio source separation and speech enhancement so far have focused on a single point of view, namely microphone array processing (Gay and Benesty, 2000; Brandstein and Ward, 2001; Loizou, 2007; Cohen *et al.*, 2010), CASA (Divenyi, 2004; Wang and Brown, 2006), BSS (O'Grady *et al.*, 2005; Makino *et al.*, 2007; Virtanen *et al.*, 2015), or machine learning (Vincent *et al.*, 2010, 2014). These are complemented by books on general sensor array processing and BSS (Hyvärinen *et al.*, 2001; Van Trees, 2002; Cichocki *et al.*, 2009; Haykin and Liu, 2010; Comon and Jutten, 2010), which do not specifically focus on speech and audio, and books on general speech processing (Benesty *et al.*, 2007; Wölfel and McDonough, 2009; Virtanen *et al.*, 2012; Li *et al.*, 2015), which do not specifically focus on separation and enhancement. A few books and surveys have attempted to cross the boundaries between these points of view (Benesty *et al.*, 2005; Cohen *et al.*, 2009; Gannot *et al.*, 2017; Makino, 2018), but they do not cover all state-of-the-art approaches and all application scenarios. We designed this book to provide the most comprehensive, up-to-date overview of the state of the art and allow readers to acquire a wide understanding of these topics.

1.3.3 Typology of Approaches

With the merging of the three historical paths introduced above, a new categorization of source separation and speech enhancement methods has become necessary. One of the most relevant ones today is based on the use of training data to estimate the model parameters and on the nature of this data. This categorization differs from the one in Section 1.2.5: it does not relate to the problem posed, but to the way it is solved. Both categorizations are essentially orthogonal. We distinguish four categories of approaches:

- *learning-free* methods do not rely on any training data: all parameters are either fixed manually by the user or estimated from the test mixture $\mathbf{x}(n,f)$ (e.g., frequency-domain ICA in Section 13.2);
- *unsupervised source modeling* methods train a model for each source from unannotated isolated signals of that source type, i.e. without using any information about each training signal besides the source type (e.g., so-called "supervised NMF" in Section 8.1.3);
- *supervised source modeling* methods train a model for each source from annotated isolated signals of that source type, i.e. using additional information about each training signal (e.g., isolated notes annotated with pitch information in the case of music, see Section 16.2.2.1);
- *separation based training* methods (e.g., deep neural network (DNN) based methods in Section 7.3) train a separation mechanism or jointly train models for all sources from mixture signals given the underlying true source signals.

In all cases, development data whose conditions are similar to the test mixture can be used to tune a small number of hyperparameters. Certain methods borrow ideas from several categories of approaches. For instance, "semi-supervised" NMF in Section 8.1.4 is halfway between learning-free and unsupervised source modeling based separation.

Other terms were used in the literature, such as generative vs. discriminative methods. We do not use these terms in the following and prefer the finer-grained categories above, which are specific to source separation and speech enhancement.

1.4 Outline

This book is structured in four parts.

Part I introduces the basic concepts of time-frequency processing in Chapter 2 and sound propagation in Chapter 3, and highlights the spectral and spatial properties of the sources. Chapter 4 provides additional background material on source activity detection and localization. These chapters are mostly designed for beginners and can be skipped by experienced readers.

Part II focuses on single-channel separation and enhancement based on the spectral properties of the sources. We first define the concept of spectral filtering in Chapter 5. We then explain how suitable spectral filters can be derived from various models and present algorithms to estimate the model parameters in Chapters 6 to 9. Most of these algorithms are not restricted to a given application area.

Part III addresses multichannel separation and enhancement based on spatial and/ or spectral properties. It follows a similar structure to Part II. We first define the concept of spatial filtering in Chapter 10 and proceed with several models and algorithms in Chapters 11 to 14. Chapter 15 focuses on dereverberation. Again, most of the algorithms reviewed in this part are not restricted to a given application area.

Readers interested in single-channel audio should focus on Part II, while those interested in multichannel audio are advised to read both Parts II and III since most single-channel algorithms can be employed or extended in a multichannel context. In either case, Chapters 5 and 10 must be read first, since they are are prerequisites to the other chapters. Chapters 6 to 9 and 11 to 15 are independent of each other and can be read separately, except Chapter 9 which relies on Chapter 8. Reading all chapters in either part is strongly recommended, however. This will provide the reader with a more complete view of the field and allow him/her to select the most appropriate algorithm or develop a new algorithm for his own use case.

Part IV presents the challenges and opportunities associated with the use of these algorithms in specific application areas: music in Chapter 16, speech in Chapter 17, and hearing instruments in Chapter 18. These chapters are independent of each other and may be skipped or not depending on the reader's interest. We conclude by discussing several research perspectives in Chapter 19.

Bibliography

Benesty, J., Makino, S., and Chen, J. (eds) (2005) *Speech Enhancement*, Springer.

Benesty, J., Sondhi, M.M., and Huang, Y. (eds) (2007) *Springer Handbook of Speech Processing and Speech Communication*, Springer.

Brandstein, M.S. and Ward, D.B. (eds) (2001) *Microphone Arrays: Signal Processing Techniques and Applications*, Springer.

Bregman, A.S. (1994) *Auditory scene analysis: The perceptual organization of sound*, MIT Press.

Cartwright, M., Pardo, B., Mysore, G.J., and Hoffman, M. (2016) Fast and easy crowdsourced perceptual audio evaluation, in *Proceedings of IEEE International Conference on Audio, Speech and Signal Processing*, pp. 619–623.

Cherry, E.C. (1953) Some experiments on the recognition of speech, with one and with two ears. *Journal of the Acoustical Society of America*, **25** (5), 975–979.

Cichocki, A., Zdunek, R., Phan, A.H., and Amari, S. (2009) *Nonnegative Matrix and Tensor Factorizations: Applications to Exploratory Multi-way Data Analysis and Blind Source Separation*, Wiley.

Cohen, I., Benesty, J., and Gannot, S. (2009) *Speech processing in modern communication: Challenges and perspectives*, vol. 3, Springer.

Cohen, I., Benesty, J., and Gannot, S. (eds) (2010) *Speech Processing in Modern Communication: Challenges and Perspectives*, Springer.

Comon, P. and Jutten, C. (eds) (2010) *Handbook of Blind Source Separation, Independent Component Analysis and Applications*, Academic Press.

Divenyi, P. (ed.) (2004) *Speech Separation by Humans and Machines*, Springer.

Emiya, V., Vincent, E., Harlander, N., and Hohmann, V. (2011) Subjective and objective quality assessment of audio source separation. *IEEE Transactions on Audio, Speech, and Language Processing*, **19** (7), 2046–2057.

Falk, T.H., Zheng, C., and Chan, W.Y. (2010) A non-intrusive quality and intelligibility measure of reverberant and dereverberated speech. *IEEE Transactions on Audio, Speech, and Language Processing*, **18** (7), 1766–1774.

Gannot, S., Burshtein, D., and Weinstein, E. (2001) Signal enhancement using beamforming and nonstationarity with applications to speech. *IEEE Transactions on Signal Processing*, **49** (8), 1614–1626.

Gannot, S., Vincent, E., Markovich-Golan, S., and Ozerov, A. (2017) A consolidated perspective on multi-microphone speech enhancement and source separation. *IEEE/ACM Transactions on Audio, Speech, and Language Processing*, **25** (4), 692–730.

Gay, S.L. and Benesty, J. (eds) (2000) *Acoustic Signal Processing for Telecommunication*, Kluwer.

Hänsler, E. and Schmidt, G. (2004) *Acoustic Echo and Noise Control: A Practical Approach*, Wiley.

Havelock, D., Kuwano, S., and Vorländer, M. (eds) (2008) *Handbook of Signal Processing in Acoustics*, vol. 2, Springer.

Haykin, S. and Liu, K.R. (eds) (2010) *Handbook on Array Processing and Sensor Networks*, Wiley.

Hyvärinen, A., Karhunen, J., and Oja, E. (2001) *Independent Component Analysis*, Wiley.

ITU-T (2001) Recommendation P.862. perceptual evaluation of speech quality (PESQ): An objective method for end-to-end speech quality assessment of narrow-band telephone networks and speech codecs.

ITU-T (2003) Recommendation P.835: Subjective test methodology for evaluating speech communication systems that include noise suppression algorithm.

ITU-T (2016) Recommendation P.807. subjective test methodology for assessing speech intelligibility.

Li, J., Deng, L., Haeb-Umbach, R., and Gong, Y. (2015) *Robust Automatic Speech Recognition*, Academic Press.

Loizou, P.C. (2007) *Speech Enhancement: Theory and Practice*, CRC Press.

Makino, S. (ed.) (2018) *Audio Source Separation*, Springer.

Makino, S., Lee, T.W., and Sawada, H. (eds) (2007) *Blind Speech Separation*, Springer.

O'Grady, P.D., Pearlmutter, B.A., and Rickard, S.T. (2005) Survey of sparse and non-sparse methods in source separation. *International Journal of Imaging Systems and Technology*, **15**, 18–33.

Pohlmann, K.C. (1995) *Principles of Digital Audio*, McGraw-Hill, 3rd edn.

Taal, C.H., Hendriks, R.C., Heusdens, R., and Jensen, J. (2011) An algorithm for intelligibility prediction of time-frequency weighted noisy speech. *IEEE Transactions on Audio, Speech, and Language Processing*, **19** (7), 2125–2136.

Van Trees, H.L. (2002) *Optimum Array Processing*, Wiley.

Vincent, E., Araki, S., Theis, F.J., Nolte, G., Bofill, P., Sawada, H., Ozerov, A., Gowreesunker, B.V., Lutter, D., and Duong, N.Q.K. (2012) The Signal Separation Evaluation Campaign (2007–2010): Achievements and remaining challenges. *Signal Processing*, **92**, 1928–1936.

Vincent, E., Bertin, N., Gribonval, R., and Bimbot, F. (2014) From blind to guided audio source separation: How models and side information can improve the separation of sound. *IEEE Signal Processing Magazine*, **31** (3), 107–115.

Vincent, E., Gribonval, R., and Févotte, C. (2006) Performance measurement in blind audio source separation. *IEEE Transactions on Audio, Speech, and Language Processing*, **14** (4), 1462–1469.

Vincent, E., Gribonval, R., and Plumbley, M.D. (2007) Oracle estimators for the benchmarking of source separation algorithms. *Signal Processing*, **87** (8), 1933–1950.

Vincent, E., Jafari, M.G., Abdallah, S.A., Plumbley, M.D., and Davies, M.E. (2010) Probabilistic modeling paradigms for audio source separation, in *Machine Audition: Principles, Algorithms and Systems*, IGI Global, pp. 162–185.

Virtanen, T., Gemmeke, J.F., Raj, B., and Smaragdis, P. (2015) Compositional models for audio processing: Uncovering the structure of sound mixtures. *IEEE Signal Processing Magazine*, **32** (2), 125–144.

Virtanen, T., Singh, R., and Raj, B. (eds) (2012) *Techniques for Noise Robustness in Automatic Speech Recognition*, Wiley.

Wang, D. and Brown, G.J. (eds) (2006) *Computational Auditory Scene Analysis: Principles, Algorithms, and Applications*, Wiley.

Wölfel, M. and McDonough, J. (2009) *Distant Speech Recognition*, Wiley.

2

Time-Frequency Processing: Spectral Properties

Tuomas Virtanen, Emmanuel Vincent, and Sharon Gannot

Many audio signal processing algorithms typically do not operate on raw time-domain audio signals, but rather on *time-frequency representations.* A raw audio signal encodes the amplitude of a sound as a function of time. Its Fourier spectrum represents it as a function of frequency, but does not represent variations over time. A time-frequency representation presents the amplitude of a sound as a function of both time and frequency, and is able to jointly account for its temporal and spectral characteristics (Gröchenig, 2001).

Time-frequency representations are appropriate for three reasons in our context. First, separation and enhancement often require modeling the structure of sound sources. Natural sound sources have a prominent structure both in time and frequency, which can be easily modeled in the time-frequency domain. Second, the sound sources are often mixed convolutively, and this convolutive mixing process can be approximated with simpler operations in the time-frequency domain. Third, natural sounds are more sparsely distributed and overlap less with each other in the time-frequency domain than in the time or frequency domain, which facilitates their separation.

In this chapter we introduce the most common time-frequency representations used for source separation and speech enhancement. Section 2.1 describes the procedure for calculating a time-frequency representation and converting it back to the time domain, using the *short-time Fourier transform* (STFT) as an example. It also presents other common time-frequency representations and their relevance for separation and enhancement. Section 2.2 discusses the properties of sound sources in the time-frequency domain, including sparsity, disjointness, and more complex structures such as harmonicity. Section 2.3 explains how to achieve separation by time-varying filtering in the time-frequency domain. We summarize the main concepts and provide links to other chapters and more advanced topics in Section 2.4.

2.1 Time-Frequency Analysis and Synthesis

In order to operate in the time-frequency domain, there is a need for analysis methods that convert a time-domain signal to the time-frequency domain, and synthesis methods that convert the resulting time-frequency representation back to the time domain after separation or enhancement. For simplicity, we consider the case of a single-channel signal ($I = 1$) and omit the channel index $i = 1$. In the case of multichannel signals, the time-frequency representation is simply obtained by applying the same procedure individually to each channel.

2.1.1 STFT Analysis

Our first example of time-frequency representation is the STFT. It is the most commonly used time-frequency representation for audio source separation and speech enhancement due to its simplicity and low computational complexity in comparison to the available alternatives. Figure 2.1 illustrates the process of segmenting and windowing an audio signal into frames, and calculating the *discrete Fourier transform* (DFT) spectrum in each frame. For visualization the figure uses the magnitude spectrum $|x(n, f)|$ only, and does not present the phase spectrum $\angle x(n, f)$.

The first step in the STFT analysis (Allen, 1977) is the segmentation of the input signal into fixed-length *frames*. Typical frame lengths in audio processing vary between 10 and 120 ms. Frames are usually overlapping – most commonly by 50% or 75%. After segmentation, each frame is multiplied elementwise by a window function. The segmented and windowed signal $x(n, t)$ in frame $n \in \{0, \ldots, N - 1\}$ can be defined as

$$x(n, t) = x(t + t_0 + nM)h_a(t), \quad t \in \{0, \ldots, T - 1\} \tag{2.1}$$

where N is the number of time frames, T is the number of samples in a frame, t_0 positions the first sample of the first frame, M is the hop size between adjacent frames in samples, and $h_a(t)$ is the *analysis window*.

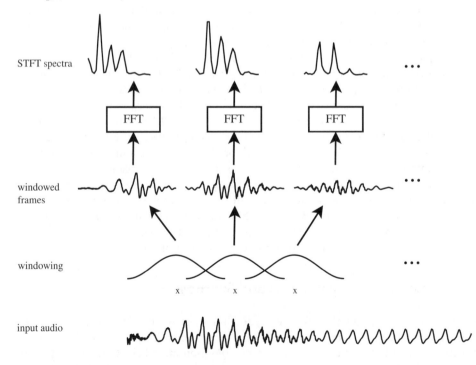

Figure 2.1 STFT analysis.

Windowing with an appropriate analysis window alleviates the spectral leakage which takes place when the DFT is applied to short frames. Spectral leakage means that energy from one frequency bin leaks to neighboring bins: even when the input frame consists of only one sinusoid, the resulting spectrum is nonzero in other bins too. The shorter the frame, the stronger the leakage. Mathematically, this can be modeled as the convolution of signal spectrum with the DFT of the window function.

For practical implementation purposes, window functions have a limited support, i.e. their values are zero outside the interval $[0, T-1]$. Typical window functions such as sine, Hamming, Hann, or Kaiser–Bessel are nonnegative, symmetric, and bell-shaped, so that the value of the window is largest at the center, and decays towards the frame boundaries. The choice of the window function is not critical, as long as a window with reasonable spectral characteristics (sufficiently narrow main lobe, and low level of sidelobes) is used. The choice of the frame length is more important, as discussed in Section 2.1.3.

After windowing, the DFT of each windowed frame is taken, resulting in complex-valued STFT coefficients

$$x(n,f) = \sum_{t=0}^{T-1} x(n,t)e^{-2j\pi tf/F}, \quad f \in \{0, \ldots, F-1\} \tag{2.2}$$

where F is the number of frequency bins, f is the discrete *frequency bin*, and j is the imaginary unit. Typically, $F = T$. We can also set F larger than the frame length T by zero-padding $x(n,t)$ by adding a desired number of zero entries $x(n,t) = 0$, $t \in \{T, \ldots, F-1\}$, to the end of the frame.

We denote the frequency in Hz associated with the positive frequency bins $f \in \{0, \ldots, \lceil F/2 \rceil\}$ as

$$v_f = \frac{f}{F}f_s \tag{2.3}$$

where f_s is the sampling frequency. The STFT coefficients for $f \in \{\lfloor F/2 \rfloor + 1, \ldots, F-1\}$ are complex conjugates of those for $f \in \{\lceil F/2 \rceil - 1, \ldots, 1\}$ and are called negative frequency bins. In the following chapters, the negative frequency bins are often implicitly discarded, nevertheless equations are always written in terms of all frequency bins $f \in \{0, \ldots, F-1\}$ for conciseness. Each term $e^{-2j\pi tf/F}$ is a complex exponential with frequency v_f, thus the DFT calculates the dot product between the windowed frame and complex basis functions with different frequencies.

The STFT has several useful properties for separation and enhancement:

- The frequency scale v_f is a linear function of the frequency bin index f.
- The resulting complex-valued STFT spectrum allows easy treatment of the *phase* $\angle x(n,f)$ and the *magnitude* $|x(n,f)|$ or the *power* $|x(n,f)|^2$ separately.
- The DFT can be efficiently calculated using the fast Fourier transform.
- The DFT is simple to invert, which will be discussed in the next section.

2.1.2 STFT Synthesis

Source separation and speech enhancement methods result in an estimate $\hat{c}(n,f)$ or $\hat{s}(n,f)$ of the target source in the STFT domain. This STFT representation is then transformed back to the time domain, at least if the signals are to be listened to. Note that we omit the source index j for conciseness.

In the STFT synthesis process, the individual STFT frames are first converted to the time domain using the inverse DFT, i.e.

$$\widehat{c}(n,t) = \frac{1}{F} \sum_{f=0}^{F-1} \widehat{c}(n,f) e^{2j\pi tf/F}, \quad t \in \{0,\dots,T-1\}. \tag{2.4}$$

The inverse DFT can also be efficiently calculated.

The STFT domain filtering used to estimate the target source STFT coefficients may introduce artifacts that affect all time samples in a given frame. These artifacts are typically most audible at the frame boundaries, and therefore the frames are again windowed by a *synthesis window* $h_s(t)$ as $\widehat{c}(n,t)h_s(t)$. The synthesis windows are also usually bell-shaped, attenuating the artifacts at the frame boundaries.

Overlapping frames are then summed to obtain the entire time domain signal $\widehat{c}(t)$, as illustrated in Figure 2.2. Together with synthesis windowing, this operation can be written as

$$\widehat{c}(t) = \sum_{n=0}^{N-1} \widehat{c}(n, t - t_0 - nM) h_s(t - t_0 - nM). \tag{2.5}$$

The above procedure is referred to as *weighted overlap-add* (Crochiere, 1980). It modifies the original overlap-add procedure of Allen (1977) by using synthesis windows to avoid artifacts at the frame boundaries. Even though in the above formula the summation extends over all time frames n, with practical window functions $h_s(t)$ that are zero outside the interval $[0, T-1]$, only those terms for which $h_s(t - t_0 - nM) \neq 0$ need to be included in the summation.

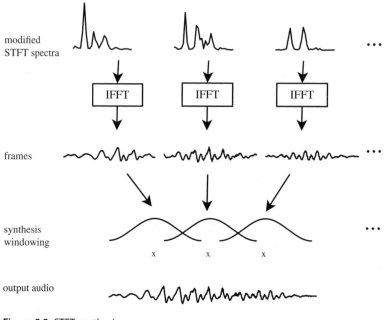

Figure 2.2 STFT synthesis.

The analysis and synthesis windows are typically chosen to satisfy the so-called *perfect reconstruction* property: when the STFT representation is not modified, i.e. $\hat{c}(n,f) = x(n,f)$, the entire analysis-synthesis procedure needs to return the original time-domain signal $\hat{c}(t) = x(t)$. Since each frame is multiplied by both the analysis and synthesis windows, perfect reconstruction is achieved if and only if condition[1] $\sum_{n=0}^{N-1} h_a(t - t_0 - nM)h_s(t - t_0 - nM) = 1$ is satisfied for all t. A commonly used analysis window is the Hamming window (Harris, 1978), which gives perfect reconstruction when no synthesis window is used (i.e., $h_s(t) = 1$). Any such analysis window that gives perfect reconstruction without a synthesis window can be transformed to an analysis-synthesis window pair by taking a square root of it, since effectively the same window becomes used twice, which cancels the square root operation.

2.1.3 Time and Frequency Resolution

Two basic properties of a time-frequency representation are its time and frequency resolution. In general, the time resolution is characterized by the window length and the hop size between adjacent windows, and the frequency resolution is characterized by the center frequencies and the bandwidths of individual frequency bins.

In the case of the STFT, the window length T is fixed over time and the hop size M can be freely chosen, as long as the perfect reconstruction condition is satisfied. The frequency scale v_f is linear so the difference between two adjacent center frequencies $v_{f+1} - v_f = f_s/F$ is constant. The bandwidth of each frequency bin depends on the used analysis window, but is always fixed over frequency and inversely proportional to the window length T. The bandwidth in which the response of a bin falls by 6 dB is on the order of $2f_s/T$ Hz for typical window functions.

From the above we can see that the frequency resolution and the time resolution are inversely proportional to each other. When the time resolution is high, the frequency resolution is low, and vice versa. It is possible to decrease the frequency difference between adjacent frequency bins by increasing the number of frequency bins F in (2.2). This operation called *zero padding* is simply achieved by concatenating a sequence of zeros after each windowed frame before calculating the DFT. It effectively results in interpolating the STFT coefficients between frequency bins, but does not affect the bandwidth of the bins, nor the capability of the representation to resolve frequency components that are close to each other.

Due to its impact on time and frequency resolution, the choice of the window length T is critical. Most of the methods discussed in this book benefit from time-frequency representations where sources to be separated exhibit little overlap in the STFT domain, and therefore the window length should depend on how stationary the sources are (see Section 2.2). Methods using multiple channels and dealing with convolutive mixtures benefit from window lengths longer from the impulse response from source to microphone, so that the convolutive mixing process is well modeled (see Section 3.4.1).In the case of separation by oracle binary masks, Vincent *et al.* (2007, fig. 5) found that a window length on the order of 50 ms (e.g., $T = 1024$ at $f_s = 16$ kHz) is suitable for speech separation, and a longer window length (e.g., $T = 4096$ at $f_s = 44.1$ kHz) for

---

1 This expression simplified from the original by Portnoff (1980) assumes that the analysis and the synthesis windows have equal lengths.

music, when the performance was measured by the signal-to-distortion ratio (SDR). For other objective evaluations of preferred window shape, window size, hop size, and zero padding see Araki *et al.* (2003) and Yılmaz and Rickard (2004).

2.1.4 Alternative Time-Frequency Representations

Alternatively to the STFT, many other time-frequency representations can be used for source separation and speech enhancement. Adaptive representations (Mallat, 1999; ISO, 2005) whose time and/or frequency resolution are automatically tuned to the signal to be processed have achieved limited success (Nesbit *et al.*, 2009). We describe below a number of time-frequency representations that differ from the STFT by the use of a fixed, nonlinear frequency scale. These representations can be either derived from the STFT or computed via a filterbank.

2.1.4.1 Nonlinear Frequency Scales

The *Mel* scale (Stevens *et al.*, 1937; Makhoul and Cosell, 1976) and the *equivalent rectangular bandwidth* (ERB) scale (Glasberg and Moore, 1990) are two nonlinear frequency scales motivated by the human auditory system.[2] The Mel scale is popular in speech processing, while the ERB scale is widely used in computational methods inspired by auditory scene analysis. A given frequency in Mel or ERB corresponds to the following frequency in Hz:

$$v(\text{Hz}) = 700 \times (e^{v(\text{Mel})/1127} - 1) \tag{2.6}$$

$$v(\text{Hz}) = 229 \times (e^{v(\text{ERB})/9.26} - 1). \tag{2.7}$$

If frequency bins or filterbank channels are linearly spaced on the Mel scale according to $v_f(\text{Mel}) = \frac{f}{F-1} v_{max}(\text{Mel}), f \in \{0, \ldots, F-1\}$, where $v_{max}(\text{Mel})$ is the maximum frequency in Mel, then their center frequencies $v_f(\text{Hz})$ in Hz are approximately linearly spaced below 700 Hz and logarithmically spaced above that frequency. The same property holds for the ERB scale, except that the change from linear to logarithmic behavior occurs at 229 Hz. The logarithmic scale (Brown, 1991; Schörkhuber and Klapuri, 2010)

$$v_f(\text{Hz}) = v_{min}(\text{Hz}) \times 2^{f/F_{oct}} \tag{2.8}$$

with $v_{min}(\text{Hz})$ the lowest frequency in Hz and F_{oct} the number of frequency bins per octave is also commonly used in music signal processing applications, since the frequencies of musical notes are distributed logarithmically. It allows easy implementation of models where change in pitch corresponds to translating the spectrum in log-frequency.

When building a time-frequency representation from the logarithmic scale (2.8), the bandwidth of each frequency bin is generally chosen so that it is proportional to the center frequency, a property known as *constant-Q* (Brown, 1991). More generally, for any nonlinear frequency scale, the bandwidth is often set to a small multiple of the frequency difference between adjacent bins. This implies that the frequency resolution is narrower at low frequencies and broader at high frequencies. Conversely, the time resolution is narrower at high frequencies and coarser at low frequencies (when the

---

2 The Mel scale measures the perceived frequency ratio between pure sinusoidal signals, while the ERB scale characterizes the frequency response of peripheral auditory filters.

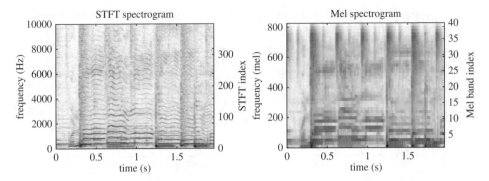

Figure 2.3 STFT and Mel spectrograms of an example music signal. High energies are illustrated with dark color and low energies with light color.

representation is calculated using a filterbank as explained in Section 2.1.4.3, not via the STFT as explained in Section 2.1.4.2). This can be seen in Figure 2.3, which shows example time-frequency representations calculated using the STFT and Mel scale.

These properties can be desirable for two reasons. First, the amplitude of natural sounds varies more quickly at high frequencies. Integrating it over wider bands makes the representation more stable. Second, there is typically more structure in sound at low frequencies, which is beneficial to model by using a higher frequency resolution for lower frequencies. By using a nonlinear frequency resolution, the number of frequency bins, and therefore the computational and memory cost of further processing, can in some scenarios be reduced by a factor of 4 to 8 without sacrificing the separation performance in a single-channel setting (Burred and Sikora, 2006). This is counter-weighted in a multichannel setting by the fact that the narrowband model of the convolutive mixing process (see Section 3.4.1) becomes invalid at high frequencies due to the increased bandwidth. Duong *et al.* (2010) showed that a full-rank model (see Section 3.4.3) is required in this case.

2.1.4.2 Computation of Power Spectrum via the STFT

The first way of computing a time-frequency representation on a nonlinear frequency scale is to derive it from the STFT. Even though there are methods that utilize STFT-domain processing to obtain complex spectra with nonlinear frequency scale, here we resort to methodology that estimates the power spectrum only. The resulting power spectrum cannot be inverted back to the time domain since it does not contain phase information. It can, however, be employed to estimate a separation filter that is then interpolated to the DFT frequency resolution and applied in the complex-valued STFT domain.

In order to distinguish the STFT and the nonlinear frequency scale representation, we momentarily index the frequency bins of the STFT by $f' \in \{0, \dots, F' - 1\}$ and the frequency bins of the nonlinear representation by $f \in \{0, \dots, F - 1\}$. The computation consists of the following steps:

1) Window the signal into frames and calculate the DFT $x(n, f')$ of each frame, similarly to the STFT analysis in Section 2.1.
2) Compute the power spectrum $|x(n, f')|^2$ in each frame.

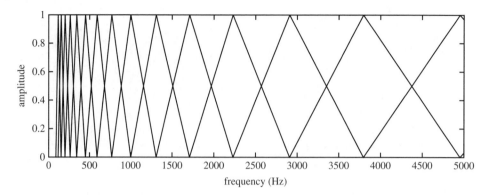

Figure 2.4 Set of triangular filter responses distributed uniformly on the Mel scale.

3) Multiply this spectrum elementwise by a set of bandpass filter responses $h(f,f')$ that are linearly spaced on the chosen frequency scale.
4) Sum over f' to obtain the nonlinear spectrum $|x(n,f)|^2 = \sum_{f'=0}^{F'-1} h(f,f')|x(n,f')|^2$ for each $f \in \{0, \dots, F-1\}$.

The Mel spectrum is usually computed using a set of triangular filter responses, as depicted in Figure 2.4. In the multichannel case, the quantity $|x(n,f')|^2$ can be replaced by $\mathbf{x}(n,f')\mathbf{x}^H(n,f')$, which results in a quadratic time-frequency representation (Gröchenig, 2001, chap. 4) as shown by Vincent (2006). In addition to to the power spectrum, this spatial covariance matrix representation contains information about the interchannel phase and level differences (IPDs and ILDs, respectively), which are useful cues in multichannel processing.

2.1.4.3 Computation via a Filterbank

Alternatively, a time-frequency representation with phase information can be obtained by filterbank analysis. A filterbank consists of a set of time-domain finite impulse response (FIR) filters[3] $h_f(\tau)$, $\tau \in \{-T_f/2, \dots, T_f/2\}$ whose center frequencies are linearly spaced on the desired scale and whose lengths T_f vary with frequency and are inversely proportional to the desired bandwidth. These filters can be generated by modulating and scaling a prototype impulse response (Burred and Sikora, 2006). The input signal $x(t)$ is convolved with each of the filters to obtain a set of complex-valued subband signals $x_f(t)$ as $x_f(t) = h_f \star x(t)$, which can then be decimated by a factor M to get $x(n,f) = x_f(nM)$. For a more detailed discussion of filterbank processing, see Vaidyanathan (1993).

The resulting representation can be approximately inverted back to the time domain by reverting the decimation operation by interpolation, convolving each subband signal by a synthesis filter, and summing the filtered signals together (Slaney *et al.*, 1994). This process of inverting the representation is approximate and causes some distortion to the signal. In many applications, the amount of distortion caused by the inversion is much smaller than the artifacts caused by the separation process, such that perfect reconstruction is not required.

---

3 In the general case infinite impulse response filters can also be used, but for simplicity we resort to FIR filters in this chapter.

Mathematically, time-frequency representations obtained via the STFT or filterbanks are very similar (see Crochiere and Rabiner (1983) for a full proof). Indeed, the STFT analysis process described in Section 2.1 can be written as $x(n,f) = \sum_{t=0}^{T-1} x(t + t_0 + nM)h_a(t)e^{-2j\pi tf/F}$ for each f. This corresponds to convolving the signal $x(t)$ with a time-reversed version of the FIR filter $h_a(t)e^{-2j\pi tf/F}$ and decimating by a factor M. Thus, STFT analysis is essentially a special case of filterbank analysis (Portnoff, 1980). Filterbanks can also be used to compute single-channel power spectra or multichannel quadratic representations by integrating the squared subband signals over time (Vincent, 2006).

2.2 Source Properties in the Time-Frequency Domain

Natural sound sources have several properties in the time-frequency domain which can be exploited in source separation or speech enhancement methods. In this section we discuss the most important properties of natural sound sources from this point of view.

2.2.1 Sparsity

Audio sources are sparse in the time-frequency domain, which means that only a small proportion of time-frequency bins have a significant amplitude. This is illustrated by the bottom left panel of Figure 2.5, which shows that the vast majority of STFT coefficients of an exemplary speech signal have very low magnitude and only a small fraction are large. This kind of distribution is termed as *sparse*.

Sparsity leads to a related phenomenon called *W-disjoint orthogonality* (Yılmaz and Rickard, 2004), which means that there is a small probability that two independent sources have significant amplitude in the same time-frequency bin. This is illustrated in the bottom right panel of Figure 2.5, which shows the bivariate histogram of the STFT magnitudes of two sources. Most observed magnitude pairs are distributed along the horizontal or vertical axis, and only about 0.2% of them have a significant amplitude in the same time-frequency bin.

W-disjoint orthogonality (Yılmaz and Rickard, 2004) is the foundation for single- and multichannel classification or clustering based methods (see Chapters 7 and 12) which predict the dominant source in each time-frequency bin and for multichannel separation methods based on nongaussian source models (see Chapter 13). Furthermore, since only one source is assumed to be dominant in each time-frequency bin, the phase of the mixture is typically close to that of the dominant source. This is one motivation for assigning the phase of the mixture to the estimated dominant source (see Chapter 5). It should be noted that reverberation decreases sparsity and therefore also W-disjoint orthogonality of sources, making these separation methods less effective in reverberant spaces.

2.2.2 Structure

Natural sounds typically have structure in both time and frequency, which translates to different entries of their time-frequency representations being dependent on each other. In the simplest case, the spectrum of a highly stationary noise source changes only very little over time, making the spectrum estimation task easier (see Chapter 6).

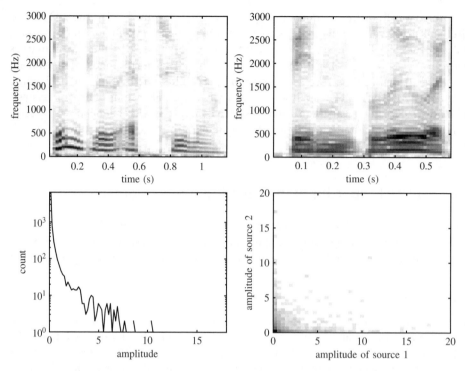

Figure 2.5 Independent sound sources are sparse in time-frequency domain. The top row illustrates the magnitude STFT spectrograms of two speech signals. The bottom left panel illustrates the histogram of the magnitude STFT coefficients of the first signal, and the bottom right panel the bivariate histogram of the coefficients of both signals.

The structure can also be much more complex, and present e.g. at different time scales. Joint modeling of different time-frequency parts of a sound allows separating sources that overlap with each other in time and frequency, allowing more accurate estimation of individual source statistics even from single-channel mixtures. For example, the amplitude of a harmonic component (see discussion below about harmonic sounds) that overlaps with another source can be predicted based on the amplitudes of other harmonics of the source.

One specific type of structure is repetition over time. Natural sound sources often consist of basic units that are present multiple times over time. For example, speech consists of phonemes, syllables, and words that are used to compose utterances. Music consists of individual notes played by different instruments that form chords, rhythmic patterns, and melodies. When processing a long audio signal, a single basic unit (e.g., phoneme, syllable, note) does not typically appear only once, but there are multiple repetitions of the unit, which are similar to each other. There exist various methods for finding repeating temporal structures (see Chapters 8, 9, 14, and 16).

Another specific type of structure within natural sound sources is *harmonicity*. Harmonic sounds have resonant modes at approximately integer multiples of the

Figure 2.6 Magnitude spectrum of an exemplary harmonic sound. The fundamental frequency is marked with a cross. The other harmonics are at integer multiples of the fundamental frequency.

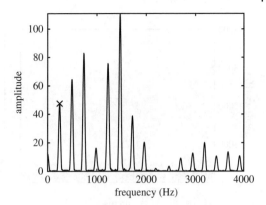

fundamental frequency of the sound, also called *pitch*, as shown in Figure 2.6. Harmonic sounds include vowels in speech, notes played by most pitched musical instruments, and many animal vocalizations. In the specific case of speech, harmonic sounds are called *voiced* and other sounds are called *unvoiced*. Harmonicity has motivated source separation and speech enhancement methods that constrain the estimated source spectra to be harmonic (see Chapter 8), and two-step methods that first track the pitch of a source over time and then predict which time-frequency bins have significant energy by considering its harmonics (see Chapter 16).

Some sound sources consist of only one type of above-discussed elementary components (noise-like, harmonic, or transient), but many natural sound sources such as speech are a combination of them, as illustrated in Figure 2.7. Each source has slightly different characteristics that make it unique and can be used to differentiate it from other sources. Source-specific models accounting for these characteristics can be trained with appropriate machine learning algorithms such as nonnegative matrix factorization (NMF), as we shall see in, for example, Chapters 8 and 9.

In order to take advantage of the structure of sounds discussed above, an appropriate representation of sound should be used. Time-domain representations often do not have as clear structure as time-frequency representations, since the frequency components of a source are generally not in phase lock and their phase is affected by the room impulse response. In the time-frequency domain the phase values are similarly quite stochastic and subject to different kinds of variabilities.

Therefore methods that exploit the structure of sounds in a single-channel setting often discard the phase and model the magnitude spectrum only. In a multichannel setting, the IPDs are extensively used since they bring essential information (see Chapters 3, 10, 11, 12, 14, and 18) and exhibit less variability in comparison to the phase values.

2.3 Filtering in the Time-Frequency Domain

Most source separation and speech enhancement methods apply a time-varying filter to the mixture signal. Since the source signals and the convolutive mixing process can both be modeled in the time-frequency domain, it is desirable to implement this filter in the same domain. In other words, the objective of source separation and speech

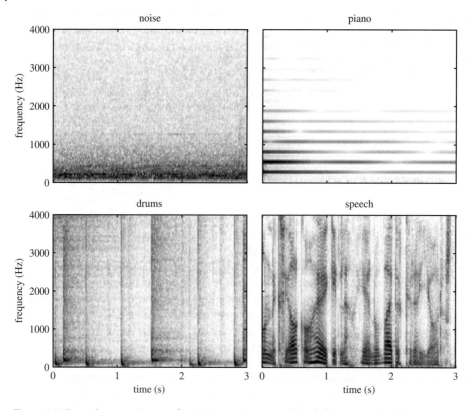

Figure 2.7 Example spectrograms of a stationary noise signal (top left), a note played by a piano (top right), a sequence of drum hits (bottom left), and a speech signal (bottom right).

enhancement methods is to estimate the target time-frequency coefficients $\hat{c}(n,f)$ and $\hat{s}(n,f)$ from the mixture coefficients $x(n,f)$.

2.3.1 Time-Domain Convolution as Interframe and Interband Convolution

Let us consider how a time-domain FIR filter $w(\tau)$, $\tau \in \{0, \ldots, L-1\}$, can be implemented in the time-frequency domain. For simplicity, we consider $w(\tau)$ to be time-invariant for the moment.

Time-domain convolution can be implemented as a complex-valued multiplication in the STFT domain via either the overlap-save method (by removing the samples that underwent circular convolution) or the overlap-add method (by properly designed zero-padding of the analysis window) (Shynk, 1992). These methods are exact if both the analysis and synthesis windows are rectangular with different lengths. If the filter length L is longer than the window length, the linear convolution can still be implemented, by partitioning the filters into blocks (Serviére, 2003; Soo and Pang, 1990). They have been used in the context of source separation and speech enhancement by, for example, Gannot *et al.* (2001), Kellermann and Buchner (2003), Serviére (2003), and Mirsamadi *et al.* (2012) but the use of rectangular analysis and synthesis windows severely limits their performance.

Using the conventional STFT with arbitrary analysis window instead, time-domain convolution translates into interframe and interband convolution (Gilloire and Vetterli, 1992; Avargel and Cohen, 2007):

$$c(n,f) = \sum_{f'=0}^{F-1} \sum_{n'=-\infty}^{\infty} w(n',f',f)x(n - n',f'). \tag{2.9}$$

This expression simply stems from the linearity of the STFT analysis operation. It is exact but computationally and memory intensive: $w(n',f',f)$ is nonzero for a few values of n' only (on the order of L/M) but for all values of f' and f. Therefore, for a given output frequency bin f, all input frequency bins f' need to be taken into account. For more discussion about this, see Chapter 19.

2.3.2 Practical Approximations

The computational complexity can be reduced by noting that $w(n',f',f)$ typically decays with increasing frequency difference $|f' - f|$, where the rate of decay is governed by the window shape. A first approximation is to assume that $w(n',f',f) \approx 0$ if $f' \neq f$, which yields the *subband filtering* operation:

$$c(n,f) = \sum_{n'=-\infty}^{\infty} w(n',f)x(n - n',f). \tag{2.10}$$

In the limit when the filter length is much shorter than the analysis window length, i.e. $L \ll T$, one can further assume that $w(n',f) \approx 0$ for $n' \neq 0$, which yields the so-called *narrowband approximation*:

$$c(n,f) = w(f)x(n,f). \tag{2.11}$$

This approximation is also valid for time-frequency representations computed by filter-bank analysis in the limit when $L \ll T_f$.

The majority of source separation and speech enhancement techniques employ the narrowband approximation even with a filter length equal to the frame size, namely $L = T$. This approach is convenient since, contrary to assuming that $L < T$, it does not confine the vector of filter coefficients $\mathbf{w}(n) = [w(n,0), \dots, w(n,F-1)]^T$ to belong to an L-dimensional subspace. However, breaching the condition $L \ll T$ typically results in cyclic convolution artifacts, namely wrapping of the frames due to the filter application. These cyclic convolution effects can be alleviated by applying frequency-domain smoothing to the frequency response of the filter, tapered analysis and synthesis windows.

It should be noted that in most source separation and speech enhancement methods, time-varying filter coefficients $w(n,f)$ are used, since the sources to be separated are nonstationary.

2.4 Summary

In this chapter, we showed how a time-domain signal can be transformed to the time-frequency domain and back to the time domain, and how the time and frequency resolution of this transform can be controlled. In addition we discussed how the time-frequency coefficients of the target source can be approximately obtained by narrowband filtering in the time-frequency domain. This will be exploited to design single-channel and multichannel filters using the various methods discussed in this book. We also reviewed the main properties of the magnitude spectra of audio sources, which will be used to derive spectral models in the remaining chapters. For advanced readers, the properties of phase spectra and interframe and/or interband filtering techniques are discussed in Chapter 19.

Bibliography

Allen, J. (1977) Short term spectral analysis, synthesis, and modification by discrete Fourier transform. *IEEE Transactions on Acoustics, Speech, and Signal Processing*, **25** (3), 235–238.

Araki, S., Mukai, R., Makino, S., Nishikawa, T., and Saruwatari, H. (2003) The fundamental limitation of frequency domain blind source separation for convolutive mixtures of speech. *IEEE Transactions on Speech and Audio Processing*, **11** (2), 109–116.

Avargel, Y. and Cohen, I. (2007) System identification in the short-time Fourier transform domain with crossband filtering. *IEEE Transactions on Audio, Speech, and Language Processing*, **15** (4), 1305–1319.

Brown, J. (1991) Calculation of a constant Q spectral transform. *Journal of the Acoustical Society of America*, **89** (1), 425–434.

Burred, J. and Sikora, T. (2006) Comparison of frequency-warped representations for source separation of stereo mixtures, in *Proceedings of the Audio Engineering Society Convention*. Paper number 6924.

Crochiere, R. (1980) A weighted overlap-add method of short-time Fourier analysis/synthesis. *IEEE Transactions on Acoustics, Speech, and Signal Processing*, **28** (1), 99–102.

Crochiere, R.E. and Rabiner, L.R. (1983) *Multirate Digital Signal Processing*, Prentice Hall.

Duong, N.Q.K., Vincent, E., and Gribonval, R. (2010) Under-determined reverberant audio source separation using local observed covariance and auditory-motivated time-frequency representation, in *Proceedings of International Conference on Latent Variable Analysis and Signal Separation*, pp. 73–80.

Gannot, S., Burshtein, D., and Weinstein, E. (2001) Signal enhancement using beamforming and nonstationarity with applications to speech. *IEEE Transactions on Signal Processing*, **49** (8), 1614–1626.

Gilloire, A. and Vetterli, M. (1992) Adaptive filtering in subbands with critical sampling: analysis, experiments, and application to acoustic echo cancellation. *IEEE Transactions on Signal Processing*, **40** (8), 1862–1875.

Glasberg, B.R. and Moore, B.C.J. (1990) Derivation of auditory filter shapes from notched-noise data. *Hearing Research*, **47**, 103–138.

Gröchenig, K. (2001) *Foundations of Time-Frequency Analysis*, Springer.

Harris, F.J. (1978) On the use of windows for harmonic analysis with the discrete Fourier transform. *Proceedings of the IEEE*, **66** (1), 51–83.

ISO (2005) Information technology — Coding of audio-visual objects — Part 3: Audio (ISO/IEC 14496-3:2005).

Kellermann, W. and Buchner, H. (2003) Wideband algorithms versus narrowband algorithms for adaptive filtering in the DFT domain, in *Proceedings of Asilomar Conference on Signals, Systems, and Computers*, pp. 1278–1282.

Makhoul, J. and Cosell, L. (1976) LPCW: An LPC vocoder with linear predictive spectral warping, in *Proceedings of IEEE International Conference on Audio, Speech and Signal Processing*.

Mallat, S. (1999) *A Wavelet Tour of Signal Processing*, Academic Press, 2nd edn.

Mirsamadi, S., Ghaffarzadegan, S., Sheikhzadeh, H., Ahadi, S.M., and Rezaie, A.H. (2012) Efficient frequency domain implementation of noncausal multichannel blind deconvolution for convolutive mixtures of speech. *IEEE Transactions on Audio, Speech, and Language Processing*, **20** (8), 2365–2377.

Nesbit, A., Vincent, E., and Plumbley, M.D. (2009) Extension of sparse, adaptive signal decompositions to semi-blind audio source separation, in *Proceedings of International Conference on Independent Component Analysis and Signal Separation*, pp. 605–612.

Portnoff, M.R. (1980) Time-frequency representation of digital signals and systems based on short-time Fourier analysis. *IEEE Transactions on Acoustics, Speech, and Signal Processing*, **28** (1), 55–69.

Schörkhuber, C. and Klapuri, A. (2010) Constant-Q transform toolbox for music processing, in *Proceedings of Sound and Music Computing Conference*.

Serviére, C. (2003) Separation of speech signals with segmentation of the impulse responses under reverberant conditions, in *Proceedings of International Conference on Independent Component Analysis and Signal Separation*, pp. 511–516.

Shynk, J. (1992) Frequency-domain and multirate and adaptive filtering. *IEEE Signal Processing Magazine*, **9** (1), 14–37.

Slaney, M., Naar, D., and Lyon, R.F. (1994) Auditory model inversion for sound separation, in *Proceedings of IEEE International Conference on Audio, Speech and Signal Processing*.

Soo, J.S. and Pang, K.K. (1990) Multidelay block frequency domain adaptive filter. *IEEE Transactions on Acoustics, Speech, and Signal Processing*, **38** (2), 373–376.

Stevens, S.S., Volkmann, J., and Newman, E.B. (1937) A scale for the measurement of the psychological magnitude pitch. *Journal of the Acoustical Society of America*, **8** (3), 185–190.

Vaidyanathan, P.P. (1993) *Multirate Systems And Filter Banks*, Prentice Hall.

Vincent, E. (2006) Musical source separation using time-frequency source priors. *IEEE Transactions on Audio, Speech, and Language Processing*, **14** (1), 91–98.

Vincent, E., Gribonval, R., and Plumbley, M.D. (2007) Oracle estimators for the benchmarking of source separation algorithms. *Signal Processing*, **87** (8), 1933–1950.

Yılmaz, Ö. and Rickard, S. (2004) Blind separation of speech mixtures via time-frequency masking. *IEEE Transactions on Signal Processing*, **52** (7), 1830–1847.

3

Acoustics: Spatial Properties

Emmanuel Vincent, Sharon Gannot, and Tuomas Virtanen

In Chapter 2, we presented the spectral properties of sound sources which can be exploited for the separation or enhancement of single-channel signals. In multichannel scenarios, the fact that the acoustic scene is observed from different positions in space can also be exploited. In this chapter, we recall basic elements of acoustics and sound engineering, and use them to model multichannel mixtures.

We consider the relationship between a source signal and its spatial image in a given channel in Section 3.1, and examine how it translates in the case of microphone recordings or artificial mixtures in Sections 3.2 and 3.3, respectively. We then introduce several possible models in Section 3.4. We summarize the main concepts and provide links to other chapters and more advanced topics in Section 3.5.

3.1 Formalization of the Mixing Process

3.1.1 General Mixing Model

Sturmel *et al.* (2012) proposed the following general two-stage model for audio mixtures. In the first stage, each single-channel point source signal $s_j(t)$ is transformed into an $I \times 1$ multichannel source spatial image signal $c_j(t)$ by means of a possibly nonlinear spatialization operation $\mathfrak{A}_j$:

$$c_j(t) = [\mathfrak{A}_j(s_j)](t). \tag{3.1}$$

In the second stage, the source spatial image signals $c_j(t)$, $j \in \{1 \dots, J\}$, of all (point and diffuse) sources are added together and passed through a possibly nonlinear post-mixing operation $\mathfrak{A}$ to obtain the $I \times 1$ multichannel mixture signal $x(t)$:

$$x(t) = \left[\mathfrak{A} \left(\sum_{j=1}^{J} c_j \right) \right](t). \tag{3.2}$$

The linear, time-invariant case is of particular interest. In that case, the spatialization operations $\mathfrak{A}_j$ boil down to linear, time-invariant filters $a_j(\tau) = [a_{1j}(\tau), \dots, a_{Ij}(\tau)]^T$

$$c_j(t) = \sum_{\tau=-\infty}^{+\infty} a_j(\tau) s_j(t - \tau) \tag{3.3}$$

Audio Source Separation and Speech Enhancement, First Edition.
Edited by Emmanuel Vincent, Tuomas Virtanen and Sharon Gannot.
© 2018 John Wiley & Sons Ltd. Published 2018 by John Wiley & Sons Ltd.
Companion Website: https://project.inria.fr/ssse/

and the post-mixing operation $\mathfrak{A}$ reduces to identity[1]

$$\mathbf{x}(t) = \sum_{j=1}^{J} \mathbf{c}_j(t). \tag{3.4}$$

The filters with coefficients $a_{ij}(\tau)$ are called *mixing filters* or *impulse responses*.

3.1.2 Microphone Recordings vs. Artificial Mixtures

To investigate models (3.1) and (3.2) further, one must consider how the mixture was obtained in practice. Two different situations arise. *Microphone recordings* refer to the situation when multiple sources which are simultaneously active are captured by a microphone *array*. Typical examples include hands-free phones, audioconferencing systems, and hearing aids. *Artificial mixtures*, by contrast, are generated by mixing individually recorded sound sources using appropriate hardware or software. Most audio media (television, music, cinema, etc.) fall into this category. Certain audio media such as classical music or documentaries result from a more complicated mixing process by which microphone recordings are first conducted and then artificially remixed in a studio. For more information about the recording and mixing strategies used by sound engineers, see Bartlett and Bartlett (2012).

3.2 Microphone Recordings

3.2.1 Acoustic Impulse Responses

In the case of a microphone recording, the mixing process is due to the propagation of sound in the recording environment. This phenomenon is linear and time-invariant provided that the sources are static (not moving), so models (3.3) and (3.4) hold (Kuttruff, 2000). Each *acoustic impulse response* $a_{ij}(\tau)$ represents the propagation of sound from one source j to one microphone i and it is causal, i.e. $a_{ij}(\tau) = 0$ for $\tau < 0$.

In *free field* that is in open air without any obstacle, sound propagation incurs a delay r_{ij}/c and an attenuation $1/\sqrt{4\pi r_{ij}}$ as a function of the distance r_{ij} from the source to the microphone. The acoustic impulse response is given by

$$a_{ij}(\tau) = \frac{1}{\sqrt{4\pi r_{ij}}} \delta\left(\tau - \frac{r_{ij}}{c} f_s\right) \tag{3.5}$$

where c is the sound speed (343 m/s at 20°C), f_s the sampling rate, and δ the Dirac function or, more generally, a fractional delay filter. The attenuation due to distance directly affects the signal-to-noise ratio (SNR) (ISO, 2003).

In practice, various obstacles such as walls and furniture must be considered. The propagation of a sound of frequency v changes depending on the size of the obstacle compared to its wavelength $\lambda = c/v$, which varies from $\lambda = 17$ mm at $v = 20$ kHz to $\lambda = 17$ m at $v = 20$ Hz. Obstacles which are substantially smaller than λ have little or no impact on the delay and attenuation. Obstacles of comparable dimension to λ result

---

1 Because of the linearity of summation, linear post-mixing, if any, is considered to be part of $\mathbf{a}_j(\tau)$.

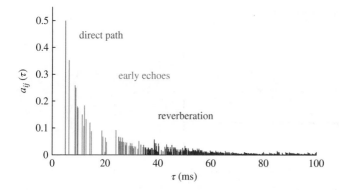

Figure 3.1 Schematic illustration of the shape of an acoustic impulse response $a_{ij}(\tau)$ for a room of dimensions $8.00 \times 5.00 \times 3.10$ m, an RT60 of 230 ms, and a source distance of $r_{ij} = 1.70$ m. All reflections are depicted as Dirac impulses.

in *diffraction*: sound takes more time to pass the obstacle and it is more attenuated than in the free field. This phenomenon is famous for binaural recordings, i.e. recordings obtained from in-ear microphones, where the torso, head, and pinna of the listener act as obstacles (Blauert, 1997). It also explains source directivity, i.e. the fact that the sound emitted by a source varies with spatial direction. Finally, surfaces of dimension larger than λ result in *reflection* of the sound wave in the opposite direction with respect to the surface normal and absorption of part of its power. Many reflections typically occur on different obstacles, which induce multiple propagation paths. The acoustic impulse response between each source and each microphone results from the summation of all those paths.

Figure 3.1 provides a schematic illustration of the shape of an acoustic impulse response and Figure 3.2 shows a real acoustic impulse response. The real response differs from the illustration as it exhibits both positive and negative values, but its magnitude follows the same overall shape. Three parts can be seen. The first peak is the line of sight, called the *direct path* (3.5). It is followed by a few disjoint reflections on the closest obstacles called *early echoes*. Many reflections then simultaneously occur and form an exponentially decreasing tail called *late reverberation* or simply *reverberation*. The boundary τ_c between early echoes and reverberation, called the *mixing time*, depends on the acoustic properties of the room. A typical value is 50 ms after the direct path. One can then decompose each acoustic impulse response $\mathbf{a}_j(\tau)$ into the sum of a direct part $\mathbf{a}_j^{\text{dir}}(\tau)$ and an indirect part due to echoes and reverberation $\mathbf{a}_j^{\text{rev}}(\tau)$. Similarly, each source spatial image can be decomposed as $\mathbf{c}_j(t) = \mathbf{c}_j^{\text{dir}}(t) + \mathbf{c}_j^{\text{rev}}(t)$. Note that early echoes are sometimes considered as part of $\mathbf{c}_j^{\text{dir}}(t)$ instead of $\mathbf{c}_j^{\text{rev}}(t)$ when defining the task of dereverberation (see Chapter 15).

3.2.2 Main Properties of Acoustic Impulse Responses

Acoustic impulse responses manifest themselves by a modification of the phase and power spectrum of the emitted signal and they smear the signal across time. Although they typically have thousands of coefficients, they can be described by three main properties. The *reverberation time* (RT60) is the duration over which the envelope of the

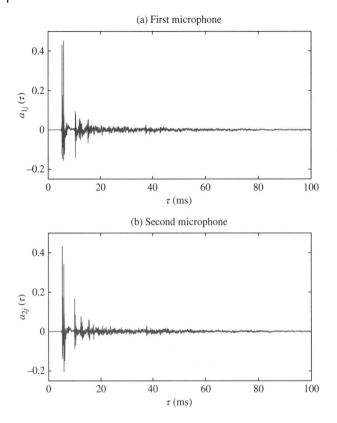

Figure 3.2 First 100 ms of a pair of real acoustic impulse responses $\mathbf{a}_j(\tau)$ from the Aachen Impulse Response Database (Jeub *et al.*, 2009) recorded in a meeting room with the same characteristics as in Figure 3.1 and sampled at 48 kHz.

reverberant tail decays by 60 decibels (dB). It depends on the size and absorption level of the room (including obstacles) and it represents the time scale of smearing. Table 3.1 reports typical RT60 values for various environments. The *direct-to-reverberant ratio* (DRR) is the ratio of the power of direct and indirect sound. It varies with the size and the absorption of the room, but also with the distance between the source and the microphone according to the curves in Figure 3.3. It governs the amount of smearing of the signal. The distance beyond which the power of indirect sound becomes larger than that of direct sound is called the *critical distance*. Finally, the *direct-to-early ratio*, which is the power of direct sound divided by the remaining power in the first τ_c samples (defined above), quantifies the modification of the power spectrum of the signal induced by early echoes. It is low when the microphone and/or the source is close to an obstacle such as a table or a window, and higher otherwise. The DRR and the direct-to-early ratio are not systematically reported when describing experiments in the literature, yet they are as important as the RT60 to characterize multichannel mixtures. Also, as we shall see, the RT60 values in Table 3.1 are larger than usually considered in the literature until recently.

Another useful property of acoustic impulse responses is the statistical dependency between impulse responses corresponding to the same source. Due to the summation

Table 3.1 Range of RT60 reported in the literature for different environments (Ribas *et al.*, 2016).

Environment		RT60 (s)
	Car	0.05
Work	Office	0.25–0.43
	Meeting room	0.23–0.70
Home	Living room	0.44–0.74
	Bedroom	0.39–0.68
	Bathroom	0.41–0.75
	Kitchen	0.41–0.83
Public spaces	Classroom	0.20–1.27
	Lecture room	0.64–1.25
	Restaurant	0.50–1.50

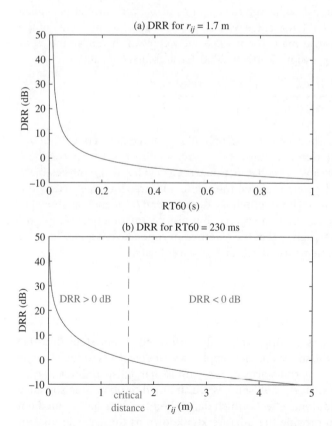

Figure 3.3 DRR as a function of the RT60 and the source distance r_{ij} based on Eyring's formula (Gustafsson *et al.*, 2003). These curves assume that there is no obstacle between the source and the microphone, so that the direct path exists. The room dimensions are the same as in Figure 3.1.

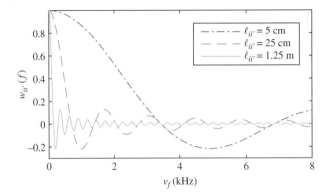

Figure 3.4 IC $\omega_{ii'}(f)$ of the reverberant part of an acoustic impulse response as a function of microphone distance $\ell_{ii'}$ and frequency v_f.

of many propagation paths, reverberation can be statistically modeled using the law of large numbers as a zero-mean Gaussian noise signal with decaying amplitude (Polack, 1993). This Gaussian noise signal is characterized by its normalized correlation, called the *interchannel coherence* (IC). On average over all possible absolute positions of the sources and the microphone array in the room, the IC between two channels i and i' has the following closed-form expression (Kuttruff, 2000; Gustafsson *et al.*, 2003):

$$\omega_{ii'}(f) = \frac{\mathbb{E}\{c_{ij}^{\text{rev}}(\cdot,f)c_{i'j}^{\text{rev}\,*}(\cdot,f)\}}{\sqrt{\mathbb{E}\{|c_{ij}^{\text{rev}}(\cdot,f)|^2\}}\sqrt{\mathbb{E}\{|c_{i'j}^{\text{rev}}(\cdot,f)|^2\}}} = \frac{\sin(2\pi v_f \ell_{ii'}/c)}{2\pi v_f \ell_{ii'}/c} \tag{3.6}$$

where $\ell_{ii'}$ denotes the distance between the microphones, v_f the center frequency of frequency bin f, and the expectation operator is taken over all directions of space. These scalar ICs can be grouped into an $I \times I$ coherence matrix $\mathbf{\Omega}(f) = [\omega_{ii'}(f)]_{ii'}$. Interestingly, the IC does not depend on the source nor on the room: it is large for small arrays and low frequencies and it decreases with microphone distance and frequency, as shown in Figure 3.4. This result holds not only on average, but also in any practical setup provided that the RT60 is large enough so that the reverberant sound field is approximately diffuse, that is for all environments listed in Table 3.1 except cars.

3.3 Artificial Mixtures

In the case of artificial mixtures, mixing is typically performed in four steps (Sturmel *et al.*, 2012). In the first step, the sound engineer applies a series of effects to each source. In the second step, the source is transformed into a multichannel spatial image $\mathbf{c}_j(t)$. In the third step, the spatial images of all sources are summed to obtain the so-called "master". In the last step, additional effects which depend on the distribution medium are applied to the master to provide the mixture $\mathbf{x}(t)$ known as the "artistic mix" or "commercial mix". Steps 1 and 2 and steps 3 and 4 are formalized in equations (3.1) and (3.2), respectively. The overall mixing process then results from the effects chosen by the sound engineer. Example effects are listed in Table 3.2.

Table 3.2 Example artificial mixing effects, from Sturmel *et al.*
(2012).

Linear instantaneous effects	Gain
	Panning (instantaneous mixing)
Linear convolutive effects	Equalization
	Reverberation
	Delay
Nonlinear effects	Dynamic compression
	Chorus
	Distortion

The inversion of nonlinear effects has been sparsely studied and shown to be difficult even when the nonlinearity is known (Gorlow and Reiss, 2013). For this reason, it is desirable to express the mixing process in linear form. It turns out that this is feasible under two conditions. First, the effects used to transform the source signal into its spatial image in step 2 must be linear. This condition often holds since panning or convolution by simulated or real reverberant impulse responses are typically used in this step and they are linear. The nonlinear effects possibly applied in step 1 can then be considered as part of the original source signal. Second, the nonlinear effects applied to the master in step 4 must be amenable to time-varying linear filtering. This condition generally holds too since dynamic compression and equalization are often the only effects applied at this stage and they can be modeled as a linear time-varying filter whose coefficients depend on the master signal. This time-varying filter may then be equivalently applied to all source images before summation. The mixture then becomes equal to the sum of the source images, as in equation (3.4), and each source image can be expressed similarly to equation (3.3), except that the mixing filters are time-varying. If convolution by reverberant impulse responses is used in step 2 and the amount of nonlinearity in step 4 is limited, the mixing filters share similar characteristics with the acoustic impulse responses reviewed above.

3.4 Impulse Response Models

Given the physical properties of mixing filters described above, we can now build models for multichannel source separation and enhancement. Throughout the rest of this book, we assume linear mixing and static sources. The additional issues raised by moving sources or time-varying mixing are discussed in Chapter 19.

Time-domain modeling of the mixing filters as finite impulse response (FIR) filters of a few thousand coefficients was popular in the early stages of research (Nguyen Thi and Jutten, 1995; Ehlers and Schuster, 1997; Gupta and Douglas, 2007) and has gained new interest recently with sparse decomposition-based approaches (Lin *et al.*, 2007; Benichoux *et al.*, 2014; Koldovský *et al.*, 2015). However, the large number of coefficients to be estimated and integration with time-frequency domain models for the sources result in costly algorithms (Kowalski *et al.*, 2010).

Most methods today model both the sources and the mixing filters in the time-frequency domain. Exact modeling using the theoretical tools in Section 2.3.1 is feasible but uncommon and it is discussed in Chapter 19. In the following, we present three approximate models which can be applied both to microphone recordings and artificial mixtures. For each model, we also explain how the parameters may be constrained in the specific case of microphone recordings. Similar constraints may be designed for artificial mixtures.

3.4.1 Narrowband Approximation

3.4.1.1 Definition
Let us denote by $c_j(n,f)$ and $s_j(n,f)$ the short-time Fourier transform (STFT) of $c_j(t)$ and $s_j(t)$, respectively. The most common model is based on the narrowband approximation. Under the conditions discussed in Section 2.3.2, time-domain filtering can be approximated by complex-valued multiplication in the STFT domain:

$$c_j(n,f) = a_j(f)\, s_j(n,f) \tag{3.7}$$

where the $I \times 1$ vector $a_j(f)$ is called the *mixing vector*. Each element $a_{ij}(f)$ of the mixing vector is the discrete Fourier transform (DFT) associated with $a_{ij}(\tau)$ called the *transfer function* or *acoustic transfer function*. The mixing vectors of all sources are sometimes concatenated into an $I \times J$ matrix $A_j(f) = [a_1(f), \dots, a_J(f)]$ called the *mixing matrix*.

3.4.1.2 Steering Vector – Near Field vs. Far Field
When the source position is known, geometrical (soft or hard) constraints can be set on $a_j(f)$ to ensure that it is close to the *steering vector* $d_j(f)$ which encodes the direct path (Parra and Alvino, 2002; Knaak *et al.*, 2007). In the case of a microphone recording, the steering vector for source j is given by

$$d_j(f) = \begin{bmatrix} \frac{1}{\sqrt{4\pi r_{1j}}} e^{-2j\pi r_{1j}v_f/c} \\ \vdots \\ \frac{1}{\sqrt{4\pi r_{Ij}}} e^{-2j\pi r_{Ij}v_f/c} \end{bmatrix} \tag{3.8}$$

where each element is the DFT of the free-field acoustic impulse response (3.5) from the source to microphone i. This expression is mainly applied in the *near field*, that is when the source-to-microphone distances r_{ij} are smaller or comparable to the microphone distances $\ell_{ii'}$. In the *far field*, the attenuation factors $1/\sqrt{4\pi r_{ij}}$ become approximately equal so the following expression of the steering vector (up to a multiplicative factor) is used instead:

$$d_j(f) = \begin{bmatrix} e^{-2j\pi r_{1j}v_f/c} \\ \vdots \\ e^{-2j\pi r_{Ij}v_f/c} \end{bmatrix}. \tag{3.9}$$

Note that, in either case, the steering vector depends both on the *direction of arrival* (DOA) and the distance of the source relative to the array.

3.4.2   Relative Transfer Function and Interchannel Cues

3.4.2.1   Definition

The transfer functions $a_{ij}(f)$ have a specific phase and amplitude for each channel i. In theory, this could be exploited to perform source localization and separation even in a single-channel or monaural setting (Blauert, 1997; Asari *et al.*, 2006) by disambiguating $a_{ij}(f)$ from $s_j(n,f)$ in 3.7. In practice, however, the phase spectrum of the source is unknown and its magnitude spectrum is rarely known to the required level of precision.[2] This has motivated researchers to disregard monaural cues and consider the differences between channels instead.

Taking the first channel as a reference, the *relative mixing vector* for source j is defined as (Gannot *et al.*, 2001; Markovich *et al.*, 2009)

$$\tilde{\mathbf{a}}_j(f) = \frac{1}{a_{1j}(f)} \mathbf{a}_j(f). \tag{3.10}$$

The elements $\tilde{a}_{ij}(f)$ of this vector are called *relative transfer functions* (RTFs). They can be interpreted as transfer functions relating the channels of the source spatial image to each other. Note that $\tilde{\mathbf{a}}_j(f)$ is defined only when $a_{1j}(f) \neq 0$, which is sometimes not true in low DRR conditions. An alternative definition was given by Affes and Grenier (1997) and Sawada *et al.* (2007):

$$\bar{\mathbf{a}}_j(f) = \frac{e^{-j\angle a_{1j}(f)}}{\|\mathbf{a}_j(f)\|_2} \mathbf{a}_j(f). \tag{3.11}$$

By taking all channels into account, this definition increases the chance that the relative mixing vector is defined and it makes it more invariant to the magnitude of the reference channel. For generalizations of this concept, see Li *et al.* (2015).

The RTFs encode the *interchannel level difference* (ILD), also known as the *interchannel intensity difference*, in decibels and the *interchannel phase difference* (IPD) in radians between pairs of microphones as a function of frequency:

$$\text{ILD}_{ij}(f) = 20 \log_{10} |\tilde{a}_{ij}(f)| \tag{3.12}$$

$$\text{IPD}_{ij}(f) = \angle \tilde{a}_{ij}(f). \tag{3.13}$$

Figure 3.5 illustrates these two quantities as a function of frequency. The ILD and the IPD appear to cluster around the ILD and the IPD associated with the direct path, but they can exhibit significant deviations due to early echoes and reverberation.

The *interchannel time difference* (ITD) in seconds is sometimes considered instead of the IPD:

$$\text{ITD}_{ij}(f) = \frac{\angle \tilde{a}_{ij}(f)}{2\pi \nu_f}. \tag{3.14}$$

Note, however, that the ITD is unambiguously defined only below the frequency c/ℓ_{i1}. With a sampling rate of 16 kHz, this requires a microphone distance ℓ_{i1} of less than

---

2 Contrary to a widespread belief, human audition relies more on head movements than monaural cues to solve ambiguous spatial percepts (Wallach, 1940; Wightman and Kistler, 1999).

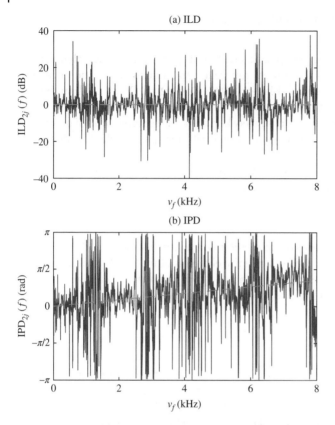

Figure 3.5 ILD and IPD corresponding to the pair of real acoustic impulse responses in Figure 3.2. Dashed lines denote the theoretical ILD and IPD in the free field, as defined by the relative steering vector in (3.15).

4.3 cm. For larger distances or higher frequencies, *spatial aliasing* occurs: several candidate ITDs correspond to a given IPD up to an integer multiple of 2π, therefore the ITD can be measured only up to an integer multiple of $1/v_f$. In a binaural setting, the ILD spans a large range and it can be exploited to disambiguate multiple ITDs corresponding to the same IPD. With free-field microphone arrays, this is hardly feasible as the ILD is smaller in the far field and varies a lot more with reverberation. One must then integrate the IPD information across frequency to recover the ITD.

These interchannel quantities play a key role in human hearing and, consequently, in hearing aids. For more details, see Chapter 18.

3.4.2.2 Relative Steering Vector
Similarly to Section 3.4.1.2, geometrical constraints can be set on $\tilde{\mathbf{a}}_j(f)$ to ensure that it is close to the *relative steering vector* $\tilde{\mathbf{d}}_j(f) = \mathbf{d}_j(f)/d_{1j}(f)$ (Yılmaz and Rickard, 2004;

Figure 3.6 Geometrical illustration of the position of a far-field source j with respect to a pair of microphones on the horizontal plane, showing the azimuth θ_j, the elevation φ_j, the angle of arrival α_j, the microphone distance ℓ_{21}, the source-to-microphone distances r_{1j} and r_{2j}, and the unit-norm vector $\mathbf{k}_j$ pointing to the source.

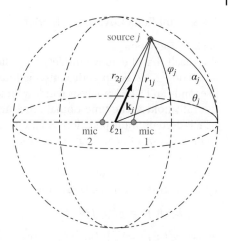

Sawada *et al.*, 2007; Reindl *et al.*, 2013), as observed in Figure 3.5. In the far field, the relative steering vector for source j is given by

$$\tilde{\mathbf{d}}_j(f) = \begin{bmatrix} 1 \\ e^{-2j\pi\Delta_{2j}\nu_f} \\ \vdots \\ e^{-2j\pi\Delta_{Ij}\nu_f} \end{bmatrix} \tag{3.15}$$

where

$$\Delta_{ij} = \frac{r_{ij} - r_{1j}}{c} \tag{3.16}$$

is the ITD in the free field called the *time difference of arrival* (TDOA). The TDOA in the far field depends only on the source DOA (not on its distance). More precisely, denoting by θ_j and φ_j the azimuth and elevation of source j with respect to the array center, as represented in Figure 3.6, and by $\mathbf{k}_j = [\cos\theta_j\cos\varphi_j, \sin\theta_j\cos\varphi_j, \sin\varphi_j]^T$ the unit-norm vector pointing to the source, the TDOA can be computed as

$$\Delta_{ij} = \frac{\mathbf{k}_j^T(\mathbf{m}_i - \mathbf{m}_1)}{c} \tag{3.17}$$

with $\mathbf{m}_i$ the Cartesian coordinates of microphone i with respect to the array center. In the case when there are two microphones or all microphones are spatially aligned with each other, it can also be expressed as

$$\Delta_{ij} = \frac{\ell_{i1}\cos\alpha_j}{c} \tag{3.18}$$

with α_j the *angle of arrival* with respect to the microphone axis. The TDOA can also be defined in the near field according to (3.16), but its expression and that of the relative steering vector also depend on the source distance in that case.

3.4.3 Full-Rank Covariance Model

3.4.3.1 Definition

We recall that the above models are valid only for point sources under the conditions in Section 2.3.2. For practical mixing filter lengths on the order of several hundred milliseconds and STFT analysis window lengths on the order of 50 ms, these conditions do not fully hold so the time-domain mixing process (3.3) is only roughly approximated by STFT-domain multiplication (3.7). One approach which partly overcomes this issue is to move from a linear (first-order) model to a second-order model of the signals.

Considering all signals of interest as wide-sense stationary processes within each time frame n, we denote by $\boldsymbol{\Sigma}_{\mathbf{c}_j}(n,f) = \mathbb{E}\{\mathbf{c}_j(n,f)\mathbf{c}_j^H(n,f)\}$ the $I \times I$ covariance matrix of $\mathbf{c}_j(n,f)$ across channels. Under the narrowband approximation, it can be shown that

$$\boldsymbol{\Sigma}_{\mathbf{c}_j}(n,f) = \sigma_{s_j}^2(n,f)\mathbf{R}_j(f) \tag{3.19}$$

where $\sigma_{s_j}^2(n,f) = \mathbb{E}\{|s_j(n,f)|^2\}$ is the variance of $s_j(n,f)$ and the $I \times I$ rank-1 matrix $\mathbf{R}_j(f) = \mathbf{a}_j(f)\mathbf{a}_j^H(f)$ is called the *spatial covariance matrix* (Févotte and Cardoso, 2005; Vincent *et al.*, 2009). The rank-1 property implies that the channels of $\mathbf{c}_j(n,f)$ are coherent, i.e. perfectly correlated.

Duong *et al.* (2010) and Sawada *et al.* (2013) proposed to consider the spatial covariance matrix $\mathbf{R}_j(f)$ as a *full-rank* matrix instead. This more flexible model applies to longer impulse responses and to diffuse sources. In such conditions, the sound emitted by each source reaches the microphones from several directions at once, such that the channels of $\mathbf{c}_j(n,f)$ become incoherent. The entries $(\mathbf{R}_j(f))_{ii'}$ of $\mathbf{R}_j(f)$ encode not only the ILD and the IPD, but also the IC[3]

$$\mathrm{IC}_{ii'j}(f) = \frac{\mathbb{E}\{c_{ij}(\cdot,f)c_{i'j}^*(\cdot,f)\}}{\sqrt{\mathbb{E}\{|c_{ij}(\cdot,f)|^2\}}\sqrt{\mathbb{E}\{|c_{i'j}(\cdot,f)|^2\}}}. \tag{3.20}$$

Indeed, they can be expressed as $\mathrm{ILD}_{ij}(f) = 10\log_{10}(|(\mathbf{R}_j(f))_{ii}|/|(\mathbf{R}_j(f))_{11}|)$, $\mathrm{IPD}_{ij}(f) = \angle(\mathbf{R}_j(f))_{i1}$, and $\mathrm{IC}_{ii'j}(f) = (\mathbf{R}_j(f))_{ii'}/\sqrt{(\mathbf{R}_j(f))_{ii}}\sqrt{(\mathbf{R}_j(f))_{i'i'}}$. The quantity $|\mathrm{IC}_{ii'j}(f)|^2$ is referred to as the magnitude squared coherence (MSC).

3.4.3.2 Parametric Covariance Models

When the source position and the room characteristics are known, geometrical (soft or hard) constraints can be set on $\mathbf{R}_j(f)$. The average value of the spatial covariance matrix over all possible absolute positions of the sources and the microphone array in the room is equal to (Duong *et al.*, 2010)

$$\mathbf{D}_j(f) = \mathbf{d}_j(f)\mathbf{d}_j^H(f) + \sigma_{\mathrm{rev}}^2(f)\boldsymbol{\Omega}(f) \tag{3.21}$$

with $\mathbf{d}_j(f)$ the steering vector in (3.8), $\boldsymbol{\Omega}(f)$ the covariance matrix of a diffuse sound field in (3.6), and $\sigma_{\mathrm{rev}}^2(f)$ the power of early echoes and reverberation. The matrix $\mathbf{D}_j(f)$ generalizes the concept of steering vector to the second-order case. Duong *et al.* (2013) showed that, for moderate or large RT60, $\mathbf{R}_j(f)$ is close to $\mathbf{D}_j(f)$. Nikunen and Virtanen (2014) alternatively constrained $\mathbf{R}_j(f)$ as the weighted sum of rank-1 matrices of the form $\mathbf{d}_j(f)\mathbf{d}_j^H(f)$ uniformly spanning all possible incoming sound directions on the 3D sphere. Ito *et al.* (2015) proposed similar linear subspace constraints for diffuse sources.

---

3 Note that the IC is defined for the full spatial image in (3.20) instead of the reverberant part only in (3.6).

3.5   Summary

In this chapter, we described the various types of mixtures encountered in audio and argued that, in most cases, they boil down to a linear mixing model. We examined the properties of impulse responses and reviewed the most common impulse response models.

These models are essentially used for multichannel separation and enhancement in Part III of the book. Specifically, the narrowband approximation and RTFs are used in Chapters 10, 11, 12, and 13, and full-rank models in Chapter 14. For specific use of binaural properties, see Chapter 18. Advanced topics such as handling moving sources or microphones, convolution in the STFT domain, and learning the manifold of impulse responses are discussed in Chapter 19.

Bibliography

Affes, S. and Grenier, Y. (1997) A signal subspace tracking algorithm for microphone array processing of speech. *IEEE Transactions on Speech and Audio Processing*, **5** (5), 425–437.

Asari, H., Pearlmutter, B.A., and Zador, A.M. (2006) Sparse representations for the cocktail party problem. *The Journal of Neuroscience*, **26** (28), 7477–7490.

Bartlett, B. and Bartlett, J. (2012) *Practical Recording Techniques: the Step-by-step Approach to Professional Recording*, Focal Press, 6th edn.

Benichoux, A., Simon, L.S.R., Vincent, E., and Gribonval, R. (2014) Convex regularizations for the simultaneous recording of room impulse responses. *IEEE Transactions on Signal Processing*, **62** (8), 1976–1986.

Blauert, J. (1997) *Spatial Hearing: The Psychophysics of Human Sound Localization*, MIT Press.

Duong, N.Q.K., Vincent, E., and Gribonval, R. (2010) Under-determined reverberant audio source separation using a full-rank spatial covariance model. *IEEE Transactions on Audio, Speech, and Language Processing*, **18** (7), 1830–1840.

Duong, N.Q.K., Vincent, E., and Gribonval, R. (2013) Spatial location priors for Gaussian model based reverberant audio source separation. *EURASIP Journal on Advances in Signal Processing*, **2013**, 149.

Ehlers, F. and Schuster, H.G. (1997) Blind separation of convolutive mixtures and an application in automatic speech recognition in a noisy environment. *IEEE Transactions on Signal Processing*, **45** (10), 2608–2612.

Févotte, C. and Cardoso, J.F. (2005) Maximum likelihood approach for blind audio source separation using time-frequency Gaussian models, in *Proceedings of IEEE Workshop on Applications of Signal Processing to Audio and Acoustics*, pp. 78–81.

Gannot, S., Burshtein, D., and Weinstein, E. (2001) Signal enhancement using beamforming and nonstationarity with applications to speech. *IEEE Transactions on Signal Processing*, **49** (8), 1614–1626.

Gorlow, S. and Reiss, J.D. (2013) Model-based inversion of dynamic range compression. *IEEE Transactions on Audio, Speech, and Language Processing*, **21** (7), 1434–1444.

Gupta, M. and Douglas, S.C. (2007) Beamforming initialization and data prewhitening in natural gradient convolutive blind source separation of speech mixtures, in *Proceedings* of *International Conference on Independent Component Analysis and Signal Separation*, pp. 462–470.

Gustafsson, T., Rao, B.D., and Trivedi, M. (2003) Source localization in reverberant environments: Modeling and statistical analysis. *IEEE Transactions on Speech and Audio Processing*, **11**, 791–803.

ISO (2003) ISO 9921. Ergonomics – Assessment of speech communication.

Ito, N., Vincent, E., Nakatani, T., Ono, N., Araki, S., and Sagayama, S. (2015) Blind suppression of nonstationary diffuse noise based on spatial covariance matrix decomposition. *Journal of Signal Processing Systems*, **79** (2), 145–157.

Jeub, M., Schäfer, M., and Vary, P. (2009) A binaural room impulse response database for the evaluation of dereverberation algorithms, in *Proceedings of IEEE International Conference on Digital Signal Processing*, pp. 1–4.

Knaak, M., Araki, S., and Makino, S. (2007) Geometrically constrained independent component analysis. *IEEE Transactions on Audio, Speech, and Language Processing*, **15** (2), 715–726.

Koldovský, Z., Malek, J., and Gannot, S. (2015) Spatial source subtraction based on incomplete measurements of relative transfer function. *IEEE/ACM Transactions on Audio, Speech, and Language Processing*, pp. 1335–1347.

Kowalski, M., Vincent, E., and Gribonval, R. (2010) Beyond the narrowband approximation: Wideband convex methods for under-determined reverberant audio source separation. *IEEE Transactions on Audio, Speech, and Language Processing*, **18** (7), 1818–1829.

Kuttruff, H. (2000) *Room acoustics*, Taylor & Francis.

Li, X., Horaud, R., Girin, L., and Gannot, S. (2015) Local relative transfer function for sound source localization, in *Proceedings of European Signal Processing Conference*, pp. 399–403.

Lin, Y., Chen, J., Kim, Y., and Lee, D.D. (2007) Blind channel identification for speech dereverberation using ℓ1-norm sparse learning, in *Proceedings of Neural Information Processing Systems*, pp. 921–928.

Markovich, S., Gannot, S., and Cohen, I. (2009) Multichannel eigenspace beamforming in a reverberant noisy environment with multiple interfering speech signals. *IEEE Transactions on Audio, Speech, and Language Processing*, **17** (6), 1071–1086.

Nguyen Thi, H.L. and Jutten, C. (1995) Blind source separation for convolutive mixtures. *Signal Processing*, **45** (2), 209–229.

Nikunen, J. and Virtanen, T. (2014) Direction of arrival based spatial covariance model for blind sound source separation. *IEEE/ACM Transactions on Audio, Speech, and Language Processing*, **22** (3), 727–739.

Parra, L.C. and Alvino, C.V. (2002) Geometric source separation: Merging convolutive source separation with geometric beamforming. *IEEE Transactions on Speech and Audio Processing*, **10** (6), 352–362.

Polack, J.D. (1993) Playing billiards in the concert hall: The mathematical foundations of geometrical room acoustics. *Applied Acoustics*, **38** (2), 235–244.

Reindl, K., Zheng, Y., Schwarz, A., Meier, S., Maas, R., Sehr, A., and Kellermann, W. (2013) A stereophonic acoustic signal extraction scheme for noisy and reverberant environments. *Computer Speech and Language*, **27** (3), 726–745.

Ribas, D., Vincent, E., and Calvo, J.R. (2016) A study of speech distortion conditions in real scenarios for speaker recognition applications, in *Proceedings of IEEE Spoken Language Technology Workshop*, pp. 13–20.

Sawada, H., Araki, S., Mukai, R., and Makino, S. (2007) Grouping separated frequency components with estimating propagation model parameters in frequency-domain blind source separation. *IEEE Transactions on Audio, Speech, and Language Processing*, **15** (5), 1592–1604.

Sawada, H., Kameoka, H., Araki, S., and Ueda, N. (2013) Multichannel extensions of non-negative matrix factorization with complex-valued data. *IEEE Transactions on Audio, Speech, and Language Processing*, **21** (5), 971–982.

Sturmel, N., Liutkus, A., Pinel, J., Girin, L., Marchand, S., Richard, G., Badeau, R., and Daudet, L. (2012) Linear mixing models for active listening of music productions in realistic studio conditions, in *Proceedings of the Audio Engineering Society Convention*. Paper 8594.

Vincent, E., Arberet, S., and Gribonval, R. (2009) Underdetermined instantaneous audio source separation via local Gaussian modeling, in *Proceedings of International Conference on Independent Component Analysis and Signal Separation*, pp. 775 –782.

Wallach, H. (1940) The role of head movements and vestibular and visual cues in sound localization. *Journal of Experimental Psychology*, **27** (4), 339–368.

Wightman, F.L. and Kistler, D.J. (1999) Resolution of front-back ambiguity in spatial hearing by listener and source movement. *Journal of the Acoustical Society of America*, **105** (5), 2841–2853.

Yılmaz, Ö. and Rickard, S.T. (2004) Blind separation of speech mixtures via time-frequency masking. *IEEE Transactions on Signal Processing*, **52** (7), 1830–1847.

4

Multichannel Source Activity Detection, Localization, and Tracking

Pasi Pertilä, Alessio Brutti, Piergiorgio Svaizer, and Maurizio Omologo

In the previous chapters, we have seen how both spectral and spatial properties of sound sources are relevant in describing an acoustic scene picked up by multiple microphones distributed in a real environment. This chapter now provides an introduction to the most common problems and methods related to source activity detection, localization, and tracking based on a multichannel acquisition setup. In Section 4.1 we start with a brief overview and with the definition of some basic notions, in particular related to time difference of arrival (TDOA) estimation and to the so-called acoustic maps. In Section 4.2, the activity detection problem will be addressed, starting from an overview of the most common methods and concluding with recent trends. Section 4.3 will examine the localization problem for both static and moving sources, with some insights into the localization of multiple sources. Section 4.4 will conclude the chapter.

Note that, although activity detection and localization apply to any kind of audio source, often the signal of interest is speech. Due to its spectral and temporal peculiarities, speech calls for specific algorithmic solutions which typically do not generalize to other sources. As an example, speech is by definition a nonstationary process, characterized by both long and very short pauses. Therefore, especially in Section 4.2 and Section 4.3.3, we will mainly focus on the specific case of speech signals.

4.1 Basic Notions in Multichannel Spatial Audio

In general, an acoustic scene consists of several contributions, including point sources (e.g., human speakers), diffuse sources, and reverberation due to reflections on obstacles and walls inside an enclosure.

Let us assume that the environment is equipped with a microphone array, or with a network of distributed microphones or small arrays. Given the possibility of observing the acoustic scene from different spatial positions, the corresponding multichannel signals can be exploited to derive the spatial position of each target sound source at each time instant and track its trajectory over time. This task is not trivial, except for simple cases corresponding to low noise and reverberation conditions, i.e. high signal-to-noise ratio (SNR) and high direct-to-reverberant ratio (DRR). In addition, it may also be more

Audio Source Separation and Speech Enhancement, First Edition.
Edited by Emmanuel Vincent, Tuomas Virtanen and Sharon Gannot.
© 2018 John Wiley & Sons Ltd. Published 2018 by John Wiley & Sons Ltd.
Companion Website: https://project.inria.fr/ssse/

or less complicated according to the directivity of the target source, the number of other sources in the scene and how stationary they are.

Our ability to sense sound source direction and even distance using only two ears demonstrates a remarkable ability that necessitates the utilization of propagation changes introduced by the head, torso, and pinnae, referred to as *head-related transfer functions* (Blauert, 1997), in combination with small head movements. Specific methods have been developed to reproduce this capability using in-ear microphones. This chapter focuses on the more common scenario of microphone arrays, which typically allow source direction to be evaluated in an unambiguous manner, for instance based on the estimation of TDOA of direct wavefronts between microphones.

The literature on sound source localization and tracking of the last two decades is very rich in references to different methodologies addressing a large variety of scenarios, from anechoic to highly reverberant conditions, and from static to moving sources. The most common problem addressed by different scientific communities and application fields (e.g., speaker diarization and meeting transcription, videoconferencing, robotics, etc.) is that of localizing a human speaker.

A relevant aspect that deserves to be mentioned is the fact that localizing a sound source makes sense only for temporal segments when that source is active. Therefore, the problems of activity detection and localization are inherently coupled, and the performance of either can be of crucial importance for the other. For this reason, the two topics are treated together in this chapter.

4.1.1 TDOA Estimation

Algorithms for the estimation of TDOA have been widely investigated in many application fields in which the signals captured by multiple sensors need to be related temporally to each other. In general, optimal estimators take advantage of the knowledge about the statistical properties of the source signals in order to detect the correlation between different channels even under very low SNR conditions. Most techniques are based on computing the *generalized cross-correlation* (GCC) between pairs of channels, whose maximum, in ideal conditions, identifies the TDOA (Carter, 1987). In real conditions, the problem is complicated by the presence of noise and reverberation (Chen *et al.*, 2006). GCC associated with direct wavefronts is obfuscated by the cross-contributions due to the reflected wavefronts and other sources. As a consequence, TDOA estimation may become infeasible (e.g., if a talker is speaking while turning his/her back to the microphones). The most frequently used technique for TDOA estimation in noisy and reverberant conditions is the GCC with *phase transform* (GCC-PHAT) (Knapp and Carter, 1976), which has been shown to be quite robust to the adverse acoustic conditions of real environments, especially for broadband signals such as speech.

An alternative method uses the *multiple signal classification* algorithm, initially proposed by Schmidt (1986), for high-resolution spectral estimation, which has been widely studied and extended to deal with broadband sources. It is based on the eigendecomposition of the covariance matrix of the observation vectors.

Another approach to TDOA estimation is based on blind identification of the acoustic impulse responses (Section 3.2.1) from the source to the microphones. The *adaptive eigenvalue decomposition* algorithm (Benesty, 2000) finds the eigenvector associated with the smallest eigenvalue of the covariance matrix of two microphone signals and derives an adaptive estimate of the two impulse responses. The time delay estimate is determined as the difference between the lags of the two main peaks, which are assumed to be associated with the direct paths.

TDOA estimation across multiple channels can be jointly achieved by adopting multichannel algorithms that exploit the mutual interchannel correlation expressed in a covariance matrix (Chen *et al.*, 2003) or in other representations related to multichannel propagation, e.g. in a demixing matrix estimated by means of blind source separation (BSS) (Nesta and Omologo, 2012).

The *Cramér–Rao lower bound* represents the lower bound on the variance of an unbiased estimator. In case of TDOA estimation, if the presence of outliers is excluded, the Cramér–Rao lower bound has been shown to be inversely related to the time-bandwidth product and the SNR, therefore an increase in any one of these parameters will decrease the TDOA variance. Specific bounds have been developed to detect SNR conditions where the Cramér–Rao lower bound applies, refer to Sadler and Kozick (2006) for an overview.

4.1.2 GCC-PHAT

The GCC-PHAT between the signals of a microphone pair (i, i') as a function of the lag τ is obtained from the inverse Fourier transform of the cross-power spectrum normalized with respect to its amplitude, such that only its phase information is kept (Omologo and Svaizer, 1994). The GCC-PHAT at frame n is computed by

$$\psi_{ii'}(n, \tau) = \frac{1}{F} \sum_{f=0}^{F-1} \frac{x_i(n,f)x_{i'}^*(n,f)}{|x_i(n,f)||x_{i'}(n,f)|} e^{2j\pi\tau f/F}, \tag{4.1}$$

where $x_i(n,f)$ and $x_{i'}(n,f)$ are the short-time Fourier transforms (STFTs) of the microphone signals. In the ideal case without noise and reverberation, the phase of the cross-power spectrum is perfectly linear and the lag $\hat{\tau}$ of the peak of the GCC-PHAT is equal to the TDOA $\Delta_{ii'}$. Noise and reverberation corrupt the phase information in (4.1) (Champagne *et al.*, 1996) but, unless the conditions are too critical, the position of the peak of $\psi_{ii'}(n, \tau)$ still indicates the correct TDOA of the direct wavefront (Omologo and Svaizer, 1997). GCC-PHAT can be shown to achieve the best linear phase interpolation of the cross-spectrum phase, although the transformation itself is not linear and therefore TDOA information gets partially distorted in the presence of multiple sources, giving rise to possible outliers, i.e. spurious peaks in $\psi_{ii'}(n, \tau)$.

4.1.3 Beamforming and Acoustic Maps

Based on the mutual TDOAs across a set of microphones, it is possible to achieve selective sound acquisition, i.e. to reinforce the sound coming from a given direction/position while mitigating the components generated elsewhere. This operation, known as beamforming, also provides a method to localize active sources.

Assuming a source in spatial position **p**, the corresponding TDOA of channels $i \in \{2, \dots, I\}$ with respect to the reference channel $i' = 1$ must be compensated to

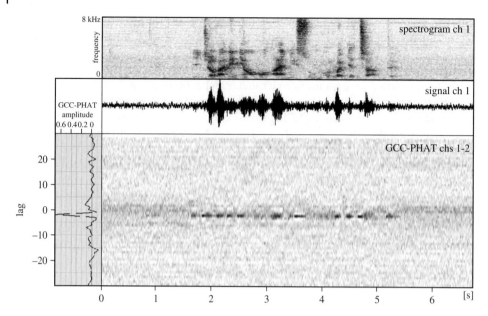

Figure 4.1 Example of GCC-PHAT computed for a microphone pair and represented with gray levels in the case of a speech utterance in noisy and reverberant environment. The left part highlights the GCC-PHAT at a single frame, with a clear peak at lag $\hat{\tau} = -2$.

produce a beamformed signal. The output of the delay-and-sum (DS) beamformer is given by

$$\hat{s}(n,f) = \mathbf{d}^H(\mathbf{p},f)\mathbf{x}(n,f), \qquad (4.2)$$

with $\mathbf{d}(\mathbf{p},f)$ the steering vector defined in (3.8). Chapter 10 will provide more details on beamforming and on other variants, e.g. minimum variance distortionless response (MVDR). The power $|\hat{s}(n,f)|^2$ of the beamformed signal at time frame n can be evaluated for any position $\mathbf{p}$ of interest. The result, interpreted as a function of the coordinates $\mathbf{p}$, is an acoustic map $\mathcal{M}^{\text{SRP}}(n, \mathbf{p})$ called the *steered response power* (SRP) (Brandstein and Ward, 2001) or power field (Alvarado, 1990), whose maximum ideally corresponds to the position of sound emission:

$$\mathcal{M}^{\text{SRP}}(n, \mathbf{p}) = \sum_{f=0}^{F-1} \mathbf{d}^H(\mathbf{p},f)\mathbf{x}(n,f)\mathbf{x}^H(n,f)\mathbf{d}(\mathbf{p},f). \qquad (4.3)$$

Acoustic maps can also be built based on the spatial correlation of the received signals rather than on beamforming power. One of the most widely used acoustic maps exploiting the spatial correlation is the *global coherence field*, introduced by De Mori (1998). For a given frame n and position $\mathbf{p}$, the global coherence field is defined as the acoustic map:

$$\mathcal{M}^{\text{SRP-PHAT}}(n, \mathbf{p}) = \sum_{ii' \in \mathcal{I}} \psi_{ii'}(n, \Delta_{ii'}(\mathbf{p})), \qquad (4.4)$$

where $\mathcal{I}$ is the set of microphone pairs. Basically, for any spatial position or direction of arrival (DOA) $\mathbf{p}$, the map value is obtained by summing the correlation of each

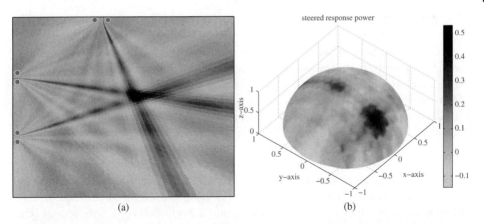

steered response power

Figure 4.2 Two examples of global coherence field acoustic maps. (a) 2D spatial localization using a distributed network of three microphone pairs, represented by circles. Observe the hyperbolic high correlation lines departing from the three pairs and crossing in the source position. (b) DOA likelihood for upper hemisphere angles using an array of five microphones in the (*x,y*) plane. High likelihood region exists around azimuth $\theta = -90°$ and elevation $\varphi = 45°$, i.e. unit-norm vector $\mathbf{k} = [0, -1/\sqrt{2}, 1/\sqrt{2}]^T$, see Figure 3.6.

microphone pair for the geometrically derived time delay $\Delta_{ii'}(\mathbf{p})$ if the source is in $\mathbf{p}$. Note that in the case of a compact microphone array, the set $\mathcal{I}$ can consist of all possible microphone pairs, resulting in a double summation over terms i and i'. Conversely, in the case of distributed arrays, considering all possible microphone combinations may be infeasible (due to large distances) or even detrimental (due to uneven microphone distributions). Figure 4.2 shows two examples of global coherence field-based acoustic maps applied to 2D source localization and DOA estimation.

The global coherence field is also known as the SRP-PHAT (Brandstein and Ward, 2001) since it can be obtained from the SRP by applying the PHAT whitening. Replacing (4.1) in (4.4) and considering summation over all microphone pairs we obtain this alternative formulation:

$$\mathcal{M}^{\text{SRP-PHAT}}(n, \mathbf{p}) = \sum_{f=0}^{F-1} \mathbf{d}^H(\mathbf{p},f) \frac{\mathbf{x}(n,f)}{|\mathbf{x}(n,f)|} \frac{\mathbf{x}^H(n,f)}{|\mathbf{x}(n,f)|} \mathbf{d}(\mathbf{p},f). \tag{4.5}$$

Due to their efficacy and easy implementation, acoustic maps have been widely investigated in the literature and several variants have been presented. Also worth mentioning are the oriented global coherence field (Brutti *et al.*, 2005; Abad *et al.*, 2007), which allows estimating the source orientation, and the use of a product in (4.4) instead of a sum (Pertilä *et al.*, 2008).

The global coherence field (4.5) uses amplitude normalization, which results in equal contribution from all frequency bands, while only frequencies emitted by the speaker are relevant for determining his/her position. Traditional SNR estimation techniques (Valin *et al.*, 2007) and, more recently, convolutional neural networks (Pertilä and Cakir, 2017) have been combined with GCC-PHAT in order to reduce the contribution of nonspeech frequencies in the resulting acoustic map.

4.2 Multi-Microphone Source Activity Detection

Given the basic notions above, we now consider the problem of detecting temporal segments during which one or more sound sources are active. This task is generally referred to as *acoustic event detection*.

Many approaches have been described in the literature, ranging from the detection of an isolated acoustic event under quiet conditions to the more challenging detection of an event immersed in background noise, and given the possible presence of overlapping events. Humans have the ability to perform this analysis, also under the most hostile noisy conditions, while the development of accurate automatic systems still represents an open challenge. Related to these goals, let us recall the recent DCASE challenges (Stowell *et al.*, 2015) and the earlier CLEAR evaluation task (Temko *et al.*, 2009).

In general, the methods used to achieve acoustic event detection have strong similarities with those adopted for the specific problem of *voice activity detection* (VAD), also known as speech activity detection. VAD aims to find the beginning and the end of each speech sequence inside an audio signal.[1] The VAD problem was explored initially for speech enhancement and voice communication (Kondoz, 2004; Vary and Martin, 2006; Loizou, 2013), and for speech recognition and related applications (De Mori, 1998; Rabiner and Juang, 1993; Ramírez *et al.*, 2007). Early VAD algorithms were conceived for close-talking telephone input signals and they were particularly effective for saving bandwidth, complexity, and battery in early cellular phones. Their performance[2] was found to be satisfactory under moderate to high SNR conditions. However, the problem becomes very complex as soon as ideal conditions do not hold anymore, for instance with low SNRs, in the presence of nonstationary noise and with reverberation in the case of far-field input, also referred to as "distant speech" (Wölfel and McDonough, 2009). One of the main reasons for this loss of performance is that in moderate or high SNR conditions the energy distribution is generally bimodal, which makes it easy to distinguish noise segments from speech. In low SNR conditions and in the presence of reverberation, however, this statistical distribution becomes more confused. For this reason, the choice of a good set of acoustic features, and the application of normalization and other processing techniques, are crucial in the development of a robust VAD solution.

4.2.1 Single-Channel Methods and Acoustic Features

In general, a VAD system consists of a simple automaton, i.e. a finite state machine with two states that correspond to "speech" and "nonspeech", respectively. A simple way to process the input signal is to derive its log-energy in the full band, or to derive an estimate of the SNR, on a frame-by-frame basis, and to compare these features to one or several thresholds, in order to move from one state to the other. Additional states can

---

1 Another similar task is the estimation of the speech presence probability (SPP) for each time-frequency bin, which can be used as preprocessing for different tasks. SPP estimation will be addressed in Chapter 6.
2 The performance of a VAD algorithm depends on the back-end application. In general the evaluation is conducted by measuring the match between the derived boundaries and ground-truth ones. Receiver operating characteristic curves are commonly used to represent miss vs. false alarm rates. In the case of ASR applications, the error rate of the recognizer can represent a significant indicator of VAD performance. Latency and computational cost are also important features in real-time applications.

be added to the state machine, for instance to distinguish short or long pauses, the short ones being often related to stop closures or to a very short silence between words. Also hangover schemes are often applied to reduce misdetection of speech, especially in the case of low energy segments.

It is also worth noting that most speech applications require real-time VAD process-ing, with a prompt detection of a new utterance, for an immediate delivering of the entire speech sequence to the back-end, e.g. to the decoding step in case of speech recognition applications. Hence, a compromise between latency, sensitivity, accuracy, and computa-tional cost is often necessary in a real-world application. Moreover, in some cases VAD is tightly connected with other processing techniques having critical real-time constraints, as in the case of acoustic echo cancellation and suppression (e.g., in the automotive case), in which discriminating between speech and nonspeech segments is of fundamental importance. One of the most relevant functionalities involving the latter techniques is the so-called *barge-in* in conversational interfaces, with which the user can interrupt the system while a prompt is being played, in order to allow for a more natural and efficient interaction.

As mentioned above, energy and SNR estimates represent acoustic features that are intuitively suitable for a VAD task. However, the resulting detectors are generally very sensitive to variations in noisy conditions. To improve VAD robustness, many approaches were explored in the past, e.g. based on adaptive thresholds and energies in subbands. In this regard, tightly connected to VAD is the noise estimation problem, which aims to model nonspeech segments (see Chapter 6) for an effective discrim-ination between speech and silence. Actually, some VAD algorithms were proposed that include a noise subtraction stage. This can be particularly effective in the case of stationary background noise.

Some of the acoustic features that were investigated to tackle the adverse conditions typically observed in the presence of noise and/or reverberation are zero-crossing rate, pitch, periodicity measures (Tucker, 1992), linear predictive coding parameters, higher order statistics of the linear predictive coding residual, long-term spectral envelopes, Mel-frequency cepstral coefficients (MFCCs), and possible combinations of such fea-tures. Early statistical approaches, including likelihood ratios, Gaussian mixture mod-els (GMMs), and hidden Markov models (HMMs), were also investigated in order to further improve the robustness of VAD algorithms (Sohn *et al.*, 1999). Finally, a vari-ety of approaches have been proposed which exploit the spectral and temporal struc-ture of speech (e.g., based on long-term features), even when immersed in noise and reverberation.

4.2.2 Multichannel Methods

When two or more microphones are available, additional cues can be considered in the development of a VAD algorithm. The simplest approach consists of exploiting redundant information that is jointly available in the microphone signals. Besides the use of energy, SNR, or other acoustic features such as those listed above, one can also adopt cross-correlation or other spatial features. An example of this approach is the use of the maximum of the GCC-PHAT function (4.1), generally corresponding to the TDOA at the microphone pair, as a feature discriminating speech from silence (Armani *et al.*, 2003). Figure 4.1 shows the effectiveness of this simple approach when processing noisy and reverberant signals. Note that the frame length, the frame shift, as well as a possible windowing of the two signals may have an impact on the accuracy of this analysis. Moreover, it is worth mentioning that the GCC-PHAT analysis becomes effective in highlighting speech activity only when the microphone pair is impinged by a sufficiently high energy of the direct path, i.e. with a high DRR, which may also be influenced by the speaker's head orientation.

Many dual-microphone methods described in the literature are based on this technique, or based on the use of the normalized cross-correlation between the microphone signals in the time-domain. Most of the alternative approaches rely on the correlation of the information in the two signals, for instance Rubio *et al.* (2007) exploited the homogeneity of the DOA estimates by profiting from the concentration of speech energy in time, frequency, and space.

So far, we have mostly referred to the detection of one active speaker. However, many applications currently being explored are characterized by the presence of two or more speakers, sometimes talking at the same time. In this case, the challenge is also related to the need for tracking an unpredictable number of speakers and sound sources that may be active simultaneously, although not all of them may be contributing in the same way at every time instant. As discussed in the following chapters, the case of multiple sound sources can be addressed in an effective way by adopting a BSS framework. In fact, most BSS techniques implicitly extract the activity of each source over time (Nesta *et al.*, 2011).

4.2.3 Deep Learning based Approaches

As is happening in many other application fields, for the VAD problem the current trend is also to adopt deep learning. A huge variety of novel methods and techniques have been reported in the recent literature (Carlin and Elhilali, 2015; Zhang and Wu, 2013), often relying on supervised and data-driven approaches based on recurrent (RNN) and deep neural networks (DNNs).

Although machine learning based VAD is far from being fully explored, it has several advantages. The integration between VAD and a speech recognition system is immediate, and it can be addressed as a unique processing framework. A more rigorous statistical formulation of the VAD task is provided thanks to the theoretical foundations of machine learning. Machine learning gives the chance to fuse together features of different natures (Zhang and Wu, 2013), which are often addressed in traditional VAD algorithms by resorting to the use of heuristics.

4.3 Source Localization

Given the acoustic measurements/observations introduced in Section 4.1 (i.e., TDOA, GCC-PHAT, SRP-PHAT), in this section we see how the spatial location of the emitting source can be derived from multichannel recordings. This problem, known as *source localization* or *speaker localization*, has been widely addressed in the literature for decades. Early solutions were mostly developed for localization of a static continuous emitter, focusing on closed-form solutions. However, most naturally occurring sound sources are typically neither persistently active nor stationary. To address such issues, the complexity of the localization solutions has increased over the years from a single static source scenario to multiple moving time-overlapping sources. Such solutions typically utilize activity detection and data-association with (multiple) target tracking methods. In this section we first list the most common approaches to infer the position of a single persistent acoustic source with framewise processing, then we present solutions for tracking the positions of moving intermittent sources and, finally, give a quick overview of how localization can be tackled in presence of multiple sources.

In the following we formulate the problem as that of estimating the 3D spatial position of the emitting source. However, the approaches reviewed here are also applicable to the estimation of the source direction in far-field scenarios using a compact microphone array.

4.3.1 Single-Frame Localization of a Static Source

This section introduces different localization methods that utilize the data from a single time frame to obtain an instantaneous location estimate.

Given a set of observed TDOAs $\widehat{\mathbf{\Delta}} = \{\widehat{\Delta}_{ii'}\}_{ii'\in\mathcal{I}}$, where $\mathcal{I}$ is the set of microphone pairs, the source position can be estimated by maximizing likelihood: $\widehat{\mathbf{p}} = \mathrm{argmax}_{\mathbf{p}}\, p(\mathbf{p} \mid \widehat{\mathbf{\Delta}})$, which, assuming independent Gaussian error between TDOA measurements with uniform variance, can be reformulated as the least squares problem:

$$\widehat{\mathbf{p}} = \underset{\mathbf{p}}{\mathrm{argmin}} \sum_{ii'\in\mathcal{I}} |\Delta_{ii'}(\mathbf{p}) - \widehat{\Delta}_{ii'}|^2. \tag{4.6}$$

This formulation holds both in the far-field and the near-field, depending on the definition of $\Delta_{ii'}(\mathbf{p})$. In the far-field case, the 3D DOA vector $\mathbf{k}$ in Cartesian coordinates is linearly related to the TDOA measurement (see (3.17)), and an exact closed-form solution can be obtained if the microphone positions span a 3D space (Yli-Hietanen *et al.*, 1996). However, the near-field TDOA equation is nonlinearly related to the source position and must be solved iteratively, e.g. with gradient descent. To obtain a reasonable initial guess or to ensure a computationally lightweight solution, approximate or multiple step closed-form procedures can be used. See Li *et al.* (2016) for a recent review of closed-form TDOA-based localization.

In contrast to the TDOA-based methods, direct localization does not require the preliminary estimation of the TDOA between microphone pairs. This intermediate estimation is especially critical when localizing weak sources for which a TDOA estimation step might produce erroneous values. Instead, they directly aggregate the evidence from acoustic measurements into the acoustic map described in Section 4.1.3 (e.g., SRP-PHAT). The position of the emitting source is estimated by maximizing the acoustic map: $\widehat{\mathbf{p}} = \mathrm{argmax}_{\mathbf{p}}\mathcal{M}(n, \mathbf{p})$. The downside is that the direct maximization of the acoustic map requires a strategy that sets the bounds and the granularity of the search, and an exhaustive search might not always be computationally feasible in practice.

The localization problem can also be solved using a maximum likelihood (ML) formulation in the frequency domain using the multichannel power spectrum as an observation vector, assuming that microphone noise and reverberation have a Gaussian distribution (Zhang *et al.*, 2007). A similar approach was used by Mungamuru and Aarabi (2004) where the source orientation was also estimated.

Recently, dictionary based solutions (Malioutov *et al.*, 2005; Asaei *et al.*, 2016; Gretsistas and Plumbley, 2010; Yin and Chen, 2011) have gained interest in the research community as they include some awareness of the environment and of the microphones. Typically, the unknown propagation channel is approximated as a sparse linear combination of an overcomplete set of measured impulse responses. These methods are tightly related to the manifold learning methods reviewed in Section 19.3.2.

Finally, after some pioneering works in the past (Arslan and Sakarya, 2000), supervised learning using artificial neural networks has recently been applied for DOA estimation of a speech source (Xiao *et al.*, 2015) and position and orientation estimation (Nakano *et al.*, 2009). Unlike conventional approaches based on geometrical or physical properties, these methods learn the acoustic properties of a given environment from training data. An emerging trend is manifold learning, where a mapping between a set of measured transfer functions and any source position is learned, either in a room-dependent or room-independent way. Further details on these new research trends are reported in Chapter 19.

4.3.2 Effect of Microphone Array Geometry

Independently of the algorithm, the accuracy of the estimates is strongly influenced by the geometry of the microphone array and by the position of the source. Error analysis can be used when designing the array geometry. Increasing the distance between microphones increases the spatial resolution at the cost of a lower correlation between the captured signals due to spatial aliasing and reverberation, see (3.6). Note that increasing the resolution using interpolation is sometimes useful, but it does not increase the underlying fundamental information. The function that maps source position into TDOA values is nonlinear. Therefore, the errors in TDOA measurement will propagate to the source position estimate in a way that depends on the source and microphone positions, i.e. on the partial derivative of the nonlinear near-field TDOA function with respect to the source position.

Three different configurations consisting of two separate microphone pairs with 20 cm spacing are used to quantify positioning errors by assuming a TDOA measurement error standard deviation of $\sigma_\Delta = \pm 1$ sample at 48 kHz sampling rate. Position error covariance is here obtained similarly to Kaplan and Hegarty (2006, chap. 7). The resulting position error is depicted as a confidence region around the true source position in Figure 4.3, where the ellipses cover a 1-σ position confidence interval. In Figure 4.3a the error distributes in very narrow ellipses, whose length increases as the source moves in a lateral position or far away from the microphone pairs. In Figure 4.3b, the error in the depth dimension is substantially reduced by increasing the distance between the pairs. In Figure 4.3c the error is rather uniformly distributed along the two dimensions and increases when the source is distant from both pairs. Note also the very poor results in the bottom left corner where the source is lateral to both pairs. The key observation is that if a shift in position results in a small change in the TDOA measurements, the accuracy decreases. For further study on the sensitivity of the location estimation refer to Brandstein (1995).

4.3.3 Localization of Moving and Intermittent Sources

In Section 4.3.1 we addressed the localization problem assuming that the target source is both continuously emitting and static. When considering speaker localization over

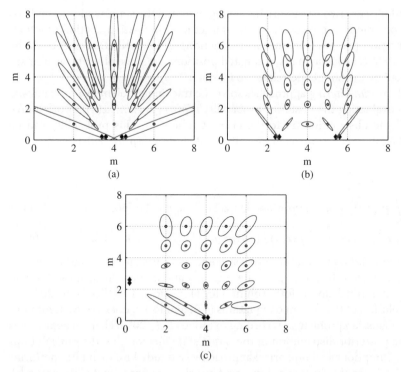

Figure 4.3 1-σ position error distributions (68.3% confidence) resulting from TDOA errors ($\sigma = 1$ sample) for two microphone pairs in three different geometries.

time, neither the temporal persistence of the emitter nor the speaker stationarity can be assumed.

Localization can be applied only to frames with detected speech activity to handle speech inactivity. To guarantee continuous output of the location system a simple threshold-and-hold approach can be used, where observations whose reliability (e.g., probability of activity, or the acoustic map peak height) is below a given threshold are discarded and the previous output is presented. Further constraints can be introduced on the spatial consistency of two consecutive estimates to remove outlier measurements. The *state-space tracking* framework offers a more elegant and solid solution to jointly track a moving source and handle outliers.

Let us denote as $\bar{\mathbf{p}}(n) = [\mathbf{p}(n), \dot{\mathbf{p}}(n)]^T$, where $\dot{\mathbf{p}}$ is the first derivative of $\mathbf{p}$, the target space at time frame n that captures the relevant kinematic properties of the target in a stacked column vector form. In addition, other parameters related to the properties of the target such as orientation can be included. Under the assumption that the target motion process is Markovian, i.e. the current state depends only on the previous state and not on the earlier trajectory, given the observation vector $\mathbf{z}(n)$, the system can be modeled as:

$$\bar{\mathbf{p}}(n) = g(\bar{\mathbf{p}}(n-1), \mathbf{v}(n)), \tag{4.7}$$

$$\mathbf{z}(n) = h(\bar{\mathbf{p}}(n), \boldsymbol{\eta}(n)), \tag{4.8}$$

where $\boldsymbol{v}(n)$ and $\boldsymbol{\eta}(n)$ are two noise variables independent of the state, while $g(\cdot)$ and $h(\cdot)$ are two functions modeling, respectively, the target motion (state transition function) and the observation process (measurement function). The observation $\mathbf{z}(n)$ could be TDOA, GCC-PHAT, acoustic maps, estimated positions, or some other measurement that is generated by the target and related to its state.

Sequential Bayesian filtering approaches solve the tracking problem in a probabilistic framework by evaluating the posterior probability $p(\bar{\mathbf{p}}(n) \mid \mathbf{z}(1), \ldots, \mathbf{z}(n))$ of the source state given all the observation up to the current time step. This probability can be obtained by Bayesian recursion given the initial distribution $p(\bar{\mathbf{p}}(0) \mid \mathbf{z}(0))$ (Stone *et al.*, 2013, chap. 3):

$$p(\bar{\mathbf{p}}(n) \mid \mathbf{z}(1), \ldots, \mathbf{z}(n - 1))$$

$$= \int p(\bar{\mathbf{p}}(n) \mid \bar{\mathbf{p}}(n - 1))p(\bar{\mathbf{p}}(n - 1) \mid \mathbf{z}(1), \ldots, \mathbf{z}(n - 1))\mathrm{d}\bar{\mathbf{p}}(n - 1) \quad (4.9)$$

$$p(\bar{\mathbf{p}}(n) \mid \mathbf{z}(1), \ldots, \mathbf{z}(n)) \propto p(\mathbf{z}(n) \mid \bar{\mathbf{p}}(n))p(\bar{\mathbf{p}}(n) \mid \mathbf{z}(1), \ldots, \mathbf{z}(n - 1)). \quad (4.10)$$

If the observation and target motion process contains Gaussian noise, and the observation is linearly related to the target state (i.e., $g(\cdot)$ and $h(\cdot)$ are linear), the *Kalman filter* provides an optimal closed-form solution for (4.9) and (4.10) (Klee *et al.*, 2006). A closed-form solution of the system of equations (4.9) and (4.10) does not exist for nonlinear and nonGaussian problems. According to Stone *et al.* (2013, chap. 3) approaches to evaluate the posterior distribution of the target state include closed-form solutions with relaxed assumptions and numerical approaches. Extended Kalman filter performs relaxation of the linear observation requirement by using an approximated linear model, while the unscented Kalman filter estimates the parameters of the involved Gaussian distributions by measuring the nonGaussian target state and measurement errors using a discrete set of points.

The likelihood and the state-space can be represented by discrete grids, and (4.9) is evaluated with discrete transition probabilities. The approach does not rely on any linearity or Gaussianity assumption, but one is left to select the grid size and form to capture all relevant situations, or to utilize a dynamic grid adjustment. In *particle filtering* or *sequential Monte Carlo* approximation, the posterior at time n is represented by a discrete set of K samples, or particles, $\bar{\mathbf{p}}_k(n)$, with associated weights $\gamma_k(n)$:

$$p(\bar{\mathbf{p}}(n) \mid \mathbf{z}(1), \ldots, \mathbf{z}(n)) = \sum_{k=1}^{K} \gamma_k(n)\delta_{\bar{\mathbf{p}}_k(n)}(\bar{\mathbf{p}}(n)) \quad (4.11)$$

with $\delta_{\bar{\mathbf{p}}_k(n)}$ the Dirac function centered in $\bar{\mathbf{p}}_k(n)$. At each time frame n the implementation of the particle filter consists of the following three steps: i) particles are propagated according to the motion model, ii) weights are computed according to the measurement likelihood, e.g. using an acoustic map or considering the least squares function in (4.6), and iii) particles are resampled using the posterior. Figure 4.4 graphically summarizes the steps of a particle filter based tracking approach. A detailed overview is available from Arulampalam *et al.* (2002).

As mentioned above, state-space tracking allows dealing with minor gaps in the information stream due to small pauses, spectral content of the emitted sound, interfering sounds, etc. However, a tracker should be activated to provide estimates only when a source is actually present in the scene and then released when the source

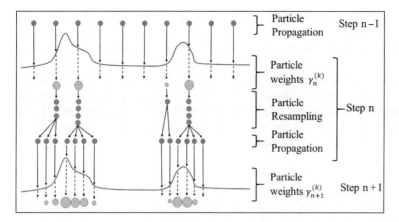

Figure 4.4 Graphical example of the particle filter procedure.

is silent. This activation can be achieved using, in parallel to the tracking, one of the VAD approaches described in Section 4.2 so that tracking starts only when the source activity exceeds a threshold. However, a too high threshold can result in missed weak targets, while a low value will cause false detections. Alternatively, activity detection can be embedded in the tracking scheme, extending the target state with the source activity information. Lehmann and Johansson (2006) presented an integrated solution where a measure for source activity is used to weight the likelihood. Similarly, *track-before-detect* approaches have been proposed to integrate target detection into tracking: Boers and Driessen (2003) present a track-before-detect particle filter that uses a likelihood ratio obtained from particle weights for source presence detection.

A variety of other tracking methods are available, for instance using HMMs (Roman and Wang, 2008; Kepesi *et al.*, 2007) or employing concepts similar to those of particle filtering such as the probability hypothesis density filter. Note that the tracking framework presented here uses only past information to estimate the current target state, which is referred to as filtering. If minor delays are acceptable in the application context, the future observations could also be used to enhance the tracking accuracy, which is referred to as *smoothing*, e.g. using the forward-backward algorithm (Caljon *et al.*, 2005).

To conclude, note that tracking can be performed in other domains instead of space, such as multidimensional TDOA (Brutti and Nesta, 2013) and relative time of arrival (Pertilä and Tinakari, 2013).

4.3.4 Towards Localization of Multiple Active Sources

Localization of multiple moving and intermittent sources (such as multiple simultaneous speakers) is a hard problem for which an established solution is yet to emerge. In general this problem is addressed by taking advantage of the speech sparsity in time, frequency, and spatial domains. Observations can be obtained through clustering in the spatio-temporal domain (Lathoud and Odobez, 2007), frequency domain (Blandin *et al.*, 2012), and eventually using expectation-maximization (EM) (Mandel *et al.*, 2010; Schwartz and Gannot, 2014; Dorfan and Gannot, 2015). This is particularly challenging in the presence of noise and reverberation. Therefore, GCC-PHAT deemphasis (Brutti *et al.*, 2010) or beamformer nulling (Valin *et al.*, 2004) could be employed to highlight all sources. In this direction, solutions originally developed for BSS were recently used to solve the multiple source localization problem, for instance based on independent component analysis (ICA) (Loesch *et al.*, 2009; Sawada *et al.*, 2003) or blind multiple-input multiple-output system identification (Buchner *et al.*, 2007).

Given measurements from multiple sources, traditional approaches for multiple target tracking can be divided into two steps (Stone *et al.*, 2013): i) assigning measurements to targets and ii) performing single target tracking separately. Often several reasonable alternatives to perform the association exist. The *multiple hypothesis tracking* method deals with this issue by creating different hypotheses, each of which assigns measurements to targets in a unique way. However, the number of hypotheses grows exponentially in time, and must be limited in practice (Pulford, 2005). Also, the hard association can be relaxed into a probabilistic one (Potamitis *et al.*, 2004). Lee *et al.* (2010) use multiple assignment hypotheses of TDOA measurements to targets and perform localization using a particle representation.

The requirement of having multiple acoustic measurements from each time step for the association process can be dealt with in several ways. Assuming that low correlation is associated with clutter or noise, simple thresholding of measurements is sufficient. For a TDOA-based tracking, evaluating all pairwise time delay measurements (e.g., GCC-PHAT) with all possible delays and selecting the highest peaks is both simple and often computationally tractable due to the single dimension of the search space. While the evaluation of all possible locations to obtain the most likely source positions is still straightforward for direct localization approaches (see Section 4.3.1, e.g. SRP-PHAT), it comes at the cost of computation burden depending on the dimension of the space to evaluate. As an alternative, the track-before-detect particle filtering paradigm allows the use of direct localization approaches for multiple target tracking with a mechanism for detecting appearing sources without thresholding the measurements (Pertilä and Hämäläinen, 2010; Brutti and Nesta, 2013; Valin *et al.*, 2007; Fallon and Godsill, 2010).

4.4 Summary

This chapter provided an overview of the basic notions and the most popular algorithms for speech activity detection and acoustic source localization in multichannel recordings. We focused on established solutions, giving insights into emerging trends and most recent developments. The accompanying web page includes audio samples, exercises, and Matlab code related to specific didactic cases.

Bibliography

Abad, A., Segura, C., Nadeu, C., and Hernando, J. (2007) Audio-based approaches for head orientation estimation in a smart room, in *Proceedings of Interspeech*, pp. 590–593.

Alvarado, V.M. (1990) *Talker Localization and Optimal Placement of Microphones for a Linear Microphone Array Using Stochastic Region Contraction*, Ph.D. thesis, Brown University.

Armani, L., Matassoni, M., Omologo, M., and Svaizer, P. (2003) Use of a CSP-based voice activity detector for distant-talking ASR, in *Proceedings of European Conference on Speech Communication and Technology*, vol. 2, pp. 501–504.

Arslan, G. and Sakarya, F.A. (2000) A unified neural-network-based speaker localization technique. *IEEE Transactions on Neural Networks*, **11** (4), 997–1002.

Arulampalam, M.S., Maskell, S., Gordon, N., and Clapp, T. (2002) A tutorial on particle filters for on-line nonlinear/non-gaussian Bayesian tracking. *IEEE Transactions on Signal Processing*, **50** (2), 174–178.

Asaei, A., Bourlard, H., Taghizadeh, M.J., and Cevher, V. (2016) Computational methods for underdetermined convolutive speech localization and separation via model-based sparse component analysis. *Speech Communication*, **76** (C), 201–217.

Benesty, J. (2000) Adaptive eigenvalue decomposition algorithm for passive acoustic source localization. *Journal of the Acoustical Society of America*, **107** (1), 384–391.

Blandin, C., Ozerov, A., and Vincent, E. (2012) Multi-source TDOA estimation in reverberant audio using angular spectra and clustering. *Signal Processing*, **92** (8), 1950–1960.

Blauert, J. (1997) *Spatial Hearing: the Psychophysics of Human Sound Localization*, MIT Press.

Boers, Y. and Driessen, H. (2003) A particle-filter-based detection scheme. *IEEE Signal Processing Letters*, **10** (10), 300–302.

Brandstein, M.S. (1995) *A Framework for Speech Source Localization Using Sensor Arrays*, Ph.D. thesis, Brown University.

Brandstein, M.S. and Ward, D. (2001) *Microphone Arrays – Signal Processing Techniques and Applications*, Springer.

Brutti, A. and Nesta, F. (2013) Tracking of multidimensional TDOA for multiple sources with distributed microphone pairs. *Computer Speech and Language*, **27** (3), 660–682.

Brutti, A., Omologo, M., and Svaizer, P. (2010) Multiple source localization based on acoustic map de-emphasis. *EURASIP Journal on Audio, Speech, and Music Processing*, **2010**, 11:1–11:17.

Brutti, A., Svaizer, P., and Omologo, M. (2005) Oriented global coherence field for the estimation of the head orientation in smart rooms equipped with distributed microphone arrays, in *Proceedings of Interspeech*, pp. 2337–2340.

Buchner, H., Aichner, R., and Kellermann, W. (2007) TRINICON-based blind system identification with application to multiple-source localization and separation, in *Blind Speech Separation*, Springer, pp. 101–147.

Caljon, T., Enescu, V., Schelkens, P., and Sahli, H. (2005) An offline bidirectional tracking scheme, in *Proceedings of International Conference on Advanced Concepts for Intelligent Vision Systems*, pp. 587–594.

Carlin, M.A. and Elhilali, M. (2015) A framework for speech activity detection using adaptive auditory receptive fields. *IEEE/ACM Transactions on Audio, Speech, and Language Processing*, **23** (12), 2422–2433.

Carter, G.C. (1987) Coherence and time delay estimation. *Proceedings of the IEEE*, **75** (2), 236–255.

Champagne, B., Bedard, S., and Stephenne, A. (1996) Performance of time-delay estimation in the presence of room reverberation. *IEEE Transactions on Speech and Audio Processing*, **4** (2), 148–152.

Chen, J., Benesty, J., and Huang, Y. (2003) Robust time delay estimation exploiting redundancy among multiple microphones. *IEEE Transactions on Speech and Audio Processing*, **11** (6), 549–557.

Chen, J., Benesty, J., and Huang, Y. (2006) Time delay estimation in room acoustic environments: An overview. *EURASIP Journal on Applied Signal Processing*, **2006**, 170–170.

De Mori, R. (ed.) (1998) *Spoken Dialogues with Computers*, Academic Press.

Dorfan, Y. and Gannot, S. (2015) Tree-based recursive expectation-maximization algorithm for localization of acoustic sources. *IEEE/ACM Transactions on Audio, Speech, and Language Processing*, **23** (10), 1692–1703.

Fallon, M.F. and Godsill, S. (2010) Acoustic source localization and tracking using track before detect. *IEEE Transactions on Audio, Speech, and Language Processing*, **18** (6), 1228–1242.

Gretsistas, A. and Plumbley, M. (2010) A multichannel spatial compressed sensing approach for direction of arrival estimation, in *Proceedings of International Conference on Latent Variable Analysis and Signal Separation*, pp. 458–465.

Kaplan, E.D. and Hegarty, C.J. (eds) (2006) *Understanding GPS: Principles and Applications*, Artech House, 2nd edn.

Kepesi, M., Pernkopf, F., and Wohlmayr, M. (2007) Joint position-pitch tracking for 2-channel audio, in *Proceedings of International Workshop on Content-Based Multimedia Indexing*, pp. 303–306.

Klee, U., Gehrig, T., and McDonough, J. (2006) Kalman filters for time delay of arrival-based source localization. *EURASIP Journal on Applied Signal Processing*, **2006**, 167–167.

Knapp, C. and Carter, G. (1976) The generalized correlation method for estimation of time delay. *IEEE Transactions on Acoustics, Speech, and Signal Processing*, **24** (4), 320–327.

Kondoz, A. (2004) *Digital Speech - Coding for Low Bit Rate Communication Systems*, Wiley.

Lathoud, G. and Odobez, J.M. (2007) Short-term spatio-temporal clustering applied to multiple moving speakers. *IEEE Transactions on Audio, Speech, and Language Processing*, **15** (5), 1696–1710.

Lee, Y., Wada, T.S., and Juang, B.H. (2010) Multiple acoustic source localization based on multiple hypotheses testing using particle approach, in *Proceedings of IEEE International Conference on Audio, Speech and Signal Processing*, pp. 2722–2725.

Lehmann, E.A. and Johansson, A.M. (2006) Particle filter with integrated voice activity detection for acoustic source tracking. *EURASIP Journal on Advances in Signal Processing*, **2007** (1), 1–11.

Li, X., Deng, Z.D., Rauchenstein, L.T., and Carlson, T.J. (2016) Contributed review: Source-localization algorithms and applications using time of arrival and time difference of arrival measurements. *Review of Scientific Instruments*, **87** (4), 041502.

Loesch, B., Uhlich, S., and Yang, B. (2009) Multidimensional localization of multiple sound sources using frequency domain ICA and an extended state coherence transform, in *Proceedings of IEEE Workshop on Statistical Signal Processing*, pp. 677–680.

Loizou, P.C. (2013) *Speech Enhancement: Theory and Practice*, CRC Press.

Malioutov, D., Cetin, M., and Willsky, A.S. (2005) A sparse signal reconstruction perspective for source localization with sensor arrays. *IEEE Transactions on Signal Processing*, **53** (8), 3010–3022.

Mandel, M.I., Weiss, R.J., and Ellis, D.P.W. (2010) Model-based expectation-maximization source separation and localization. *IEEE Transactions on Audio, Speech, and Language Processing*, **18** (2), 382–394.

Mungamuru, B. and Aarabi, P. (2004) Enhanced sound localization. *IEEE Transactions on Systems, Man, and Cybernetics—Part B: Cybernetics*, **34** (3), 1526–1540.

Nakano, A.Y., Nakagawa, S., and Yamamoto, K. (2009) Automatic estimation of position and orientation of an acoustic source by a microphone array network. *Journal of the Acoustical Society of America*, **126** (6), 3084–3094.

Nesta, F. and Omologo, M. (2012) Generalized state coherence transform for multidimensional TDOA estimation of multiple sources. *IEEE/ACM Transactions on Audio, Speech, and Language Processing*, **20** (1), 246–260.

Nesta, F., Svaizer, P., and Omologo, M. (2011) Convolutive BSS of short mixtures by ICA recursively regularized across frequencies. *IEEE/ACM Transactions on Audio, Speech, and Language Processing*, **19** (3), 624–639.

Omologo, M. and Svaizer, P. (1994) Acoustic event location using a crosspower-spectrum phase based technique, in *Proceedings of IEEE International Conference on Audio, Speech and Signal Processing*, vol. 2, pp. 273–276.

Omologo, M. and Svaizer, P. (1997) Use of the crosspower-spectrum phase in acoustic event location. *IEEE Transactions on Speech and Audio Processing*, **5** (3), 288–292.

Pertilä, P. and Hämäläinen, M.S. (2010) A track before detect approach for sequential Bayesian tracking of multiple speech sources, in *Proceedings of IEEE International Conference on Audio, Speech and Signal Processing*, pp. 4974–4977.

Pertilä, P. and Tinakari, A. (2013) Time-of-arrival estimation for blind beamforming, in *Proceedings of IEEE International Conference on Digital Signal Processing*, pp. 1–6.

Pertilä, P. and Cakir, E. (2017) Robust direction estimation with convolutional neural networks based steered response power, in *Proceedings of IEEE International Conference on Audio, Speech and Signal Processing*, pp. 6125 –6129.

Pertilä, P., Korhonen, T., and Visa, A. (2008) Measurement combination for acoustic source localization in a room environment. *EURASIP Journal on Audio, Speech, and Music Processing*, **2008** (1), 1–14.

Potamitis, I., Chen, H., and Tremoulis, G. (2004) Tracking of multiple moving speakers with multiple microphone arrays. *IEEE Transactions on Audio, Speech, and Language Processing*, **12** (5), 520–529.

Pulford, G.W. (2005) Taxonomy of multiple target tracking methods. *IEE Proceedings on Radar, Sonar and Navigation*, **152** (5), 291–304.

Rabiner, L. and Juang, B.H. (1993) *Fundamentals of Speech Recognition*, Prentice-Hall.

Ramírez, J., Górriz, M., and Segura, J.C. (2007) Voice activity detection. fundamentals and speech recognition system robustness, in *Robust Speech Recognition and Understanding*, InTech.

Roman, N. and Wang, D. (2008) Binaural tracking of multiple moving sources. *IEEE Transactions on Audio, Speech, and Language Processing*, **16** (4), 728–739.

Rubio, J., Ishizuka, K., Sawada, H., Araki, S., Nakatani, T., and Fujimoto, M. (2007) Two-microphone voice activity detection based on the homogeneity of the direction of arrival estimates, in *Proceedings of IEEE International Conference on Audio, Speech and Signal Processing*, vol. 4, pp. 385–388.

Sadler, B.M. and Kozick, R.J. (2006) A survey of time delay estimation performance bounds, in *Proceedings of IEEE Workshop on Sensor Array and Multichannel Processing*, pp. 282–288.

Sawada, H., Mukai, R., and Makino, S. (2003) Direction of arrival estimation for multiple source signals using independent component analysis, in *Proceedings of International Symposium on Signal Processing and its Applications*, vol. 2, pp. 411–414.

Schmidt, R. (1986) Multiple emitter location and signal parameter estimation. *IEEE Transactions on Antennas and Propagation*, **34** (3), 276–280.

Schwartz, O. and Gannot, S. (2014) Speaker tracking using recursive EM algorithms. *IEEE/ACM Transactions on Audio, Speech, and Language Processing*, **22** (2), 392–402.

Sohn, J., Kim, N.S., and Sung, W. (1999) A statistical model-based voice activity detection. *IEEE Signal Processing Letters*, **6** (1), 1–3.

Stone, L.D., Streit, R.L., Corwin, T.L., and Bell, K.L. (2013) *Bayesian Multiple Target Tracking*, Artech House, 2nd edn.

Stowell, D., Giannoulis, D., Benetos, E., Lagrange, M., and Plumbley, M.D. (2015) Detection and classification of acoustic scenes and events. *IEEE Transactions on Multimedia*, **17** (10), 1733–1746.

Temko, A., Nadeu, C., Macho, D., Malkin, R., Zieger, C., and Omologo, M. (2009) Acoustic event detection and classification, in *Computers in the Human Interaction Loop*, Springer, pp. 61–73.

Tucker, R. (1992) Voice activity detection using a periodicity measure. *IEE Proceedings-I*, **139** (4), 377–380.

Valin, J.M., Michaud, F., Hadjou, B., and Rouat, J. (2004) Localization of simultaneous moving sound sources for mobile robot using a frequency-domain steered beamformer approach, in *Proceedings of IEEE International Conference on Robotics and Automation*, vol. 1, pp. 1033–1038.

Valin, J.M., Michaud, F., and Rouat, J. (2007) Robust localization and tracking of simultaneous moving sound sources using beamforming and particle filtering. *Robotics and Autonomous Systems*, **55** (3), 216–228.

Vary, P. and Martin, R. (2006) *Digital Speech Transmission - Enhancement, Coding, and Error Concealment*, Wiley.

Wölfel, M. and McDonough, J. (2009) *Distant Speech Recognition*, Wiley.

Xiao, X., Zhao, S., Zhong, X., Jones, D.L., Chng, E.S., and Li, H. (2015) A learning-based approach to direction of arrival estimation in noisy and reverberant environments, in *Proceedings of IEEE International Conference on Audio, Speech and Signal Processing*, pp. 2814–2818.

Yin, J. and Chen, T. (2011) Direction-of-arrival estimation using a sparse representation of array covariance vectors. *IEEE Transactions on Signal Processing*, **59** (9), 4489–4493.

Yli-Hietanen, J., Kalliojarvi, K., and Astola, J. (1996) Low-complexity angle of arrival estimation of wideband signals using small arrays, in *Proceedings of IEEE Workshop on Statistical Signal and Array Processing*, pp. 109–112.

Zhang, C., Zhang, Z., and Florêncio, D. (2007) Maximum likelihood sound source localization for multiple directional microphones, in *Proceedings of IEEE International Conference on Audio, Speech and Signal Processing*, vol. 1, pp. 125–128.

Zhang, X.L. and Wu, J. (2013) Deep belief networks based voice activity detection. *IEEE/ACM Transactions on Audio, Speech, and Language Processing*, **21** (4), 697–710.

Part II

Single-Channel Separation and Enhancement

5

Spectral Masking and Filtering
Timo Gerkmann and Emmanuel Vincent

In this chapter and the following ones, we consider the case of a single-channel input signal ($I = 1$). We denote it as $x(t)$ and omit the channel index $i = 1$ for legibility.

As discussed in Chapter 3, spatial diversity cannot be exploited to separate such a signal due to the difficulty of disambiguating the transfer function from the spectrum of each source. Therefore, single-channel separation and enhancement must rely on the spectral diversity and exploit properties of the sources such as those listed in Chapter 2. Disregarding phase, one can then separate or enhance the sources using real-valued filters in the time-frequency domain known as *time-frequency masks*.

In the following, we define the concept of time-frequency masking in Section 5.1. We introduce different models to derive a mask from the signal statistics in Section 5.2 and modify them in order to improve perceptual quality in Section 5.3. We summarize the main findings and provide links to forthcoming chapters and more advanced topics in Section 5.4.

5.1 Time-Frequency Masking

5.1.1 Definition and Types of Masks

Following the discussion in Chapter 2, filtering is performed in the time-frequency domain (Ephraim and Malah, 1984; Roweis, 2001; Benaroya *et al.*, 2006). Denoting by $x(n,f)$ the complex-valued time-frequency coefficients of the input signal, separation and enhancement can be achieved by

$$\widehat{c}_j(n,f) = w_j(n,f)\,x(n,f) \tag{5.1}$$

or

$$\widehat{s}_j(n,f) = w_j(n,f)\,x(n,f), \tag{5.2}$$

depending whether one wishes to estimate the spatial image $c_j(n,f)$ of source j, or its direct path component, which is a delayed and attenuated version of the original source

Audio Source Separation and Speech Enhancement, First Edition.
Edited by Emmanuel Vincent, Tuomas Virtanen and Sharon Gannot.
© 2018 John Wiley & Sons Ltd. Published 2018 by John Wiley & Sons Ltd.
Companion Website: https://project.inria.fr/ssse/

signal $s_j(n,f)$. The filter $w_j(n,f)$ is generally assumed to be real-valued and it is often additionally assumed to satisfy the following constraints for all n, f:

$$0 \le w_j(n,f) \le 1 \quad \text{and} \quad \begin{cases} \sum_{j=1}^{J} w_j(n,f) = 1 & \text{in (5.1)} \\ \sum_{j=1}^{J} w_j(n,f) \le 1 & \text{in (5.2).} \end{cases} \tag{5.3}$$

Such a filter is called a time-frequency mask, a spectral mask, or a masking filter because it operates by selectively hiding unwanted time-frequency areas. The constraints ensure that the sum of the estimated source spatial images $\sum_{j=1}^{J} \widehat{c}_j(n,f)$ is equal to the mixture $x(n,f)$ as per (3.4), and that the sum of the estimated direct path components $\sum_{j=1}^{J} \widehat{s}_j(n,f)$ is smaller than $x(n,f)$ due to the reduction of reverberation.

Masks can be broadly categorized depending on the value range of $w_j(n,f)$. *Binary masks* take binary values $w_j(n,f) \in \{0,1\}$. They have enjoyed some popularity in the literature due to their ability to effectively improve speech intelligibility in the presence of noise or multiple talkers despite their simplicity (Wang, 2005; Li and Loizou, 2008). *Soft masks* or *ratio masks*, by contrast, can take any value in the range $[0, 1]$.

5.1.2 Oracle Mask

In order to understand the potential of time-frequency masking, it is useful to consider the notion of *ideal or oracle mask*, i.e. the best possible mask for a given signal. This mask can be computed only on development data for which the target signal is known. It provides an upper bound on the separation or enhancement performance achievable.

For most time-frequency representations, the oracle mask cannot easily be computed due to the nonorthogonality of the transform. In practice, this issue is neglected and the oracle mask is computed in each time-frequency bin separately. For the separation of $c_j(n,f)$, for instance, the oracle mask is defined as

$$\widehat{w}_j(n,f) = \underset{w_j(n,f)}{\operatorname{argmin}} |c_j(n,f) - w_j(n,f)\, x(n,f)|^2. \tag{5.4}$$

In order to solve this optimization problem under the constraints in (5.3), it is useful to define the real part of the ratio of time-frequency coefficients of the source and the mixture: $r_j(n,f) = \Re(c_j(n,f)/x(n,f))$. In the simplest case when there are only two sources ($J = 2$), the oracle binary masks for the two sources are given by[1]

$$\widehat{w}_j^{\text{bin}}(n,f) = \begin{cases} 1 & \text{if } r_j(n,f) > \frac{1}{2}, \\ 0 & \text{otherwise,} \end{cases} \tag{5.5}$$

and the oracle soft masks by

$$\widehat{w}_j^{\text{soft}}(n,f) = \begin{cases} 0 & \text{if } r_j(n,f) < 0, \\ 1 & \text{if } r_j(n,f) > 1, \\ r_j(n,f) & \text{otherwise.} \end{cases} \tag{5.6}$$

---

1 When $r_j(n,f) = \frac{1}{2}$ for both sources, $\widehat{w}_j^{\text{bin}}(n,f)$ can be arbitrarily set to 1 for either source.

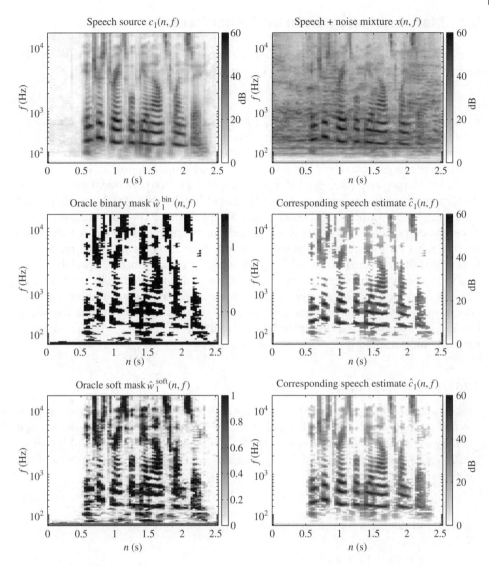

Figure 5.1 Separation of speech from cafe noise by binary vs. soft masking. The masks shown in this example are oracle masks.

See Vincent *et al.* (2007) for a proof of this result and for the general solution with three or more sources.

These two types of masks are displayed in Figure 5.1. We see that time-frequency masking can potentially lead to very good separation performance. Also, soft masking appears to perform slightly better than binary masking. As a matter of fact, it has been shown that soft masking improves both speech intelligibility (Jensen and Hendriks, 2012; Madhu *et al.*, 2013) and the maximum achievable signal-to-distortion ratio (SDR) by 3 dB compared to binary masking (Vincent *et al.*, 2007).

5.2 Mask Estimation Given the Signal Statistics

In this section we discuss different ways of obtaining filter masks that separate the desired signal from competing sources. We consider the following signal model

$$x(n,f) = c(n,f) + u(n,f) \tag{5.7}$$

where $c(n,f)$ is the target signal and $u(n,f)$ is an uncorrelated interference. For instance, $c(n,f) = c_j(n,f)$ may be the spatial image of the target source and $u(n,f) = \sum_{j' \neq j} c_{j'}(n,f)$ the superposition of all other sources. Alternatively, $c(n,f)$ may be the direct path component of the target source and late reverberation may be modeled as additive, uncorrelated to the target, and comprised of $u(n,f)$ (Lebart *et al.*, 2001). While the assumption that late reverberation is uncorrelated to the target is debatable, it yields powerful and robust estimators in practice (Lebart *et al.*, 2001; Habets, 2007; Cauchi *et al.*, 2015). Then, the derivation of the filters is rather general, meaning that we can use the same spectral mask estimators for signal enhancement, dereverberation, and source separation. The difference in spectral masking based signal enhancement, dereverberation, and source separation rather lies in the way the signals are statistically modeled and how the corresponding parameters, e.g. the power spectra of target and interference, are estimated. Due to the fact that spectral masking is applied in each time-frequency bin and for each source independently, we will drop indices j, n, f, in the following sections unless needed.

5.2.1 Spectral Subtraction

Probably the simplest and earliest method for interference reduction is the concept of *spectral subtraction* (Boll, 1979; Berouti *et al.*, 1979). In its simplest form, the average interference magnitude spectrum $|\bar{u}| = \frac{1}{N} \sum_{n=0}^{N-1} |u(n,f)|$ is subtracted from the magnitude spectral coefficients $|x|$ of the mixture and combined with the phase $\angle x$ of the mixture (Boll, 1979):

$$\hat{c} = (|x| - |\bar{u}|)e^{j\angle x}. \tag{5.8}$$

This spectral subtraction rule can be represented by means of a mask w_{SS} as

$$\hat{c} = \underbrace{\left(1 - \frac{|\bar{u}|}{|x|}\right)}_{w_{SS}} x = w_{SS}x. \tag{5.9}$$

Equations (5.8) and (5.9) present the simplest forms of spectral subtraction and are somewhat heuristically motivated. It is important to realize that even though the complex spectral coefficients of the target and the interference are additive, neither their magnitudes nor the averages or the expected values of their magnitudes are additive (with the exception of trivial phases):

$$|c| \neq |x| - |u| \tag{5.10}$$
$$|c| \neq |x| - |\bar{u}| \tag{5.11}$$
$$\mathbb{E}\{|c|\} \neq \mathbb{E}\{|\hat{c}|\} = \mathbb{E}\{|x|\} - \mathbb{E}\{|u|\}. \tag{5.12}$$

Thus, from a mathematical perspective, the simple spectral amplitude subtraction rule is not optimal.

This is somewhat improved when spectral subtraction is defined on power spectral coefficients, leading to power spectral subtraction:

$$\hat{c} = (|x|^2 - |\bar{u}|^2)^{\frac{1}{2}} e^{j\angle x} \tag{5.13}$$

$$= \underbrace{\left(1 - \frac{|\bar{u}|^2}{|x|^2}\right)^{\frac{1}{2}}}_{w_{\text{PSS}}} x. \tag{5.14}$$

When the temporal average is interpreted as an estimate of the noise power spectrum, i.e. $|\bar{u}|^2 = \hat{\sigma}_u^2$, the quantity $|x|^2 - |\bar{u}|^2$ can be interpreted as an estimate of the target power spectrum $\sigma_c^2 = \mathbb{E}\{|c|^2\}$ (Hendriks *et al.*, 2013). Under the additive signal model (5.7) with zero-mean uncorrelated target and interference, we have $\mathbb{E}\{cx^*\} = 0$ and

$$|c|^2 \neq |x|^2 - |\bar{u}|^2, \text{ but} \tag{5.15}$$

$$\mathbb{E}\{|c|^2\} = \mathbb{E}\{|x|^2\} - \mathbb{E}\{|\bar{u}|^2\} = \mathbb{E}\{|\hat{c}|^2\}, \tag{5.16}$$

i.e. the power subtraction rule (5.14) represents an unbiased estimator of the target power spectrum $\mathbb{E}\{|c|^2\}$. The zero-mean assumption stems from the fact that the phase is assumed to be uniformly distributed. From a practical viewpoint, the application of power spectral subtraction only requires an estimate of the interference power spectrum $\hat{\sigma}_u^2 = |\bar{u}|^2$.

5.2.2 Wiener Filtering

A more rigorous way of finding a spectral mask w is based on minimizing the mean square error (MSE) between the target c and the estimate $\hat{c} = w^*x$. Similarly to the problem statement for the oracle mask (5.4), this can be written as

$$w_{\text{SWF}} = \underset{w}{\arg\min} \, \mathbb{E}\{|c - w^*x|^2\}. \tag{5.17}$$

The resulting spectral mask is called the *single-channel Wiener filter*. In this expression, both the target spectral coefficients c and the mixture spectral coefficients x are considered as random variables while the spectral mask w is considered to be deterministic and independent of x. In contrast to (5.4), we now consider the possibility that w is complex-valued. Such masking filters are referred to as linearly constrained filters because the estimate $\hat{c}$ is expressed as a linear function of the mixture x. Thus, the Wiener filter is the linear *minimum mean square error* (MMSE) estimator.

Using the linearity of expectation, we can rephrase the cost in (5.17) as

$$\mathbb{E}\{|c - w^*x|^2\} = \mathbb{E}\{|c|^2\} + |w|^2 \mathbb{F}\{|x|^2\} - 2\Re(w^*\mathbb{E}\{c^*x\}) \tag{5.18}$$

First, let us look at the phase of w. Obviously, the phase of w only influences the last term $-2\Re(w^*\mathbb{E}\{c^*x\})$. Thus, to minimize (5.18) we need to maximize the real part of $w^*\mathbb{E}\{c^*x\}$. It is easy to show that this happens when the phase of w is the same as that of $\mathbb{E}\{c^*x\}$ such that the product $w^*\mathbb{E}\{c^*x\}$ is real-valued:

$$\angle w_{\text{SWF}} = \angle\mathbb{E}\{c^*x\}. \tag{5.19}$$

Secondly, let us look at the magnitude of w. The optimal magnitude can be found by inserting the optimal phase (5.19) and equating the derivative with respect to $|w|$ to zero

$$0 = \frac{\partial}{\partial|w|}(\mathbb{E}\{|c|^2\} + |w|^2\mathbb{E}\{|x|^2\} - 2|w||\mathbb{E}\{c^*x\}|) \tag{5.20}$$

$$= 2|w|\mathbb{E}\{|x|^2\} - 2|\mathbb{E}\{c^*x\}|. \tag{5.21}$$

Solving for $|w|$, we obtain

$$|w_{\text{SWF}}| = \frac{|\mathbb{E}\{c^*x\}|}{\mathbb{E}\{|x|^2\}}. \tag{5.22}$$

Combining the optimal magnitude (5.22) and the optimal phase (5.19), we obtain the optimal masking filter

$$w_{\text{SWF}} = \frac{\mathbb{E}\{c^*x\}}{\mathbb{E}\{|x|^2\}}. \tag{5.23}$$

With the assumption that the target c and the interference u are zero-mean and mutually uncorrelated, we have $\mathbb{E}\{c^*x\} = \mathbb{E}\{|c|^2\}$ and $\mathbb{E}\{|x|^2\} = \mathbb{E}\{|c|^2\} + \mathbb{E}\{|u|^2\}$ such that the spectral mask is given by

$$w_{\text{SWF}} = \frac{\mathbb{E}\{|c|^2\}}{\mathbb{E}\{|c|^2\} + \mathbb{E}\{|u|^2\}} = \frac{\sigma_c^2}{\sigma_c^2 + \sigma_u^2}. \tag{5.24}$$

The Wiener filter w_{SWF} turns out to be real-valued and to satisfy the constraints in (5.3). From a practical viewpoint, in contrast to power spectral subtraction, estimates of both the target and the interference power spectra are needed.

Note that, depending on the problem formulation, the Wiener filter can also be complex-valued. For instance, if we aim to find the source signal s instead of its spatial image c, we would replace c by s in (5.17). Assuming a narrowband model $x = as + u$, the Wiener filter solution (5.23) would result in

$$w_{\text{SWF}} = \frac{a\sigma_s^2}{|a|^2\sigma_s^2 + \sigma_u^2}. \tag{5.25}$$

As explained in Chapter 2, for commonly chosen frame sizes, the narrowband model does not properly account for late reverberation. An alternative way to perform dereverberation is to model late reverberation as additive and uncorrelated with the target. Then, the late reverberant power spectrum can be incorporated into the interference power spectrum σ_u^2 in (5.24). This approach can yield robust results in practice (Lebart *et al.*, 2001; Habets, 2007; Cauchi *et al.*, 2015).

5.2.3 Bayesian Estimation of Gaussian Spectral Coefficients

In the previous section we showed that if we constrain the estimate to be a linear function of the mixture, the MMSE estimator is given by the Wiener filter (5.24). For its derivation, we did not assume any underlying distribution of the target or interference spectral magnitude coefficients, but only that the target and interference spectral coefficients are zero-mean and mutually uncorrelated. Hence, this result is very compact and general. However, the question arises if we can get an even better result if we allow for a nonlinear relationship between the input x and the output $\hat{c}(x)$ of the filter. For this, we need to optimize the more general problem

$$\hat{c}^{\text{Bayes}} = \underset{\hat{c}}{\text{argmin}} \; \mathbb{E}\{|c - \hat{c}(x)|^2\} \tag{5.26}$$

where the estimate $\hat{c}(x)$ is any, possibly nonlinear, function of the mixture x.

It can be shown that solving this general MMSE problem is equivalent to finding the *posterior mean* (Schreier and Scharf, 2010, sec. 5.2), i.e.

$$\hat{c}^{\text{Bayes}} = \mathbb{E}\{c \mid x\} = \int c \, p(c \mid x) \mathrm{d}c. \tag{5.27}$$

Thus, the MMSE estimator is also referred to as the posterior mean estimator. The formulation as a posterior mean estimator in (5.27) allows us to elegantly use the concepts of *Bayesian statistics* to find the unconstrained MMSE estimator, i.e. to solve (5.26).

To find the posterior mean (5.27), we need a model for the conditional probability distribution of the searched quantity c, referred to as *posterior* in Bayesian estimation. While finding a model for the posterior is often difficult, using Bayes' theorem the posterior $p(c \mid x)$ can be expressed as a function of the *likelihood* $p(x \mid c)$ and the *prior* $p(c)$ as

$$p(c \mid x) = \frac{p(c, x)}{p(x)} = \frac{p(c, x)}{\int p(c, x) \mathrm{d}c} = \frac{p(c)p(x \mid c)}{\int p(c)p(x \mid c)\mathrm{d}c}. \tag{5.28}$$

This means that instead of the posterior, we now need models for the likelihood and the prior over c in order to solve (5.27).

If we have an additive signal model then, as the target signal is given, the randomness in the likelihood $p(x \mid c)$ is given only by the interference signal. A common assumption is that the interference signal is zero-mean *complex Gaussian* distributed with variance $\sigma_u^2 = \mathbb{E}\{|u|^2\}$. As a consequence, the likelihood is complex Gaussian with mean c and variance σ_u^2:

$$p(x \mid c) = \frac{1}{\pi\sigma_u^2} \exp\left(-\frac{|x - c|^2}{\sigma_u^2}\right). \tag{5.29}$$

This Gaussian interference model is the most popular. Other interference models, e.g. Laplacian, have also been discussed in the literature (Martin, 2005; Benaroya *et al.*, 2006).

While the likelihood requires defining a statistical model for the interference coefficients, the prior corresponds to a statistical model of the target spectral coefficients. The simplest assumption is a zero-mean complex Gaussian model with variance σ_c^2:

$$p(c) = \frac{1}{\pi\sigma_c^2} \exp\left(-\frac{|c|^2}{\sigma_c^2}\right). \tag{5.30}$$

For uncorrelated zero-mean Gaussian target and interference, the sum $x = c + u$ is zero-mean complex Gaussian with variance $\sigma_c^2 + \sigma_u^2$, i.e. the resulting *evidence* model is

$$p(x) = \frac{1}{\pi(\sigma_c^2 + \sigma_u^2)} \exp\left(-\frac{|x|^2}{\sigma_c^2 + \sigma_u^2}\right). \tag{5.31}$$

Using the Gaussian likelihood (5.29) and prior (5.30) in the numerator and the evidence (5.31) in the denominator of (5.28), we obtain

$$p(c \mid x) = \frac{\dfrac{1}{\pi\sigma_c^2} \exp\left(-\dfrac{|c|^2}{\sigma_c^2}\right) \dfrac{1}{\pi\sigma_u^2} \exp\left(-\dfrac{|x-c|^2}{\sigma_u^2}\right)}{\dfrac{1}{\pi(\sigma_c^2 + \sigma_u^2)} \exp\left(-\dfrac{|x|^2}{\sigma_c^2 + \sigma_u^2}\right)},$$ (5.32)

$$= \frac{1}{\pi\dfrac{\sigma_c^2\sigma_u^2}{\sigma_c^2 + \sigma_u^2}} \exp\left(-\frac{\dfrac{\sigma_c^2}{\sigma_c^2 + \sigma_u^2}|x-c|^2 + \dfrac{\sigma_u^2}{\sigma_c^2 + \sigma_u^2}|c|^2 - \dfrac{\sigma_c^2\sigma_u^2}{(\sigma_c^2 + \sigma_u^2)^2}|x|^2}{\dfrac{\sigma_c^2\sigma_u^2}{\sigma_c^2 + \sigma_u^2}}\right).$$ (5.33)

Substituting $\lambda = \dfrac{\sigma_c^2\sigma_u^2}{\sigma_c^2 + \sigma_u^2}$, using $|x - c|^2 = |x|^2 + |c|^2 - 2\Re(xc^*)$ we obtain

$$p(c \mid x) = \frac{1}{\pi\lambda} \exp\left(-\frac{|c|^2 + |x|^2\left(\dfrac{\sigma_c^2}{\sigma_c^2 + \sigma_u^2}\right)^2 - 2\Re(xc^*)\dfrac{\sigma_c^2}{\sigma_c^2 + \sigma_u^2}}{\lambda}\right)$$ (5.34)

$$= \frac{1}{\pi\lambda} \exp\left(-\frac{\left|c - \dfrac{\sigma_c^2}{\sigma_c^2 + \sigma_u^2}x\right|^2}{\lambda}\right).$$ (5.35)

This result is interesting in many ways. First of all, we see that for a Gaussian likelihood and prior, the posterior is also Gaussian. Secondly, we may directly see that the mean of the posterior is given by

$$\mathbb{E}\{c \mid x\} = \frac{\sigma_c^2}{\sigma_c^2 + \sigma_u^2}x = w_{\text{SWF}}\, x = \hat{c},$$ (5.36)

which is identical to the Wiener solution found as the MMSE linearly constrained filter (5.24). In other words, for Gaussian target and interference, the optimal estimate in the MMSE sense is the Wiener filter no matter whether we constrain the filter to be linear or not. Note, however, that for nongaussian target or interference this is not necessarily true and the MMSE estimate is generally a nonlinear function of the observation. Finally, we also see that the posterior exhibits the variance $\lambda = \frac{\sigma_c^2\sigma_u^2}{\sigma_c^2 + \sigma_u^2}$, which can also be seen as a measure of uncertainty for the Wiener estimate.

The concept of Bayesian statistics and estimation is a general and powerful tool. In Table 5.1 an overview of the probability distributions relating to the searched quantity θ (so far we considered $\theta = c$) and the observation x is given. Based on these conditional distributions, other estimators can be defined too. For instance, the θ that maximizes the likelihood is referred to as the maximum likelihood (ML) estimate, while the θ that maximizes the posterior is referred to as the *maximum a posteriori* (MAP) estimate

Table 5.1 Bayesian probability distributions for the observation x and the searched quantity θ.

$p(\theta \mid x)$	Posterior
$p(x \mid \theta)$	Likelihood
$p(\theta)$	Prior
$p(x)$	Evidence

Table 5.2 Criteria for the estimation of θ.

$\widehat{\theta}^{\mathrm{ML}} = \underset{\theta}{\mathrm{argmax}}\, p(x \mid \theta)$	ML estimator
$\widehat{\theta}^{\mathrm{MAP}} = \underset{\theta}{\mathrm{argmax}}\, p(\theta \mid x)$	MAP estimator
$\widehat{\theta}^{\mathrm{MMSE}} = \mathbb{E}\{\theta \mid x\}$	MMSE estimator

(see Table 5.2). Using Bayes' theorem, the MAP estimator can be expressed as a function of the likelihood and the prior as

$$\widehat{\theta}^{\mathrm{MAP}} = \underset{\theta}{\mathrm{argmax}}\, p(\theta \mid x) \tag{5.37}$$

$$= \underset{\theta}{\mathrm{argmax}}\, \frac{p(x \mid \theta)\, p(\theta)}{p(x)} \tag{5.38}$$

$$= \underset{\theta}{\mathrm{argmax}}\, p(x \mid \theta)\, p(\theta) \tag{5.39}$$

Here the normalization by the evidence model $p(x)$ is not necessary as it is not a function of the searched quantity θ. Hence, whenever prior information about the searched quantity is given, it can be used to extend the likelihood and to obtain an improved estimator by means of a MAP estimator or, using (5.28), the MMSE estimator.

If the respective conditional distributions are known and unimodal, the ML and MAP estimators defined in Table 5.2 can be obtained by equating the derivative of the conditional distributions with respect to the searched quantity θ to zero and solving for θ. In the Gaussian case considered so far, finding the ML and MAP estimates of c is rather simple. The likelihood (5.29) is maximum when the argument of its exponential function is as small as possible, i.e. when $c = x$. In other words, the ML estimate is equal to the mixture spectral coefficient

$$\widehat{c}^{\mathrm{ML}} = x = w_{\mathrm{ML}} x. \tag{5.40}$$

The resulting masking filter is $w_{\mathrm{ML}} = 1$ for all time-frequency bins and thus does not result in any interference reduction.

As the Gaussian distribution is not only unimodal but also symmetric, the mean of the posterior is identical to its mode, meaning that for any unimodal and symmetric posterior, the MAP estimator is equivalent to the MMSE estimator, i.e. for a Gaussian target and interference model the estimate is given by

$$\widehat{c}^{\mathrm{MAP}} = \frac{\sigma_c^2}{\sigma_c^2 + \sigma_u^2} x = w_{\mathrm{MAP}}\, x. \tag{5.41}$$

The MAP solution is thus identical to the MMSE solution, and $w_{\mathrm{MAP}} = w_{\mathrm{SWF}}$.

From the above results, it may seem as if the Bayesian theory is not of much help, as under a Gaussian signal model either a trivial filter arises as the ML estimator in (5.40), or simply the Wiener masking filter arises as the unconstrained MMSE estimator in (5.36) or the MAP estimator in (5.41). However, taking the linear approach (5.17) the Wiener solution is found without any assumptions on the distribution. So why would the Bayesian concept be of importance? The answer is simple: for many random variables the Gaussian assumption is either invalid or impossible to verify and alternative distributions can be assumed. Bayesian estimation then provides a very general concept to find optimal estimators for these nongaussian quantities that may outperform the simple Wiener filter.

5.2.4 Estimation of Magnitude Spectral Coefficients

A prominent example where nongaussianity matters is the estimation of nonnegative quantities such as magnitude or power spectral coefficients. For instance, estimating spectral magnitudes rather than complex spectral coefficients is thought to be perceptually more meaningful (Ephraim and Malah, 1984, 1985) (see also Section 5.3). We now argue that for such nonnegative quantities the Wiener filter is not the optimal solution, neither in the linearly constrained MMSE sense nor in the Bayesian sense. For this, it is important to note that nonnegative quantities are not zero-mean and hence a multiplicative linear filter as defined in (5.17) may result in a biased estimate, i.e. $\mathbb{E}\{\hat{\theta}\} \neq \mathbb{E}\{\theta\}$ where θ is the nonnegative searched quantity. To find an unbiased linear estimator that minimizes the MSE, the problem statement in (5.17) must be extended by adding a term m such that the MSE and the bias can both be controlled. Let r be a nonzero-mean mixture and θ be the nonzero-mean searched quantity. Without loss of generality $r = |x|$ could be the amplitude of the mixture while $\theta = |c|$ could be the magnitude of the target spectral coefficients. The problem then becomes

$$\min_{w,m} \mathbb{E}\{|\theta - (wr + m)|^2\}. \tag{5.42}$$

Optimizing for both w and m, the resulting estimator boils down to subtracting the mean of the nonnegative mixture $r' = r - \mathbb{E}\{r\}$ and the nonnegative searched quantity $\theta' = \theta - \mathbb{E}\{\theta\}$, applying a filter similar to (5.23) and adding the mean back of the desired quantity $\mathbb{E}\{\theta\}$ as

$$\hat{\theta} = \frac{\mathbb{E}\{\theta' r'\}}{\mathbb{E}\{r'^2\}} r' + \mathbb{E}\{\theta\}. \tag{5.43}$$

However, in contrast to (5.24), just as for Bayesian estimators, the relationship between $\mathbb{E}\{\theta r\}$, $\mathbb{E}\{r^2\}$, and $\mathbb{E}\{\theta\}$ depends on the distributions of the random variables and not only on their variances (Hendriks *et al.*, 2013). Therefore, we now put aside linearly constrained estimators and focus on the potentially stronger unconstrained Bayesian estimators.

Besides the insight that nonnegative quantities are not zero-mean, it is also important to realize that they are not Gaussian. This is intuitively clear when considering that Gaussian distributed random variables take values between $-\infty$ and $+\infty$ per definition, while nonnegative random variables can only take values between 0 and $+\infty$ per definition. As a result, the Wiener filter (5.24) obtained as the MMSE and MAP solutions (5.41) under a Gaussian signal model is not the optimal estimator for nonnegative quantities.

As a concrete example, we consider the Bayesian estimation of spectral magnitudes, i.e. $\theta = |c|$, obtained from complex Gaussian distributed spectral coefficients c following (5.30). The prior $p(\theta) = p(|c|)$ can be obtained by transforming $p(c)$ (5.30) into polar coordinates and integrating out the phase. The result is the well known *Rayleigh* distribution

$$p(|c|) = \frac{2|c|}{\sigma_c^2} \exp\left(-\frac{|c|^2}{\sigma_c^2}\right). \tag{5.44}$$

The posterior could then be obtained from this prior and the likelihood (5.29) using Bayes theorem and integration as in (5.28). Alternatively, the posterior $p(\theta \mid x) = p(|c| \mid x)$ can be obtained by directly transforming $p(c \mid x)$ (5.35) into polar coordinates and integrating out the phase. It is well known that the resulting magnitude posterior follows a *Rician* distribution (Wolfe and Godsill, 2003)

$$p(|c| \mid x) = \frac{2|c|}{\lambda} \exp\left(-\frac{|c|^2 + w_{\text{SWF}}^2 |x|^2}{\lambda}\right) I_0\left(\frac{2|x|\,|c|\,w_{\text{SWF}}}{\lambda}\right), \tag{5.45}$$

with λ defined below (5.33), w_{SWF} defined in (5.36), and $I_0(\cdot)$ the modified zeroth-order Bessel function of the first kind (Gradshteyn and Ryzhik, 2000). The MMSE optimal estimator of spectral magnitudes is thus given as the mean of the Rician distribution (5.45). Just as for other common distributions, the mean of the Rician distribution can nowadays easily be found in the literature. In the context of speech enhancement the resulting estimator has been proposed by Ephraim and Malah (1984) and referred to as the *short-time spectral amplitude* estimator:

$$\mathbb{E}\{|c| \mid x\} = \Gamma(1.5) \sqrt{\lambda} \; \Phi\left(-0.5, 1; -w_{\text{SWF}}\frac{|x|^2}{\sigma_u^2}\right), \tag{5.46}$$

where $\Gamma(\cdot)$ is the gamma function (Gradshteyn and Ryzhik, 2000, (8.31)), $\Gamma(1.5) = \sqrt{\pi}/2$, and $\Phi(\cdot, \cdot; \cdot)$ is the confluent hypergeometric function (Gradshteyn and Ryzhik, 2000, (9.210)). An estimate of the target complex coefficients is then obtained by combining the amplitude estimate (5.46) with the phase of the mixture. A spectral masking filter can be obtained by dividing the estimated target coefficients by the mixture coefficients. However, the relationship between the mixture and the estimate remains nonlinear and the resulting spectral masking filter may not satisfy the constraints in (5.3).

Furthermore, as the Rician distribution is not symmetric, the mode and the mean of the posterior are not identical anymore. Thus, in contrast to the Gaussian case discussed in Section 5.2.3, the MAP estimator of spectral magnitudes is different from the MMSE estimate (5.46). This is also illustrated in Figure 5.2, where the Rician posterior (5.45) is shown along with the MMSE and the MAP estimates. While analytically finding the mode of the Rician distribution is difficult, Wolfe and Godsill (2003) proposed an approximate closed-form solution for the MAP estimator of target magnitudes which is easier to implement than the MMSE estimator in (5.46):

$$|c|^{\text{MAP}} \approx \left(\frac{1}{2} w_{\text{SWF}} + \sqrt{\left(\frac{1}{2} w_{\text{SWF}}\right)^2 + \frac{\lambda}{4|x|^2}}\right) |x| \tag{5.47}$$

The larger the argument of the Bessel function in (5.45), the better this approximation holds. In Figure 5.2 an example for the approximate MAP estimate is given.

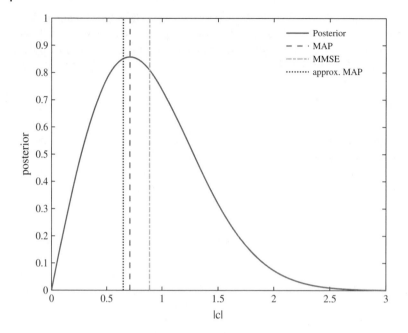

Figure 5.2 Illustration of the Rician posterior $p(|c| \mid x)$ (5.45) for $\sigma_c^2 = 1$, $\sigma_u^2 = 10$, and $|x| = \sqrt{11}$. The red dashed line shows the mode of the posterior and thus the MAP estimate of target spectral magnitudes $|c|$. The purple dotted line corresponds to the approximate MAP estimate (5.47), and the yellow dash-dotted line corresponds to the posterior mean (5.46) and thus the MMSE estimate of $|c|$.

5.2.5 Heavy-Tailed Priors

Another situation when Bayesian estimation helps is when considering nongaussian distributions for the source spectral coefficients. While we previously argued that the Gaussian model is a very useful and generic model, alternative distributions can also be assumed (Martin, 2005). Indeed, it is impossible to know the true distribution of the source spectral coefficients due to the fact that it is nonstationary, i.e. it varies from one time-frequency bin to another, and it cannot be estimated from the observation in a single time-frequency bin. One attempt to estimate the distribution of clean speech is to compute the histogram for a narrow range of estimated speech powers $\hat{\sigma}_c^2$ (Martin, 2005). A second approach is to normalize the speech spectral coefficients by the square-root of the estimated speech power $\sqrt{\hat{\sigma}_c^2}$ and to compute the histogram over time-frequency bins where speech is active (Gerkmann and Martin, 2010). In both cases, the found distributions also depend on the chosen spectral transformation, frame size, and power spectrum estimator (Gerkmann and Martin, 2010). However, whatever the choices made, the obtained histogram typically follows a *heavy-tailed* distribution, also known as a *super-Gaussian* or *sparse* distribution. This means that small and large values are more likely and medium values are less likely compared to a Gaussian distribution. In Figure 5.3, an example of the histogram of normalized speech coefficients is shown (Gerkmann and Martin, 2010). Here, the histogram of the real part of complex speech coefficients $\Re(c)$ is compared to a Gaussian and a *Laplacian* distribution. Clearly, the histogram is more similar to the heavy-tailed Laplacian

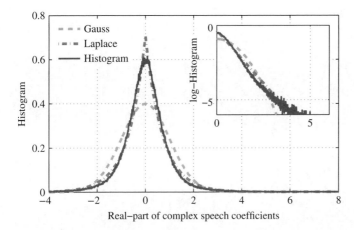

Figure 5.3 Histogram of the real part of complex speech coefficients (Gerkmann and Martin, 2010).

distribution. We emphasize that the Laplacian distribution is only an example of a heavy-tailed distribution and sparser distributions may often provide a better fit (Vincent, 2007).

If the target spectral coefficients follow a heavy-tailed distribution, nonlinear estimators can be derived that outperform the simple Wiener filter. This idea has already been discussed by Porter and Boll (1984), who proposed to obtain the optimal filter based on training data. Martin proposed a closed-form solution when the target follows a gamma (Martin, 2002) or Laplacian (Martin and Breithaupt, 2003) prior distribution and showed that improved performance can be achieved.

As a consequence, in the last decade many more proposals and improvements using heavy-tailed priors were proposed. For this, parameterizable speech priors were proposed in order to optimize the achieved results by means of instrumental measures or listening experiments. One example of such a parameterizable speech prior is the χ *distribution* for spectral amplitudes $|c|$ as

$$p(|c|) = \frac{2}{\Gamma(\mu)} \left(\frac{\mu}{\sigma_c^2} \right)^{\mu} |c|^{2\mu-1} \exp\left(-\frac{\mu}{\sigma_c^2} |c|^2 \right) \tag{5.48}$$

with $\Gamma(\cdot)$ as defined below (5.46). The so-called shape parameter μ controls the heavy-tailedness of the target prior. While for $\mu = 1$ (5.48) corresponds to the Rayleigh distribution (5.44) thus implying complex Gaussian coefficients c, for $0 < \mu < 1$ a heavy-tailed speech prior results. Using different heavy-tailed priors, both MAP and MMSE estimators were derived for complex spectral coefficients, spectral amplitudes, and compressed spectral amplitudes (Martin, 2005; Lotter and Vary, 2005; Benaroya *et al.*, 2006; Erkelens *et al.*, 2007; Breithaupt *et al.*, 2008b). In Figure 5.4 it can be seen that for a large input $x/\sigma_c \gg 1$, a heavy-tailed prior results in a larger output than when a Gaussian prior is used. This is because using heavy-tailed target priors, outliers in the input are more likely attributed to the target signal. This behavior of super-Gaussian estimators results in less speech attenuation and thus less target distortion in the processed signal. However, this behavior may also increase the amount of undesired outliers in the processed signal that may be perceived as annoying musical tones.

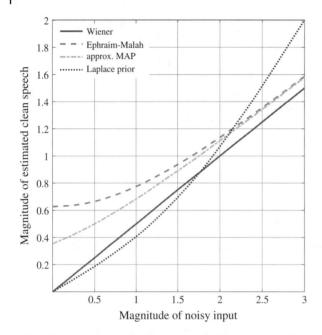

Figure 5.4 Input-output characteristics of different spectral filtering masks. In this example $\sigma_c^2 = \sigma_u^2 = 1$. "Wiener" refers to the Wiener filter, "Ephraim–Malah" to the short-time spectral amplitude estimator of Ephraim and Malah (1984), and "approx. MAP" to the approximate MAP amplitude estimator (5.47) of Wolfe and Godsill (2003). While "Wiener", "Ephraim–Malah", and "approx. MAP" are based on a Gaussian speech model, "Laplace prior" refers to an estimator of complex speech coefficients with a super-Gaussian speech prior (Martin and Breithaupt, 2003). Compared to the linear Wiener filter, amplitude estimators tend to apply less attenuation for low inputs, while super-Gaussian estimators tend to apply less attenuation for high inputs.

5.2.6 Masks Based on Source Presence Statistics

While in deriving linear MMSE estimators via (5.17) a spectral masking filter was explicitly estimated, for the Bayesian estimators considered so far, the targets were (functions of) the target coefficients c. A spectral masking filter could then be obtained by dividing input and output as $w = \hat{c}/x$.

A different way to obtain a spectral masking filter using Bayesian statistics is to estimate the source presence probability. For this, we define $\mathcal{H}_1$ as the hypothesis that the target is active in the considered time-frequency bin, while $\mathcal{H}_0$ denotes the hypothesis that it is inactive. A spectral masking filter can then be obtained by computing the posterior probability that the target is present. Using Bayes' theorem, similar to (5.28), this posterior probability can be expressed as

$$w = P(\mathcal{H}_1 \mid x) = \frac{p(x \mid \mathcal{H}_1)P(\mathcal{H}_1)}{p(x \mid \mathcal{H}_1)P(\mathcal{H}_1) + p(x \mid \mathcal{H}_0)P(\mathcal{H}_0)}. \tag{5.49}$$

As opposed to the previously described mask estimators, the optimality criterion is not on the separation or enhancement of sources anymore but merely on the estimation of their presence. The source presence probability is a particularly powerful tool, e.g. to estimate parameters such as the interference power spectrum. More details about speech presence probability (SPP) estimation can be found in Chapter 6.

5.3 Perceptual Improvements

Whenever applying a spectral mask, undesired artifacts may occur. When the mask in a given time-frequency bin is lower than it should be, target distortion occurs. When the mask is larger, interference reduction is limited. The artifacts are even more disturbing when the mask exhibits large values in isolated time-frequency regions that result in isolated outliers in the estimated signal. In the time domain, these isolated spectral peaks result in sinusoidal components of short duration and are often perceived as annoying musical noise. An early proposal to reduce musical noise artifacts is to *overestimate* the interference power spectrum in order to reduce spectral outliers in the mask (Berouti *et al.*, 1979) at the costs of increased target distortion. Another way is to apply a lower limit, a so-called *spectral floor* (Berouti *et al.*, 1979), to the spectral masking filter at the cost of lesser interference reduction. Using this spectral floor, both musical noise and target distortion are controlled and perceptually convincing results can be obtained.

Porter and Boll (1984) observed that when spectral masking filters are derived by estimating compressed spectral amplitudes, musical noise can be reduced. As a compression of amplitudes is related to the way we perceive the loudness of sounds, also the estimation of compressed spectral coefficients is considered to be perceptually more meaningful than an amplitude estimation without compression. Ephraim and Malah (1985) were the first to derive a closed-form solution for a Bayesian estimator of logarithmically compressed amplitudes under a Gaussian prior and likelihood. You *et al.* (2005) derived a more general estimator for powers of spectral amplitudes as $\mathbb{E}\{|c|^\beta \mid x\}$ that also generalizes the square root ($\beta = 1/2$) and logarithmic ($\beta \to 0$) compression. Breithaupt *et al.* (2008b) again generalized this result for the parameterizable prior (5.48), thus enabling the estimation of compressed spectral coefficients under heavy-tailed priors. This flexible estimator results in

$$\mathbb{E}\{|c|^\beta \mid x\} = \left(\frac{\sigma_c^2 \sigma_u^2}{\sigma_c^2 + \mu\sigma_u^2} \right)^{\frac{\beta}{2}} \frac{\Gamma(\mu + \beta/2)}{\Gamma(\mu)} \frac{\Phi(1 - \mu - \beta/2, 1; -\nu)}{\Phi(1 - \mu, 1; -\nu)} \tag{5.50}$$

with $\nu = |x|^2 \sigma_c^2 / (\sigma_u^4 \mu + \sigma_u^2 \sigma_c^2)$ and $\Gamma(\cdot), \Phi(\cdot)$ as defined below (5.46). Compression is obtained for $0 < \beta < 1$.

Another way to reduce processing artifacts is to apply smoothing methods to the spectral masking filter or its parameters (Vincent, 2010). This has to be done with great care in order not to introduce smearing artifacts. In the single-channel case, simple non-adaptive temporal smoothing often does not lead to satisfactory results and adaptive smoothing methods are used instead (Ephraim and Malah, 1984; Cappé, 1994; Martin and Lotter, 2001). Good results can also be achieved by carefully smoothing over both time and frequency (Cohen and Berdugo, 2001; Gerkmann *et al.*, 2008), or smoothing in perceptually motivated filter bands (Esch and Vary, 2009; Brandt and Bitzer, 2009). An elegant way to incorporate typical speech spectral structures in the smoothing process is to apply so-called *cepstral smoothing* (Breithaupt *et al.*, 2007), as illustrated in Figure 5.5. The *cepstrum* is defined as the spectral transform of the logarithmic amplitude spectrum. In this domain speech-like spectral structures are compactly represented by few lower cepstral coefficients that represent the speech spectral envelope, and a peak in the upper cepstrum that represents the spectral fine structure of voiced speech. Thus, in the cepstral domain the speech related coefficients can be preserved while smoothing is mainly applied to the remaining coefficients that represent spectral structures that are

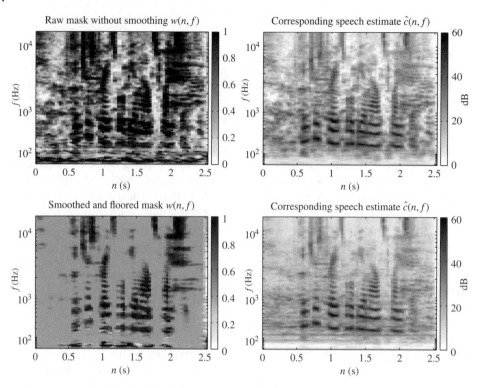

Figure 5.5 Examples of estimated filters for the noisy speech signal in Figure 5.1. The filters were computed in the STFT domain but are displayed on a nonlinear frequency scale for visualization purposes.

not speech-like. This method can be applied directly to spectral masks (Breithaupt *et al.*, 2007), or to the target and interference spectra from which the masks are computed (Breithaupt *et al.*, 2008a; Gerkmann *et al.*, 2010).

5.4 Summary

In this chapter, we introduced the concept of spectral masking for signal enhancement and separation. We reviewed different ways of deriving time-frequency masks, from spectral subtraction and Wiener filtering to more general Bayesian estimation of source spectral coefficients or activity masks. We argued that when functions of complex coefficients, such as magnitudes, are targeted better estimators than spectral subtraction and Wiener filtering can be derived using Bayesian estimation. We finally discussed methods to control the perceptual quality of the output by heuristic tweaks, estimation of compressed spectral coefficients, and time-frequency smoothing.

The reviewed estimators depend on the statistics of the source signals, namely the source activities $P(\mathcal{H}_1 \mid x)$ and $P(\mathcal{H}_0 \mid x)$ (zeroth-order statistics) or the source variances σ_c^2 and σ_u^2 (second-order statistics). In the case of three or more sources, these boil down to estimation of the activities or the variances $\sigma_{c_j}^2$ of all sources. The estimation of these

Table 5.3 Overview of the discussed estimation schemes.

Method	Pros	Cons
Spectral subtraction	Simple	Somewhat heuristic
Bayesian estimation	Very flexible, well defined optimality criteria	Models needed, closed-form solutions not guaranteed to exist
Wiener filtering	Simple	Not optimal for nonnegative or nongaussian quantities
Heavy-tailed priors	Less target distortion	Often more musical noise
Source presence statistics	Powerful tool to estimate parameters such as the interference power spectrum	Optimality not defined in terms of the separated/enhanced signals
Perceptual improvements	Better sound quality	Often more interference

quantities in the single-channel case is covered in Chapters 6, 7, 8, and 9. An overview of the discussed mask estimators is given in Table 5.3.

In recent years, improved estimators that go beyond the time-frequency masking paradigm have been proposed. For instance, researchers showed that the correlation of neighboring time-frequency bins and the spectral phase can be estimated and exploited for enhancement with reduced distortion. These techniques are reviewed in Chapter 19.

Bibliography

Benaroya, L., Bimbot, F., and Gribonval, R. (2006) Audio source separation with a single sensor. *IEEE Transactions on Audio, Speech, and Language Processing*, **14** (1), 191–199.

Berouti, M., Schwartz, R., and Makhoul, J. (1979) Enhancement of speech corrupted by acoustic noise, in *Proceedings of IEEE International Conference on Audio, Speech and Signal Processing*, pp. 208–211.

Boll, S.F. (1979) Suppression of acoustic noise in speech using spectral subtraction. *IEEE Transactions on Acoustics, Speech, and Signal Processing*, **27** (2), 113–120.

Brandt, M. and Bitzer, J. (2009) Optimal spectral smoothing in short-time spectral attenuation (STSA) algorithms: Results of objective measures and listening tests, in *Proceedings of European Signal Processing Conference*, pp. 199–203.

Breithaupt, C., Gerkmann, T., and Martin, R. (2007) Cepstral smoothing of spectral filter gains for speech enhancement without musical noise. *IEEE Signal Processing Letters*, **14** (12), 1036–1039.

Breithaupt, C., Gerkmann, T., and Martin, R. (2008a) A novel a priori SNR estimation approach based on selective cepstro-temporal smoothing, in *Proceedings of IEEE International Conference on Audio, Speech and Signal Processing*, pp. 4897–4900.

Breithaupt, C., Krawczyk, M., and Martin, R. (2008b) Parameterized MMSE spectral magnitude estimation for the enhancement of noisy speech, in *Proceedings of IEEE International Conference on Audio, Speech and Signal Processing*, pp. 4037–4040.

Cappé, O. (1994) Elimination of the musical noise phenomenon with the Ephraim and Malah noise suppressor. *IEEE Transactions on Speech and Audio Processing*, **2** (2), 345–349.

Cauchi, B., Kodrasi, I., Rehr, R., Gerlach, S., Jukic, A., Gerkmann, T., Doclo, S., and Goetze, S. (2015) Combination of MVDR beamforming and single-channel spectral processing for enhancing noisy and reverberant speech. *EURASIP Journal on Advances in Signal Processing*, **2015** (61), 1–12.

Cohen, I. and Berdugo, B. (2001) Speech enhancement for non-stationary noise environments. *Signal Processing*, **81** (11), 2403–2418.

Ephraim, Y. and Malah, D. (1984) Speech enhancement using a minimum mean-square error short-time spectral amplitude estimator. *IEEE Transactions on Acoustics, Speech, and Signal Processing*, **32** (6), 1109–1121.

Ephraim, Y. and Malah, D. (1985) Speech enhancement using a minimum mean-square error log-spectral amplitude estimator. *IEEE Transactions on Acoustics, Speech, and Signal Processing*, **33** (2), 443–445.

Erkelens, J.S., Hendriks, R.C., Heusdens, R., and Jensen, J. (2007) Minimum mean-square error estimation of discrete Fourier coefficients with generalized Gamma priors. *IEEE Transactions on Audio, Speech, and Language Processing*, **15** (6), 1741–1752.

Esch, T. and Vary, P. (2009) Efficient musical noise suppression for speech enhancement systems, in *Proceedings of IEEE International Conference on Audio, Speech and Signal Processing*, pp. 4409–4412.

Gerkmann, T., Breithaupt, C., and Martin, R. (2008) Improved a posteriori speech presence probability estimation based on a likelihood ratio with fixed priors. *IEEE Transactions on Audio, Speech, and Language Processing*, **16** (5), 910–919.

Gerkmann, T., Krawczyk, M., and Martin, R. (2010) Speech presence probability estimation based on temporal cepstrum smoothing, in *Proceedings of IEEE International Conference on Audio, Speech and Signal Processing*, pp. 4254–4257.

Gerkmann, T. and Martin, R. (2010) Empirical distributions of DFT-domain speech coefficients based on estimated speech variances, in *Proceedings of International Workshop on Acoustic Echo and Noise Control*.

Gradshteyn, I.S. and Ryzhik, I.M. (2000) *Table of Integrals Series and Products*, Academic Press, 6th edn.

Habets, E.A.P. (2007) *Single- and Multi-Microphone Speech Dereverberation using Spectral Enhancement*, Ph.D. thesis, Technische Universiteit Eindhoven.

Hendriks, R.C., Gerkmann, T., and Jensen, J. (2013) *DFT-Domain Based Single-Microphone Noise Reduction for Speech Enhancement: A Survey of the State of the Art*, Morgan & Claypool.

Jensen, J. and Hendriks, R.C. (2012) Spectral magnitude minimum mean-square error estimation using binary and continuous gain functions. *IEEE Transactions on Audio, Speech, and Language Processing*, **20** (1), 92–102.

Lebart, K., Boucher, J.M., and Denbigh, P.N. (2001) A new method based on spectral subtraction for speech dereverberation. *Acta Acustica*, **87**, 359–366.

Li, N. and Loizou, P.C. (2008) Factors influencing intelligibility of ideal binary-masked speech: Implications for noise reduction. *Journal of the Acoustical Society of America*, **123** (3), 1673–1682.

Lotter, T. and Vary, P. (2005) Speech enhancement by MAP spectral amplitude estimation using a super-Gaussian speech model. *EURASIP Journal on Advances in Signal Processing*, **2005** (7), 1110–1126.

Madhu, N., Spriet, A., Jansen, S., Koning, R., and Wouters, J. (2013) The potential for speech intelligibility improvement using the ideal binary mask and the ideal Wiener filter in single channel noise reduction systems: Application to auditory prostheses. *IEEE Transactions on Audio, Speech, and Language Processing*, **21** (1), 61–70.

Martin, R. (2002) Speech enhancement using MMSE short time spectral estimation with Gamma distributed speech priors, in *Proceedings of IEEE International Conference on Audio, Speech and Signal Processing*, pp. 253–256.

Martin, R. (2005) Speech enhancement based on minimum mean-square error estimation and supergaussian priors. *IEEE Transactions on Speech and Audio Processing*, **13** (5), 845–856.

Martin, R. and Breithaupt, C. (2003) Speech enhancement in the DFT domain using Laplacian speech priors, in *Proceedings of International Workshop on Acoustic Echo and Noise Control*, pp. 87–90.

Martin, R. and Lotter, T. (2001) Optimal recursive smoothing of non-stationary periodograms, in *Proceedings of International Workshop on Acoustic Echo and Noise Control*, pp. 167–170.

Porter, J.E. and Boll, S.F. (1984) Optimal estimators for spectral restoration of noisy speech, in *Proceedings of IEEE International Conference on Audio, Speech and Signal Processing*, pp. 18A.2.1–18A.2.4.

Roweis, S.T. (2001) One microphone source separation, in *Proceedings of Neural Information Processing Systems*, pp. 793–799.

Schreier, P.J. and Scharf, L.L. (2010) *Statistical Signal Processing of Complex-valued Data: The Theory of Improper and Noncircular Signals*, Cambridge University Press.

Vincent, E. (2007) Complex nonconvex l_p norm minimization for underdetermined source separation, in *Proceedings of International Conference on Independent Component Analysis and Signal Separation*, pp. 430–437.

Vincent, E. (2010) An experimental evaluation of Wiener filter smoothing techniques applied to under-determined audio source separation, in *Proceedings of International Conference on Latent Variable Analysis and Signal Separation*, pp. 157–164.

Vincent, E., Gribonval, R., and Plumbley, M.D. (2007) Oracle estimators for the benchmarking of source separation algorithms. *Signal Processing*, **87** (8), 1933–1950.

Wang, D.L. (2005) On ideal binary mask as the computational goal of auditory scene analysis, in *Speech Separation by Humans and Machines*, Springer, pp. 181–197.

Wolfe, P.J. and Godsill, S.J. (2003) Efficient alternatives to the Ephraim and Malah suppression rule for audio signal enhancement. *EURASIP Journal on Applied Signal Processing*, **10**, 1043–1051.

You, C.H., Koh, S.N., and Rahardja, S. (2005) β-order MMSE spectral amplitude estimation for speech enhancement. *IEEE Transactions on Speech and Audio Processing*, **13** (4), 475–486.

6

Single-Channel Speech Presence Probability Estimation and Noise Tracking

Rainer Martin and Israel Cohen

The single-channel enhancement filters reviewed in Chapter 5 require knowledge of the power spectra of the target and the interference signals. Since the target and interfering signals are not available their power spectra must be estimated from the mixture signal. In most acoustic scenarios the power spectra of both the target and the interfering signals are time-varying and therefore require online tracking. All together, this constitutes a challenging estimation problem, especially when the interference is highly nonstationary and when it occupies the same frequency bands as the target signal.

Most algorithms in this domain rely on specific statistical differences between speech as the target signal and interfering noise signals. The methods presented in this chapter are developed for a single speaker mixed with short-time stationary environmental noise such as car noise and multiple-speaker babble noise. These methods will most likely fail when the interference is a single competing speaker. In the latter case, there are in general no speaker-independent statistical differences that could be exploited. Single-channel speaker separation methods must then be utilized which typically require trained models of specific speakers. These methods are outside the scope of this chapter and are discussed in Chapters 7, 8, and 9. Speech and noise power spectrum estimation is closely related to voice activity detection (VAD, see Chapter 4) and the estimation of *speech presence probability* (SPP). The interplay between speech and noise power tracking and SPP estimation therefore constitutes a central aspect of this chapter.

The structure of the chapter is as follows. We review and discuss the basic principles of single-channel SPP estimation in Section 6.1, noise power spectrum tracking in Section 6.2, and the corresponding evaluation methods in Section 6.3. Throughout the chapter we present several practical algorithms and discuss their performance. We conclude with a summary in Section 6.4.

6.1 Speech Presence Probability and its Estimation

There is obviously a close coupling between voice activity and noise power estimation. While in early approaches a voice activity detector was used for the identification of

Audio Source Separation and Speech Enhancement, First Edition.
Edited by Emmanuel Vincent, Tuomas Virtanen and Sharon Gannot.
© 2018 John Wiley & Sons Ltd. Published 2018 by John Wiley & Sons Ltd.
Companion Website: https://project.inria.fr/ssse/

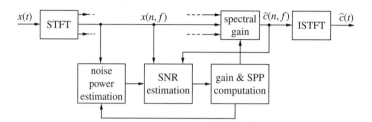

Figure 6.1 State-of-the-art single-channel noise reduction system operating in the STFT domain. The application of a spectral gain function to the output of the STFT block results in an estimate $\hat{c}(n, f)$ of clean speech spectral coefficients $c(n, f)$. The spectral gain is controlled by the estimated SNR, which in turn requires the noise power tracker as a central component.

noise-only segments, most modern approaches avoid global decisions but employ a statistical model in the time-frequency domain. In this section, we will explain the most common statistical model and the resulting interactions of SPP estimation and noise power spectrum tracking.

To set the stage for our discussion of algorithms, Figure 6.1 depicts the block diagram of a typical single-channel noise reduction algorithm implemented in the short-time Fourier transform (STFT) domain. Throughout this chapter the target signal is a speech signal $c(t)$ and the interference is an additive noise signal $u(t)$. Using the same notations as in Chapter 5, we may write the mixed signal x in the STFT domain as

$$x(n, f) = c(n, f) + u(n, f) \tag{6.1}$$

where n and f denote the time frame and the frequency bin index in the STFT representation, respectively.

In order to compute a spectral gain function (see Chapter 5), an accurate estimate of the *signal-to-noise ratio* (SNR) in each frequency bin is required. Instead of a direct SNR estimation, it is often more practical to first estimate the noise power spectrum, then derive the power of the target signal and subsequently the SNR. This is especially true when the variations of the noise power are slower than the power variations of the speech signal. Furthermore, noise power estimation is largely facilitated when an estimate of the probability of speech presence is available. It is therefore common practice to feed back an SPP estimate into the control of the noise power tracking process. Therefore, both estimation processes are linked and need to be discussed in the context of an appropriate statistical model.

6.1.1 Speech Presence Probability

Discrete Fourier transform (DFT) coefficients and the quantities derived thereof have a number of statistical properties that facilitate the design of noise power and SPP estimation algorithms. For large transform lengths, the complex Fourier coefficients are known to be asymptotically complex Gaussian distributed (Brillinger, 1981) with zero mean. Under this model $x(n, f)$ is distributed in each time-frequency bin as

$$p(x(n, f)) = \frac{1}{\pi \sigma_x^2(n, f)} \exp\left(-\frac{|x(n, f)|^2}{\sigma_x^2(n, f)}\right) \tag{6.2}$$

where $\sigma_x^2(n, f) = \mathbb{E}\{|x(n, f)|^2\}$ denotes the signal power in that time-frequency bin. The power $\sigma_x^2(n, f)$ is equal to the noise power $\sigma_u^2(n, f)$ during speech pauses and to

$\sigma_c^2(n,f) + \sigma_u^2(n,f)$ during speech activity, where the common assumption of statistical independence between speech and noise components has been utilized. This model and its supergaussian variants are discussed in greater detail in Section 5.2.

Given an appropriate statistical model we may now formulate the hypothesis of speech presence as $\mathcal{H}_1(n,f)$ and the hypothesis of speech absence as $\mathcal{H}_0(n,f)$ in each time-frequency bin. Accordingly, the probabilities of speech presence and speech absence given the observed noisy signal are denoted as $P(\mathcal{H}_1(n,f) \mid x(n,f))$ and $P(\mathcal{H}_0(n,f) \mid x(n,f)) = 1 - P(\mathcal{H}_1(n,f) \mid x(n,f))$, respectively. Using Bayes theorem we obtain

$$P(\mathcal{H}_1(n,f) \mid x(n,f))$$

$$= \frac{p(x(n,f) \mid \mathcal{H}_1(n,f))P(\mathcal{H}_1(n,f))}{p(x(n,f) \mid \mathcal{H}_1(n,f))P(\mathcal{H}_1(n,f)) + p(x(n,f) \mid \mathcal{H}_0(n,f))P(\mathcal{H}_0(n,f))}$$

$$= \frac{\rho(n,f)}{1 + \rho(n,f)} \tag{6.3}$$

where $P(\mathcal{H}_1(n,f))$ and $P(\mathcal{H}_0(n,f)) = 1 - P(\mathcal{H}_1(n,f))$ denote the *prior probabilities of speech presence or absence*, respectively. The term $\rho(n,f)$ denotes the *generalized likelihood ratio* and is defined as

$$\rho(n,f) = \frac{P(\mathcal{H}_1(n,f))}{P(\mathcal{H}_0(n,f))} \frac{p(x(n,f) \mid \mathcal{H}_1(n,f))}{p(x(n,f) \mid \mathcal{H}_0(n,f))} . \tag{6.4}$$

Obviously, a large likelihood ratio results in an SPP close to one. For the complex Gaussian model of the transformed signal we have for both hypotheses

$$p(x(n,f) \mid \mathcal{H}_1(n,f))$$

$$= \frac{1}{\pi(\sigma_c^2(n,f) + \sigma_u^2(n,f))} \exp\left(-\frac{|x(n,f)|^2}{\sigma_c^2(n,f) + \sigma_u^2(n,f)}\right) \tag{6.5}$$

$$p(x(n,f) \mid \mathcal{H}_0(n,f)) = \frac{1}{\pi\sigma_u^2(n,f)} \exp\left(-\frac{|x(n,f)|^2}{\sigma_u^2(n,f)}\right) . \tag{6.6}$$

Therefore, the generalized likelihood ratio is established as

$$\rho(n,f) = \frac{P(\mathcal{H}_1(n,f))}{P(\mathcal{H}_0(n,f))} \frac{1}{1 + \xi(n,f)} \exp\left(\gamma(n,f)\frac{\xi(n,f)}{1 + \xi(n,f)}\right) \tag{6.7}$$

where

$$\xi(n,f) = \frac{\sigma_c^2(n,f)}{\sigma_u^2(n,f)} \tag{6.8}$$

is commonly known as the *a priori SNR* and

$$\gamma(n,f) = \frac{|x(n,f)|^2}{\sigma_u^2(n,f)} \tag{6.9}$$

as the *a posteriori SNR* (Ephraim and Malah, 1984). We note that these SNR quantities are closely related as

$$\xi(n,f) = \mathbb{E}\{\gamma(n,f) - 1\} . \tag{6.10}$$

Obviously, the estimation of the likelihood ratio and thus of the SPP requires estimates of the noise power spectrum, of the a posteriori and a priori SNRs, and of the prior

SPPs. While noise power spectrum estimation will be discussed in detail in Section 6.2, we will discuss the latter three estimation tasks briefly below. To this end we assume for the remainder of this section that a sufficiently accurate noise power spectrum estimate $\hat{\sigma}_u^2(n,f)$ is available.

6.1.2   Estimation of the a Posteriori SNR

For the Gaussian signal model in (6.2), $|x(n,f)|^2$ is exponentially distributed and therefore also $\gamma(n,f)$ follows an *exponential distribution*, i.e.

$$p(\gamma(n,f)) = \frac{1}{1 + \xi(n,f)} \exp\left(-\frac{\gamma(n,f)}{1 + \xi(n,f)}\right) . \tag{6.11}$$

This distribution is parameterized by the a priori SNR $\xi(n,f)$. Furthermore, we note that during speech absence we have $\xi(n,f) = 0$. The above expression reduces to a standard exponential distribution with unit mean and unit variance.

As the squared magnitude of the noisy signal x in the STFT domain is immediately available, an estimate of the a posteriori SNR is obtained by replacing the true noise power by its most recent unbiased estimate $\hat{\sigma}_u^2(n,f)$. Hence, we may write

$$\hat{\gamma}(n,f) = \frac{|x(n,f)|^2}{\hat{\sigma}_u^2(n,f)} . \tag{6.12}$$

6.1.3   Estimation of the a Priori SNR

To estimate the a priori SNR we may now rewrite (6.11) as a likelihood $p(\gamma \mid \xi)$. Its derivative with respect to ξ yields the maximum likelihood (ML) estimate of the a priori SNR

$$\hat{\xi}_{\mathrm{ML}}(n,f) = \gamma(n,f) - 1 \approx \hat{\gamma}(n,f) - 1 . \tag{6.13}$$

Thus, both SNR quantities are closely related. Since this estimate makes use of the instantaneous squared magnitude of the noisy signal, it has a relatively large variance. A reduction of the variance therefore requires smoothing.

A very effective form of smoothing has been proposed by Ephraim and Malah (1984), which relies on the estimated target signal STFT $\hat{c}(n - 1,f)$ of the previous frame and the ML estimate of (6.13):

$$\hat{\xi}_{\mathrm{DD}}(n,f) = \max\left(\lambda \frac{|\hat{c}(n - 1,f)|^2}{\hat{\sigma}_u^2(n,f)} + (1 - \lambda)(\hat{\gamma}(n,f) - 1), \, \mathrm{SNR}_{\min}\right) \tag{6.14}$$

where $\lambda \in [0, 1]$ is a smoothing parameter and $\mathrm{SNR}_{\min} > 0$ is a parameter that allows a lower positive limit to be set on the SNR estimates. This limit, typically in the range of [0.01,0.1], has been found to be effective for the avoidance of musical noise artifacts in the processed output signal (Cappé, 1994).

The above is known as the *decision-directed* a priori SNR estimation method and is widely used. Its name is derived from an analogy to the decision-feedback equalizer known in communications technology (Proakis and Salehi, 2007). Just as the decision feedback equalizer uses bit decisions for the improvement of the channel estimate, the SNR estimator in (6.14) feeds back the estimated signal of the previous frame into the update of the current SNR value. Typically, this process leads to a delay of one frame in the estimate with the consequence that speech onsets are not instantly reproduced in the processed signal. However, for values of λ close to one, the large variance of the ML estimator is significantly reduced and thus musical noise is largely avoided (Cappé, 1994). In order to counteract the delay, Plapous *et al.* (2004) proposed a two-stage procedure which first estimates a preliminary gain function and then re-estimates the a priori SNR and the final gain. It should also be noted that the performance of the decision-directed a priori SNR estimator depends on the SNR and on the method employed for the estimation of the target signal STFT $\hat{c}(n,f)$ (Breithaupt and Martin, 2011).

6.1.4 Estimation of the Prior Speech Presence Probability

It is common either to set the prior probability $P(\mathcal{H}_1(n,f))$ to a fixed value in the range of [0.5,0.8] (McAulay and Malpass, 1980; Ephraim and Malah, 1984) or to track it adaptively, as proposed, for example, by Malah *et al.* (1999), Cohen and Berdugo (2001), and Cohen (2003). For example, the probability of speech presence can be obtained from a test on the a priori SNR ξ where the a priori SNR is compared to a preset threshold $\xi_{\min}$. Since the a priori SNR parameterizes the statistics of the a posteriori SNR in terms of the exponential distribution, a test can be devised which relies exclusively on the a posteriori SNR (Malah *et al.*, 1999). In this test, Malah *et al.* (1999) compare the a posteriori SNR to a threshold $\gamma_{\min}$ using the indicator variable

$$\mathbb{1}_\gamma(n,f) = \begin{cases} 1 & \text{if } \hat{\gamma}(n,f) > \gamma_{\min} \\ 0 & \text{otherwise.} \end{cases} \tag{6.15}$$

When speech is present, the decisions are smoothed over time in each frequency bin by *recursive averaging*, also known as *exponential smoothing*:

$$\hat{P}(\mathcal{H}_1(n,f)) = \lambda_p \, \hat{P}(\mathcal{H}_1(n-1,f)) + (1-\lambda_p) \, \mathbb{1}_\gamma(n,f) . \tag{6.16}$$

During speech pauses, the estimated prior probability $\hat{P}(\mathcal{H}_1(n,f))$ is set to a fixed low value. Thus, in conjunction with a reliable voice activity detector the adaptive tracking method enables good discrimination between time-frequency bins which are mostly occupied by speech or by noise.

Cohen and Berdugo (2001) and Cohen (2003) introduced estimators for the prior SPP, which are computationally efficient and characterized by the ability to quickly follow abrupt changes in the noise spectrum. The estimators are controlled by the minimum values of the smoothed power spectrum without a hard distinction between speech absence and presence, thus continuously updating the prior SPP even during weak speech activity. The smoothing of the noisy power spectrum is carried out in both time and frequency, which accounts for the strong correlation of speech presence in neighboring frequency bins of consecutive frames.

6.1.5 SPP Estimation with a Fixed SNR Prior

While all of the above methods contribute to successful SPP estimation, the adjustment of the a priori SPP to the actual speech presence in each time-frequency bin may appear to be somewhat unnatural. However, without this adaptation of the prior the discrimination between bins with speech activity and speech absence is less effective. In particular, when the SPP prior is set to a fixed value we note that the generalized likelihood ratio in (6.7) approaches the ratio $P(\mathcal{H}_1)/P(\mathcal{H}_0)$ during speech absence, i.e. for $\sigma_c^2 = 0$. As a consequence the SPP will be equal to $P(\mathcal{H}_1)$ in this case. When the SPP value is used, e.g. to construct a spectral mask for noise reduction, and the prior is set to $P(\mathcal{H}_1) = 0.5$, the attenuation of the interference is limited to 6 dB. In order to achieve lower values of the SPP during speech pause, a modification of the generalized likelihood ratio has been proposed by Gerkmann *et al.* (2008). In this work the authors use not only a fixed SPP prior but also set the a priori SNR to a fixed value. This fixed SNR value is selected as the typical value that would be observed when speech were present in the noisy signal. Among other factors, this value depends on the amount of smoothing applied to the estimated a posteriori SNR. Minimizing the false alarm and missed hit probabilities, an optimal value of 8 dB is found for a specific example where 15 adjacent bins of the time-frequency plane are included in the smoothing process (Gerkmann *et al.*, 2008).

In Figure 6.2 we finally provide a comparison of the traditional SPP (6.3) and the SPP estimate with a fixed a priori SNR. In order to allow a direct comparison we do not smooth the a posteriori SNR across time or frequency, although this is recommended in general. As pointed out before, smoothing the a posteriori SNR will reduce the number of outliers but also requires a reformulation of the likelihood ratio in terms of a χ^2 statistics. In contrast to the traditional estimate, which is lower bounded by $P(\mathcal{H}_1)/P(\mathcal{H}_0)$, we

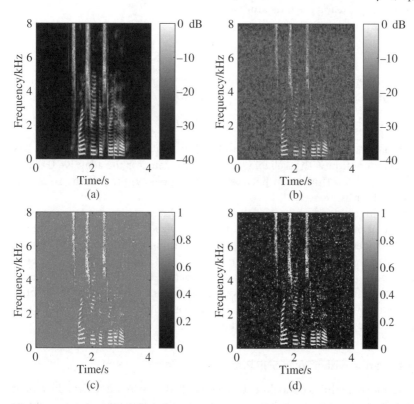

Figure 6.2 Spectrograms of (a) a clean speech sample, (b) clean speech mixed with traffic noise at 5 dB SNR, (c) SPP based on the Gaussian model, and (d) SPP based on the Gaussian model with a fixed a priori SNR prior. Single bins in the time-frequency plane are considered. The fixed a priori SNR was set to 15 dB and $P(\mathcal{H}_1) = 0.5$.

find that the SPP estimate using a fixed a priori SNR covers the full range of values in the interval [0, 1]. Therefore, the latter method is most useful in applications of spectral masking but its utility has been also shown in conjunction with noise power spectrum tracking (Gerkmann and Hendriks, 2012). However, a direct comparison of all the above methods and possible refinements thereof with respect to the noise power tracking application remains to be conducted.

6.2 Noise Power Spectrum Tracking

All of the above procedures assume that a noise power spectrum estimate is available. It is the objective of this section to review the most common *noise power tracking* methods. A fundamental assumption of single-channel noise power tracking algorithms is that certain statistical properties of the noise signals are different from those of the target speech signal. In the early days of noise power tracking, but to some extent also to this day, many algorithms rely on the assumption that the noise signal is more stationary than the target signal. The search for discriminating properties between the target and the interference has also led to specialized algorithms that work well for some specific types of noise signals but perform much worse on other types of noise.

6.2.1 Basic Approaches

Many early approaches to noise power estimation used either a voice activity detector or some simple form of noise power tracking. Often a comparison between the noise power estimate of the previous frame and the short-time power of the noisy input signal controls the update decision. To this end the short-time power spectrum of the noisy signal x is often estimated by recursive averaging

$$\hat{\sigma}_x^2(n,f) = \lambda_x \hat{\sigma}_x^2(n-1,f) + (1-\lambda_x)|x(n,f)|^2 . \tag{6.17}$$

Then, the computationally efficient method by Doblinger (1995), for instance, updates the estimated noise power spectrum using either a smoothed version of the difference $\hat{\sigma}_x^2(n,f) - \beta\hat{\sigma}_x^2(n-1,f)$ or the estimated short-time power spectrum $\hat{\sigma}_x^2(n,f)$ directly,

$$\hat{\sigma}_u^2(n,f) = \begin{cases} \lambda_u\hat{\sigma}_u^2(n-1,f) + \frac{1-\lambda_u}{1-\beta}(\hat{\sigma}_x^2(n,f) - \beta\hat{\sigma}_x^2(n-1,f)) \\ \qquad\qquad\qquad\qquad \text{if } \hat{\sigma}_u^2(n-1,f) < \hat{\sigma}_x^2(n,f) \\ \hat{\sigma}_x^2(n,f) \qquad\qquad\quad \text{otherwise,} \end{cases} \tag{6.18}$$

where λ_u is the smoothing parameter of the update recursion and β controls the update term. We note that first, this algorithm does not allow the noise power estimate to become larger than the short-time power of the noisy signal. This is obviously a sensible constraint which also provides fast tracking of decreasing noise power levels. Second, the magnitude of the noise power update depends on the slope of the short-time power of the noisy signal. This might be reasonable in some scenarios but might also cause leakage of speech power into the noise power estimate in high SNR conditions.

A similarly efficient approach to tracking the noise power uses a slope parameter ϵ which defines the adaptation speed and is typically set to 5 dB/s. Again, a minimum operation restricts the estimated noise power to values smaller than or equal to the short-time power of the noisy signal (Paul, 1981; Hänsler and Schmidt, 2005):

$$\hat{\sigma}_u^2(n,f) = \min(\hat{\sigma}_x^2(n,f), \hat{\sigma}_u^2(n-1,f))(1+\epsilon) . \tag{6.19}$$

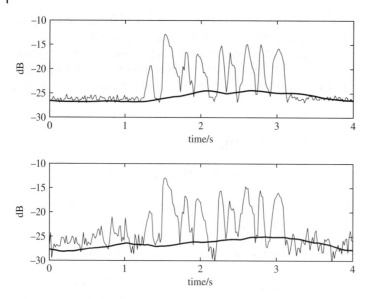

Figure 6.3 Power of the noisy speech signal (thin solid line) and estimated noise power (thick solid line) using the noise power tracking approach according to (6.19). The slope parameter is set such that the maximum slope is 5 dB /s. Top: stationary white Gaussian noise; bottom: nonstationary multiple-speaker babble noise.

The general difficulty of this approach is finding the appropriate slope ϵ. On the one hand, the slope should be large to cope with varying noise powers. On the other hand, a large slope will also track speech components and thus will lead to speech leakage. Therefore, this method is suitable only for slowly varying noise types. When a voiced/unvoiced decision is available, the performance can be improved when the slope is set to smaller values for voiced speech segments (Paul, 1981). Figure 6.3 shows examples for stationary white Gaussian noise and for babble noise using (6.19). It can be observed that this estimator has a tendency to overestimate the average noise power during speech activity while it underestimates the average noise power during speech absence. In noise reduction systems the underestimation during speech pause is often counteracted by means of a noise overestimation factor. A constant overestimation factor, however, will aggravate the noise overestimation during speech activity and hence should be avoided.

Another basic approach has been described by Hirsch and Ehrlicher (1995) using a recursively smoothed noise magnitude $\overline{|u(n,f)|}$, where the update of the recursion is based on the magnitude $|x(n,f)|$ of the noisy signal and is controlled by a hard decision. Updates are computed whenever $(|x(n,f)| - \beta\overline{|u(n,f)|}) \leq 0$ where β ranges in $[1.5, 2.5]$. A second method is proposed in the same paper where this threshold decision is used to build a truncated histogram of past magnitudes of the noisy signal. The most frequently occurring bin then constitutes the noise magnitude estimate. The histogram is built from signal magnitudes across time segments of about 400 ms. It is reported that this second method is significantly more accurate than the first.

6.2.2   The Minimum Statistics Approach

While the approaches discussed above provide sufficient performance for nearly stationary noise types, they also suffer from speech leakage and other robustness issues
when the noise signal has a higher variability. Therefore, significant efforts have been
made to design more robust estimators. In order to approach these issues, we explain
the principle of the well-known *minimum statistics* algorithm for noise power spectrum
estimation and discuss the most crucial measures which are necessary to provide a stable
and accurate performance (Martin, 2001).

The minimum statistics method assumes that the noise power spectrum $\sigma_u^2(n,f) =$
$\mathbb{E}\{|u(n,f)|^2\}$ may be found by tracking power minima in each frequency band, irrespec-
tive of speech activity. To this end we consider the minimum of D successive and in
general time-varying power values of the noisy input signal x as

$$\sigma_{x,\min}^2(n,f) = \min_{n' \in \{n-D+1,\ldots,n\}} \sigma_x^2(n',f) \tag{6.20}$$

and similarly of the noise signal u as

$$\sigma_{u,\min}^2(n,f) = \min_{n' \in \{n-D+1,\ldots,n\}} \sigma_u^2(n',f). \tag{6.21}$$

If D is sufficiently large to span across the periods of syllables or words we may assume
that

$$\sigma_{u,\min}^2(n,f) \approx \sigma_{x,\min}^2(n,f). \tag{6.22}$$

Thus, the minimum power of the noisy signal provides access to the minimum of the
noise power. As the minimum is smaller than the mean the bias $\mathbb{E}\{\sigma_{u,\min}^2(n,f)\} - \sigma_u^2(n,f)$
must be compensated. Then, a noise power estimate is obtained from (6.22) as

$$\hat{\sigma}_u^2(n,f) = \kappa_{\min}(n,f)\sigma_{x,\min}^2(n,f) \tag{6.23}$$

where $\kappa_{\min}(n,f)$ is a multiplicative (and thus scale-invariant) bias correction that
accounts for the difference between the mean minimum value and the desired true
noise power. With respect to the approaches discussed in the previous section, the
improved robustness can be attributed to the (nonlinear) minimum operation that
effectively decouples the noise power estimate from the speech power. However, in
scenarios with highly nonstationary noise signals and when the bias compensation
factor $\kappa_{\min}(n,f)$ is adjusted to more stationary noise types the minimum operation may
lead to an underestimation of the average noise power. In conjunction with a speech
enhancement system this may result in less noise reduction but also less target signal
distortion.

In practice, we estimate $\sigma_x^2(n,f)$ by temporal smoothing of successive instantaneous
power values. This may be achieved either by nonrecursive averaging using a sliding
window or by recursive averaging

$$\hat{\sigma}_x^2(n,f) = \lambda(n,f)\hat{\sigma}_x^2(n-1,f) + (1 - \lambda(n,f))|x(n,f)|^2 \tag{6.24}$$

where $\lambda(n,f)$ denotes an adaptive smoothing parameter. Although the multiplicative
compensation $\kappa_{\min}(n,f)$ of the bias $B(n,f)$ is scale-invariant it still depends on the probability distribution of the smoothed power $\hat{\sigma}_x^2(n,f)$ and as such on the duration and

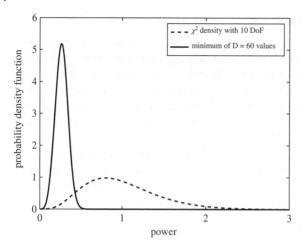

Figure 6.4 Probability distribution of short-time power (χ^2 distribution with 10 degrees of freedom (DoF)) and the corresponding distribution of the minimum of $D = 60$ independent power values.

the shape of the effective smoothing window. To achieve the required level of accuracy the bias compensation has to consider the probability distribution of the minimum. Therefore, this noise power estimator is known as the minimum statistics method. The distribution of short-time smoothed data and the distribution of the minimum of 60 values drawn from it are illustrated in Figure 6.4. Both distributions clearly have different mean values. For χ^2 independently and identically distributed (i.i.d.) power spectra, e.g. obtained from complex Gaussian i.i.d. spectral coefficients, the ratio of these mean values can be computed analytically. For the more practical case of recursive averaging and temporally dependent data, the bias has to be estimated from training data (Martin, 2006). Furthermore, it is important to choose an adaptive smoothing parameter $\lambda(n,f)$ such that less smoothing is applied in high SNR conditions and more smoothing in low SNR conditions or speech absence. A fixed small smoothing parameter would lead to a relatively high variance of the estimated short-time power and thus less reliable power estimates. A fixed large smoothing parameter would result in a long period until the short-time power has decayed from the peaks of speech activity to the noise floor. In high SNR conditions this would require a larger length D of the minimum search window and would reduce the tracking speed of the algorithm. Under the assumption of speech absence, an appropriate smoothing parameter may be derived by minimizing (Martin, 2001)

$$\mathbb{E}\{(\hat{\sigma}_x^2(n,f) - \sigma_u^2(n,f))^2 \mid \hat{\sigma}_x^2(n-1,f)\} \tag{6.25}$$

and is given by (see Figure 6.5)

$$\lambda(n,f) = \frac{1}{1 + (\hat{\sigma}_x^2(n,f)/\sigma_u^2(n,f) - 1)^2} . \tag{6.26}$$

Accordingly, also the bias compensation needs to be adjusted to the time-varying statistics of the smoothed power $\hat{\sigma}_x^2(n,f)$. Finally, we note that besides the minimum, other order statistics have been explored. Stahl *et al.* (2000) designed a noise reduction preprocessor for an automatic speech recognition (ASR) task and obtained the best performance when using the median of the temporal sequence of power values as a noise power estimate. The use of the median, however, requires that a noisy speech file contains a sufficiently large fraction of speech pauses.

Figure 6.5 Optimal smoothing parameter $\lambda(n,f)$ as a function of the power ratio $\hat{\sigma}_x^2(n,f)/\sigma_u^2(n,f)$.

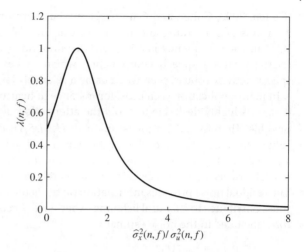

6.2.3 Minima Controlled Recursive Averaging

A common noise estimation technique is to recursively average past spectral power values of the noisy measurement during periods of speech absence, and hold the estimate during speech presence. Specifically, we may write

$$\mathcal{H}_0(n,f) : \quad \bar{\sigma}_u^2(n+1,f) = \lambda\,\bar{\sigma}_u^2(n,f) + (1-\lambda)\,|x(n,f)|^2 \tag{6.27}$$

$$\mathcal{H}_1(n,f) : \quad \bar{\sigma}_u^2(n+1,f) = \bar{\sigma}_u^2(n,f)\,. \tag{6.28}$$

Under speech presence uncertainty, we can employ the SPP and carry out the recursive averaging by

$$\bar{\sigma}_u^2(n+1,f) = P_1(n,f)\,\bar{\sigma}_u^2(n,f)$$
$$+ (1-P_1(n,f))[\lambda\,\bar{\sigma}_u^2(n,f)P_1(n,f) + (1-\lambda)\,|x(n,f)|^2] \tag{6.29}$$

where $P_1(n,f)$ is an estimator for the SPP $P(\mathcal{H}_1(n,f)\mid x(n,f))$. Equivalently, the recursive averaging can be obtained by

$$\bar{\sigma}_u^2(n+1,f) = \tilde{\lambda}(n,f)\,\bar{\sigma}_u^2(n,f) + (1-\tilde{\lambda}(n,f))\,|x(n,f)|^2 \tag{6.30}$$

where

$$\tilde{\lambda}(n,f) = \lambda + (1-\lambda)\,P_1(n,f) \tag{6.31}$$

represents a time-varying frequency-dependent smoothing parameter. Note that the recursive averaging employs a variable time segment in each subband, which takes the probability of speech presence into account. The time segment is longer in subbands that contain frequent speech portions and shorter in subbands that contain frequent silence portions. This feature has been considered a desirable characteristic of noise estimation, which improves its robustness and tracking capability.

The estimator $P_1(n,f)$ requires an estimate of the prior, $\hat{P}(\mathcal{H}_1(n,f))$, as implied from (6.3). In this method, $\hat{P}(\mathcal{H}_1(n,f))$ is obtained by tracking the minima values of a smoothed power spectrum of the noisy signal. The smoothing of the noisy power spectrum is carried out in both time and frequency. This accounts for the strong correlation of speech presence in neighboring frequency bins of consecutive frames

(Cohen, 2003; Cohen and Gannot, 2008). Furthermore, the procedure comprises two iterations of smoothing and minimum tracking. The first iteration provides a rough VAD in each frequency band. Then, the smoothing in the second iteration excludes relatively strong speech components, which makes the minimum tracking during speech activity robust, even when using a relatively large smoothing window.

In many application scenarios, for instance in hearing aids and when the SNR is positive, it is highly desirable to avoid the attenuation of speech components as much as possible. Therefore, the prior estimator $\hat{P}(\mathcal{H}_1(n,f))$ is deliberately biased toward higher values, since deciding speech is absent when speech is present ultimately results in the attenuation of speech components and a reduction of speech intelligibility. The alternative false decision, up to a certain extent, merely introduces some level of residual noise. The residual noise provides information on the acoustic environment to the listener and thus contributes to a natural listening experience. Accordingly, a bias compensation factor is included in the noise estimator

$$\hat{\sigma}_u^2(n,f) = \kappa_0 \bar{\sigma}_u^2(n,f) \tag{6.32}$$

such that the factor $\kappa_0 \geq 1$ compensates the bias when speech is absent:

$$\kappa_0 = \left. \frac{\sigma_u^2(n,f)}{\mathbb{E}\{\bar{\sigma}_u^2(n,f)\}} \right|_{c(n,f)=0} . \tag{6.33}$$

The value of κ_0 is completely determined by the particular prior estimator, and can be derived analytically (Cohen, 2003). The advantage of introducing an upward bias in $\hat{P}(\mathcal{H}_1(n,f))$ and compensating the bias in $\hat{\sigma}_u^2(n,f)$ is that transient speech components are less attenuated, while pseudo-stationary noise components are further suppressed.

Let λ_x $(0 < \lambda_x < 1)$ be a smoothing parameter, and let $h_a(f)$ denote a normalized window function of length $2F' + 1$, i.e., $\sum_{f=-F'}^{F'} h_a(f) = 1$. The frequency smoothing of the noisy power spectrum in each frame is then defined by

$$\bar{\sigma}_x^2(n,f) = \sum_{f'=-F'}^{F'} h_a(f') \, |x(n, f - f')|^2 . \tag{6.34}$$

Subsequently, temporal smoothing is performed by a first-order recursive averaging:

$$\hat{\sigma}_x^2(n,f) = \lambda_x \, \hat{\sigma}_x^2(n-1,f) + (1 - \lambda_x) \, \bar{\sigma}_x^2(n,f) . \tag{6.35}$$

In accordance with the minimum statistics approach, the minima values of $\hat{\sigma}_x^2(n,f)$ are picked within a finite window of length D for each frequency bin:

$$\hat{\sigma}_{x,\min}^2(n,f) = \min_{n' \in \{n-D+1,\dots,n\}} \hat{\sigma}_x^2(n',f). \tag{6.36}$$

It follows (Martin, 2001) that there exists a constant factor $\kappa_{\min}$, independent of the noise power spectrum, such that

$$\mathbb{E}\{\hat{\sigma}_{x,\min}^2(n,f) \mid \mathcal{H}_0\} = \kappa_{\min}^{-1} \sigma_u^2(n,f) . \tag{6.37}$$

The factor $\kappa_{\min}$ compensates the bias of a minimum noise estimate, and generally depends on the values of D, λ_x, F' and the spectral analysis parameters (type, length and overlap of the analysis windows). The value of $\kappa_{\min}$ can be estimated by generating white Gaussian noise, and computing the inverse of the mean of $\hat{\sigma}_{x,\min}^2(n,f)$. This also accounts for the time-frequency correlation of the noisy squared magnitude $|x(n,f)|^2$.

Let $\gamma_{\min}(n,f)$ and $\zeta(n,f)$ be defined by

$$\gamma_{\min}(n,f) = \frac{|x(n,f)|^2}{\kappa_{\min}\,\hat{\sigma}^2_{x,\min}(n,f)} \tag{6.38}$$

$$\zeta(n,f) = \frac{\hat{\sigma}^2_x(n',f)}{\kappa_{\min}\,\hat{\sigma}^2_{x,\min}(n,f)} \tag{6.39}$$

where the denominator $\kappa_{\min}\,\hat{\sigma}^2_{x,\min}(n,f)$ represents an estimate of $\sigma^2_u(n,f)$ based on the minimum statistics method. Let γ_0 and ζ_0 denote threshold parameters that satisfy

$$P(\gamma_{\min} \geq \gamma_0 \mid \mathcal{H}_0) < \epsilon \tag{6.40}$$

$$P(\zeta \geq \zeta_0 \mid \mathcal{H}_0) < \epsilon \tag{6.41}$$

for a certain significance level $\epsilon \ll 1$. Cohen (2003) suggested that in speech enhancement applications, it is useful to replace the prior SPP with the following expression

$$\hat{P}(\mathcal{H}_1(n,f)) = \begin{cases} 0 & \text{if } \gamma_{\min}(n,f) \leq 1 \text{ and } \zeta(n,f) < \zeta_0 \\ \frac{\gamma_{\min}(n,f)-1}{\gamma_0-1} & \text{if } 1 < \gamma_{\min}(n,f) < \gamma_0 \text{ and } \zeta(n,f) < \zeta_0 \\ 1 & \text{otherwise.} \end{cases} \tag{6.42}$$

It assumes that speech is present $(\hat{P}(\mathcal{H}_1(n,f)) = 1)$ whenever $\zeta(n,f) \geq \zeta_0$ or $\gamma_{\min}(n,f) \geq \gamma_0$. That is, whenever the smoothed power spectrum (local measured power), $\hat{\sigma}^2_x(n,f)$, or the instantaneous power, $|x(n,f)|^2$, are relatively high compared to the noise spectrum $\kappa_{\min}\,\hat{\sigma}^2_{x,\min}(n,f) \approx \sigma^2_u(n,f)$. It assumes that speech is absent $(\hat{P}(\mathcal{H}_1(n,f)) = 0)$ whenever both the smoothed and instantaneous measured power spectra are relatively low compared to the noise power spectrum $(\gamma_{\min}(n,f) \leq 1$ and $\zeta(n,f) < \zeta_0)$. In between, there is a soft transition between speech absence and speech presence, based on the value of $\gamma_{\min}(n,f)$.

The main objective of combining conditions on both $\gamma_{\min}(n,f)$ and $\zeta(n,f)$ is to prevent an increase in the estimated noise during weak speech activity, especially when the input SNR is low. Weak speech components can often be extracted using the condition on $\zeta(n,f)$. Sometimes, speech components are so weak that $\zeta(n,f)$ is smaller than ζ_0. In that case, most of the speech power is still excluded from the averaging process using the condition on $\gamma_{\min}(n,f)$. The remaining speech components can hardly affect the noise estimator, since their power is relatively low compared to that of the noise. The steps of noise estimation in this method known as *minima controlled recursive averaging* are summarized in Table 6.1.

6.2.4 Harmonic Tunneling and Subspace Methods

Another approach to noise power spectrum estimation explicitly exploits the harmonic structure of voiced speech (Ealey *et al.*, 2001). The *harmonic tunneling* method first estimates the spectral peaks in the STFT domain and uses an estimate of the fundamental frequency to classify each peak as speech or noise. After a further analysis of the harmonic pattern in the time-frequency plane the peaks of speech energy are consolidated. Then, minimum values between peaks are found and these values are used for an interpolation of the noise power across the full spectrum. The method is especially useful for nonstationary noise types and has been used in conjunction with an ASR task (Pearce and Hirsch, 2000).

Table 6.1 Summary of noise estimation in the minima controlled recursive averaging approach (Cohen, 2003).

Compute the smoothed power spectrum $\hat{\sigma}_x^2(n,f)$ using (6.34) and (6.35).

Update its running minimum $\hat{\sigma}_{x,\min}^2(n,f)$ using (6.36).

Compute $\gamma_{\min}(n,f)$ and $\zeta(n,f)$ using (6.38) and (6.39).

Compute $\hat{P}(\mathcal{H}_1(n,f))$ using (6.42).

Update the SPP $P(\mathcal{H}_1(n,f) \mid x(n,f))$ using (6.3)

Compute $\bar{\sigma}_u^2(n+1,f)$ using (6.29).

Update the noise spectrum estimate $\hat{\sigma}_u^2(n,f)$ using (6.32).

In a more general approach, which also provides a continuous estimate of the noise power spectrum, Hendriks *et al.* (2008) proposed the decomposition of the temporal sequence $x(n',f)$, $n' \in \{n - L + 1, \ldots, n\}$, of L noisy STFT coefficients via a *subspace method*. The method is based on the assumption that the speech correlation matrix is of low rank in each frequency bin and thus a noise-only subspace can be identified from the correlation matrix of the noisy sequence in each frequency bin. In a first step, the sample correlation matrix and the eigenstructure of the prewhitened correlation matrix are computed. Then, an ML estimate of the noise-only subspace is constructed using an average of the smallest eigenvalues. The method requires both an estimate of the subspace dimensions and a compensation of the bias which may result from a systematic overestimation or underestimation of the noise-only subspace dimensions. The method performs well in moderately nonstationary noise but is computationally complex as the eigenvalue decomposition must be performed in each frequency bin. In addition, the presence of deterministic noise components requires further measures as this noise type may be assigned to the signal subspace and is thus not properly estimated in the subspace procedure.

6.2.5 MMSE Noise Power Estimation

A systematic approach to designing an SPP-controlled noise power spectrum estimator is achieved using soft decision minimum mean square error (MMSE) estimation (Gerkmann and Hendriks, 2012). The MMSE noise power spectrum estimate is given by the conditional expectation $\mathbb{E}\{|u(n,f)|^2 \mid x(n,f)\}$ and may be expanded in terms of the hypotheses of speech presence and absence as

$$
\begin{aligned}
\hat{\sigma}_u^2(n,f) &= \mathbb{E}\{|u(n,f)|^2 \mid x(n,f)\} \\
&= \mathbb{E}\{|u(n,f)|^2 \mid x(n,f), \mathcal{H}_1(n,f)\}P(\mathcal{H}_1(n,f)) \\
&\quad + \mathbb{E}\{|u(n,f)|^2 \mid x(n,f), \mathcal{H}_0(n,f)\}P(\mathcal{H}_0(n,f)) \\
&= \mathbb{E}\{|u(n,f)|^2 \mid x(n,f), \mathcal{H}_1(n,f)\}P(\mathcal{H}_1(n,f)) \\
&\quad + \mathbb{E}\{|u(n,f)|^2 \mid x(n,f), \mathcal{H}_0(n,f)\}(1 - P(\mathcal{H}_1(n,f))) .
\end{aligned} \tag{6.43}
$$

When speech is absent, we may write $\mathbb{E}\{|u(n,f)|^2 \mid x(n,f), \mathcal{H}_0(n,f)\} = \mathbb{E}\{|x(n,f)|^2\}$ where again the expected value might be approximated by smoothing successive values in time. In a framewise processing scheme and during speech absence, even the instantaneous noise power spectrum estimate $|x(n,f)|^2 = |u(n,f)|^2$ will provide close-to-optimal performance. During speech activity, however, the computation of $\mathbb{E}\{|u(n,f)|^2 \mid x(n,f), \mathcal{H}_1(n,f)\}$ is less obvious. Since the noise power is not easily accessible it is common to substitute $\mathbb{E}\{|u(n,f)|^2 \mid x(n,f), \mathcal{H}_1(n,f)\} = \hat{\sigma}_u^2(n-1,f)$. Therefore, during speech activity, the noise power is not or only slowly updated. In total, we arrive at a first-order recursive system for smoothing the instantaneous power spectrum independently in each frequency bin f as

$$\hat{\sigma}_u^2(n,f) = \lambda(n,f)\hat{\sigma}_u^2(n-1,f) + (1 - \lambda(n,f))|x(n,f)|^2 \qquad (6.44)$$

where the SPP acts as an adaptive smoothing parameter $\lambda(n,f) = P(\mathcal{H}_1(n,f) \mid x(n,f))$. Whenever the SPP is close to one, this simple recursion will rely on the noise power spectrum estimate of the previous signal frame. When speech is not present the power spectrum is updated using the instantaneous power $|x(n,f)|^2$ of the mixed signal. Obviously, the success of this method hinges on the accuracy of the SPP estimate as the estimated noise power is based on a linear combination of instantaneous power spectra of the noisy signal.

To improve the above procedure, the MMSE estimator for the instantaneous power (Accardi and Cox, 1999) of the noise signal may be used (Hendriks *et al.*, 2010; Gerkmann and Hendriks, 2012), i.e.

$$\mathbb{E}\{|u(n,f)|^2 \mid x(n,f), \mathcal{H}_1(n,f)\}$$
$$= \left(\frac{1}{1 + \xi(n,f)}\right)|x(n,f)|^2 + \frac{\xi(n,f)}{1 + \xi(n,f)}\sigma_u^2(n,f) . \qquad (6.45)$$

However, this estimator requires estimations of the noise power spectrum and the a priori SNR. The substitution of the true values by corresponding estimated values gives rise to a bias which can be analytically computed and compensated (Hendriks *et al.*, 2010). As shown by Gerkmann and Hendriks (2012) the bias is avoided when soft decision weights as in (6.43) are employed. The overall system then provides a reasonable tradeoff between tracking speed and estimation errors. Figure 6.6 depicts

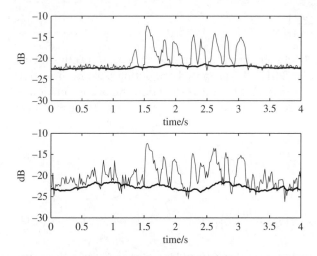

Figure 6.6 Power of noisy speech signal (thin solid line) and estimated noise power (thick solid line) using the MMSE noise power tracking approach (Hendriks *et al.*, 2010). Top: stationary white Gaussian noise; bottom: nonstationary multiple-speaker babble noise.

two examples where the power spectra of the noisy signal and of the MMSE noise power estimate (Hendriks *et al.*, 2010) have been averaged across all frequency bins. Compared to Figure 6.3 we note less overestimation during speech activity and less underestimation during speech absence.

6.3 Evaluation Measures

The evaluation of noise power estimators is not straightforward as in most cases the estimator is embedded into a speech enhancement system. In this context many of the commonly used instrumental measures of speech quality and speech intelligibility may be used. This includes the measures reviewed in Section 1.2.6 or the mutual information k-nearest neighbor measure (Taghia and Martin, 2014). Furthermore, the selection of appropriate test data with different degrees of temporal and spectral variability and a wide range of SNRs is necessary. To demonstrate the tracking properties, often real-world noise signals and artificially modulated noise signals are used. However, the above measures do not provide the full picture. Therefore, in addition, listening experiments are useful to evaluate quality and intelligibility aspects of the overall system. Noise power estimators which have a tendency to underestimate the true noise power are likely to generate a higher level of residual noise in the processed signal. Estimators which allow the tracking of rapidly varying noise power levels might also have a tendency to overestimate the true noise power level and therefore might lead to an attenuation of the target signal. Both of these effects are hard to evaluate using the above general instrumental measures only.

Therefore, besides the above application-driven evaluation procedures, there are also a number of measures for direct assessment of the noise power estimate. Typical evaluation measures for noise tracking algorithms are the *log-error* (Hendriks *et al.*, 2008) and the log-error variance (Taghia *et al.*, 2011). The log-error distortion measure is defined as

$$\mathrm{LE} = \frac{1}{NF} \sum_{nf} \left| 10 \log \frac{\hat{\sigma}_u^2(n,f)}{\sigma_u^2(n,f)} \right| \tag{6.46}$$

where N and F define the overall number of time and frequency bins, respectively. Furthermore, Gerkmann and Hendriks (2012) also proposed computing separate distortion measures $\mathrm{LE}_{\mathrm{under}}$ and $\mathrm{LE}_{\mathrm{over}}$ for errors related to underestimation and overestimation as these error types cause different perceptual effects, as discussed above. These measures are defined as

$$\mathrm{LE}_{\mathrm{under}} = \frac{1}{NF} \sum_{nf} \left| \min\left(0, 10 \log \frac{\hat{\sigma}_u^2(n,f)}{\sigma_u^2(n,f)} \right) \right| \tag{6.47}$$

and

$$\mathrm{LE}_{\mathrm{over}} = \frac{1}{NF} \sum_{nf} \left| \max\left(0, 10 \log \frac{\hat{\sigma}_u^2(n,f)}{\sigma_u^2(n,f)} \right) \right| . \tag{6.48}$$

Gerkmann and Hendriks (2012) also show that for babble noise and traffic noise, the rapidly tracking MMSE estimators significantly reduce the errors associated with noise power underestimation in comparison to the minimum statistics methods, but increase

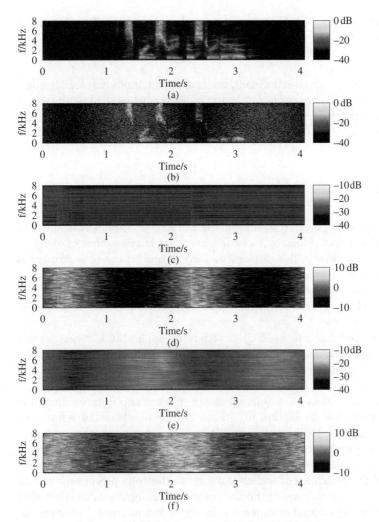

Figure 6.7 Spectrogram of (a) clean speech, (b) clean speech plus additive amplitude-modulated white Gaussian noise, (c) minimum statistics noise power estimate, (d) log-error for the minimum statistics estimator, (e) MMSE noise power estimate, and (f) log-error for the MMSE estimator. For the computation of these noise power estimates we concatenated three identical phrases of which only the last one is shown in the figure. Note that in (d) and (f) light color indicates noise power overestimation while dark color indicates noise power underestimation.

the errors associated with noise power overestimation. This can be also observed in Figure 6.7, where we use a short speech phrase contaminated with amplitude-modulated white Gaussian noise as an example. Here, the modulation frequency is 0.5 Hz and the amplitude of the modulating sinusoid is 3 dB. We find that the minimum statistics tracker cannot track these fast variations as the size of the window for minimum search has a duration similar to the inverse modulation frequency. Therefore, the minimum statistics approach underestimates the noise power. For this specific noisy speech file the total log-error distortion amounts to LE = 4.53 where underestimation errors contribute a large share of LE_{under} = 4.04 and overestimation errors amount to only

$LE_{over} = 0.49$. By contrast, the MMSE approach is well capable of tracking the rising noise power with a small delay. Now, the total log-error distortion is only $LE = 2.50$, however $LE_{under} = 1.34$ is attributed to underestimation errors and $LE_{over} = 1.16$ to overestimation errors.

Thus, in the context of speech enhancement, fast tracking methods may lead to additional speech distortions. A comparison of several state-of-the-art methods with respect to the above measures is provided, e.g. by Taghia *et al.* (2011) and Yong and Nordholm (2016), where in the latter work further refinements of the above method are described as well.

6.4   Summary

In the past 20 years, noise power tracking for the purpose of speech enhancement has been a very active area of research. As a consequence, a large number of methods and publications are available. In this chapter, we have described the basic principles in general terms and have summarized some of the better known methods in greater detail. Furthermore, we emphasize again the close relation between noise power spectrum tracking and SPP estimation, which especially under the MMSE paradigm, go hand-in-hand.

Methods for noise tracking are used in a variety of applications, most notably in single-channel speech enhancement where the computation of the gain function requires a noise power spectrum estimate or an SNR estimate as described in Chapter 5. Whenever a large number of frequency bands are used in these applications the computational complexity and memory requirements are of great importance. Therefore, the basic methods described in Section 6.2.1 have also seen substantial refinement recently (Baasch *et al.*, 2014; Heese and Vary, 2015), where improvements stem from tracking in the log-magnitude domain and from adaptive slope parameters.

A general challenge resides in the tradeoff between a fast noise power update in nonstationary noise and the avoidance of speech leaking into the noise power estimate. In fact, it has been observed that some of the fast reacting tracking methods also inflict more speech distortions when used in a speech enhancement framework. An improved overall solution therefore must take the accuracy of the noise power estimate and a distortion measure on the target signal into account. This perspective calls for novel solutions which will require the rapid online estimation of target signal distortions. These solutions rely on trained noise templates (Rosenkranz and Puder, 2012; Heese *et al.*, 2014) and are also suitable for transient noise types.

Bibliography

Accardi, A. and Cox, R. (1999) A modular approach to speech enhancement with an application to speech coding, in *Proceedings of IEEE International Conference on Audio, Speech and Signal Processing*, vol. 1, pp. 201–204.

Baasch, C., Rajan, V.K., Krini, M., and Schmidt, G. (2014) Low-complexity noise power spectral density estimation for harsh automobile environments, in *Proceedings of International Workshop on Acoustic Signal Enhancement*, pp. 218–222.

Breithaupt, C. and Martin, R. (2011) Analysis of the decision-directed SNR estimator for speech enhancement with respect to low-SNR and transient conditions. *IEEE Transactions on Audio, Speech, and Language Processing*, **19** (2), 277–289.

Brillinger, D.R. (1981) *Time Series: Data Analysis and Theory*, Holden-Day.

Cappé, O. (1994) Elimination of the musical noise phenomenon with the Ephraim and Malah noise suppressor. *IEEE Transactions on Speech and Audio Processing*, **2** (2), 345–349.

Cohen, I. (2003) Noise estimation in adverse environments: Improved minima controlled recursive averaging. *IEEE Transactions on Speech and Audio Processing*, **11** (5), 466–475.

Cohen, I. and Berdugo, B. (2001) Speech enhancement for non-stationary noise environments. *Signal Processing*, **81** (11), 2403–2418.

Cohen, I. and Gannot, S. (2008) Spectral enhancement methods, in *Springer Handbook of Speech Processing*, Springer, chap. 44, pp. 873–901.

Doblinger, G. (1995) Computationally efficient speech enhancement by spectral minima tracking in subbands, in *Proceedings of European Conference on Speech Communication and Technology*, vol. 2, pp. 1513–1516.

Ealey, D., Kelleher, H., and Pearce, D. (2001) Harmonic tunnelling: tracking non-stationary noises during speech, in *Proceedings of European Conference on Speech Communication and Technology*, p. 437–440.

Ephraim, Y. and Malah, D. (1984) Speech enhancement using a minimum mean-square error short-time spectral amplitude estimator. *IEEE Transactions on Acoustics, Speech, and Signal Processing*, **32** (6), 1109–1121.

Gerkmann, T., Breithaupt, C., and Martin, R. (2008) Improved a posteriori speech presence probability estimation based on a likelihood ratio with fixed priors. *IEEE Transactions on Audio, Speech, and Language Processing*, **16** (5), 910–919.

Gerkmann, T. and Hendriks, R. (2012) Unbiased MMSE-based noise power estimation with low complexity and low tracking delay. *IEEE Transactions on Audio, Speech, and Language Processing*, **20** (4), 1383–1393.

Hänsler, E. and Schmidt, G. (2005) *Acoustic Echo and Noise Control: A Practical Approach*, Wiley.

Heese, F., Nelke, C.M., Niermann, M., and Vary, P. (2014) Selflearning codebook speech enhancement, in *Proceedings of ITG Symposium on Speech Communication*, pp. 1–4.

Heese, F. and Vary, P. (2015) Noise PSD estimation by logarithmic baseline tracing, in *Proceedings of IEEE International Conference on Audio, Speech and Signal Processing*, pp. 4405–4409.

Hendriks, R., Jensen, J., and Heusdens, R. (2008) Noise tracking using DFT domain subspace decompositions. *IEEE Transactions on Audio, Speech, and Language Processing*, pp. 541–553.

Hendriks, R.C., Heusdens, R., and Jensen, J. (2010) MMSE based noise PSD tracking with low complexity, in *Proceedings of IEEE International Conference on Audio, Speech and Signal Processing*, pp. 4266–4269.

Hirsch, H.G. and Ehrlicher, C. (1995) Noise estimation techniques for robust speech recognition, in *Proceedings of IEEE International Conference on Audio, Speech and Signal Processing*, vol. 1, pp. 153–156.

Malah, D., Cox, R., and Accardi, A. (1999) Tracking speech-presence uncertainty to improve speech enhancement in non-stationary noise environments, in *Proceedings of IEEE International Conference on Audio, Speech and Signal Processing*, pp. 789–792.

Martin, R. (2001) Noise power spectral density estimation based on optimal smoothing and minimum statistics. *IEEE Transactions on Speech and Audio Processing*, **9** (5), 504–512.

Martin, R. (2006) Bias compensation methods for minimum statistics noise power spectral density estimation. *Signal Processing*, **86** (6), 1215–1229.

McAulay, R. and Malpass, M. (1980) Speech enhancement using a soft-decision noise suppression filter. *IEEE Transactions on Acoustics, Speech, and Signal Processing*, **28** (2), 137–145.

Paul, D. (1981) The spectral envelope estimation vocoder. *IEEE Transactions on Acoustics, Speech, and Signal Processing*, **29** (4), 786–794.

Pearce, D. and Hirsch, H. (2000) The Aurora experimental framework for the performance evaluation of speech recognition systems under noisy conditions, in *Proceedings of the International Conference on Spoken Language Processing*, pp. 29–32.

Plapous, C., Marro, C., Mauuary, L., and Scalart, P. (2004) A two-step noise reduction technique, in *Proceedings of IEEE International Conference on Audio, Speech and Signal Processing*, pp. 289–292.

Proakis, J. and Salehi, M. (2007) *Digital Communications*, McGraw-Hill, 5th edn.

Rosenkranz, T. and Puder, H. (2012) Improving robustness of codebook-based noise estimation approaches with delta codebooks. *IEEE Transactions on Audio, Speech, and Language Processing*, **20** (4), 1177–1188.

Stahl, V., Fischer, A., and Bippus, R. (2000) Quantile based noise estimation for spectral subtraction and Wiener filtering, in *Proceedings of IEEE International Conference on Audio, Speech and Signal Processing*, vol. 3, pp. 1875–1878.

Taghia, J. and Martin, R. (2014) Objective intelligibility measures based on mutual information for speech subjected to speech enhancement processing. *IEEE/ACM Transactions on Audio, Speech, and Language Processing*, **22** (1), 6–16.

Taghia, J., Taghia, J., Mohammadiha, N., Sang, J., Bouse, V., and Martin, R. (2011) An evaluation of noise power spectral density estimation algorithms in adverse acoustic environments, in *Proceedings of IEEE International Conference on Audio, Speech and Signal Processing*, pp. 4640–4643.

Yong, P.C. and Nordholm, S. (2016) An improved soft decision based noise power estimation employing adaptive prior and conditional smoothing, in *Proceedings of International Workshop on Acoustic Signal Enhancement*.

7

Single-Channel Classification and Clustering Approaches

Felix Weninger, Jun Du, Erik Marchi, and Tian Gao

The separation of sources from single-channel mixtures is particularly challenging. If two or more microphones are available, information on relative amplitudes or relative time delays can be used to identify the sources and help to perform the separation (see Chapter 12). Yet, with only one microphone, this information is not available. Instead, information about the structure of the source signals must be exploited to identify and separate the different components.

Methods for single-channel source separation can be roughly grouped into two categories: clustering and classification/regression. *Clustering* algorithms are based on grouping similar time-frequency bins. This particularly includes *computational auditory scene analysis* (CASA) approaches, which rely on psychoacoustic cues in a learning-free mode, i.e. no models of individual sources are assumed, but rather generic properties of acoustic signals are exploited. In contrast, *classification* and *regression* algorithms are used in separation-based training to predict the source belonging to the target class or classify the type of source that dominates each time-frequency bin. *Factorial hidden Markov models* (HMMs) are a generative model explaining the statistics of a mixture based on statistical models of individual source signals, and hence rely on source-based unsupervised training, i.e. training a model for each source from isolated signals of that source. In particular, deep learning-based approaches are discriminatively trained to learn a complex nonlinear mapping from mixtures to sources, which can be regarded as mixture-based supervised training: train a separation mechanism or jointly train models for all sources from mixture signals, with knowledge of the ground truth underlying source signals.

In this chapter, we first introduce conventional source separation methods, including CASA and spectral clustering in Section 7.1 and factorial HMMs in Section 7.2. Then, we move on to recently proposed classification and regression methods based on deep learning in Section 7.3. We provide a summary in Section 7.4.

Audio Source Separation and Speech Enhancement, First Edition.
Edited by Emmanuel Vincent, Tuomas Virtanen and Sharon Gannot.
© 2018 John Wiley & Sons Ltd. Published 2018 by John Wiley & Sons Ltd.
Companion Website: https://project.inria.fr/ssse/

7.1   Source Separation by Computational Auditory Scene Analysis

In this section, we introduce one broad class of single-channel source separation, namely CASA (Brown and Cooke, 1994; Wang and Brown, 2006). Auditory scene analysis (Bregman, 1994) refers to the ability of human listeners to form perceptual representations of the components sources in an acoustic source mixture. CASA is the study of auditory scene analysis in a "computational manner". CASA systems are essentially machine listening systems that aim to separate mixtures of sound sources in the same way as human listeners. Some CASA systems focus on the known stages of auditory processing, whereas others adopt a model-based approach. This section covers CASA studies aimed at feature-based grouping, focusing on segregation of a target source from acoustic interference using auditory features such as fundamental frequency. The knowledge-based grouping relying on learned patterns or trained models is introduced in Section 7.2, exemplified by factorial-max HMMs. In the current section, we first introduce the concept of auditory scene analysis. Then, we discuss the feature-based CASA system for source separation, and its two important components, namely segmentation and grouping. Finally, spectral clustering is presented as an application of CASA to source separation.

7.1.1   Auditory Scene Analysis

In many naturalistic scenes, different sound sources are usually active at the same time, producing an auditory stream, as in the well-known "cocktail party" effect. With a number of voices speaking at the same time or with background noises, humans are able to follow a particular voice even though other voices and background noises are present. How can this complex source mixture be parsed to one isolated source? Bregman (1994) answered this question. He contends that humans perform an auditory scene analysis, which can be conceptualized as a two-stage process. In the first stage, called *segmentation*, the source mixture is decomposed into groups of contiguous segments. A segment can be regarded as an atomic part of a single source. Then, a *grouping* process combines segments that are likely originated from the same source, forming a perceptual representation called a stream. The grouping process relies on intrinsic source cues such as fundamental frequency and temporal continuity.

7.1.2   CASA System for Source Separation

Research in auditory scene analysis has inspired a series of studies of CASA (Brown and Cooke, 1994; Wang and Brown, 2006). The structure of a typical feature-based CASA system is closely related to Bregman's conceptual model, as shown in Figure 7.1. CASA retrieves one source from a mixture using grouping rules based on psychological cues. It is based on two main stages as auditory scene analysis: segmentation and grouping. The segmentation stage contains peripheral processing and feature extraction. The subsequent grouping process integrates the components of the same source in time. The result of grouping is a segregated target source stream and segregated interference.

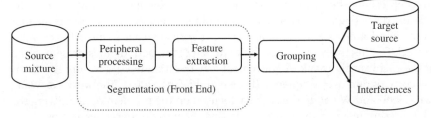

Figure 7.1 An implementation of a feature-based CASA system for source separation.

7.1.2.1 Segmentation

The first step of a CASA system is usually a time-frequency analysis of the mixture signal to generate *mid-level representations*, as described in Chapter 2. In the context of CASA, a *gammatone* filterbank is often used, which is a bank of bandpass filters designed to approximate the impulse responses of the auditory nerve fibers (Cooke, 2005). Other time-frequency representations that are sometimes employed for CASA, but are not oriented on the human auditory system, include short-time Fourier transform (STFT) and discrete wavelet transform (Nakatani and Okuno, 1999; Nix *et al.*, 2003).

We will now describe the gammatone filterbank in more detail. The filterbank usually uses 128 gammatone filters (or 64 filters) with center frequencies ranging from 80 to 5000 Hz. The impulse response of this filter has the following form:

$$h_f(\tau) = \begin{cases} a\tau^{n-1}e^{-2\pi b_f \tau}\cos(2\pi\tau v_f/f_s + \phi) & \text{for } t > 0 \\ 0 & \text{otherwise} \end{cases} \tag{7.1}$$

where v_f (in Hz) is the center frequency, ϕ (in radians) is the phase of the carrier, a is the amplitude, n is the filter's order, b_f (in Hz) is the filter's bandwidth, and t is time. A non-linear frequency resolution is used in the gammatone filter motivated by the distribution of sounds at high frequencies: natural signals are rather stochastic at high frequencies, and measuring the amplitude or phase of sound within a narrow band that is similar across all the frequencies would give a rather unreliable representation at high frequencies. So, in the application of the gammatone filterbank, the bandwidth of each filter varies according to its center frequency. In particular, the filter bandwidths increase from low to high frequency regions. Thus, in low-frequency regions, the individual harmonics of complex sounds such as speech can be resolved. In high-frequency regions, stochasticity is introduced by using broader bandwidths.

Most CASA systems further extract features from the peripheral time-frequency representation to explicitly encode properties which are implicit in the source signal. These features are useful for the grouping stage. One of the most effective features for signal-channel source separation is fundamental frequency (F0). An important class of F0 estimation algorithms is based on a temporal model of pitch perception proposed by Licklider (1951). Weintraub (1985) first implemented Licklider's theory in a computational manner, and referred to it as an auto-coincidence representation. Slaney and Lyon (1990) further introduced the *correlogram*, which is computed through an autocorrelation using the output of cochlear filter analysis,

$$\sigma_x(n,f,\tau) = \sum_{t=0}^{T-1} x_f(t + t_0 + nM)x_f(t + t_0 + nM - \tau)h_a(t) \tag{7.2}$$

where $x_f(t)$ represents the cochlear filter response for channel f at time t, τ is the autocorrelation delay (lag), $h_a(t)$ is an analysis window of length T, t_0 positions the first sample of the first frame, and M is the hop size between adjacent frames. The correlogram can track F0 effectively due to its ability to detect the periodicities present in the output of the cochlear filterbank. A convenient way to estimate F0 is to sum the channels of the correlogram over frequency (Brown and Wang, 2005). The works of Wang and Brown (1999) and Wu *et al.* (2003) have focused on the problem of identifying the multiple F0s present in a mixture source and using them to do source separation.

7.1.2.2 Grouping

After the segmentation stage, we obtain a time-frequency representation in order to extract features that are useful for grouping. The problem addressed in the grouping stage is to determine which segments should be grouped together and identified as the same source. Principal features that are used to this end include F0 (Hu and Wang, 2006). The segments are split into groups based on the similarity of their source and location attributes. These groups represent the separated sources.

Grouping can be classified into *simultaneous grouping* (across frequency) and *sequential grouping* (across time), also known as *streaming*. Simultaneous grouping aims to group segments overlapped in time (Bregman, 1994; Wang and Brown, 1999; Hu and Wang, 2006; Shao and Wang, 2006) and results in a collection of streams, each of which is temporally continuous. The streams are still separated in time and interleaved with segments from other sources. Given the results from simultaneous grouping, sequential grouping links together the continuous streams originating from the same source that are separated in time. For speech separation, the task is to group successive speech segments of the same speaker.

7.1.3 Application: Spectral Clustering for Source Separation

Spectral clustering was applied to single-channel source separation problems as an application of CASA, by casting the problem as one of segmentation of the spectrogram (Bach and Jordan, 2006, 2009). It works within a spectrogram, and exploits the sparsity of source signals in this representation. In the spectral learning algorithm, the training examples are obtained by mixing separately normalized source signals. Then a set of psychoacoustic cues such as continuity, common fate cues, pitch and timbre (Bach and Jordan, 2009) are extracted for the spectral segmentation. With these cues, parameterized similarity matrices are built that can be used to define a spectral segmenter. A set of basis similarity matrices are first defined for each cue, and then combined with learned weights because a single cue is not sufficient and the combination of multiple cues can further improve the performance. The algorithms for learning the similarity matrix and weights are discussed by Ng *et al.* (2002) and Bach and Jordan (2006).

The problems of the above spectral clustering algorithms are high computational cost and shallow learning. *Deep neural network* (DNN) based algorithms to spectral clustering are a promising direction to address these issues. See Section 19.1.3 for details.

7.2   Source Separation by Factorial HMMs

We have introduced a feature-based CASA system for source separation in the above sections. The approaches have focused on substantial knowledge of the human auditory system and its psychoacoustical cues, relying mainly on the independence between different signals. Factorial HMMs, on the other hand, have been investigated for the separation problem by utilizing the statistical knowledge about the signals to be separated (Roweis, 2001, 2003; Ozerov *et al.*, 2005). The approaches are classified as source-based unsupervised training, which train specific models for specific sources to discover source regularities from a large amount of source data and then use the learned models to compute the masks in order to perform time-frequency masking as defined in Chapter 5.

The target source can be estimated by performing time-frequency masking as follows,

$$\hat{s}_j(n,f) = w_j(n,f)x(n,f) \tag{7.3}$$

where, $w_j(n,f)$ is the mask for target source $c_j(n,f)$ at time-frequency domain, $x(n,f)$ is the source mixture, and j is the source index in the mixture. When the masks are well chosen, separation is indeed possible. So, the essential problem is how to automatically compute the $w_j(n,f)$ from a single input mixture. The goal is to group together regions of the spectrogram that belong to the same source object. For generative method, two problems can be observed. The first problem is to find a signal model that is most likely to have generated a set of observations for each source. The second problem is to estimate the signal model parameters from mixed observations where the interaction between the sources is considered an important issue. Specific to the factorial HMMs, there are three steps:

1) Train one HMM per source with Gaussian mixture model (GMM) observations and build a factorial HMM architecture.
2) Decode the MAP mixture states.
3) Perform time-frequency masking.

7.2.1   GMM-HMM and Factorial-Max Architecture

In addressing the first step, all sources are modeled with separate GMM-HMMs, which are assumed to be trained separately with corresponding data, as shown in Figure 7.2. For each source a GMM-HMM is fitted using log-power spectrograms as observations. A GMM-HMM describes two dependent random processes, an observable process and a hidden Markov process. The observation sequence is assumed to be generated by each hidden state according to a GMM. A GMM-HMM is parameterized by vector state prior probabilities, the state transition probability matrix, and by a set of state-dependent parameters in GMMs. The emission probability distributions of the GMM model the typical spectral patterns, while the transition probabilities of the HMM encourage spectral continuity. The highly efficient expectation-maximization

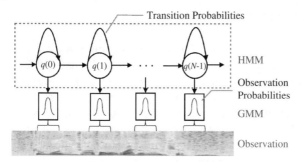

Figure 7.2 Architecture of a GMM-HMM. The GMM models the typical spectral patterns produced by each source, while the HMM models the spectral continuity.

(EM) algorithm (Baum and Petrie, 1966) is the key point to estimate model parameters. This maximum likelihood (ML) method, often called the *Baum-Welch algorithm*, was a principal way of training the GMM-HMM (Young *et al.*, 2006).

Factorial-max HMM is an extension of factorial-max vector quantization (Roweis, 2003) through time regarding the max approximation to log-spectrograms of mixtures. Considering a two-source mixture $x(n,f) = c_1(n,f) + c_2(n,f)$, the *log-max approximation* assumption states that the log-power spectrum $\log(|x(n,f)|^2)$ can be approximated by the maximum of the individual log-power spectra

$$\log(|x(n,f)|^2) \approx \max(\log(|c_1(n,f)|^2), \log(|c_2(n,f)|^2)). \tag{7.4}$$

Formally, factorial-max vector quantization is a latent variable probabilistic model. The model consists of several vector quantizers, each with a number of codebook vectors. Latent variables control which codebook vector each vector quantizer selects. Given these selections, the final output vector is generated as a noisy version of the elementwise maximum of the selected codewords. Specific to factorial-max HMM, Markov chains and mean vectors of Gaussian distribution are defined as the vector quantizers and codebook vectors, respectively.

We consider only two Markov chains in the factorial HMM, whose states are denoted as $q_1(n)$ and $q_2(n)$. For each state, only a single Gaussian was used. So at each frame, one chain proposes an $F \times 1$ mean vector $\boldsymbol{\mu}_{q_1(n)}$ and the other proposes $\boldsymbol{\mu}_{q_2(n)}$. The likelihood of an input log-power spectrogram feature $\mathbf{o}(n) = [\log(|x(n,0)|^2), \ldots, \log(|x(n,F-1)|^2)]^T$ depends on the states of all the chains. The elementwise maximum rule is taken to combine them to generate the observation $\mathbf{o}(n)$. This maximum operation reflects the observation that the log-magnitude spectrogram of a mixture of sources is close to the elementwise maximum of the individual spectrograms. The full generative model for factorial-max HMM can be written simply as:

$$\mathbf{o}(n) \mid q_1(n), q_2(n) \sim \mathcal{N}(\mathbf{o}(n) \mid \max(\boldsymbol{\mu}_{q_1(n)}, \boldsymbol{\mu}_{q_2(n)}), \boldsymbol{\Sigma}) \tag{7.5}$$

which is a Gaussian distribution with mean $\max(\boldsymbol{\mu}_{q_1(n)}, \boldsymbol{\mu}_{q_2(n)})$ and covariance $\boldsymbol{\Sigma}$ (shared across states and Markov chains). $\max(\cdot)$ is the elementwise maximum operation on two vectors.

7.2.2 MAP Decoding for HMM State Sequence

Given an input waveform, the observation sequence $\mathbf{O} = [\mathbf{o}(0), \ldots, \mathbf{o}(N-1)]$ is created from the log-power spectrogram. Separation can be done by first inferring a joint underlying state sequence $\{\hat{\mathbf{q}}_1, \hat{\mathbf{q}}_2\} = \{\hat{q}_1(n), \hat{q}_2(n)\}_n$ of the two Markov chains in the model using the maximum a posteriori (MAP) estimation as follows,

$$\hat{\mathbf{q}}_1, \hat{\mathbf{q}}_2 = \underset{\mathbf{q}_1, \mathbf{q}_2}{\operatorname{argmax}} P(\mathbf{q}_1, \mathbf{q}_2 \mid \mathbf{O}). \tag{7.6}$$

7.2.3 Mask Estimation given State Sequences

After the inferring of state sequences $\hat{q}_1(n)$ and $\hat{q}_2(n)$, the corresponding mean vectors $\mu_{q_1(n)}$ and $\mu_{q_2(n)}$ are used to compute a binary mask (assuming $q_1(n)$ is the Markov chain of the target source):

$$\hat{w}_1(n,f) = \begin{cases} 1 & \text{if } \mu_{q_1(n)}(f) > \mu_{q_2(n)}(f) \\ \epsilon & \text{otherwise} \end{cases} \tag{7.7}$$

where $\hat{w}_1(n,f)$ is the estimated mask of target source $c_1(n,f)$ at frame n and frequency f, and ϵ is the spectral floor, i.e. a small value close to zero to reduce musical noise artifacts. Finally, the estimated signal of the target source can be obtained by performing time-frequency masking.

7.3 Separation Based Training

In contrast to the previous sections, the source separation problem is now formulated as minimizing the error of a mapping from mixtures to a representation of the corresponding source signal(s). The mapping is defined as a model that can be trained using machine learning methods, e.g. a DNN. The inference needed in the separation stage is limited to evaluating the mapping at the point of the mixture signal. This is very efficient compared to iterative methods such as nonnegative matrix factorization (NMF) (see Chapter 8). The category of separation-based training approaches is mostly applied to separating a desired source from interference(s), e.g. extracting the leading voice from polyphonic music or separating speech from background noise (speech enhancement).

7.3.1 Prerequisites for Separation-Based Training

It is straightforward to derive a supervised training scheme for source separation, by training a system to predict a representation of a wanted signal from features of a mixed signal. To train single-channel source separation, a *training set* of mixture signals $x(t)$ along with the source signals $c_j(t)$ is assumed to be available. The former are used to generate features for training, usually a spectral representation $x(n,f)$, and the latter are used to generate training targets, such as the ideal binary mask (cf. below for details). This is also called *stereo data training*.[1] The easiest way to obtain stereo training data is to add isolated source signals in the time domain to generate mixture signals. In this way, the assumption of stereo data is reduced to the assumption of availability of training signals for all sources, which is also made by other approaches such as unsupervised source modeling methods.

---

1 Note that the meaning of "stereo" here is totally different from its classical meaning in audio, i.e. two-channel.

The generalization of trained models to unseen test data is of crucial importance in any machine learning task. In contrast to model-inspired methods such as those introduced in earlier chapters, which have only a few parameters tuned by experts, machine learning methods often determine millions of model parameters automatically, and hence they might memorize the correct labeling of the training patterns instead of deriving a general solution to the problem at hand (*overfitting*). In the following, we briefly recapitulate the basics of *cross-validation*, which is the standard method to measure generalization. The interested reader is referred to the machine learning literature for details (Schuller, 2013; Goodfellow *et al.*, 2016).

Besides a training corpus, one usually employs a *validation set* that is disjointed from the training corpus. One purpose of the validation corpus is to assess the generalization capability of the model during model training by periodically evaluating the performance on the validation corpus. A popular strategy is to adjust the learning parameters (e.g., the learning rate in the case of the gradient descent algorithm) when the error on the validation corpus increases, or to stop training altogether once the performance on the validation corpus stops improving (*early stopping*). Another purpose of the validation corpus is to tune the hyperparameters of the training. This particularly includes the model size (number of parameters) – if too low, the model might not be capable enough of modeling a complex problem (*underfitting*); if too high, the model is prone to overfitting (cf. above) – and the learning parameters.

After training and parameter tuning, the final performance of the model should be evaluated on a *test set* that is disjointed from both the training and validation corpus. This is to evaluate the performance on data that is unknown at the time of model training.

The sizes of the validation and test corpora are chosen empirically. If too small, performance measurements might be noisy and differences not statistically significant; if too large, the amount of training data is decreased considerably, usually leading to worse generalization, and the computational cost of evaluating the performance becomes a burden.

7.3.2 Deep Neural Networks

In principle, there is no restriction on the type of machine learning model to use with separation-based training. In particular, it has been shown that binary classification by decision trees (Gonzalez and Brookes, 2014) and support vector machines (Le Roux *et al.*, 2013) can be effectively used for this task. However, as in many other areas of audio processing, there is currently an increasing trend towards DNN-based source separation (Narayanan and Wang, 2013; Huang *et al.*, 2014; Weninger *et al.*, 2014a).

DNNs have a few convenient properties that can be exploited for training the task of source separation. First, the source representation (e.g., the filter $w(n,f)$ for a desired source) for all frequency bins f can be represented in a single model, which allows for efficient computation at test time. In addition, nonlinearities in the feature representation can be handled effectively, thus allowing for (e.g., logarithmic) compression of the spectral magnitudes, which is considered useful in speech processing. Finally, it is easy to incorporate various discriminative costs into DNN training by the backpropagation algorithm, since only the gradient computation of the cost function with respect to the source representation output by the DNN needs to be changed. In this chapter, a functional description of several DNN architectures along with the basics of training is given. For an in-depth discussion the reader is referred to the standard machine learning literature (Goodfellow *et al.*, 2016; Montavon *et al.*, 2012).

DNNs are very generic models which are – in their basic form – not specifically designed for source separation, but rather for handling a wide variety of feature types and recognition tasks. Still, they have been proven very successful in source separation, particularly of speech, music and environmental sounds (Lu *et al.*, 2013; Xia and Bao, 2013; Huang *et al.*, 2014; Simpson *et al.*, 2015; Narayanan and Wang, 2015; Zöhrer *et al.*, 2015).

A H-layer DNN computes a nonlinear function

$$g_{\hat{z}}(\mathbf{y}_0) = g_H(\mathbf{Z}_H g_{H-1}(\mathbf{Z}_{H-1} \cdots g_1(\mathbf{Z}_1 \mathbf{y}_0))). \tag{7.8}$$

There, we use the notations $\mathbf{Z}_h$ for the weight matrices, $\mathbf{y}_0$ for *input features*, and g_h, $h \in \{1, \ldots, H\}$ for *activation functions*. In our notation, the *hidden layers* of the DNN correspond to the operations with indices $h \in \{1, \ldots, H-1\}$, so that the DNN (7.8) is said to have $H-1$ hidden layers. The hidden layer activations in layer h are defined as $\mathbf{y}_h = g_h(\bar{\mathbf{y}}_h)$ with $\bar{\mathbf{y}}_h = \mathbf{Z}_h \mathbf{y}_{h-1}$ a $K_h \times 1$ vector with K_h being the number of *neurons* in layer h. The layer with index H is called the *output layer* with activations $g_{\hat{z}}(\mathbf{y}_0) = \mathbf{y}_H$. In the following, the shorthand notation $\mathbf{y}$ will be used for the network outputs. Typically, the activations $\mathbf{y}_h$, $h \in \{0, \ldots, H-1\}$ are augmented by a constant input (usually 1), so that the rightmost column of $\mathbf{Z}_h$ forms a bias vector. In the following, the constant input is left out for the sake of readability. Figure 7.3 shows an exemplary DNN with two hidden layers and three neurons per layer.

The input features usually correspond to a time-frequency representation of the input signal, such as (logarithmic) power, magnitude, or Mel spectra. To provide acoustic *context* to the DNN, consecutive feature frames are concatenated (stacked) in a sliding window approach (Huang *et al.*, 2014). For the sake of readability, this will be neglected in the notations used by this chapter, and it is easy to see that all the methods presented in this section can be easily generalized to sliding windows of input features. An alternative to input feature stacking is to use networks with feedback loops (see Section 7.3.2.1 below).

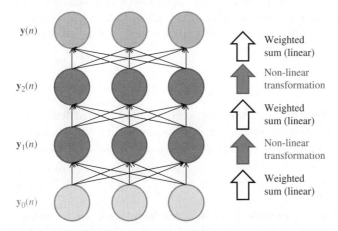

Figure 7.3 Schematic depiction of an exemplary DNN with two hidden layers and three neurons per layer.

The activation functions g_h are usually chosen from the following (where the layer index h is omitted for readability):

- identity: $g(\bar{\mathbf{y}}) = \bar{\mathbf{y}}$;
- rectifier: $g([\bar{y}_1, \ldots, \bar{y}_K]^T) = [\max(\bar{y}_1, 0), \ldots, \max(\bar{y}_K, 0)]^T$;
- *logistic*: $g([\bar{y}_1, \ldots, \bar{y}_K]^T) = [\sigma(\bar{y}_1), \ldots, \sigma(\bar{y}_K)]^T$ with $\sigma(x) = 1/(1 + e^{-x})$;
- hyperbolic tangent: $g([\bar{y}_1, \ldots, \bar{y}_K]^T) = [\tanh(\bar{y}_1), \ldots, \tanh(\bar{y}_K)]^T$
  with $\tanh(x) = 2\sigma(2x) - 1$.

Consequently, in case of identity and rectifier activation functions, the corresponding hidden layer units are often referred to as linear and *rectified linear*. When using DNNs for source separation, the choice of the output layer activation function is particularly interesting. When using the logistic function, the outputs of the DNN are in the interval $[0, 1]$ and can hence be interpreted as a time-frequency mask $w(n,f)$. For the reasons pointed out by LeCun *et al.* (1998), the hyperbolic tangent function, which is a rescaled version of the logistic function, typically yields better convergence. In this case, the outputs need to be linearly rescaled to $[0, 1]$ to yield a time-frequency mask. When using the halfwave function, the outputs of the DNN are nonnegative and can hence be understood as magnitude or power spectra of a source $c_j(n,f)$. When using the identity, the outputs are unrestricted and can hence correspond to other features of a source. Note that the so-called softmax activation function, which is typically used for multiclass classification with DNNs (see Section 17.2.1.4), is not considered in the context of source separation, since it forces the output layer activations to sum to one.

To determine the weights $\mathcal{Z} = \{\mathbf{Z}_h\}_h$ of a DNN, the following cost function is minimized:

$$C(\mathcal{Z}) = \sum_n C_n(\mathcal{Z}), \tag{7.9}$$

where the sum is taken over all times frames n in the training set and the cost $C_n(\mathcal{Z})$ on time frame n quantifies the discrepancy between the network outputs $\mathbf{y}(n) = [y(n,f)]_f$ and the training targets $\mathbf{y}^{\text{tgt}}(n) = [y^{\text{tgt}}(n,f)]_f$ in that frame. In the case of source separation, the training targets usually correspond to a representation of a desired source to be extracted from the input signal (cf. below). The sum of squared errors criterion can be considered as the most generic cost function, and corresponds to the sum of squared deviations of the outputs from the training targets,

$$C^{\text{SSE}}(\mathcal{Z}) = \sum_{nf} (y^{\text{tgt}}(n,f) - y(n,f))^2. \tag{7.10}$$

As will be discussed in more detail below, other choices of C can be motivated depending on the application but also on the network structure, in particular the activation function g_H of the output layer. The minimization of the cost function C (7.9) is usually achieved by gradient descent,

$$\mathcal{Z}^{(m+1)} = \mathcal{Z}^{(m)} - \eta \nabla C(\mathcal{Z}^{(m)}) \tag{7.11}$$

with training epoch m and learning rate $\eta > 0$. When minimizing this and other similar costs with gradient descent, it must be taken into account that their scale depends on the training set size N, and hence the choice of η depends on N. Hence, in practice, the cost function is often normalized by the factor $1/N$. In this particular case, this leads to the *mean square error* (MSE) cost:

$$C^{\text{MSE}}(\mathcal{Z}) = \frac{1}{N} \sum_{nf} (y^{\text{tgt}}(n,f) - y(n,f))^2. \tag{7.12}$$

Gradient descent is furthermore often augmented by a *momentum* term (Polyak, 1964), which serves to avoid oscillations of the cost function. The momentum method is based on the intuition that the weight update $\Delta\mathcal{Z}$ corresponds to the current "velocity" of the gradient descent process (Sutskever *et al.*, 2013):

$$\Delta\mathcal{Z}^{(m+1)} = \mu\Delta\mathcal{Z}^{(m)} - \eta\nabla C(\mathcal{Z}^{(m)}), \tag{7.13}$$

$$\mathcal{Z}^{(m+1)} = \mathcal{Z}^{(m)} + \Delta\mathcal{Z}^{(m+1)}, \tag{7.14}$$

with $\Delta\mathcal{Z}^{(0)} = \mathbf{0}$ and $0 < \mu \leq 1$ being the momentum coefficient.

Especially with large training sets, standard gradient descent is replaced by *stochastic gradient descent*, where weight updates are performed on *minibatches* $B \subset \{0, \dots, N-1\}$ of size $|B|$, based on the assumption that each minibatch is a representative sample of the whole training set. This leads to a set of cost functions for minibatches B,

$$C_B(\mathcal{Z}) = \sum_{n \in B} C_n(\mathcal{Z}), \tag{7.15}$$

such that $\sum_B C_B = C$. Hence, the minibatch update becomes

$$\mathcal{Z}^{(m+1)} = \mathcal{Z}^{(m)} - \frac{\eta}{|B|}\nabla C_B(\mathcal{Z}^{(m)}). \tag{7.16}$$

In computing the gradient of the cost function with respect to the weights, one observes that due to the chain rule, the function composition in the computation of the DNN output (7.8) translates to the multiplication of the weight gradients per layer. Thus, the iteration in computing the DNN output directly corresponds to an iterative algorithm (*backpropagation*) that computes the gradient with respect to the layer activations and weights starting from the gradient of the error function with respect to the network outputs.

Note that the term DNN is commonly associated with a couple of "tricks of the trade" that make gradient descent-based training of large models on large amounts of data practicable – an overview is given by Deng *et al.* (2013).

For example, *layerwise training* eases the high-dimensional optimization problem, especially when the amount of training data is low, by avoiding the training of many weight parameters at once from a random initialization. A straightforward, yet effective iterative implementation of layerwise training of a H-layer DNN uses the weights of an $h - 1$ layer DNN as initialization for the weights of an h-layer DNN, $h \in \{2, \dots, H\}$, with only the weights in the hth layer being randomly initialized (Yu *et al.*, 2011). This is also known as discriminative pretraining, as opposed to generative pretraining (Hinton, 2002), which is an example for a model-inspired method.

Yet another crucial point for the success of DNNs is that training on large amounts of data is becoming more and more practicable due to the increased prevalence of multicore central processing unit and graphical processing unit architectures. This is as large parts of DNN training can be easily parallelized on such architectures. In particular, within a single forward and backward pass, all training vectors can be processed in parallel using matrix-matrix multiplications and elementwise operations; sequential computation is only required between layers.

The stochastic gradient descent algorithm, however, still incurs significant sequential computation, as no two sets of minibatches can be computed in parallel. To remedy this problem, *asynchronous gradient descent* (Dean *et al.*, 2012) slightly reformulates the update to allow a time delay Δm between the current estimate of the weights and the estimate of the weights that the weight update calculation is based on.

As DNNs are generic models that are capable of modeling highly nonlinear relationships, they are more likely to overfit than other model types. Besides the general strategies explained in Section 7.3.1, other heuristics frequently employed to combat overfitting during training are (a) *input noise* or *weight noise*, where white noise is added to the input features or weights during inference (incorporating the model constraint that the network's output should be invariant to small variations of the input features and/or weights, and hence regularizing the regression/decision function); (b) *dropout*, where activations are randomly reset to zero, effectively computing and averaging the gradient for many small variations of the network structure, where random sets of neurons are removed; and (c) training data *shuffling*, where the order of training instances is determined randomly – this is important for stochastic gradient descent, as it effectively randomizes the order in which gradient steps corresponding to minibatches are taken, and supposedly reduces the susceptibility to local minima in the error function. Optimal DNN training is still an active area of research, and more "recipes" to improve generalization are presented by Montavon *et al.* (2012).

7.3.2.1 Recurrent Neural Networks

Due to the sequential nature of audio, it is unsurprising that recently sequence learners such as *recurrent neural networks* (RNNs) have seen a resurgence in popularity for speech and music processing tasks, in particular source separation (Weninger *et al.*, 2014a; Huang *et al.*, 2014), as well as the related tasks of polyphonic piano note transcription (Böck and Schedl, 2012) and noise-robust speech recognition (Weninger *et al.*, 2014b). The combination of deep structures with temporal recurrence for sequence learning yields so-called *deep recurrent neural networks* (DRNNs) (Graves *et al.*, 2013).

In contrast to a nonrecurrent DNN, the input of a DRNN is a sequence of input features $\mathbf{y}_0(n)$, $n \in \{0, \dots, N-1\}$ and the output is a sequence $\mathbf{y}(n)$. The function g_z computed by a H-layer DRNN can be defined by the following iteration (forward pass) for $h \in \{1, \dots, H-1\}$ and $n \in \{0, \dots, N-1\}$:

$$\mathbf{y}_h(n) = g_h(\mathbf{Z}_{h-1,h}\mathbf{y}_{h-1}(n) + \mathbf{Z}_{h,h}\mathbf{y}_h(n-1)), \tag{7.17}$$

$$\mathbf{y}(n) = g_H(\mathbf{Z}_H\mathbf{y}_{H-1}(n)). \tag{7.18}$$

In the above, $\mathbf{y}_h(-1)$ is initialized to zero for all h.

The hidden layer weight matrices $\mathbf{Z}_h$, $h \in \{1, \dots, H-1\}$, are now split into feedforward parts $\mathbf{Z}_{h-1,h}$, which connect the layers to the previous layers, and recurrent parts $\mathbf{Z}_{h,h}$, which contain the feedback weights within the layers.

Figure 7.4 shows a sketch of a DRNN with a recurrent layer at position $h = 2$. On the left, the DRNN is shown with feedback connections (in green), indicated by "-1" for a time delay of 1 step. The feedforward connections between layers are depicted by black arrows. For clarity, not all the feedback connections are drawn. On the right, the DRNN is unfolded into a very deep DNN (showing all the feedback connections

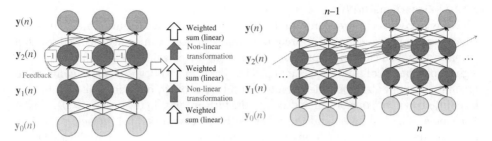

Figure 7.4 DRNN and unfolded DRNN.

in green). The unfolding is exemplified by the steps $n - 1$ and n. As can be seen, the unfolded DRNN is equivalent to a DNN that takes a sequence of features as input and outputs a sequence of features $\mathbf{y}(n)$ that undergo Hn nonlinear transformations from the input, making the DRNN a very powerful model.

To train RNNs, *backpropagation through time* is typically used, which is conceptually similar to performing backpropagation after unfolding the DNN to a nonrecurrent DNN, as exemplified in Figure 7.4. However, for larger N, the optimization of RNNs with stochastic gradient descent becomes difficult, as the gradients tend to either vanish or explode (Bengio *et al.*, 1994). As a result, RNNs are often not able to outperform DNNs in speech processing tasks, including source separation (Huang *et al.*, 2014). One of the oldest, yet most effective solutions proposed to remedy this problem is to add structure to the RNN following the *long short-term memory* (LSTM) principle as defined by Hochreiter and Schmidhuber (1997) and Gers *et al.* (2000).

The central idea of LSTM is to introduce a cell state variable $\mathbf{y}_h^{\text{cell}}(n)$ alongside the RNN hidden activation $\mathbf{y}_h^{\text{RNN}}(n)$. In contrast to the RNN hidden activation, the cell state is updated additively (not multiplicatively) and has an identity activation function:

$$\mathbf{y}_h^{\text{cell}}(n) = \mathbf{y}_h^{\text{forget}}(n) \circ \mathbf{y}_h^{\text{cell}}(n - 1) + \mathbf{y}_h^{\text{input}}(n) \circ \mathbf{y}_h^{\text{RNN}}(n). \tag{7.19}$$

Here, $\mathbf{y}_h^{\text{forget}}(n)$ and $\mathbf{y}_h^{\text{input}}(n)$ denote the forget and input gate activations, which are computed similarly to $\mathbf{y}_h^{\text{RNN}}(n)$, following (7.17), and $\circ$ denotes elementwise multiplication. The hidden activation of the LSTM cell is transformed by an activation function g_h and is scaled by the value of a third gate unit, the output gate $\mathbf{y}_h^{\text{output}}(n)$:

$$\mathbf{y}_h(n) = \mathbf{y}_h^{\text{output}}(n) \circ g_h(\mathbf{y}_h^{\text{cell}}(n)). \tag{7.20}$$

In LSTM, g_h is typically chosen as the hyperbolic tangent while the gate activation functions are chosen as logistic (Graves *et al.*, 2013).

From (7.19), it is obvious that unless the forget gate is used to reset the state of the cell, i.e. $\mathbf{y}_h^{\text{forget}}(n) = \mathbf{0}$, the gradient of the cell state with respect to previous time steps, $\partial \mathbf{y}_h^{\text{cell}}(n)/\partial \mathbf{y}_h^{\text{cell}}(n - n')$ for $n' > 0$, will not vanish, allowing the LSTM-RNN to exploit an unbounded amount of context. It can be shown that the LSTM approach allows for effectively training DRNNs using gradient descent, and for learning long-term dependencies (Hochreiter and Schmidhuber, 1997). In Section 7.3.6 we will show that using LSTM-DRNNs leads to improved source separation performance compared to DNNs with fixed context windows.

7.3.2.2 Bidirectional RNNs

RNNs as introduced above can exploit context from previous feature frames. In cases where real-time processing is not required, future context can be used as well. A fixed amount of lookahead can be provided to an RNN by delaying the training targets by a given number n' of time steps, i.e. replacing $y^{\text{tgt}}(n,f)$ by $y^{\text{tgt}}(n-n',f)$. In contrast, to take into account an unbounded amount of future context, one can split each hidden layer into two parts, one of which executes the forward pass in the order $n = 0, \ldots, N-1$ as above (forward layer) and the other in the reverse order, i.e. replacing $n-1$ by $n+1$ for the recurrent connections and iterating over $n = N-1, \ldots, 0$ (backward layer). The forward layer and backward layer have separate weight matrices, $\mathbf{Z}_h^{\text{forw}}$ and $\mathbf{Z}_h^{\text{back}}$. This yields a *bidirectional RNN* or, if the hidden units are designed as LSTM units, a bidirectional LSTM-DRNN.

For each time step n, the activations of the hth forward ($\mathbf{y}_h^{\text{forw}}(n)$) and backward ($\mathbf{y}_h^{\text{back}}(n)$) layers are concatenated to a single vector $\mathbf{y}_h(n)$. Both the forward and backward layers in the next level ($h+1$) process this concatenated vector as input. Thus, conceptually, the bidirectional RNN forward pass is equivalent to processing the sequence in both directions, collecting the activations and using them as input sequence for a bidirectional forward pass on the next level, etc.

Backpropagation for bidirectional RNNs can be easily implemented due to the untying of the forward layer and backward layer weights. The weights $\mathbf{Z}_h^{\text{forw}}$ are obtained in analogy to the unidirectional RNN case, as described above, based on the activations $\mathbf{y}_h^{\text{forw}}(n)$. Conversely, the weights $\mathbf{Z}_h^{\text{back}}$ are obtained by backpropagation through time with reversed temporal dependencies (replacing $n+1$ by $n-1$), and using the activations $\mathbf{y}_h^{\text{back}}(n)$ accordingly.

7.3.2.3 Other Architectures

Finding optimal DNN architectures for source separation is still an active area of research. Simpson *et al.* (2015) proposed the usage of *convolutional neural networks*, which have the advantage of being invariant to pitch variations in the input features. Furthermore, new DNN architectures for source separation can be derived by exploiting model-inspired constraints such as nonnegativity (see Chapter 8). Following this paradigm, nonnegative DNNs were introduced by Le Roux *et al.* (2015).

7.3.3 Learning Source Separation as Classification

Source separation by binary masks can be formulated as a binary classification task, where each time-frequency bin is to be classified as to whether the desired source $c_j(t)$ dominates (1) or not (0). To train a model for this classification task, the residual signals $u(t) = x(t) - c_j(t)$ are computed given a training corpus of mixtures $x(t)$ and corresponding sources (stereo data, cf. above), where j is the index of the desired source to be extracted. Then, the STFT magnitude spectrograms $|c_j(n,f)|$ and $|u(n,f)|$ of $c(t)$ and $u(t)$ are computed. From this, an ideal ratio mask (cf. Chapter 5) for the source j is obtained as follows:

$$\hat{y}^{\text{tgt}}(n,f) = \frac{|c_j(n,f)|}{|c_j(n,f)| + |u(n,f)|}. \tag{7.21}$$

Then, $\hat{y}^{\text{tgt}}(n,f)$ is transformed into a binary mask $y^{\text{tgt}}(n,f)$ according to Chapter 5.

Now, supervised training can be applied to obtain a classifier for each frequency bin f, learning the mapping from $\mathbf{y}_0(n)$ to $y^{\text{tgt}}(n,f)$ on the training data. For example, support vector machines and decision trees can be used to train separate classifiers for each f (Le Roux *et al.*, 2013; Gonzalez and Brookes, 2014). In contrast, DNNs can be used as a joint classifier for all frequency bins. At test time, the classifier is applied to each (feature) frame of the mixture, $\mathbf{y}_0(n)$. The output of the classifier is then used to generate a binary mask for the test mixture and separate the desired source (cf. Chapter 5). Alternatively, if the classifier outputs class posteriors instead of hard decisions (which is the case for all the classifier types mentioned above), these can be used as a ratio mask for separation (cf. Chapter 5).

For training DNNs to perform classification, the logistic function is used for the output layer, and the cross-entropy cost function can be used instead of the MSE cost function:

$$C^{\text{CE}}(\mathcal{Z}) = -\frac{1}{N} \sum_{nf} y^{\text{tgt}}(n,f) \log y(n,f) + (1 - y^{\text{tgt}}(n,f)) \log(1 - y(n,f)) \qquad (7.22)$$

7.3.4 Learning Source Separation as Regression

Instead of using a binary mask as training targets, i.e. hard targets, soft targets can be used in the form of ratio masks for supervised training. This yields a straightforward regression approach to source separation, using ratio masks to separate sources at test time, as indicated above (Huang *et al.*, 2014).

The cost functions can be chosen in the same way as for classification (MSE or cross-entropy) – the only difference being that the targets $y^{\text{tgt}}(n,f)$ are not binary. However, in the case of regression, it is also possible to derive a *discriminative training* cost that is closely related to the SNR of the reconstruction of a source c_j in the magnitude time-frequency domain. This optimal reconstruction cost can be written as a distance function for the time-frequency mask $\mathbf{y}$ output by a DNN with logistic output activation function:

$$C^{\text{DT}}(\mathcal{Z}) = \frac{1}{N} \sum_{nf} (y(n,f)|x(n,f)| - |c_j(n,f)|)^2. \qquad (7.23)$$

The cost (7.23) can be easily minimized by a DNN through backpropagation. Then,

$$\frac{\partial C^{\text{DT}}}{\partial y(n,f)} = (y(n,f)|x(n,f)| - |c_j(n,f)|)|x(n,f)| \qquad (7.24)$$

is the gradient of the cost function with respect to the network outputs. The weight updates are then simply determined by backpropagation to the output and hidden layers, as outlined above. It has been shown by Weninger *et al.* (2014c) that this can improve performance compared to the standard MSE cost (7.12) (cf. Section 7.3.6). This is in line with the results obtained by Wang *et al.* (2014).

In addition, it is also easy to derive an optimal reconstruction cost for the case of Mel-domain separation, where a time-frequency mask is computed in the Mel-domain and applied to a full-resolution STFT spectrum (Weninger *et al.*, 2014c). It was shown by Weninger *et al.* (2014c) that the latter cost delivers better performance than both the previously mentioned ones. Finally, Erdogan *et al.* (2015) proposed a phase-sensitive extension of the above cost, resulting in further performance gains (cf. Section 7.3.6):

$$C^{\text{PS}}(\mathcal{Z}) = \frac{1}{N} \sum_{nf} |y(n,f)x(n,f) - c_j(n,f)|^2. \qquad (7.25)$$

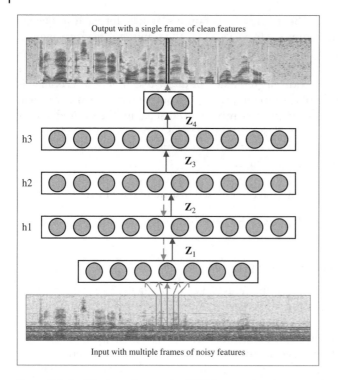

Figure 7.5 Architecture of regression DNN for speech enhancement.

In contrast to the above (7.23), the reconstruction error is measured in the complex STFT domain. Note that the network does not predict the phase, but still predicts a mask. The goal of the complex domain phase-sensitive cost function is to make the network learn to shrink the mask estimates when the noise is high. The exact shrinking amount is the cosine of the angle between the phases of the noisy and clean signals, which is known during training but unknown during testing (Erdogan *et al.*, 2015; Weninger *et al.*, 2015).

Apart from the prediction of time-frequency masks, learning approaches are adopted to predict source spectra directly for source separation. In the specific applications, Xu *et al.* (2014b) proposed a DNN-based speech enhancement framework in which DNN was regarded as a regression model to predict the clean spectra features from noisy spectra, as shown in Figure 7.5. In the training stage, a regression DNN model is trained from a collection of stereo data by minimizing the MSE cost (7.12), where the network outputs $y(n,f)$ correspond to estimated features and the training targets $y^{tgt}(n,f)$ correspond to reference features. Liu *et al.* (2014) investigated the detailed configurations of DNN, and it was shown that other cost functions, such as the Kullback–Leibler (KL) divergence or the Itakura–Saito (IS) divergence, all performed worse than the MSE. In the enhancement stage, the well-trained DNN model is fed with the features of noisy speech in order to generate the enhanced features. The additional phase information is calculated from the original noisy speech.

Xu *et al.* (2015a) updated the regression-based speech enhancement framework by taking advantage of the abundant acoustic context information and large training data, and showed that it achieves better generalization to new speakers, different SNRs, and even other languages, etc. One challenge of the speech enhancement task is the performance degradation in low SNR environments. Faced with this problem, Gao *et al.* (2015b) employed a joint framework combining speech enhancement with voice activity detection (VAD) to increase the speech intelligibility in low SNR environments, and proposed a SNR-based progressive learning framework (Gao *et al.*, 2016) that aims to decompose the complicated regression problem into a series of subproblems for enhancing system performances and reducing model complexities.

Du *et al.* (2014) applied the above regression framework to single-channel speech separation, and demonstrated its effectiveness over state-of-the-art approaches of separating a single target speaker from mixtures of two voices in both the supervised and semi-supervised modes. Tu *et al.* (2014) modified the architecture with dual outputs for learning both target source spectra and interfering source spectra. A semi-supervised mode to separate the speech of the target speaker from an unknown interfering speaker, i.e. selecting a specific model only for the target speaker and using a generic model for the interfering speakers, was discussed by Du *et al.* (2016). In many real applications, the potential of separating the target speaker from the interferer is limited by the fact that the mixed speech signals are further corrupted with background noises. Gao *et al.* (2015a) presented a speaker-dependent framework for reduction of background noise and interfering sources.

7.3.5 Generalization Capabilities

One key point of learning approaches is the generalization capability to unseen and nonstationary noise conditions (Chen *et al.*, 2015; Vincent *et al.*, 2017). To enhance this capability, a set of noise types and dynamic noise-aware training approaches were used (Xu *et al.*, 2014a). Kim and Smaragdis (2015) proposed a fine-tuning scheme at the test stage to improve the performance of a well-trained DNN. Multitask learning has been proved to improve the generalization of the target task by leveraging the other tasks. A multiobjective framework was proposed by Xu *et al.* (2015b) to improve the generalization capability of regression DNN. Note that Chapter 12 covers generalization in the multichannel setting.

7.3.6 Benchmark Performances

A benchmark of recent developments in DNN-based separation of speech from nonstationary noise is given by Weninger *et al.* (2015). The noise comprises music, interfering speakers, and environmental noise such as babble, street, and domestic noise. Methods are evaluated on the corpus of the CHiME-2 Challenge (Track 2) (Vincent *et al.*, 2013). To take into account generalization, the corpus is split speaker-independently into a training, validation, and test set, the background noise in the development and test sets is disjoint from the noise in the training set, and a different room impulse response is used to convolve the dry utterances. In the CHiME-2 setup, the speaker is positioned at

Table 7.1 SDR (dB) achieved on the CHiME-2 dataset using supervised training (2ch: average of two input channels).

Enhancement	Mel	Cost	Input SNR (dB)						Avg
			−6	−3	0	3	6	9	
Noisy			−2.27	−0.58	1.66	3.40	5.20	6.60	2.34
Sparse NMF (Chapter 8)	–	–	5.48	7.53	9.19	10.88	12.89	14.61	10.10
DNN	–	MSE	6.89	8.82	10.53	12.25	14.13	15.98	11.43
DNN	–	discrim.	7.89	9.64	11.25	12.84	14.74	16.61	12.16
DNN	√	discrim.	8.36	10.00	11.65	13.17	15.02	16.83	12.50
LSTM-DRNN	√	discrim.	10.14	11.60	13.15	14.48	16.19	17.90	**13.91**
2ch-LSTM-DRNN	√	discrim.	10.46	11.85	13.40	14.86	16.34	18.07	14.17
2ch-LSTM-DRNN	√	phase-sens.	10.97	12.28	13.76	15.13	16.57	18.26	14.49
2ch-bidir. LSTM-DRNN	√	phase-sens.	11.30	12.74	14.18	15.46	16.96	18.67	**14.88**
Ideal ratio mask	–	–	14.53	15.64	16.95	18.09	19.65	21.24	17.68

an approximate azimuth angle of 0°, i.e. facing the microphone. This means that adding the left and right channels corresponds to simple delay-and-sum (DS) beamforming, and is expected to improve the outcome.

Supervised training for DNNs was considered using the MSE (7.12), discriminative (7.23), and phase-sensitive (7.25) costs. The targets for supervised training are derived from the parallel noise-free and multicondition training sets of the CHiME-2 data. The D(R)NN topology and training parameters were as set by Weninger *et al.* (2014c) and Erdogan *et al.* (2015). The evaluation measure employed for speech separation is signal-to-distortion ratio (SDR) (Vincent *et al.*, 2006).

Table 7.1 shows a comparison of selected speech enhancement systems on the CHiME-2 test set at input SNRs from −6 to 9 dB. In the evaluation, STFT as well as Mel magnitude spectra were investigated as feature domains. Comparing the results obtained with STFT magnitudes, it is observed that the DNN significantly outperforms NMF (Chapter 8), and discriminative training of DNN according to (7.23) leads to a significant performance improvement. Switching to the Mel magnitude domain further improves the results for the DNN. The gains by using the LSTM-DRNN network architecture are complementary to the gains by discriminative training, leading to the best result of 13.91 dB average SDR. Further gains are obtained by a simple 2-channel (2ch) front-end (adding both channels), the phase-sensitive cost (7.25) as well as bidirectional LSTM networks (cf. Section 7.3.2.2). We can observe that the potential gain over the noisy baseline by these supervised methods is very large (up to 12.5 dB), and that NMF is outperformed by a large margin of 4.78 dB SDR. However, there is still a certain gap to the performance attainable by the ideal ratio mask (17.68 dB).

An additional benchmark of single-channel source separation methods is provided by Huang *et al.* (2015). The experiments were conducted on three databases (TSP, MIR-1K, and TIMIT) with the aim of focusing on joint mask optimization and DRNNs for single-channel speech separation, singing voice separation, and speech denoising. The results shows that RNN-based source separation methods outperformed NMF baselines by a margin of up to 4.98 dB SDR, in line with the results obtained above.

7.4 Summary

In this chapter, learning-free, unsupervised and supervised source modeling, as well as separation-based training approaches for single-channel source separation were presented. Single-channel source separation has many applications, including front-end processing for automatic speech recognition (ASR; see Chapter 17).

In principle, the main advantage of learning-free as well as source modeling approaches is a higher generalization capability to a variety of acoustic environments and source types compared to separation-based training. In contrast, separation-based training approaches, when trained and tested in similar conditions and on similar source types (e.g., for speech enhancement in a car or domestic environment), can yield best performance, but generalization remains a concern. Today, generalization of separation-based training is mostly approached by training data augmentation, which comes at the price of increasing the computational effort at training time.

Factorial models combine the advantages of exploiting knowledge about sources with increased generalization capabilities, as models for various sources can be combined modularly. In contrast, to integrate a new type of source into a separation-based training approach, costly retraining of the entire model is needed. Furthermore, with factorial models, models of sources that are unknown at the training stage can be inferred during the separation stage, which is not directly possible with separation-based training approaches. In Chapter 8, NMF will be introduced as a powerful extension of factorial models that has the same advantages.

Conversely, separation-based training approaches are cheap to evaluate in the separation stage (inference mode), as only a deterministic function needs to be evaluated at the point of the current signal frame, while factorial models require probabilistic inference. A promising perspective is to unite the benefits of source modeling and separation-based training approaches, as discussed in Section 19.1.

Bibliography

Bach, F.R. and Jordan, M.I. (2006) Learning spectral clustering, with application to speech separation. *Journal of Machine Learning Research*, 7, 1963–2001.

Bach, F.R. and Jordan, M.I. (2009) Spectral clustering for speech separation, in *Automatic Speech and Speaker Recognition: Large Margin and Kernel Methods*, Wiley, pp. 221–253.

Baum, L.E. and Petrie, T. (1966) Statistical inference for probabilistic functions of finite state Markov chains. *Annals of Mathematical Statistics*, **37** (6), 1554–1563.

Bengio, Y., Simard, P., and Frasconi, P. (1994) Learning long-term dependencies with gradient descent is difficult. *IEEE Transactions on Neural Networks*, **5** (2), 157–166.

Böck, S. and Schedl, M. (2012) Polyphonic piano note transcription with recurrent neural networks, in *Proceedings of IEEE International Conference on Audio, Speech and Signal Processing*, pp. 121–124.

Bregman, A.S. (1994) *Auditory Scene Analysis: The Perceptual Organization of Sound*, MIT Press.

Brown, G. and Wang, D. (2005) Separation of speech by computational auditory scene analysis, in *Speech Enhancement*, Springer, pp. 371–402.

Brown, G.J. and Cooke, M. (1994) Computational auditory scene analysis. *Computer Speech and Language*, **8** (4), 297–336.

Chen, J., Wang, Y., and Wang, D. (2015) Noise perturbation improves supervised speech separation, in *Proceedings of International Conference on Latent Variable Analysis and Signal Separation*, pp. 83–90.

Cooke, M. (2005) *Modelling Auditory Processing and Organisation*, Cambridge University Press.

Dean, J., Corrado, G., Monga, R., Chen, K., Devin, M., Mao, M., Ranzato, M., Senior, A., Tucker, P., Yang, K., Le, Q.V., and Ng, A.Y. (2012) Large scale distributed deep networks, in *Proceedings of Neural Information Processing Systems*, pp. 1223–1231.

Deng, L., Hinton, G., and Kingsbury, B. (2013) New types of deep neural network learning for speech recognition and related applications: An overview, in *Proceedings of IEEE International Conference on Audio, Speech and Signal Processing*, pp. 8599–8603.

Du, J., Tu, Y., Xu, Y., Dai, L.R., and Lee, C.H. (2014) Speech separation of a target speaker based on deep neural networks, in *Proceedings of International Conference on Speech Processing*, pp. 473–477.

Du, J., Tu, Y.H., Dai, L.R., and Lee, C.H. (2016) A regression approach to single-channel speech separation via high-resolution deep neural networks. *IEEE/ACM Transactions on Audio, Speech, and Language Processing*, **24** (8), 1424–1437.

Erdogan, H., Hershey, J.R., Watanabe, S., and Le Roux, J. (2015) Phase-sensitive and recognition-boosted speech separation using deep recurrent neural networks, in *Proceedings of IEEE International Conference on Audio, Speech and Signal Processing*.

Gao, T., Du, J., Dai, L.R., and Lee, C.H. (2016) SNR-based progressive learning of deep neural network for speech enhancemen, in *Proceedings of Interspeech*.

Gao, T., Du, J., Xu, L., Liu, C., Dai, L.R., and Lee, C.H. (2015a) A unified speaker-dependent speech separation and enhancement system based on deep neural networks, in *Proceedings of ChinaSIP*, pp. 687–691.

Gao, T., Du, J., Xu, Y., Liu, C., Dai, L.R., and Lee, C.H. (2015b) Improving deep neural network based speech enhancement in low SNR environments, in *Proceedings of International Conference on Latent Variable Analysis and Signal Separation*, pp. 75–82.

Gers, F., Schmidhuber, J., and Cummins, F. (2000) Learning to forget: Continual prediction with LSTM. *Neural Computation*, **12** (10), 2451–2471.

Gonzalez, S. and Brookes, M. (2014) Mask-based enhancement for very low quality speech, in *Proceedings of IEEE International Conference on Audio, Speech and Signal Processing*, pp. 7079–7083.

Goodfellow, I., Bengio, Y., and Courville, A. (2016) *Deep Learning*, MIT Press.

Graves, A., Mohamed, A., and Hinton, G. (2013) Speech recognition with deep recurrent neural networks, in *Proceedings of IEEE International Conference on Audio, Speech and Signal Processing*, pp. 6645–6649.

Hinton, G.E. (2002) Training products of experts by minimizing contrastive divergence. *Neural Computation*, **14**, 1771–1800.

Hochreiter, S. and Schmidhuber, J. (1997) Long short-term memory. *Neural Computation*, **9** (8), 1735–1780.

Hu, G. and Wang, D. (2006) An auditory scene analysis approach to monaural speech segregation, in *Topics in Acoustic Echo and Noise Control*, Springer, pp. 485–515.

Huang, P.S., Kim, M., Hasegawa-Johnson, M., and Smaragdis, P. (2014) Deep learning for monaural speech separation, in *Proceedings of IEEE International Conference on Audio, Speech and Signal Processing*, pp. 1562–1566.

Huang, P.S., Kim, M., Hasegawa-Johnson, M., and Smaragdis, P. (2015) Joint optimization of masks and deep recurrent neural networks for monaural source separation. *IEEE Transactions on Audio, Speech, and Language Processing*, **23** (12), 2136–2147.

Kim, M. and Smaragdis, P. (2015) Adaptive denoising autoencoders: A fine-tuning scheme to learn from test mixtures, in *Proceedings of International Conference on Latent Variable Analysis and Signal Separation*, pp. 100–107.

Le Roux, J., Hershey, J.R., and Weninger, F. (2015) Deep NMF for speech separation, in *Proceedings of IEEE International Conference on Audio, Speech and Signal Processing*, pp. 66–70.

Le Roux, J., Watanabe, S., and Hershey, J. (2013) Ensemble learning for speech enhancement, in *Proceedings of IEEE Workshop on Applications of Signal Processing to Audio and Acoustics*.

LeCun, Y., Bottou, L., Orr, G., and Müller, K. (1998) Efficient backprop, in *Neural Networks: Tricks of the Trade*.

Licklider, J.C.R. (1951) A duplex theory of pitch perception. *Journal of the Acoustical Society of America*, **23** (1), 147–147.

Liu, D., Smaragdis, P., and Kim, M. (2014) Experiments on deep learning for speech denoising, in *Proceedings of Interspeech*, pp. 2685–2689.

Lu, X., Tsao, Y., Matsuda, S., and Hori, C. (2013) Speech enhancement based on deep denoising autoencoder, in *Proceedings of Interspeech*, pp. 3444–3448.

Montavon, G., Orr, G.B., and Müller, K.R. (eds) (2012) *Neural Networks: Tricks of the Trade*, Springer.

Nakatani, T. and Okuno, H.G. (1999) Harmonic sound stream segregation using localization and its application to speech stream segregation. *Speech Communication*, **27** (3), 209–222.

Narayanan, A. and Wang, D.L. (2013) Ideal ratio mask estimation using deep neural networks for robust speech recognition, in *Proceedings of IEEE International Conference on Audio, Speech and Signal Processing*, pp. 7092–7096.

Narayanan, A. and Wang, D.L. (2015) Improving robustness of deep neural network acoustic models via speech separation and joint adaptive training. *IEEE Transactions on Audio, Speech, and Language Processing*, **23** (1), 92–101.

Ng, A.Y., Jordan, M.I., and Weiss, Y. (2002) On spectral clustering: Analysis and an algorithm, vol. 2, pp. 849–856.

Nix, J., Kleinschmidt, M., and Hohmann, V. (2003) Computational auditory scene analysis by using statistics of high-dimensional speech dynamics and sound source direction., in *Proceedings of Interspeech*, pp. 1441–1444.

Ozerov, A., Philippe, P., Gribonval, R., and Bimbot, F. (2005) One microphone singing voice separation using source-adapted models, in *Proceedings of IEEE Workshop on Applications of Signal Processing to Audio and Acoustics*, pp. 90–93.

Polyak, B.T. (1964) Some methods of speeding up the convergence of iteration methods. *Computational Mathematics and Mathematical Physics*, **4** (5), 1–17.

Roweis, S.T. (2001) One microphone source separation, in *Proceedings of Neural Information Processing Systems*, vol. 13, pp. 793–799.

Roweis, S.T. (2003) Factorial models and refiltering for speech separation and denoising, in *Proceedings of Interspeech*.

Schuller, B. (2013) *Intelligent Audio Analysis*, Springer.

Shao, Y. and Wang, D. (2006) Model-based sequential organization in cochannel speech. *IEEE Transactions on Audio, Speech, and Language Processing*, **14** (1), 289–298.

Simpson, A.J., Roma, G., and Plumbley, M.D. (2015) Deep karaoke: Extracting vocals from musical mixtures using a convolutional deep neural network, in *Proceedings of International Conference on Latent Variable Analysis and Signal Separation*, pp. 429–436.

Slaney, M. and Lyon, R.F. (1990) A perceptual pitch detector, in *Proceedings of IEEE International Conference on Audio, Speech and Signal Processing*, IEEE, pp. 357–360.

Sutskever, I., Martens, J., Dahl, G., and Hinton, G. (2013) On the importance of initialization and momentum in deep learning, in *Proceedings of International Conference on Machine Learning*.

Tu, Y., Du, J., Xu, Y., Dai, L.R., and Lee, C.H. (2014) Speech separation based on improved deep neural networks with dual outputs of speech features for both target and interfering speakers, in *Proceedings of International Symposium on Chinese Spoken Language Processing*, pp. 250–254.

Vincent, E., Barker, J., Watanabe, S., Le Roux, J., Nesta, F., and Matassoni, M. (2013) The second 'CHiME' speech separation and recognition challenge: Datasets, tasks and baselines, in *Proceedings of IEEE International Conference on Audio, Speech and Signal Processing*, pp. 126–130.

Vincent, E., Gribonval, R., and Févotte, C. (2006) Performance measurement in blind audio source separation. *IEEE Transactions on Audio, Speech, and Language Processing*, **14**(4), 1462–1469.

Vincent, E., Watanabe, S., Nugraha, A.A., Barker, J., and Marxer, R. (2017) An analysis of environment, microphone and data simulation mismatches in robust speech recognition. *Computer Speech and Language*, **46**, 535–557.

Wang, D. and Brown, G.J. (2006) *Computational Auditory Scene Analysis: Principles, Algorithms, and Applications*, Wiley-IEEE Press.

Wang, D.L. and Brown, G.J. (1999) Separation of speech from interfering sounds based on oscillatory correlation. *IEEE Transactions on Neural Networks*, **10** (3), 684–697.

Wang, Y., Narayanan, A., and Wang, D. (2014) On training targets for supervised speech separation. *IEEE/ACM Transactions on Audio, Speech, and Language Processing*, **22** (12), 1849–1858.

Weintraub, M. (1985) *A Theory and Computational Model of Monaural Auditory Sound Separation*, Ph.D. thesis, Stanford Univ.

Weninger, F., Erdogan, H., Watanabe, S., Vincent, E., Le Roux, J., Hershey, J.R., and Schuller, B. (2015) Speech enhancement with LSTM recurrent neural networks and its application to noise-robust ASR, in *Proceedings of International Conference on Latent Variable Analysis and Signal Separation*, pp. 91–99.

Weninger, F., Eyben, F., and Schuller, B. (2014a) Single-channel speech separation with memory-enhanced recurrent neural networks, in *Proceedings of IEEE International Conference on Audio, Speech and Signal Processing*, pp. 3737–3741.

Weninger, F., Geiger, J., Wöllmer, M., Schuller, B., and Rigoll, G. (2014b) Feature enhancement by deep LSTM networks for ASR in reverberant multisource environments. *Computer Speech and Language*, **28** (4), 888–902.

Weninger, F., Hershey, J.R., Le Roux, J., and Schuller, B. (2014c) Discriminatively trained recurrent neural networks for single-channel speech separation, in *Proceedings of GlobalSIP*, pp. 740–744.

Wu, M., Wang, D., and Brown, G.J. (2003) A multipitch tracking algorithm for noisy speech. *IEEE Transactions on Speech and Audio Processing*, **11** (3), 229–241.

Xia, B.Y. and Bao, C.C. (2013) Speech enhancement with weighted denoising auto-encoder, in *Proceedings of Interspeech*, pp. 436–440.

Xu, Y., Du, J., Dai, L.R., and Lee, C.H. (2014a) Dynamic noise aware training for speech enhancement based on deep neural networks, in *Proceedings of Interspeech*, pp. 2670–2674.

Xu, Y., Du, J., Dai, L.R., and Lee, C.H. (2014b) An experimental study on speech enhancement based on deep neural networks. *IEEE Signal Processing Letters*, **21** (1), 65–68.

Xu, Y., Du, J., Dai, L.R., and Lee, C.H. (2015a) A regression approach to speech enhancement based on deep neural networks. *IEEE/ACM Transactions on Audio, Speech, and Language Processing*, **23** (1), 7–19.

Xu, Y., Du, J., Huang, Z., Dai, L.R., and Lee, C.H. (2015b) Multi-objective learning and mask-based post-processing for deep neural network based speech enhancement, in *Proceedings of Interspeech*, pp. 1508–1512.

Young, S., Evermann, G., Gales, M., Hain, T., Kershaw, D., Liu, X., Moore, G., Odell, J., Ollason, D., Povey, D. *et al.* (2006) The HTK book (for HTK version 3.4). *Cambridge University Engineering Department*, **2** (2), 2–3.

Yu, D., Deng, L., Seide, F., and Li, G. (2011) Discriminative pretraining of deep neural networks, US Patent 13/304643.

Zöhrer, M., Peharz, R., and Pernkopf, F. (2015) Representation learning for single-channel source separation and bandwidth extension. *IEEE/ACM Transactions on Audio, Speech, and Language Processing*, **23** (12), 2398–2409.

8

Nonnegative Matrix Factorization

Roland Badeau and Tuomas Virtanen

Nonnegative matrix factorization (NMF) refers to a set of techniques that have been used to model the spectra of sound sources in various audio applications, including source separation. Sound sources have a structure in time and frequency: music consists of basic units like notes and chords played by different instruments, speech consists of elementary units such as phonemes, syllables or words, and environmental sounds consist of sound events produced by various sound sources. NMF models this structure by representing the spectra of sounds as a sum of components with fixed spectrum and time-varying gain, so that each component in the model represents these elementary units in the sound.

Modeling this structure is beneficial in source separation, since inferring the structure makes it possible to use contextual information for source separation. NMF is typically used to model the magnitude or power spectrogram of audio signals, and its ability to represent the structure of audio sources makes separation possible even in single-channel scenarios.

This chapter presents the use of NMF-based single-channel techniques. In Section 8.1 we introduce the basic NMF model used in various single-channel source separation scenarios. In Section 8.2, several deterministic and probabilistic frameworks for NMF are presented, along with various NMF algorithms. Then several methods that can be used to learn NMF components by using suitable training material are presented in Section 8.3. In Section 8.4, some advanced NMF models are introduced, including regularizations and nonstationary models. Finally, Section 8.5 summarizes the key concepts introduced in this chapter.

8.1 NMF and Source Separation

Let $\hat{\mathbf{V}} = [\hat{v}(n,f)]_{fn}$ denote the $F \times N$ nonnegative time-frequency representation of a signal $x(t)$, where $n \in \{0, \ldots, N-1\}$ is the time frame index, and $f \in \{0, \ldots, F-1\}$ is the frequency index. For instance, if $\mathbf{X}$ is the $F \times N$ complex-valued short-time Fourier transform (STFT) of x, then $\hat{\mathbf{V}}$ can be the magnitude spectrogram $|\mathbf{X}|$ or the power spectrogram $|\mathbf{X}|^2$ (Smaragdis and Brown, 2003). Other choices may include perceptual

Audio Source Separation and Speech Enhancement, First Edition.
Edited by Emmanuel Vincent, Tuomas Virtanen and Sharon Gannot.
© 2018 John Wiley & Sons Ltd. Published 2018 by John Wiley & Sons Ltd.
Companion Website: https://project.inria.fr/ssse/

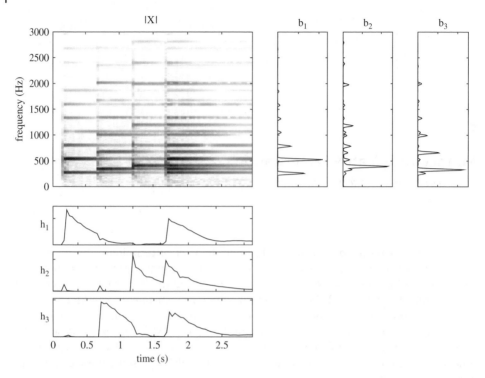

Figure 8.1 An example signal played by piano consists of a sequence of notes C, E, and G, followed by the three notes played simultaneously. The basic NMF models the magnitude spectrogram of the signal $\hat{\mathbf{V}} = |\mathbf{X}|$ (top left) as a sum of components having fixed spectra $\mathbf{b}_k$ (rightmost panels) and activation coefficients $h_k(n)$ (lowest panels). Each component represents parts of the spectrogram corresponding to an individual note.

frequency scales, such as constant-Q (Fuentes *et al.*, 2013) or equivalent rectangular bandwidth (ERB) (Vincent *et al.*, 2010) representations (see Chapter 2 for a discussion about different time-frequency representations).

NMF (Lee and Seung, 1999) approximates the nonnegative matrix $\hat{\mathbf{V}}$ with another nonnegative matrix $\mathbf{V} = [v(n,f)]_{fn}$ with entries $v(n,f) = \sigma_x^2(n,f)$, defined as the product

$$\mathbf{V} = \mathbf{BH} \tag{8.1}$$

of an $F \times K$ nonnegative matrix $\mathbf{B}$ and a $K \times N$ nonnegative matrix $\mathbf{H}$ of lower rank $K < \min(F, N)$. This factorization can also be written $\mathbf{V} = \sum_k \mathbf{V}_k$, where $\mathbf{V}_k = [v_k(n,f)]_{fn} = \mathbf{b}_k \mathbf{h}_k^T$, for all $k \in \{1, \dots, K\}$, is the kth rank-1 matrix component. The kth column vector $\mathbf{b}_k = [b_k(f)]_f$ can be interpreted as its spectrum, and the kth row vector $\mathbf{h}_k^T = [h_k(n)]_n$ comprises its *activation coefficients* over time (cf. Figure 8.1). We also write $v_k(n,f) = b_k(f)h_k(n)$. All the parameters of the model, as well as the observed magnitude or power spectra, are elementwise nonnegative.

8.1.1 NMF Masking

The rationale of using NMF in source separation is that each source typically has a characteristic spectral structure that is different from other sources, and therefore ideally each source can be modeled using a distinct set of *basis spectra*. In other words, an individual source in a mixture is modeled by using a subset of basis spectra. If we denote $\mathcal{K}_j$ the subset of basis spectra of source j, the model for the magnitude or power spectrum vector of the source in frame n can be written as $\sum_{k \in \mathcal{K}_j} \mathbf{b}_k h_k(n)$. Assuming that we have a method for estimating the basis vectors and activation coefficients in a mixture, we can generate a time-frequency mask vector, for example as

$$\mathbf{w}_j(n) = \frac{\sum_{k \in \mathcal{K}_j} \mathbf{b}_k h_k(n)}{\sum_{k=1}^{K} \mathbf{b}_k h_k(n)} \in [0, 1] \tag{8.2}$$

which can be used to filter source j from the mixture as described in Chapter 5. In (8.2), the divisions must be understood elementwise. The entire process of generating the mask is illustrated in Figure 8.2.

The criteria used for estimating the NMF model parameters are presented in Section 8.2.1. It should be noted that other models exist that similarly represent an observation as a linear combination of basis vectors weighted by activations. However, those models are based on completely different estimation criteria, such as independent component analysis (ICA) (Brown and Smaragdis, 2004; Casey and Westner, 2000), which does not lead to nonnegative components. There are also models which combine

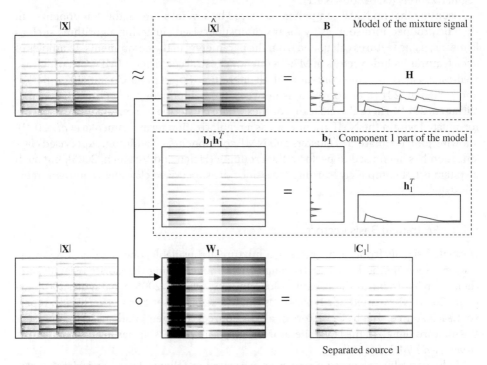

Figure 8.2 The NMF model can be used to generate time-frequency masks in order to separate sources from a mixture. Top row: the spectrogram $\hat{\mathbf{V}} = |\mathbf{X}|$ of the mixture signal in Figure 8.1 is modeled with an NMF. Middle row: the model for an individual source can be obtained using a specific set of components. In this case, only component 1 is used to represent an individual note in the mixture. Bottom row: the mixture spectrogram is elementwise multiplied by the time-frequency mask matrix $\mathbf{W}_1$, resulting in a separated source spectrogram $|\mathbf{C}_1|$.

additive nonnegative summation of sources and their components with linear modeling of source variation in the log-spectral domain (Vincent and Rodet, 2004), which can be viewed as a more realistic model to characterize the variability of natural sources.

8.1.2 Learning-Free Separation

In the early NMF-based approaches by Smaragdis and Brown (2003) and Virtanen (2003), mixture signals were separated into sources without any training stage. In this context, both the basis vectors and their activations are estimated from a mixture spectrogram. In this type of *unsupervised NMF* processing, there is no prior knowledge on the indexes of the components that represent a given source. Actually, one NMF component may even be shared by different sources, which is detrimental to source separation. With this type of entirely learning-free methods, a good separation performance is obtained only in some simple scenarios where the characteristics of source components are clearly different from each other and well modeled with NMF, such as in the example in Figure 8.1. Therefore a significant part of the NMF literature has focused on producing NMF components that characterize one source only, e.g. by using pretrained basis vectors (cf. Section 8.1.3) or by introducing various types of regularizations (cf. Section 8.4.1).

Assuming that no NMF component is shared by several sources, the components can then be grouped into sources by means of unsupervised clustering algorithms such as k-means, using features calculated from the parameters of the components. Examples of such features include vectors of Mel-frequency cepstral coefficients (MFCCs), which are well-established in audio content analysis literature (Spiertz and Gnann, 2009; Barker and Virtanen, 2013), or correlations between the component activations (Barker and Virtanen, 2013). There also exist approaches where clustering in a certain feature space is formulated within the NMF framework as a regularization term (Kameoka *et al.*, 2012), allowing joint estimation of clusters and NMF components. In addition, supervised classification has been used to perform the grouping (Helén and Virtanen, 2005), but such a system is not completely learning-free, since the supervised classifier requires a training stage.

8.1.3 Pretrained Basis Vectors

If isolated signals from each source in a mixture are available, one can develop powerful separation methods by first training source models using the isolated signals, and then applying them to the mixture (Schmidt and Olsson, 2006; Smaragdis, 2007; Raj *et al.*, 2010). These methods typically estimate a separate set of basis vectors to represent each source. The basis vectors of all the sources are then jointly used to represent the mixture signal. In the NMF literature, this type of processing has been referred to as *supervised NMF*.

Mathematically, the processing can be expressed as follows. First, the isolated material of source j is used to estimate a set $\mathcal{K}_j$ of basis vectors to represent the source. This source-specific *dictionary* can also be expressed as a basis matrix $\mathbf{B}_j$. There exist various dictionary learning methods that are discussed in more detail in Section 8.3 of this chapter. Common approaches are based, for example, on the NMF of the training data, on vector quantization, or on random sampling.

The magnitude spectrogram $\hat{\mathbf{V}} = |\mathbf{C}_j|$ (or the power spectrogram $\hat{\mathbf{V}} = |\mathbf{C}_j|^2$) of source j within the mixture is now modeled with the basic NMF model as

$$\hat{\mathbf{V}}_j \approx \mathbf{B}_j \mathbf{H}_j, \tag{8.3}$$

where $\mathbf{H}_j$ is the activation matrix of source j. For the mixture of all the sources, the model can therefore be written as

$$\hat{\mathbf{V}} \approx \sum_{j=1}^{J} \hat{\mathbf{V}}_j$$

$$\approx \sum_{j=1}^{J} \mathbf{B}_j \mathbf{H}_j,$$

$$= \mathbf{B}\mathbf{H} \tag{8.4}$$

where the basis matrix $\mathbf{B} = [\mathbf{B}_1, \ldots, \mathbf{B}_J]$ consists of the dictionaries of all sources and the activation matrix $\mathbf{H} = [\mathbf{H}_1^T, \ldots, \mathbf{H}_J^T]^T$ contains all the activations. It should be noted that in the above formulation we assume that the magnitude (or power) spectrogram of the mixture equals the sum of the magnitude (or power) spectrograms of the individual sources, which is in general only an approximation.

Assuming the additivity of power spectrograms is theoretically justified for noisy signals, since the variances of independent random variables can be summed. In practice, however, audio signals are not only made of noise. Experimental studies show that assuming the additivity of magnitude spectrograms is a better fit if the performance is measured in terms of signal-to-distortion ratio (SDR) or signal-to-artifacts ratio (SAR), and using a magnitude exponent even smaller than one can be beneficial for some metrics, when certain cost functions (see Section 8.2.1) are used (King *et al.*, 2012). The assumption of additivity of magnitude or power spectrograms holds relatively well, but the resulting unavoidable inaccuracies can lead to some artifacts such as musical noise or interference.

As can be seen from (8.4), the model for the mixture spectrogram now corresponds to the basic NMF model, with the constraint that the basis and activation matrices consist of source-specific submatrices. In the parameter estimation, the pretrained basis vectors are kept fixed, and only their activations are estimated. Once the activations have been estimated, source j can be separated from the mixture by applying a mask, as explained earlier in this chapter. The overall processing is illustrated in Figure 8.3 in the case of a two-source mixture.

It should be noted that in addition to the separate estimation of source-specific basis matrices, there exist methods that estimate the basis matrices jointly for all sources (Grais and Erdogan, 2013) using material from all the sources at the same time. These approaches can also include the use of mixtures to optimize their parameters, as long as isolated reference signals for each source are available (Weninger *et al.*, 2014).

8.1.4 Combining Pretrained Basis Vectors and Learning-Free Separation

In practical source separation scenarios, it is typical that we have prior information about some sources in a mixture, but there are also sources whose exact characteristics are not known. For example, in a speech enhancement scenario we know that the

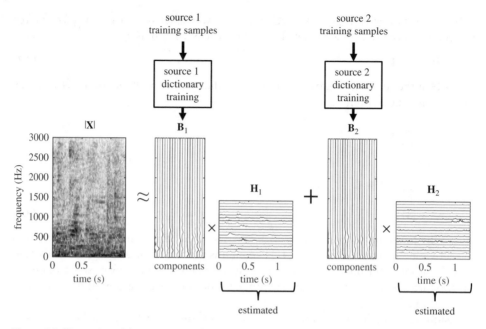

Figure 8.3 Illustration of the separation of two sources, where source-specific models are obtained in the training stage. Dictionaries $\mathbf{B}_1$ and $\mathbf{B}_2$ consisting of source-specific basis vectors are obtained for sources 1 and 2 using isolated samples of each source. The mixture spectrogram $\widehat{\mathbf{V}} = |\mathbf{X}|$ is modeled as a weighted linear combination of all the basis vectors. Matrices $\mathbf{H}_1$ and $\mathbf{H}_2$ contain activations of basis vectors in all frames.

target source is speech, but we may not know the exact type of background noise. For the target source we can train basis vectors from isolated training material, but there is a need for estimating the basis vectors for other sources from the mixture signals. In the NMF literature (Joder *et al.*, 2012; Mysore and Smaragdis, 2011), this processing has been referred to as *semi-supervised NMF* (see Chapter 1 for typology of different source separation paradigms).

The model formulation in this case is very similar to the previous case where basis vectors are trained for all sources in advance. For each source $j \in \{1, \dots, J\}$ that is known in advance and for which isolated training material is available, we will estimate a set $\mathcal{K}_j$ of basis vectors to represent this source. Again, we will express this as a basis matrix $\mathbf{B}_j$. The rest of the sources in the mixture (those for which we have not trained the basis vectors) are modeled with a set $\mathcal{K}_{J+1}$ of basis vectors (also expressed with a basis matrix $\mathbf{B}_{J+1}$ that is estimated from the mixture).

Let us denote the spectrogram of each source $j \in \{1, \dots, J\}$ within the mixture as $\widehat{\mathbf{V}}_j$, and the spectrogram of the unknown sources as $\widehat{\mathbf{V}}_{J+1}$. The spectrogram of each source is modeled with the NMF model $\widehat{\mathbf{V}}_j \approx \mathbf{B}_j \mathbf{H}_j, j \in \{1, \dots, J+1\}$. The model for the mixture is then

$$\widehat{\mathbf{V}} \approx \sum_{j=1}^{J+1} \mathbf{B}_j \mathbf{H}_j = \mathbf{B}\mathbf{H}. \tag{8.5}$$

Similarly to (8.4), matrices $\mathbf{B}$ and $\mathbf{H}$ consist of submatrices, each representing one source.

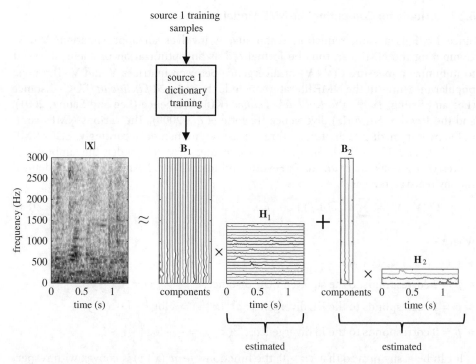

Figure 8.4 Illustration of the model where a basis vector matrix $\mathbf{B}_1$ is obtained for the target source at the training stage and kept fixed, and a basis vector matrix $\mathbf{B}_2$ that represents other sources in the mixture is estimated from the mixture. The two activation matrices $\mathbf{H}_1$ and $\mathbf{H}_2$ are both estimated from the mixture.

Similarly to the previous model, we also estimate activations for all the sources. The part of the basis matrix $\mathbf{B}$ corresponding to sources $j \in \{1, \dots, J\}$ is also kept fixed. However, the basis matrix $\mathbf{B}_{J+1}$ that represents the unknown sources is estimated from the mixture, i.e. the basis vectors representing the unknown sources are not fixed but estimated from the mixture. This model is illustrated in Figure 8.4. In order to ensure that the basis vectors estimated from the mixture do not model the target sources, their number must be rather low and their activations must be regularized, as will be discussed in Section 8.4.1. Note also that in addition to the two types of basis vectors discussed above (those that are estimated at the training stage and kept fixed, and those that are estimated from the test mixture without prior information), there are scenarios where some basis vectors are estimated in advance, but are adapted to the mixture signal, in order to compensate differences between training and test data.

8.2 NMF Theory and Algorithms

In this section, we first present the standard criteria for computing the NMF model parameters (Section 8.2.1), then we introduce probabilistic frameworks for NMF (Section 8.2.2), and we describe several algorithms designed for computing an NMF (Section 8.2.3).

8.2.1 Criteria for Computing the NMF Model Parameters

Since NMF is a rank reduction technique, it involves an approximation: $\mathbf{V} \approx \hat{\mathbf{V}}$. Computing the NMF can thus be formalized as an optimization problem: we want to minimize a measure $C(\hat{\mathbf{V}} \mid \mathbf{V})$ of divergence between matrices $\hat{\mathbf{V}}$ and $\mathbf{V}$. The most popular measures in the NMF literature include the squared *Euclidean* (EUC) distance (Lee and Seung, 1999), the *Kullback–Leibler* (KL) divergence (Lee and Seung, 2001), and the *Itakura–Saito* (IS) divergence (Févotte *et al.*, 2009). The various NMFs computed by minimizing each of these three measures are named accordingly: EUC-NMF, KL-NMF, and IS-NMF. Actually, these three measures fall under the umbrella of *β-divergences* (Nakano *et al.*, 2010; Févotte and Idier, 2011). Formally, they are defined for any real-valued β as

$$C^{\beta}(\hat{\mathbf{V}} \mid \mathbf{V}) = \sum_{nf} d^{\beta}(\hat{v}(n,f) \mid v(n,f)), \tag{8.6}$$

where

- $\forall \beta \notin \{0, 1\}$, $d^{\beta}(x \mid y) = \frac{1}{\beta(\beta-1)}(x^{\beta} + (\beta - 1)y^{\beta} - \beta x y^{\beta-1})$;
- $\beta = 2$ corresponds to the squared EUC distance: $d^{\text{EUC}}(x \mid y) = \frac{1}{2}|x - y|^2$;
- $\beta = 1$ corresponds to the KL divergence: $d^{\text{KL}}(x \mid y) = x \log\left(\frac{x}{y}\right) - x + y$;
- $\beta = 0$ corresponds to the IS divergence: $d^{\text{IS}}(x \mid y) = \frac{x}{y} - \log\left(\frac{x}{y}\right) - 1$.

It can be easily proved that $\forall x > 0$, the function $y \mapsto d^{\beta}(x \mid y)$ is convex with respect to y if and only if $\beta \in [1, 2]$ (Févotte and Idier, 2011). This means that minimizing $C^{\beta}(\hat{\mathbf{V}} \mid \mathbf{BH})$ with respect to $\mathbf{H}$ with $\mathbf{B}$ fixed, or conversely with respect to $\mathbf{B}$ with $\mathbf{H}$ fixed, is a convex optimization problem if and only if $\beta \in [1, 2]$. This convexity property is particularly convenient in a pretrained framework, where matrix $\mathbf{B}$ is fixed and only matrix $\mathbf{H}$ is estimated from the observed data (cf. Section 8.1.3). Optimization algorithms are then insensitive to initialization, which might explain the better success of KL-NMF and EUC-NMF compared with IS-NMF in the NMF literature.

However, in the context of learning-free separation (cf. Section 8.1.2), whatever the value of β, minimizing $C^{\beta}(\hat{\mathbf{V}} \mid \mathbf{BH})$ jointly with respect to $\mathbf{B}$ and $\mathbf{H}$ is not a convex optimization problem. Indeed, this factorization is not unique, since we also have $\mathbf{V} = \mathbf{B'H'}$ with $\mathbf{B'} = \mathbf{BA\Pi}$ and $\mathbf{H'} = \mathbf{\Pi}^T \mathbf{\Lambda}^{-1} \mathbf{H}$, where $\mathbf{\Lambda}$ can be any $K \times K$ diagonal matrix with positive diagonal entries, and $\mathbf{\Pi}$ can be any $K \times K$ permutation matrix. Note that the nonuniqueness of the model is actually ubiquitous in source separation and is generally not considered as a problem: sources are recovered up to a scale factor and a permutation. In the case of NMF, however, other kinds of indeterminacies may also exist (Laurberg *et al.*, 2008). Due to the existence of local minima, optimization algorithms become sensitive to initialization (cf. Section 8.2.3). In practice, with a random initialization there is no longer any guarantee to converge to a solution that is helpful for source separation. For this reason, advanced NMF models have been proposed to enforce some specific desired properties in the decomposition (cf. Section 8.4).

8.2.2 Probabilistic Frameworks for NMF

Computing an NMF can also be formalized as a parametric estimation problem based on a probabilistic model that involves both observed and latent (hidden) random variables. Typically, observed variables are related to matrix $\hat{\mathbf{V}}$, whereas latent variables are related to matrices $\mathbf{B}$ and $\mathbf{H}$. The main advantages of using a probabilistic framework are the facility of exploiting some a priori knowledge that we may have about matrices $\mathbf{B}$ and $\mathbf{H}$, and the existence of well-known statistical inference techniques, such as the *expectation-maximization* (EM) algorithm.

Popular probabilistic models of nonnegative time-frequency representations include Gaussian models that are equivalent to EUC-NMF (Schmidt and Laurberg, 2008) (Section 8.2.2.1), and count models, such as the celebrated *probabilistic latent component analysis* (PLCA) (Shashanka *et al.*, 2008) (Section 8.2.2.2) and the Poisson NMF model based on the Poisson distribution (Virtanen *et al.*, 2008) (Section 8.2.2.3), which are both related to KL-NMF. However these probabilistic models do not account for the fact that matrix $\hat{\mathbf{V}}$ has been generated from a time-domain signal $x(t)$. As a result, they can be used to estimate a nonnegative time-frequency representation, but they are not able to account for the phase that is necessary to reconstruct a time-domain signal.

Other probabilistic frameworks focus on power or magnitude spectrograms, and intend to directly model the STFT $\mathbf{X}$ instead of the nonnegative time-frequency representation $\hat{\mathbf{V}}$, in order to permit the resynthesis of time-domain signals. The main advantage of this approach is the ability to account for the phase and, in source separation applications, to provide a theoretical ground for using time-frequency masking techniques. Such models include Gaussian models that are equivalent to IS-NMF (Févotte *et al.*, 2009) (Section 8.2.2.4) and the Cauchy NMF model based on the Cauchy distribution (Liutkus *et al.*, 2015), which both fall under the umbrella of α-stable models (Liutkus and Badeau, 2015) (Section 8.2.2.5).

8.2.2.1 Gaussian Noise Model

A simple probabilistic model for EUC-NMF was presented by Schmidt and Laurberg (2008): $\hat{\mathbf{V}} = \mathbf{BH} + \mathbf{U}$, where matrices $\mathbf{B}$ and $\mathbf{H}$ are seen as deterministic parameters, and the entries of matrix $\mathbf{U}$ are Gaussian, independent and identically distributed (i.i.d.): $u(n,f) \sim \mathcal{N}(u(n,f) \mid 0, \sigma_u^2)$. Then the log-likelihood of matrix $\hat{\mathbf{V}}$ is $\log p(\hat{\mathbf{V}} \mid \mathbf{B}, \mathbf{H}) = -\frac{1}{2\sigma_u^2}\|\hat{\mathbf{V}} - \mathbf{BH}\|_F^2 + \text{cst} = -\frac{1}{\sigma_u^2}C^2(\hat{\mathbf{V}} \mid \mathbf{V}) + \text{cst}$ (where cst denotes a constant additive term), as defined in (8.6) with $\beta = 2$. Therefore the maximum likelihood (ML) estimation of $(\mathbf{B}, \mathbf{H})$ is equivalent to EUC-NMF. The main drawback of this generative model is that it does not enforce the nonnegativity of $\hat{\mathbf{V}}$, whose entries might take negative values.

8.2.2.2 Probabilistic Latent Component Analysis

PLCA (Shashanka *et al.*, 2008) is a count model that views matrix $\mathbf{V}$ as a probability distribution (normalized so that $\sum_{nf} v(n,f) = 1$). The observation model is the following one: the probability distribution $P(n,f) = v(n,f)$ is sampled M times to produce M independent time-frequency pairs (n_m, f_m), $m \in \{1, \dots, M\}$. Then matrix $\hat{\mathbf{V}}$ is generated as a histogram: $\hat{v}(n,f) = \frac{1}{M}\sum_m \delta_{(n_m, f_m)}(n,f)$, which also satisfies $\sum_{nf} \hat{v}(n,f) = 1$. The connection with NMF is established by introducing a latent variable k that is also sampled M times to produce k_m, $m \in \{1, \dots, M\}$. More precisely, it is assumed that (k_m, n_m) are first sampled together according to distribution $P(k, n) = h_k(n)$, and that f_m is then sampled given k_m according to distribution $P(f \mid k) = b_k(f)$, resulting in the joint distribution $P(n, f, k) = P(k, n)P(f \mid k) = v_k(n, f)$. Then $P(n, f)$ is the marginal distribution resulting from the joint distribution $P(n,f,k)$: $P(n,f) = \sum_k P(n,f,k) = v(n,f)$. Finally, note that another convenient formulation of PLCA, which is used in Chapter 9, is to simply state that $\hat{\mathbf{v}}(n) \sim \mathcal{M}(\hat{\mathbf{v}}(n) \mid \|\hat{\mathbf{v}}(n)\|_1, \mathbf{Bh}(n))$ where $\hat{\mathbf{v}}(n)$ and $\mathbf{h}(n)$ are the nth columns of matrices $\hat{\mathbf{V}}$ and $\mathbf{H}$, respectively, $\mathcal{M}$ denotes the multinomial distribution, and $\mathbf{h}(n)$ and the columns of matrix $\mathbf{B}$ are vectors that sum to 1.

In Section 8.2.3, it will be shown that this probabilistic model is closely related to KL-NMF. Indeed, the update rules obtained by applying the EM algorithm are formally equivalent to KL-NMF multiplicative update rules (cf. Section 8.2.3.3).

8.2.2.3 Poisson NMF Model

The Poisson NMF model (Virtanen *et al.*, 2008) is another count model that assumes that the observed nonnegative matrix $\hat{\mathbf{V}}$ is generated as the sum of K independent, nonnegative latent components $\hat{\mathbf{V}}_k$. The entries $\hat{v}_k(n,f)$ of matrix $\hat{\mathbf{V}}_k$ are assumed independent and *Poisson*-distributed: $\hat{v}_k(n,f) \sim \mathcal{P}(\hat{v}_k(n,f) \mid v_k(n,f))$. The Poisson distribution is defined for any positive integer v as $\mathcal{P}(v \mid \lambda) = \frac{e^{-\lambda}\lambda^v}{v!}$, where λ is the intensity parameter and $v!$ is the factorial of v. A nice feature of the Poisson distribution is that the sum of K independent Poisson random variables with intensity parameters λ_k is a Poisson random variable with intensity parameter $\lambda = \sum_k \lambda_k$. Consequently, $\hat{v}(n,f) \sim \mathcal{P}(\hat{v}(n,f) \mid v(n,f))$. The NMF model $\mathbf{V} = \mathbf{BH}$ can thus be computed by maximizing $P(\hat{\mathbf{V}} \mid \mathbf{B}, \mathbf{H}) = \prod_{nf} \mathcal{P}(\hat{v}(n,f) \mid v(n,f))$. It can be seen that $\log P(\hat{\mathbf{V}} \mid \mathbf{B}, \mathbf{H}) = -C^1(\hat{\mathbf{V}} \mid \mathbf{V})$, as defined in (8.6) with $\beta = 1$. Therefore ML estimation of $(\mathbf{B}, \mathbf{H})$ is equivalent to KL-NMF.

8.2.2.4 Gaussian Composite Model

The Gaussian *composite model* introduced by Févotte *et al.* (2009) exploits a feature of the Gaussian distribution that is similar to that of the Poisson distribution: a sum of K independent Gaussian random variables of means μ_k and variances σ_k^2 is a Gaussian random variable of mean $\mu = \sum_k \mu_k$ and variance $\sigma^2 = \sum_k \sigma_k^2$. The main difference with the Poisson NMF model is that, instead of modeling the non-negative time-frequency representation $\hat{\mathbf{V}}$, IS-NMF aims to model the complex STFT $\mathbf{X}$, such that $\hat{\mathbf{V}} = |\mathbf{X}|^2$. The observed complex matrix $\mathbf{X}$ is thus generated as the sum of K independent complex latent components $\mathbf{X}_k$ (cf. Figure 8.5). The entries of matrix $\mathbf{X}_k$ are assumed independent and complex Gaussian distributed: $x_k(n,f) \sim \mathcal{N}_c(x_k(n,f) \mid 0, v_k(n,f))$. Here the complex Gaussian distribution is defined as $\mathcal{N}_c(x \mid \mu, \sigma^2) = \frac{1}{\pi\sigma^2} \exp(-\frac{|x-\mu|^2}{\sigma^2})$, where μ and σ^2 are the mean and variance parameters. Consequently, $x(n,f) = \sum_k x_k(n,f) \sim \mathcal{N}_c(x(n,f) \mid 0, v(n,f))$. The NMF model $\mathbf{V} = \mathbf{BH}$ can thus be computed by maximizing $p(\mathbf{X} \mid \mathbf{B}, \mathbf{H}) = \prod_{nf} \mathcal{N}_c(x(n,f) \mid 0, v(n,f))$. It can be seen that $\log p(\mathbf{X} \mid \mathbf{B}, \mathbf{H}) = -C^0(\hat{\mathbf{V}} \mid \mathbf{V})$, as defined in (8.6) with $\beta = 0$. Therefore ML estimation of $(\mathbf{B}, \mathbf{H})$ is equivalent to IS-NMF.

In a source separation application, the main practical advantage of this Gaussian composite model is that the minimum mean square error (MMSE) estimates of the sources are obtained by time-frequency masking, in a way that is closely related to Wiener filtering. Indeed, following the developments in Chapter 5, suppose now that the observed signal $x(t)$ is the sum of J unknown source signals $s_j(t)$, so that $x(n,f) = \sum_j s_j(n,f)$, and that each source follows an IS-NMF model: $s_j(n,f) \sim \mathcal{N}_c(s_j(n,f) \mid 0, v_j(n,f))$ where

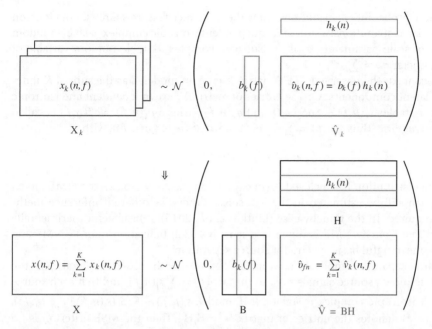

Figure 8.5 Gaussian composite model (IS-NMF) by Févotte *et al.* (2009).

$v_j(n,f) = \sigma_{s_j}^2(n,f)$ denotes the entries of matrix $\mathbf{V}_j = \mathbf{B}_j\mathbf{H}_j$. Then the minimum of the mean square error (MSE) criterion $\sum_{nf}\mathbb{E}\{|s_j(n,f) - \hat{s}_j(n,f)|^2 \mid x(n,f)\}$ is reached when

$$\forall n,f, \quad \hat{s}_j(n,f) = \mathbb{E}\{s_j(n,f) \mid x(n,f)\} = w_j(n,f)x(n,f), \tag{8.7}$$

where the time-frequency mask $w_j(n,f)$ is defined as

$$w_j(n,f) = \frac{v_j(n,f)}{\sum_{j'} v_{j'}(n,f)}. \tag{8.8}$$

8.2.2.5 α-Stable NMF Models

Despite its nice features, the Gaussian model introduced in the previous section presents two drawbacks. First, it amounts to assuming the additivity of the source power spectrograms, whereas several experimental studies have shown that the additivity of magnitude spectrograms is a better fit (see Liutkus and Badeau (2015) and references therein). Second, the IS divergence is not convex, which leads to increased optimization issues due to the existence of local minima. In order to circumvent these problems, a generalization of this model was introduced by Liutkus and Badeau (2015) based on isotropic complex α-*stable* distributions denoted $S\alpha S_c$, which stands for complex symmetric α-stable. This is a family of heavy-tailed probability distributions defined for any $\alpha \in [0, 2]$ that do not have a closed-form expression, except in the particular cases $\alpha = 2$, which corresponds to the complex Gaussian distribution, and $\alpha = 1$, which corresponds to the isotropic complex *Cauchy* distribution. In the general case, the distribution is defined by its characteristic function: $x \sim S\alpha S_c(x \mid \sigma) \Leftrightarrow \phi_z(\theta) = \mathbb{E}\{e^{j\Re(\theta^* x)}\} = e^{-\sigma^\alpha|\theta|^\alpha}$ for any complex-valued θ, where $\sigma > 0$ is the scale parameter (which corresponds to the standard deviation in the Gaussian case). These probability distributions enjoy the same nice feature shared by the Poisson and Gaussian distributions: a sum of K independent isotropic complex α-stable random variables of scale parameters σ_k is an isotropic complex α-stable random variable of scale parameter $\sigma^\alpha = \sum_k \sigma_k^\alpha$.

In this context, the observed STFT matrix $\mathbf{X}$ is again modeled as the sum of K independent latent components $\mathbf{X}_k$. The entries of matrix $\mathbf{X}_k$ are independent and isotropic complex α-stable: $x_k(n,f) \sim S\alpha S_c(x_k(n,f) \mid \sigma_k(n,f))$, where $v_k(n,f) = \sigma_k^\alpha(n,f)$ is called an α-spectrogram. Thus $x(n,f) = \sum_k x_k(n,f) \sim S\alpha S_c(x(n,f) \mid \sigma(n,f))$, with

$$v(n,f) = \sigma^\alpha(n,f) = \sum_k \sigma_k^\alpha(n,f) = \sum_k v_k(n,f). \tag{8.9}$$

When the distribution has a closed-form expression (i.e., $\alpha = 1$ or 2), the NMF model $\mathbf{V} = \mathbf{BH}$ can still be estimated in the ML sense, otherwise different inference methods are required. In the Cauchy case (Liutkus *et al.*, 2015), it has been experimentally observed that Cauchy NMF is much less sensitive to initialization than IS-NMF and produces meaningful basis spectra for source separation.

In a source separation application, we again suppose that the observed signal $x(t)$ is the sum of J unknown source signals $s_j(t)$, so that $x(n,f) = \sum_j s_j(n,f)$, and that each source follows an isotropic complex α-stable NMF model: $s_j(n,f) \sim S\alpha S_c(s_j(n,f) \mid v_j^{1/\alpha}(n,f))$, where $v_j(n,f)$ denotes the entries of matrix $\mathbf{V}_j = \mathbf{B}_j\mathbf{H}_j$. Then the MSE criterion is no longer defined for all $\alpha \in (0, 2]$, but in any case, the posterior mean $\hat{s}_j(n,f) = \mathbb{E}\{s_j(n,f) \mid x(n,f)\}$ is still well-defined, and admits the same expression as in (8.7) and (8.8).

8.2.2.6 Choosing a Particular NMF Model

When choosing a particular NMF model for a given source separation application, several criteria may be considered, including the following ones:

Robustness to initialization: Cauchy NMF has proved to be more robust to initialization than all other probabilistic NMF models. In addition, in the context of pretrained source separation, the Gaussian noise model (related to EUC-NMF) and the PLCA/Poisson NMF models (related to KL-NMF) lead to a convex optimization problem with a unique minimum, which is not the case of the Gaussian composite model (related to IS-NMF).

Source reconstruction: Only the α-stable NMF models, including IS-NMF and Cauchy NMF, provide a theoretical ground for using Wiener filtering in order to reconstruct time-domain signals.

Existence of closed-form update rules: ML estimation of the model parameters is tractable for all NMF models except α-stable models with $\alpha \neq 1, 2$.

8.2.3 Algorithms for NMF

In the literature, various algorithms have been designed for computing an NMF, including the famous multiplicative update rules (Lee and Seung, 2001), the alternated least squares method (Finesso and Spreij, 2004), and the projected gradient method (Lin, 2007). In Section 8.2.3.1 we present the multiplicative update rules that form the most celebrated NMF algorithm and we summarize their convergence properties. Then in the following sections we present some algorithms dedicated to the probabilistic frameworks introduced in Section 8.2.2.

8.2.3.1 Multiplicative Update Rules

The basic idea of *multiplicative update* rules is that the nonnegativity constraint can be easily enforced by updating the previous values of the model parameters by multiplication with a nonnegative scale factor. A heuristic way of deriving these updates consists in decomposing the gradient of the cost function $C(\hat{\mathbf{V}} \mid \mathbf{V})$, e.g. the β-divergence introduced in (8.6), as the difference of two nonnegative terms: $\nabla_{\mathbf{B}} C(\hat{\mathbf{V}} \mid \mathbf{V}) = \nabla_{\mathbf{B}}^+ C(\hat{\mathbf{V}} \mid \mathbf{V}) - \nabla_{\mathbf{B}}^- C(\hat{\mathbf{V}} \mid \mathbf{V})$, where $\nabla_{\mathbf{B}}^+ C(\hat{\mathbf{V}} \mid \mathbf{V}) \geq 0$ and $\nabla_{\mathbf{B}}^- C(\hat{\mathbf{V}} \mid \mathbf{V}) \geq 0$, meaning that all the entries of these two matrices are nonnegative. Then matrix $\mathbf{B}$ can be updated as $\mathbf{B} \leftarrow \mathbf{B} \circ (\nabla_{\mathbf{B}}^- C(\hat{\mathbf{V}} \mid \mathbf{V}) / \nabla_{\mathbf{B}}^+ C(\hat{\mathbf{V}} \mid \mathbf{V}))^{\eta}$, where $\circ$ denotes elementwise matrix product, $/$ denotes elementwise matrix division, the matrix exponentiation must be understood elementwise, and $\eta > 0$ is a stepsize similar to that involved in a gradient descent (Badeau *et al.*, 2010). The same update can be derived for matrix $\mathbf{H}$, and then matrices $\mathbf{B}$ and $\mathbf{H}$ can be updated in turn, until convergence.[1] Note that the decomposition of the gradient as a difference of two nonnegative terms is not unique, and different choices can be made, leading to different multiplicative update rules. In the case of the β-divergence, the standard multiplicative update rules are expressed as follows (Févotte and Idier, 2011):

$$\mathbf{B} \leftarrow \mathbf{B} \circ \left(\frac{(\hat{\mathbf{V}} \circ (\mathbf{BH})^{\beta-2}) \mathbf{H}^T}{(\mathbf{BH})^{\beta-1} \mathbf{H}^T} \right)^{\eta} \tag{8.10}$$

$$\mathbf{H} \leftarrow \mathbf{H} \circ \left(\frac{\mathbf{B}^T (\hat{\mathbf{V}} \circ (\mathbf{BH})^{\beta-2})}{\mathbf{B}^T (\mathbf{BH})^{\beta-1}} \right)^{\eta} \tag{8.11}$$

where matrix division and exponentiation must be understood elementwise. By using the auxiliary function approach, Nakano *et al.* (2010) proved that the cost function $C^{\beta}(\hat{\mathbf{V}} \mid \mathbf{V})$ is nonincreasing under these updates when the stepsize η is given by $\eta = \frac{1}{2-\beta}$ for $\beta < 1$, $\eta = 1$ for $1 \leq \beta \leq 2$, and $\eta = \frac{1}{\beta-1}$ for $\beta > 2$. In addition, Févotte and Idier (2011) proved that the same cost function is nonincreasing under (8.10)–(8.11) for $\eta = 1$ and for all $\beta \in [0, 2]$ (which includes the popular EUC, KL and IS-NMF). They also proved that these updates correspond to a majorization-minimization (MM) algorithm when the stepsize η is expressed as a given function of β, which is equal to 1 for all $\beta \in [1, 2]$, and that they correspond to a majorization-equalization algorithm for $\eta = 1$ and $\beta = 0$. However, contrary to a widespread belief (Lee and Seung, 2001), the decrease in the cost function is not sufficient to prove the convergence of the algorithm to a local or global minimum. Badeau *et al.* (2010) analyzed the convergence of multiplicative update rules by means of Lyapunov's stability theory. In particular, it was proved that:

- there is $\eta^{\max} > 0$ such that these rules are exponentially or asymptotically stable for all $\eta \in (0, \eta^{\max})$; moreover, $\forall \beta$, the upper bound $\eta^{\max}$ is such that $\eta^{\max} \in (0, 2]$, and if $\beta \in [1, 2]$, $\eta^{\max} = 2$;
- these rules are unstable if $\eta \notin [0, 2]$, $\forall \beta$.

In practice, the step size η permits us to control the convergence rate of the algorithm.

Note that, due to the nonuniqueness of NMF, there is a scaling and permutation ambiguity between matrices $\mathbf{B}$ and $\mathbf{H}$ (cf. Section 8.2.1). Therefore, when $\mathbf{B}$ and $\mathbf{H}$ are to be

---

1 This iterative algorithm can stop, for example, when the decrease in the β-divergence, or the distance between the successive iterates of matrices $\mathbf{B}$ and $\mathbf{H}$, goes below a given threshold.

updated in turn, numerical stability can be improved by renormalizing the columns of **B** (resp. the rows of **H**), and scaling the rows of **H** (resp. the columns of **B**) accordingly, so as to keep the product **BH** unchanged.

Finally, a well-known drawback of most NMF algorithms is the sensitivity to initialization, that is due to the multiplicity of local minima of the cost function (cf. Section 8.2.1). Many initialization strategies were thus proposed in the literature (Cichocki *et al.*, 2009). In the case of IS-NMF multiplicative update rules, a *tempering* approach was proposed by Bertin *et al.* (2009). The basic idea is the following: since the β-divergence is convex for all $\beta \in [1, 2]$, but not for $\beta = 0$, the number of local minima is expected to increase when β goes from 2 to 0. Therefore a simple solution for improving the robustness to initialization consists of making parameter β vary from 2 to 0 over the iterations of the algorithm. Nevertheless, the best way of improving the robustness to initialization in general is to select a robust NMF criterion, such as that involved in Cauchy NMF (cf. Section 8.2.2.5).

8.2.3.2 The EM Algorithm and its Variants

As mentioned in Section 8.2.2, one advantage of using a probabilistic framework for NMF is the availability of classical inference techniques, whose convergence properties are well known. Classical algorithms used in the NMF literature include the EM algorithm (Shashanka *et al.*, 2008), the space-alternating generalized EM algorithm (Févotte *et al.*, 2009), variational Bayesian (VB) inference (Badeau and Drémeau, 2013), and *Markov chain Monte Carlo* methods (Simsekli and Cemgil, 2012a).

Below, we introduce the basic principles of the space-alternating generalized EM algorithm (Fessler and Hero, 1994), which includes the regular EM algorithm as a particular case. We then apply the EM algorithm to the PLCA framework described in Section 8.2.2.2, and the space-alternating generalized EM algorithm to the Gaussian composite model described in Section 8.2.2.4.

Consider a random observed dataset $\mathcal{X}$, whose probability distribution is parameterized by a parameter set θ, which is partitioned as $\theta = \{\theta_k\}_k$. The space-alternating generalized EM algorithm aims to estimate parameters θ_k iteratively, while guaranteeing that the likelihood $p(\mathcal{X} \mid \theta)$ is nondecreasing over the iterations. It requires choosing for each subset θ_k a *hidden data* space which is complete for this particular subset, i.e. a latent dataset $\mathcal{X}_k$ such that $p(\mathcal{X}, \mathcal{X}_k \mid \theta) = p(\mathcal{X} \mid \mathcal{X}_k, \{\theta_{k'}\}_{k' \neq k})p(\mathcal{X}_k \mid \theta)$. The algorithm iterates over both the iteration index and over k. For each iteration and each k, it is composed of an expectation step (*E-step*) and a maximization step (*M-step*):

- E-step: evaluate $Q_k(\theta_k) = \mathbb{E}\{\log p(\mathcal{X}_k \mid \theta_k, \{\theta_{k'}\}_{k' \neq k}) \mid \mathcal{X}, \theta\}$;
- M-step: compute $\theta_k = \mathrm{argmax}_{\theta_k} Q_k(\theta_k)$.

The regular EM algorithm corresponds to the particular case $K = 1$, where $\mathcal{X}$ is a deterministic function of the complete data space.

8.2.3.3 Application of the EM Algorithm to PLCA

Shashanka *et al.* (2008) applied the EM algorithm to the PLCA model described in Section 8.2.2.2. The observed dataset is $\mathcal{X} = \{n_m, f_m\}_m$, the parameter set is $\theta = \{\mathbf{B}, \mathbf{H}\}$, and the complete data space is $\{n_m, f_m, k_m\}_m$. Then:

- The E-step consists in computing $P(k \mid n, f) = \frac{P(n,f,k)}{\sum_{k'} P(n,f,k')} = \frac{v_k(n,f)}{v(n,f)}$, that appears in the expression $Q(\theta) = \sum_{nf} \hat{v}(n,f) \sum_k P(k \mid n,f) \log(b_k(f)h_k(n))$.

- The M-step consists in maximizing $Q(\theta)$ with respect to $b_k(f)$ and $h_k(n)$, subject to $\forall k$, $\sum_f b_k(f) = 1$ and $\sum_{kn} h_k(n) = 1$. Given that $\sum_{nf} \hat{v}(n,f) = 1$, we get:

$$h_k(n) \leftarrow \frac{\sum_f \hat{v}(n,f) P(k \mid n,f)}{\sum_{k'n'f} \hat{v}(n',f) P(k' \mid n',f)} = h_k(n) \sum_f b_k(f) \frac{\hat{v}(n,f)}{v(n,f)}, \tag{8.12}$$

$$b_k(f) \leftarrow \frac{\sum_n \hat{v}(n,f) P(k \mid n,f)}{\sum_{nf'} \hat{v}(n,f') P(k \mid n,f')} = \frac{\tilde{b}_k(f)}{\sum_{f'} \tilde{b}_k(f')}, \tag{8.13}$$

where $\tilde{b}_k(f) = b_k(f) \sum_n h_k(n) \frac{\hat{v}(n,f)}{v(n,f)}$.

It is easy to check that this algorithm is identical to the multiplicative update rules for KL divergence, as described in (8.10)–(8.11) with $\eta = \beta = 1$, up to a scaling factor in $\mathbf{H}$ due to the normalization of matrix $\hat{\mathbf{V}}$ (Shashanka *et al.*, 2008).

8.2.3.4 Application of the Space-Alternating Generalized EM Algorithm to the Gaussian Composite Model

Févotte *et al.* (2009) applied the *space-alternating generalized EM* algorithm to the Gaussian composite model described in Section 8.2.2.4. The observed dataset is $\mathcal{X} = \mathbf{X}$, the kth parameter set is $\theta_k = \{\mathbf{b}_k, \mathbf{h}_k\}$, and the kth complete latent dataset is $\mathcal{X}_k = \mathbf{X}_k$. Then:

- The E-step consists of computing $\hat{\mathbf{V}}_k = \frac{\mathbf{V}_k^2}{\mathbf{V}^2} \circ \hat{\mathbf{V}} + \frac{\mathbf{V}_k \circ (\mathbf{V} - \mathbf{V}_k)}{\mathbf{V}}$, which appears in the expression $Q_k(\theta_k) = -C^0(\hat{\mathbf{V}}_k \mid \mathbf{V}_k)$, where criterion C^0 was defined in (8.6) with $\beta = 0$.
- The M-step computes $h_k(n) \leftarrow \frac{1}{F} \sum_f \frac{\hat{v}_k(n,f)}{b_k(f)}$ and $b_k(f) \leftarrow \frac{1}{N} \sum_n \frac{\hat{v}_k(n,f)}{h_k(n)}$.

Note that it has been experimentally observed by Févotte *et al.* (2009) that this space-alternating generalized EM algorithm converges more slowly than the IS-NMF multiplicative update rules described in (8.10)–(8.11) for $\eta = 1$ and $\beta = 0$.

8.3 NMF Dictionary Learning Methods

In NMF-based source separation, the general requirement for basis vectors is that the basis vectors of source j represent well signals originating from the source, and do not represent the other sources. When pretrained basis vectors are used, there are several alternative methods to learn the basis vectors that represent each source, using isolated material from each source. These methods also include other models in addition to the NMF itself. Learning the basis vectors is essentially similar to the problem of *dictionary learning*, which exists also in many other research fields. The key idea in NMF is that the atoms that the dictionary consists of are constrained to be elementwise nonnegative, and also the activations of the components are nonnegative.

In addition to the above general requirement for an NMF-based dictionary to result in good separation, there can be other factors affecting the choice of the dictionary. For example, in practical scenarios, the size of the dictionary, i.e. the number of basis vectors, is constrained by the memory or computational complexity requirements. Note that these requirements can be very different in different applications. For example, in real-time implementations we may want to resort to a small dictionary (tens of basis vectors), but in offline computing with graphical processing units dictionary sizes can be thousands of basis vectors. In advanced NMF models, there might be additional requirements, e.g. the capability to couple other information with each basis vector or the capability to parameterize the basis vectors somehow, e.g. to allow their adaptation.

In this section we review the most common dictionary learning methods that have been used to learn NMF dictionaries.

8.3.1 NMF Dictionaries

The most obvious choice for dictionary learning is to use NMF itself (Schmidt and Olsson, 2006). If we denote the spectrogram that includes all the training data of source j as $\hat{\mathbf{V}}_j$, NMF can be used to factorize it as

$$\hat{\mathbf{V}}_j \approx \mathbf{B}_j \mathbf{H}_j. \tag{8.14}$$

where $\mathbf{H}_j$ is the activation matrix of source j.

NMF represents the training data as a weighted sum of basis vectors, and therefore atoms in an NMF-based dictionary are part-based, i.e. each atom represents only a part of a training instance. For example, in the case of a speech source, the parts can be different formants, and in the case of polyphonic music, the parts can be individual notes played by different instruments.

NMF-based dictionaries have the same limitation as regular NMF: one cannot increase the size of the dictionary arbitrarily. If the number of components equals the number of frequencies, one can represent the training data matrix perfectly ($\hat{\mathbf{V}}_j = \mathbf{B}_j \mathbf{H}_j$) by using the $F \times F$ identity matrix $\mathbf{B}_j = \mathbf{I}_F$ as the dictionary. However, this kind of dictionary is very generic, not characteristic of the target sources. It is capable of representing any other source in the mixture, and therefore does not lead to any separation. Therefore, the limitation of NMF-based dictionaries is that they are limited to small dictionaries. The separation of sources that have a complex structure (e.g., speech) benefits from larger dictionaries and other advanced dictionary learning methods. Basic NMF can be regularized with sparsity (cf. Section 8.4.1.1), which helps to learn larger dictionaries, at least to some degree (Schmidt and Olsson, 2006).

8.3.2 Exemplar-Based Dictionaries

If there is no constraint (e.g., memory or computational complexity requirement) on the size of the dictionary, using all the training instances as dictionary atoms gives an accurate representation of each source and leads to good separation (Smaragdis *et al.*, 2009). Prior to deep neural network (DNN)-based approaches, this method led to state-of-the-art results, e.g. in noise-robust automatic speech recognition (ASR), where NMF-based source separation was used as a preprocessing step (Raj *et al.*, 2010; Gemmeke *et al.*, 2011).

The argument for using these *exemplar*-based dictionaries is as follows. The distribution of the source data is often structurally complex. The NMF model $\sum_k \mathbf{b}_k \mathbf{h}_k^T$ with $\mathbf{h}_k \geq 0$ spans a convex polyhedral cone, which can become too loose, i.e. capable of also modeling other sources, if regular NMF is used to estimate the basis vectors $\mathbf{b}_k$. When the data points themselves are used as basis vectors, the basis vectors match better with the distribution of the source data, leading to better separation.

Naturally, in practice the size of dictionaries needs to be constrained. Previous studies have found that approximately 1000 exemplars were required to achieve the performance of proper dictionary learning methods when separating two speakers (Diment and Virtanen, 2015). Current studies use up to tens of thousands of exemplars (Baby *et al.*, 2015).

It should be noted that exemplar-based dictionaries do not decompose training data instances into parts, i.e. each exemplar corresponds to the whole spectrum of a signal from the training dataset. This is advantageous in source separation scenarios where information about sources within a specific frequency band is difficult to estimate since they heavily overlap within the band. In this case, the activations of the basis vectors can be more reliably estimated by using other frequency bands where sources do not overlap, which can then be used to predict the sources within the overlapping band. On the other hand, exemplar-based dictionaries are not optimal in the case of training data consisting of additive combinations of a large number of individual events (e.g., notes in polyphonic music). Such data are combinatorial, and capturing all the combinations is difficult with a small-sized dictionary.

8.3.3 Clustering-Based Dictionary

A simple approach to reduce the size of an exemplar-based dictionary is to cluster the data (Virtanen *et al.*, 2013), and to use cluster centers as basis vectors. This approach clearly improves the separation quality when small and medium dictionary sizes (up to hundreds to basis vectors) are used (Diment and Virtanen, 2015). There also exist clustering algorithms that use divergences similar to those used in NMF (Virtanen *et al.*, 2013). Similarly to exemplar-based dictionaries, clustering does not decompose the training data into parts, but rather finds basis vectors corresponding to full spectra.

8.3.4 Discriminative Dictionaries

The above dictionary learning algorithms do not explicitly prevent the learned dictionaries from representing sources other than the target. The first approaches for including such constraints in dictionary learning used a cross-coherence cost between the dictionaries of two different sources in order to avoid the learned dictionaries of the two sources being similar to each other (Grais and Erdogan, 2013). The above criterion was included in the NMF dictionary learning framework as a regularization term.

Weninger *et al.* (2014) and Sprechmann *et al.* (2014) learned *discriminative NMF* dictionaries so as to minimize the reconstruction error of the target source recovered from a set of training mixtures. Weninger *et al.* (2014) even took into account the source reconstruction by masking in the cost function. The resulting criterion is expected to be optimal for source separation purposes. The optimization algorithm related to this kind of dictionary learning is highly nonconvex, which requires careful tuning, in order to reach a proper optimization solution (cf. Section 8.2.3.1). Weninger *et al.* (2014) circumvented the problem by using two different dictionaries, one for finding the activations and the other for source reconstruction. Sprechmann *et al.* (2014) used a stochastic gradient algorithm for the optimization.

8.3.5   Dictionary Adaptation

In practical source separation scenarios it is difficult to obtain training material from the target source whose characteristics exactly match the actual usage scenario. For example, if the target signal is speech, we can have multiple speech datasets available to develop the methods, but the actual speech signals to be processed are from an acoustic environment different from those datasets. In this kind of scenario it is possible to pretrain the basis vectors and then to adapt them to the signal to be processed.

Virtanen and Cemgil (2009) used a gamma mixture model to model the distribution of the training data based on which each basis vector was assigned a gamma prior distribution. The adaptation could then be formulated in a maximum a posteriori (MAP) estimation framework. Grais and Erdogan (2011) used a similar probabilistic framework for the adaptation, where the gamma distribution for each basis vector was obtained deterministically by allowing some variance around each pretrained atom.

Jaureguiberry *et al.* (2011) used an equalization filter to adapt the pretrained basis vectors. A single filter was shared between all the basis vectors. Since filtering corresponds to elementwise multiplication in the frequency domain, it is only needed to estimate the magnitude response vector of the equalization filter, which is used to multiply the basis vectors elementwise. This model is similar to the source-filter model discussed earlier in this chapter. The estimation of the adaptation filter response can be done using principles similar to the other NMF model parameters.

8.3.6   Regularization in Learning Source Models from a Mixture

Dictionary adaptation is closely related to the model in Section 8.1.4 where some of the basis vectors are estimated from the training data and kept fixed, and the others are estimated from the mixture, in order to model the unknown sources it contains. In the simplest scenario we can have some prior information about the activity of each source, and parts of the signal that are known to contain only the target sources can be used to, for example, extract exemplars that are added to the dictionary (Weninger *et al.*, 2011; Gemmeke and Virtanen, 2010). For example, in speech processing there are some pauses in speech, and exemplars can be extracted during those short pauses.

However, in a more challenging situation, the source for which the basis vectors are to be learned always overlaps with other sources, or there is no way to obtain information about segments where the source is present and isolated. In this case, we can use NMF algorithms with specific regularizations to learn the basis vectors (cf. Section 8.4.1). Regularizations that have been used include, for example, imposing a heavy sparsity penalty to the noise atoms and using only a small number of noise atoms (Hurmalainen *et al.*, 2013), or constraining the number of iterations that are used to update the model parameters (Germain and Mysore, 2014).

8.4   Advanced NMF Models

The basic NMF model presented in Section 8.1 has proved successful for addressing a variety of audio source separation problems. Nevertheless, the source separation performance can still be improved by exploiting prior knowledge that we may have about the source signals. For instance, we know that musical notes and voiced sounds have a harmonic spectrum (or, more generally, an inharmonic or a sparse spectrum), and that both their spectral envelope and their temporal power profile have smooth variations. On the other hand, percussive sounds rather have a smooth spectrum and a sparse temporal power profile. It may thus be desirable to impose properties such as *harmonicity*, *smoothness*, and *sparsity* on either the spectral matrix $\mathbf{B}$ or the activation matrix $\mathbf{H}$ in the NMF $\mathbf{V} = \mathbf{BH}$. For that purpose, it is possible to apply either hard constraints, e.g. by parameterizing matrix $\mathbf{B}$ or $\mathbf{H}$, or soft constraints, e.g. by adding a regularization term to the criterion (8.6) or by introducing the prior distributions of $\mathbf{B}$ or $\mathbf{H}$ in the probabilistic frameworks introduced in Section 8.2.2 (Bayesian approach). Examples of such regularizations are described in Section 8.4.1. Note that another possible way of exploiting prior information is to use a pre-defined dictionary $\mathbf{B}$ trained on a training dataset. This idea was investigated in Section 8.3.

In other respects, audio signals are known to be nonstationary, therefore it is useful to consider that some characteristics such as the fundamental frequency or the spectral envelope may vary over time. Such nonstationary models will be presented in Section 8.4.2.

8.4.1 Regularizations

In this section, we present a few examples of NMF regularizations, including sparsity (Section 8.4.1.1), group-sparsity (Section 8.4.1.2), harmonicity and spectral smoothness (Section 8.4.1.3), and inharmonicity (Section 8.4.1.4).

8.4.1.1 Sparsity
Since NMF is well suited to the problem of separating audio signals formed of a few repeated audio events, it is often desirable to enforce the sparsity of matrix $\mathbf{H}$.

The most straightforward way of doing this is to add to the NMF criterion a sparsity-promoting regularization term. Ideally, sparsity is measured by the ℓ_0 norm, which counts the number of nonzero entries in a vector. However, optimizing a criterion involving the ℓ_0 norm raises intractable combinatorial issues. In the optimization literature, the ℓ_1 norm is often preferred because it is the tightest convex relaxation of the ℓ_0 norm. Therefore the criterion $C^\beta(\widehat{\mathbf{V}} \mid \mathbf{V})$ in (8.6) may be replaced with

$$C(\widehat{\mathbf{V}} \mid \mathbf{V}) = C^\beta(\widehat{\mathbf{V}} \mid \mathbf{V}) + \lambda \sum_k \|\mathbf{h}_k\|_1, \tag{8.15}$$

where $\lambda > 0$ is a tradeoff parameter to be tuned manually, as suggested, for example, by Hurmalainen *et al.* (2015).

However, if the NMF is embedded in a probabilistic framework such as those introduced in Section 8.2.2, sparsity is rather enforced by introducing an appropriate prior distribution of matrix $\mathbf{H}$. In this case, $\mathbf{H}$ is estimated by maximizing its posterior probability given $\widehat{\mathbf{V}}$, or equivalently the maximum a posteriori (MAP) criterion $\log p(\widehat{\mathbf{V}} \mid \mathbf{B}, \mathbf{H}) + \log p(\mathbf{H})$, instead of the log-likelihood $\log p(\widehat{\mathbf{V}} \mid \mathbf{B}, \mathbf{H})$. For instance, Kameoka *et al.* (2009) consider a generative model similar to the Gaussian noise model presented in Section 8.2.2.1, where the sparsity of matrix $\mathbf{H}$ is enforced by means of a generalized Gaussian prior:

$$p(\mathbf{H}) = \prod_{kn} \frac{1}{2\Gamma\left(1 + \frac{1}{p}\right)\sigma} e^{-\frac{|h_k(n)|^p}{\sigma^p}}, \tag{8.16}$$

where $\Gamma(\cdot)$ denotes the gamma function, parameter p promotes sparsity if $0 < p < 2$, and the case $p = 2$ corresponds to the standard Gaussian distribution.

In the PLCA framework described in Section 8.2.2.2, the entries of $\mathbf{H}$ are the discrete probabilities $P(k, n)$. By noticing that the entropy $\mathbb{H}\{\mathbf{H}\}$ of this discrete probability distribution is related to the sparsity of matrix $\mathbf{H}$ (the lower $\mathbb{H}\{\mathbf{H}\}$, the sparser $\mathbf{H}$), a suitable sparsity-promoting prior is the so-called *entropic prior* (Shashanka *et al.*, 2008), defined as $p(\mathbf{H}) \propto e^{-\beta \mathbb{H}\{\mathbf{H}\}}$, where $\beta > 0$.

8.4.1.2 Group Sparsity

Now suppose that the observed signal $x(t)$ is the sum of J unknown source signals $s_j(t)$ for $j \in \{1, \dots, J\}$, whose spectrograms $\hat{\mathbf{V}}_j$ are approximated as $\mathbf{V}_j = \mathbf{B}_j \mathbf{H}_j$, as in Section 8.2.2.4. Then the spectrogram $\hat{\mathbf{V}}$ of $x(t)$ is approximated with the NMF $\sum_j \mathbf{V}_j = \mathbf{BH}$, where $\mathbf{B} = [\mathbf{B}_1, \dots, \mathbf{B}_J]$ and $\mathbf{H} = [\mathbf{H}_1^T, \dots, \mathbf{H}_J^T]^T$. In this context, it is natural to expect that if a given source j is inactive at time n, then all the entries in the nth column of $\mathbf{H}_j$ are zero. Such a property can be enforced by using *group sparsity*. A well-known group sparsity regularization term is the mixed ℓ_2-ℓ_1 norm: $\|\mathbf{H}\|_{2,1} = \sum_{jn}\|\mathbf{h}_j(n)\|_2$ where $\mathbf{h}_j(n)$ is the nth column of matrix $\mathbf{H}_j$, as suggested, for example, by Hurmalainen *et al.* (2015). Indeed, the minimization of the criterion (8.15) involving this regularization term tends to enforce sparsity over both n and j, while ensuring that the whole vector $\mathbf{h}_j(n)$ gets close to zero for most values of n and j.

Lefevre *et al.* (2011) proposed a group sparsity prior for the IS-NMF probabilistic framework described in Section 8.2.2.4. The idea is to consider a prior distribution of matrix $\mathbf{H}$ such that all vectors $\mathbf{h}_j(n)$ are independent: $p(\mathbf{H}) = \prod_{jn} p(\mathbf{h}_j(n))$. Each $p(\mathbf{h}_j(n))$ is chosen so as to promote near-zero vectors. Then, as in Section 8.4.1.1, the NMF parameters are estimated in the MAP sense: $(\mathbf{B}, \mathbf{H}) = \text{argmax}_{\mathbf{B},\mathbf{H}} \log p(\mathbf{X} \mid \mathbf{B}, \mathbf{H}) + \sum_{jn} \log p(\mathbf{h}_j(n))$, where $p(\mathbf{X} \mid \mathbf{B}, \mathbf{H})$ was defined in Section 8.2.2.4.

8.4.1.3 Harmonicity and Spectral Smoothness

Contrary to sparsity, harmonicity in matrix $\mathbf{B}$ is generally enforced as a hard constraint, by using parametric models, whose parameter set includes the fundamental frequency. For instance, Vincent *et al.* (2010) and Bertin *et al.* (2010) parameterized the spectrum vector $\mathbf{b}_k$ as a nonnegative linear combination of M narrowband, harmonic spectral patterns (cf. Figure 8.6): $b_k(f) = \sum_m e_{km} g_{km}(f)$, where all spectral patterns $\{g_{km}(f)\}_m$ share the same fundamental frequency $v_k^0 > 0$, have smooth spectral envelopes and different spectral centroids, so as to form a filterbank-like decomposition of the whole spectrum, and $\{e_{km}\}_m$ are the nonnegative coefficients of this decomposition. In this way, it is guaranteed that $b_k(f)$ is a harmonic spectrum of fundamental frequency v_k^0, with a smooth spectral envelope. If the signal of interest is a music signal, then the order K and the fundamental frequencies v_k^0 can typically be preset according to the semitone scale; otherwise they have to be estimated along with the other parameters. Two methods were proposed for estimating the coefficients e_{km} and the activations in matrix $\mathbf{H}$

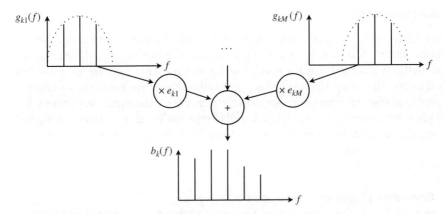

Figure 8.6 Harmonic NMF model by Vincent *et al.* (2010) and Bertin *et al.* (2010).

from the observed spectrogram: a space-alternating generalized EM algorithm based on a Gaussian model (Bertin *et al.*, 2010) (cf. Section 8.2.3.2) and multiplicative update rules (with a faster convergence speed) based either on the IS divergence (Bertin *et al.*, 2010), or more generally on the β-divergence (Vincent *et al.*, 2010).

Hennequin *et al.* (2010) proposed a similar parameterization of the spectrum vector $\mathbf{b}_k$, considering that harmonic spectra are formed of a number M of distinct partials:

$$b_k(f) = \sum_m a_k^m g_{km}(f), \tag{8.17}$$

where $a_k^m \geq 0$, $g_{km}(f) = g(v_f - v_k^m)$, $v_f = \frac{f}{F}f_s$ and $v_k^m = m\, v_k^0$, and $g(\cdot)$ is the spectrum of the analysis window used for computing the spectrogram. Multiplicative update rules based on the β-divergence were proposed for estimating this model. Since this parametric model does not explicitly enforce the smoothness of the spectral envelope, a regularization term promoting this smoothness was added to the β-divergence, resulting in a better decomposition of music spectrograms (Hennequin *et al.*, 2010).

8.4.1.4 Inharmonicity

When modeling some string musical instruments such as the piano or the guitar, the harmonicity assumption has to be relaxed. Indeed, because of the bending stiffness, the partial frequencies no longer follow an exact harmonic progression, but rather a so-called *inharmonic* progression:

$$v_k^m = m v_k^0 \sqrt{1 + Bm^2}, \tag{8.18}$$

where m is the partial index, $B > 0$ is the inharmonicity coefficient, and $v_k^0 > 0$ is the fundamental frequency of vibration of an ideal flexible string (Rigaud *et al.*, 2013). Then the spectrum vector $\mathbf{b}_k$ can be parameterized as in (8.17), and all parameters, including the inharmonicity coefficient B, can be estimated by minimizing the β-divergence criterion by means of multiplicative update rules. However, it was observed that the resulting algorithm is very sensitive to initialization (cf. Section 8.2.3.1). In order to improve the robustness to initialization, the exact parameterization of frequencies v_k^m in (8.18) was relaxed by considering these frequencies as free parameters and by adding the following regularization term to the β-divergence criterion: $\sum_{km} |v_k^m - m v_k^0 \sqrt{1 + Bm^2}|^2$.

8.4.2 Nonstationarity

In Section 8.4.1, several methods were presented for enforcing the harmonicity and the spectral smoothness of vectors $\mathbf{b}_k$ in matrix $\mathbf{B}$ by means of either hard or soft constraints. All these methods assumed that the spectra of the audio events forming the observed spectrogram are stationary. However, many real audio signals are known to be nonstationary: the fundamental frequency, as well as the spectral envelope, may vary over time. In this section we present some models that aim to represent such nonstationary signals by allowing the fundamental frequency and spectral envelope parameters to vary over time. Note that the prediction of the temporal variations of such parameters will not be addressed here, but in Chapter 9.

8.4.2.1 Time-Varying Fundamental Frequencies

In Section 8.4.1.3, a harmonic parameterization of vector $\mathbf{b}_k$ was described in (8.17). Hennequin *et al.* (2010) proposed a straightforward generalization of this model by making the spectral coefficient b_k also depend on time n: $b_k(n,f) = \sum_m a_k^m \, g(v_f - v_k^m(n))$, resulting in a a spectrogram model that is a generalization of NMF: $v(n,f) = \sum_k v_k(n,f)$ with $v_k(n,f) = b_k(n,f) \, h_k(n)$.

Multiplicative update rules based on the β-divergence were proposed for estimating this extended model, along with several regularization terms designed to better fit music spectrograms (Hennequin *et al.*, 2010).

8.4.2.2 Time-Varying Spectral Envelopes

Beyond the fundamental frequency, the spectral envelope of freely vibrating harmonic tones (such as those produced by a piano or a guitar) is not constant over time: generally, the upper partials decrease faster than the lower ones. Besides, some sounds, such as those produced by a didgeridoo, are characterized by a strong resonance in the spectrum that varies over time. Similarly, every time fingerings change on a wind instrument, the shape of the resonating body changes and the resonance pattern is different.

In order to properly model such sounds involving time-varying spectra, Hennequin *et al.* (2011) proposed making the activations in vector $\mathbf{h}_k$ not only depend on time, but also on frequency, in order to account for the temporal variations of the spectral envelope of vector $\mathbf{b}_k$. More precisely, the activation coefficient $h_k(n,f)$ is parameterized according to an autoregressive (AR) moving average model:

$$h_k(n,f) = \sigma_k^2(n) \left| \frac{1 + \sum_{n'} \beta_k(n,n') e^{-2j\pi n'f/F}}{1 - \sum_{n'} \alpha_k(n,n') e^{-2j\pi n'f/F}} \right|^2 \tag{8.19}$$

where $\sigma_k^2(n)$ is the variance parameter at time n, $\alpha_k(n,n')$ denotes the AR coefficients, and $\beta_k(n,n')$ denotes the moving average coefficients at time n. Then the NMF model is generalized in the following way: $v(n,f) = \sum_k v_k(n,f)$ with $v_k(n,f) = b_k(f) \, h_k(n,f)$. This model is also estimated by minimizing a β-divergence criterion. All parameters, including the AR moving average coefficients, are computed by means of multiplicative update rules, without any training. Note that even though the AR moving average model of $h_k(n,f)$ in (8.19) is nonnegative, the model coefficients $\alpha_k(n,n')$ and $\beta_k(n,n')$ are not necessarily nonnegative, which means that the multiplicative update rules introduced in

Section 8.2.3.1 were generalized so as to handle these coefficients appropriately (Hennequin *et al.*, 2011).

8.4.2.3 Both Types of Variations

In order to account for the temporal variations of the fundamental frequency and the spectral envelope jointly, Fuentes *et al.* (2013) proposed a model called harmonic adaptive latent component analysis. This models falls within the scope of the PLCA framework described in Section 8.2.2.2: matrix V is viewed as a discrete probability distribution $P(n,f) = v(n,f)$ and the spectrogram $\widehat{V}$ is modeled as a histogram: $\widehat{v}(n,f) = \frac{1}{M}\sum_m \delta_{(n_m,f_m)}(n,f)$, where $\{(n_m,f_m)\}_m$ are i.i.d. random vectors distributed according to $P(n,f)$.

In practice, the time-frequency transform used to compute $\widehat{V}$ is a constant-Q transform. Because this transform involves a log-frequency scale, pitch shifting can be approximated as a translation of the spectrum along this log-frequency axis. Therefore all the notes produced by source j at time n are approximately characterized by a unique template spectrum modeled by a probability distribution $P(\mu \mid j, n)$ (where μ is a frequency parameter), which does not depend on the fundamental frequency. However this distribution depends on time n in order to account for possible temporal variations of the spectral envelope. In addition, the variation of the pitch f_0 of source j over time n is modeled by a probability distribution $P(f_0 \mid j, n)$. Hence the resulting distribution of the shifted frequency $f = \mu + f_0$ is $P(f \mid j, n) = \sum_{f_0} P(f - f_0 \mid j, n)P(f_0 \mid j, n)$. Finally, the presence of source j at time n is characterized by a distribution $P(j, n)$. Therefore the resulting spectrogram corresponds to the probability distribution $P(n,f) = \sum_{jf_0} P(f - f_0 \mid j, n)P(f_0 \mid j, n)P(j, n)$.

In order to enforce both the harmonicity and the smoothness of the spectral envelope, the template spectrum $P(\mu \mid j, n)$ is modeled in the same way as in the first paragraph of Section 8.4.1.3, as a nonnegative linear combination of K narrowband, harmonic spectral patterns $P(\mu \mid k)$: $P(\mu \mid j, n) = \sum_k P(\mu \mid k)P(k \mid j, n)$, where $P(k \mid j, n)$ is the nonnegative weight of pattern k at time n for source j. Finally, the resulting harmonic adaptive latent component analysis model is expressed as

$$P(n,f) = \sum_{f_0 kj} P(f - f_0 \mid k)P(k \mid j, n)P(f_0 \mid j, n)P(j, n). \tag{8.20}$$

8.4.3 Coupled Factorizations

One of the strengths of NMF is that it can be used to build more complex models that explicitly represent some structural properties of audio or that represent audio together with some other kind of data. They can be formulated as *coupled tensor factorizations*, where the observed data is factored into a sum of components while coupling information across some data dimensions. The observed data to be factored, as well as the latent factors in the model, can be more than 2D, which is why the term "tensor" is generally used to characterize the model.

Coupled factorizations have the potential to improve source separation performance if either the underlying structure of the audio data can be modeled with a suitable factorization model, or coupling with another modality provides additional information about the underlying sources. For example, reverberation can be formulated with a specific type of factorization model, or the musical score can be used to guide source separation algorithms.

A general framework for coupled tensor factorization models can be written as (Simsekli *et al.*, 2015)

$$\hat{\mathcal{X}}_d(\mathcal{U}_d) \approx \mathcal{X}_d(\mathcal{U}_d) = \sum_{\bar{\mathcal{U}}} \prod_{\alpha=1}^{A} \mathcal{Z}_\alpha(\mathcal{V}_\alpha)^{r^{d,\alpha}}, \tag{8.21}$$

where $\hat{\mathcal{X}}_d(\mathcal{U}_d)$, $d \in \{1,\ldots,D\}$, are the data tensors to be factored (note that there can be $D \geq 1$ observed data tensors), $\mathcal{X}_d(\mathcal{U}_d)$ are their models, $\mathcal{U}_d$ is the set of indexes of the dth tensor, $\bar{\mathcal{U}}$ is the set of indexes which are not present in the dth tensor, and $\mathcal{Z}_\alpha$ are the factors themselves, each of which is indexed by the index set $\mathcal{V}_\alpha$. The number of factors is A and they are indexed by α. The binary variable $r^{d,\alpha}$ defines whether a specific factor is used in the product ($r^{d,\alpha} = 1$) or not ($r^{d,\alpha} = 0$).

In order to illustrate the model more concretely, we can, for example, rewrite the basic NMF as follows:

$$\hat{\mathcal{X}}_1(n,f) \approx \mathcal{X}_1(n,f) = \sum_{k} \mathcal{Z}_1(f,k)\mathcal{Z}_2(k,n). \tag{8.22}$$

In the basic NMF model we have only one observed tensor $\hat{\mathcal{X}}_1(n,f)$, indexed by f and n, i.e. $\mathcal{U}_1 = \{f,n\}$. The entire index set of the model consists of f, n, and k, and therefore $\bar{\mathcal{U}} = \{k\}$. We have $A = 2$ factors $\mathcal{Z}_1(f,k)$ and $\mathcal{Z}_2(k,n)$ having index sets $\mathcal{V}_1 = \{f,k\}$ and $\mathcal{V}_2 = \{k,n\}$. Both factors are present in the observed tensor, and therefore $r^{d,\alpha} = 1$ for $\alpha = 1$ and $\alpha = 2$.

The constraint that each source j is modeled with a disjoint set of basis vectors $\mathcal{K}_j$ can be formulated as

$$\hat{\mathcal{X}}_1(n,f) \approx \mathcal{X}_1(n,f) = \sum_{kj} \mathcal{Z}_1(f,k)\mathcal{Z}_2(k,n)\mathcal{Z}_3(k,j), \tag{8.23}$$

where the fixed binary tensor $\mathcal{Z}_3(k,j)$ couples the sources and components as

$$\mathcal{Z}_3(k,j) = \begin{cases} 1 & \text{if } k \in \mathcal{K}_j, \\ 0 & \text{otherwise.} \end{cases} \tag{8.24}$$

The advantage of the above formulation is that it allows an easy derivation of estimation algorithms for models having any structure, as long as they can be represented within the general model (8.21). The model is linear with respect to each factor, and it is relatively easy to derive an estimation method for each factor for several divergence measures, provided that the other parameters are kept fixed (Simsekli *et al.*, 2015). We will give some examples of these advanced models below.

The source-filter NMF model makes it possible, for example, to model differences between pretrained basis vectors and actual test data by learning a source-specific equalizing filter to accommodate for the differences. In the NMF framework we consider the magnitude response of the filter, which is used to elementwise multiply the basis vectors. This model can be formulated by adding a filter response factor $\mathcal{Z}_4(f,j)$ to (8.23):

$$\hat{\mathcal{X}}_1(n,f) \approx \mathcal{X}_1(n,f) = \sum_{kj} \mathcal{Z}_1(f,k)\mathcal{Z}_2(k,n)\mathcal{Z}_3(k,j)\mathcal{Z}_4(f,j). \tag{8.25}$$

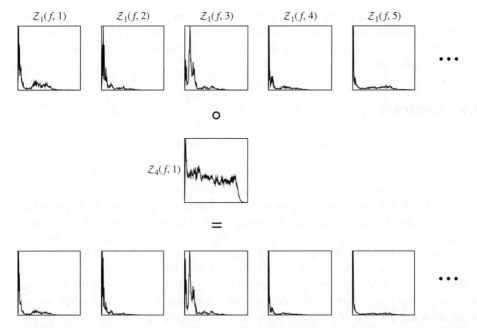

Figure 8.7 Illustration of the coupled factorization model, where basis vectors $\mathcal{Z}_1(f, k)$ acquired from training data are elementwise multiplied with an equalization filter response $\mathcal{Z}_4(f, 1)$ to better model the observed data at test time.

In this model, the basis vectors $\mathcal{Z}_1(f, k)$ would typically be estimated from training data, the coupling matrix $\mathcal{Z}_3(k, j)$ would be fixed, and the activations $\mathcal{Z}_2(k, n)$ and the filter response $\mathcal{Z}_4(f, j)$ would be estimated from the mixture signal. This model is illustrated in Figure 8.7. In addition to compensating differences between training and test data, this kind of model can be used to separately model the excitation spectrum in speech and the magnitude response of the vocal tract, or the excitation spectrum in musical instruments and the body response of the instrument.

An example of model with more than one data tensor to be factored is the model by Simsekli and Cemgil (2012b) for score-guided musical source separation (see Chapter 16 for more approaches to this problem). The model is given (in a slightly simplified form to match the notation of the chapter) as

$$\mathcal{X}_1(n, f) = \sum_k \mathcal{Z}_1(f, k)\mathcal{Z}_2(n, k)\mathcal{Z}_4(n, k), \tag{8.26}$$

$$\mathcal{X}_2(n, f) = \sum_k \mathcal{Z}_1(f, k)\mathcal{Z}_3(n, k)\mathcal{Z}_5(n, k), \tag{8.27}$$

where $\mathcal{X}_1(n, f)$ is the modeled target mixture spectrogram, $\mathcal{X}_2(n, f)$ is the modeled training data spectrogram consisting of isolated notes, $\mathcal{Z}_1(f, k)$ are the basis vectors, and $\mathcal{Z}_2(n, k)$ and $\mathcal{Z}_3(n, k)$ are their excitations in the mixture spectrogram and training data, respectively. Each basis vector is assigned a note number, and the binary factor tensor $\mathcal{Z}_4(n, k)$ indicates whether the note associated with the basis vector k is active in frame n of the mixture, based on the musical score that is temporally aligned with the mixture signal. Similarly, $\mathcal{Z}_5(n, k)$ indicates the activity of notes in the training data. The binary tensor $\mathcal{Z}_4(n, k)$ controls which basis vectors can be used to represent each frame, and the use of the isolated training data tensor $\mathcal{X}_2(n,f)$ helps to learn the basis vectors robustly. The use of both tensors contributes positively to the source separation performance. The use of coupled factorizations in music applications is discussed in more detail in Chapter 16.

8.5 Summary

In this chapter we have shown that NMF is a very powerful model for representing speech and music data. We have presented the mathematical foundations and described several probabilistic frameworks and various algorithms for computing an NMF. We have also presented some advanced NMF models that are able to more accurately represent audio signals by enforcing properties such as sparsity, harmonicity, and spectral smoothness, and by taking the nonstationarity of the data into account. We have shown that coupled factorizations make it possible to exploit some extra information we may have about the observed signal, such as the musical score. Finally, we have presented several methods that perform dictionary learning for NMF.

The benefits of NMF in comparison with other separation approaches are the capability of performing unsupervised source separation, learning source models from a relatively small amount of material (especially in comparison with DNNs), and easily implementing and adapting the source models and algorithms. The main downside is the complexity of iterative NMF algorithms. Note that beyond source separation, NMF models have also proved successful in a broad range of audio applications, including automatic music transcription (Smaragdis and Brown, 2003), multipitch estimation (Vincent *et al.*, 2010; Bertin *et al.*, 2010; Fuentes *et al.*, 2013; Benetos *et al.*, 2014), and audio inpainting (Smaragdis *et al.*, 2011).

In Chapter 9, the modeling of the temporal variations of NMF parameters will be investigated. Multichannel NMF will be addressed in Chapter 14, and many NMF examples and extensions for music will be presented in Chapter 16. Some perspectives of NMF modeling will also be presented in Chapter 19, including the complex NMF (Kameoka *et al.*, 2009) and high-resolution NMF (Badeau, 2011) models, which both focus on modeling the phase of the STFT along with its magnitude.

Bibliography

Baby, D., Virtanen, T., Gemmeke, J.F., and Van Hamme, H. (2015) Coupled dictionaries for exemplar-based speech enhancement and automatic speech recognition. *IEEE/ACM Transactions on Audio, Speech, and Language Processing*, **11** (23), 1788–1799.

Badeau, R. (2011) Gaussian modeling of mixtures of non-stationary signals in the time-frequency domain (HR-NMF), in *Proceedings of IEEE Workshop on Applications of Signal Processing to Audio and Acoustics*, pp. 253–256.

Badeau, R., Bertin, N., and Vincent, E. (2010) Stability analysis of multiplicative update algorithms and application to non-negative matrix factorization. *IEEE Transactions on Neural Networks*, **21** (12), 1869–1881.

Badeau, R. and Drémeau, A. (2013) Variational bayesian EM algorithm for modeling mixtures of non-stationary signals in the time-frequency domain (HR-NMF), in

Proceedings of IEEE International Conference on Audio, Speech and Signal Processing, pp. 6171–6175.

Barker, T. and Virtanen, T. (2013) Non-negative tensor factorisation of modulation spectrograms for monaural sound source separation, in *Proceedings of Interspeech*.

Benetos, E., Richard, G., and Badeau, R. (2014) Template adaptation for improving automatic music transcription, in *Proceedings of International Society for Music Information Retrieval Conference*, pp. 175–180.

Bertin, N., Badeau, R., and Vincent, E. (2010) Enforcing harmonicity and smoothness in Bayesian non-negative matrix factorization applied to polyphonic music transcription. *IEEE/ACM Transactions on Audio, Speech, and Language Processing*, **18** (3), 538–549.

Bertin, N., Févotte, C., and Badeau, R. (2009) A tempering approach for Itakura-Saito non-negative matrix factorization. With application to music transcription, in *Proceedings of IEEE International Conference on Audio, Speech and Signal Processing*, pp. 1545–1548.

Brown, J.C. and Smaragdis, P. (2004) Independent component analysis for automatic note extraction from musical trills. *Journal of the Acoustical Society of America*, **115** (5), 2295–2306.

Casey, M.A. and Westner, A. (2000) Separation of mixed audio sources by independent subspace analysis, in *Proceedings of International Computer Music Conference*, pp. 154–161.

Cichocki, A., Zdunek, R., Phan, A.H., and Amari, S.I. (2009) *Nonnegative Matrix and Tensor Factorizations: Applications to Exploratory Multi-way Data Analysis and Blind Source Separation*, Wiley.

Diment, A. and Virtanen, T. (2015) Archetypal analysis for audio dictionary learning, in *Proceedings of IEEE Workshop on Applications of Signal Processing to Audio and Acoustics*.

Fessler, J.A. and Hero, A.O. (1994) Space-alternating generalized expectation-maximization algorithm. *IEEE Transactions on Signal Processing*, **42** (10), 2664–2677.

Févotte, C., Bertin, N., and Durrieu, J.L. (2009) Nonnegative matrix factorization with the Itakura-Saito divergence. With application to music analysis. *Neural Computation*, **21** (3), 793–830.

Févotte, C. and Idier, J. (2011) Algorithms for nonnegative matrix factorization with the beta-divergence. *Neural Computation*, **23** (9), 2421–2456.

Finesso, L. and Spreij, P. (2004) Approximate nonnegative matrix factorization via alternating minimization, in *Proceedings of International Symposium on Mathematical Theory of Networks and Systems*.

Fuentes, B., Badeau, R., and Richard, G. (2013) Harmonic adaptive latent component analysis of audio and application to music transcription. *IEEE/ACM Transactions on Audio, Speech, and Language Processing*, **21** (9), 1854–1866.

Gemmeke, J., Virtanen, T., and Hurmalainen, A. (2011) Exemplar-based sparse representations for noise robust automatic speech recognition. *IEEE/ACM Transactions on Audio, Speech, and Language Processing*, **19** (7), 2067–2080.

Gemmeke, J.F. and Virtanen, T. (2010) Artificial and online acquired noise dictionaries for noise robust ASR, in *Proceedings of Interspeech*.

Germain, F.G. and Mysore, G.J. (2014) Stopping criteria for non-negative matrix factorization based supervised and semi-supervised source separation. *IEEE Signal Processing Letters*, **21** (10), 1284–1288.

Grais, E. and Erdogan, H. (2011) Adaptation of speaker-specific bases in non-negative matrix factorization for single channel speech-music separation, in *Proceedings of Interspeech*.

Grais, E.M. and Erdogan, H. (2013) Discriminative nonnegative dictionary learning using cross-coherence penalties for single channel source separation, in *Proceedings of Interspeech*.

Helén, M. and Virtanen, T. (2005) Separation of drums from polyphonic music using non-negative matrix factorization and support vector machine, in *Proceedings of European Signal Processing Conference*.

Hennequin, R., Badeau, R., and David, B. (2010) Time-dependent parametric and harmonic templates in non-negative matrix factorization, in *Proceedings of International Conference on Digital Audio Effects*.

Hennequin, R., Badeau, R., and David, B. (2011) NMF with time-frequency activations to model non-stationary audio events. *IEEE/ACM Transactions on Audio, Speech, and Language Processing*, **19** (4), 744–753.

Hurmalainen, A., Gemmeke, J.F., and Virtanen, T. (2013) Modelling non-stationary noise with spectral factorisation in automatic speech recognition. *Computer Speech and Language*, **28** (3), 763–779.

Hurmalainen, A., Saeidi, R., and Virtanen, T. (2015) Similarity induced group sparsity for non-negative matrix factorisation, in *Proceedings of IEEE International Conference on Audio, Speech and Signal Processing*, pp. 4425–4429.

Jaureguiberry, X., Leveau, P., Maller, S., and Burred, J.J. (2011) Adaptation of source-specific dictionaries in non-negative matrix factorization for source separation, in *Proceedings of IEEE International Conference on Audio, Speech and Signal Processing*, pp. 5–8.

Joder, C., Weninger, F., Eyben, F., Virette, D., and Schuller, B. (2012) Real-time speech separation by semi-supervised nonnegative matrix factorization, in *Proceedings of International Conference on Latent Variable Analysis and Signal Separation*, pp. 322–329.

Kameoka, H., Nakano, M., Ochiai, K., Imoto, Y., Kashino, K., and Sagayama, S. (2012) Constrained and regularized variants of non-negative matrix factorization incorporating music-specific constraints, in *Proceedings of IEEE International Conference on Audio, Speech and Signal Processing*.

Kameoka, H., Ono, N., Kashino, K., and Sagayama, S. (2009) Complex NMF: A new sparse representation for acoustic signals, in *Proceedings of IEEE International Conference on Audio, Speech and Signal Processing*, pp. 3437–3440.

King, B., Févotte, C., and Smaragdis, P. (2012) Optimal cost function and magnitude power for NMF-based speech separation and music interpolation, in *Proceedings of IEEE International Workshop on Machine Learning for Signal Processing*.

Laurberg, H., Christensen, M., Plumbley, M.D., Hansen, L.K., and Jensen, S.H. (2008) Theorems on positive data: On the uniqueness of NMF. *Computational Intelligence and Neuroscience*, **2008**. Article ID 764206, 9 pages.

Lee, D.D. and Seung, H.S. (1999) Learning the parts of objects by non-negative matrix factorization. *Nature*, **401** (6755), 788–791.

Lee, D.D. and Seung, H.S. (2001) Algorithms for non-negative matrix factorization, in *Proceedings of Neural Information Processing Systems*, pp. 556–562.

Lefevre, A., Bach, F., and Févotte, C. (2011) Itakura-saito nonnegative matrix factorization with group sparsity, in *Proceedings of IEEE International Conference on Audio, Speech and Signal Processing*, pp. 21–24.

Lin, C.J. (2007) Projected gradient methods for non-negative matrix factorization. *Neural Computation*, **19** (10), 2756–2779.

Liutkus, A. and Badeau, R. (2015) Generalized Wiener filtering with fractional power spectrograms, in *Proceedings of IEEE International Conference on Audio, Speech and Signal Processing*, pp. 266–270.

Liutkus, A., Fitzgerald, D., and Badeau, R. (2015) Cauchy nonnegative matrix factorization, in *Proceedings of IEEE Workshop on Applications of Signal Processing to Audio and Acoustics*.

Mysore, G.J. and Smaragdis, P. (2011) A non-negative approach to semi-supervised separation of speech from noise with the use of temporal dynamics, in *Proceedings of IEEE International Conference on Audio, Speech and Signal Processing*, pp. 17–20.

Nakano, M., Kameoka, H., Le Roux, J., Kitano, Y., Ono, N., and Sagayama, S. (2010) Convergence-guaranteed multiplicative algorithms for non-negative matrix factorization with beta-divergence, in *Proceedings of IEEE International Workshop on Machine Learning for Signal Processing*, pp. 283–288.

Raj, B., Virtanen, T., Chaudhure, S., and Singh, R. (2010) Non-negative matrix factorization based compensation of music for automatic speech recognition, in *Proceedings of the International Conference on Spoken Language Processing*.

Rigaud, F., David, B., and Daudet, L. (2013) A parametric model and estimation techniques for the inharmonicity and tuning of the piano. *Journal of the Acoustical Society of America*, **133** (5), 3107–3118.

Schmidt, M.N. and Laurberg, H. (2008) Non-negative matrix factorization with Gaussian process priors. *Computational Intelligence and Neuroscience*, **2008**, 1–10. Article ID 361705.

Schmidt, M.N. and Olsson, R.K. (2006) Single-channel speech separation using sparse non-negative matrix factorization, in *Proceedings of the International Conference on Spoken Language Processing*.

Shashanka, M., Raj, B., and Smaragdis, P. (2008) Probabilistic latent variable models as nonnegative factorizations. *Computational Intelligence and Neuroscience*, **2008**, 1–8. Article ID 947438.

Simsekli, U. and Cemgil, A.T. (2012a) Markov chain Monte Carlo inference for probabilistic latent tensor factorization, in *Proceedings of IEEE International Workshop on Machine Learning for Signal Processing*, pp. 1–6.

Simsekli, U. and Cemgil, A.T. (2012b) Score guided musical source separation using generalized coupled tensor factorization, in *Proceedings of European Signal Processing Conference*, pp. 2639–2643.

Simsekli, U., Virtanen, T., and Cemgil, A.T. (2015) Non-negative tensor factorization models for Bayesian audio processing. *Digital Signal Processing*, **47**, 178–191.

Smaragdis, P. (2007) Convolutive speech bases and their application to supervised speech separation. *IEEE/ACM Transactions on Audio, Speech, and Language Processing*, **15** (1), 1–12.

Smaragdis, P. and Brown, J.C. (2003) Non-negative matrix factorization for polyphonic music transcription, in *Proceedings of IEEE Workshop on Applications of Signal Processing to Audio and Acoustics*, pp. 177–180.

Smaragdis, P., Raj, B., and Shashanka, M. (2011) Missing data imputation for time-frequency representations of audio signals. *Journal of Signal Processing Systems*, **65**, 361–370.

Smaragdis, P., Shashanka, M., and Raj, B. (2009) A sparse non-parametric approach for single channel separation of known sounds, in *Proceedings of Neural Information Processing Systems*.

Spiertz, M. and Gnann, V. (2009) Source-filter based clustering for monaural blind source separation, in *Proceedings of International Conference on Digital Audio Effects*.

Sprechmann, P., Bronstein, A.M., and Sapiro, G. (2014) Supervised non-Euclidean sparse NMF via bilevel optimization with applications to speech enhancement, in *Proceedings of Joint Workshop on Hands-free Speech Communication and Microphone Arrays*, pp. 11–15.

Vincent, E., Bertin, N., and Badeau, R. (2010) Adaptive harmonic spectral decomposition for multiple pitch estimation. *IEEE/ACM Transactions on Audio, Speech, and Language Processing*, **18** (3), 528–537.

Vincent, E. and Rodet, X. (2004) Instrument identification in solo and ensemble music using independent subspace analysis, in *Proceedings of International Society for Music Information Retrieval Conference*, pp. 576–581.

Virtanen, T. (2003) Sound source separation using sparse coding with temporal continuity objective, in *Proceedings of International Computer Music Conference*, pp. 231–234.

Virtanen, T. and Cemgil, A.T. (2009) Mixtures of gamma priors for non-negative matrix factorization based speech separation, in *Proceedings of International Conference on Independent Component Analysis and Signal Separation*, pp. 646–653.

Virtanen, T., Cemgil, A.T., and Godsill, S. (2008) Bayesian extensions to non-negative matrix factorisation for audio signal modelling, in *Proceedings of IEEE International Conference on Audio, Speech and Signal Processing*, pp. 1825–1828.

Virtanen, T., Gemmeke, J., and Raj, B. (2013) Active-set Newton algorithm for overcomplete non-negative representations of audio. *IEEE/ACM Transactions on Audio, Speech, and Language Processing*, **21** (11), 2277–2289.

Weninger, F., Geiger, J., Wöllmer, M., Schuller, B., and Rigoll, G. (2011) The Munich 2011 CHiME challenge contribution: NMF-BLSTM speech enhancement and recognition for reverberated multisource environments, in *Proceedings of International Workshop on Machine Listening in Multisource Environments*, pp. 24–29.

Weninger, F., Le Roux, J., Hershey, J.R., and Watanabe, S. (2014) Discriminative NMF and its application to single-channel source separation, in *Proceedings of Interspeech*, pp. 865–869.

9

Temporal Extensions of Nonnegative Matrix Factorization

Cédric Févotte, Paris Smaragdis, Nasser Mohammadiha, and Gautham J. Mysore

Temporal continuity is one of the most important features of time series data. Our aim here is to present some of the basic as well as advanced ideas to make use of this information by modeling time dependencies in nonnegative matrix factorization (NMF). The dependencies between consecutive frames of the spectrogram can be imposed either on the basis matrix **B** or on the activations **H** (introduced in Chapter 8). The former case is known as the convolutive NMF, reviewed in Section 9.1. In this case, the repeating patterns within data are represented with multidimensional bases that are not vectors anymore, but functions that can span an arbitrary number of dimensions (e.g., time and frequency). The other case consists of imposing temporal structure on the activations **H**, in line with traditional dynamic models that have been studied extensively in signal processing. Most models considered in the NMF literature can be cast as special cases of a unifying state-space model, which will be discussed in Section 9.2. Special cases will be reviewed in subsequent sections. Continuous models are addressed in Sections 9.3 and 9.4, while Section 9.5 reviews models that involve a discrete latent state variable. Sections 9.6 and 9.7 provide quantitative and qualitative comparisons of the proposed methods, while Section 9.8 summarizes. This chapter is an extended version of the review paper by Smaragdis *et al.* (2014).

In this chapter we will denote by $\hat{\mathbf{V}} = [\hat{v}(n,f)]_{fn}$ the nonnegative spectral data, with columns $\hat{\mathbf{v}}(n) = [\hat{v}(n,f)]_f$ and coefficients $\hat{v}(n,f)$. In most cases, $\hat{\mathbf{V}}$ is either the magnitude spectrogram $|\mathbf{X}|$ or the power spectrogram $|\mathbf{X}|^2$, i.e. $\hat{v}(n,f) = |x(n,f)|$ or $|x(n,f)|^2$. We will also denote $\mathbf{V} = [v(n,f)]_{fn} = \mathbf{BH}$, with coefficients $v(n,f)$. Note that traditionally the NMF literature instead denotes the data by $\mathbf{V}$ and the approximate factorization by $\hat{\mathbf{V}}$. However, the chosen notation is here consistent with the convention used in this book, where variables with a hat denote statistics (observed quantities) and variables without a hat denote model parameters.

9.1 Convolutive NMF

Convolutive NMF (Smaragdis, 2007; O'Grady and Pearlmutter, 2008; Wang *et al.*, 2009) is a technique that is used to model each sound source as a collection of

Audio Source Separation and Speech Enhancement, First Edition.
Edited by Emmanuel Vincent, Tuomas Virtanen and Sharon Gannot.
© 2018 John Wiley & Sons Ltd. Published 2018 by John Wiley & Sons Ltd.
Companion Website: https://project.inria.fr/ssse/

time-frequency templates that each span multiple time frames and/or frequency bins (often all frequency bins). As we will show shortly, these templates often correspond to basic elements of sounds, such as phonemes of speech, notes of music, or other temporally coherent units of sound. By using this approach, we can capture a lot of the temporally important nuances that make up a given sound. However, since it directly captures a whole time-frequency patch, it can be quite inflexible and cannot model sounds of varying lengths and pitches. For example, a template of a given vowel will have difficulty modeling the same vowel of a longer length or different pitch. On the other hand, sounds of a fixed length and pitch, such as a drum hit, can be modeled quite well using this technique. These models can be seen as a deterministic way to model temporal dependencies.

9.1.1   1D Convolutive NMF

We will start by formulating the 1D version of convolutive NMF and build up from there. Recall that traditional NMF performs the approximation:

$$\hat{\mathbf{V}} \approx \mathbf{BH} \tag{9.1}$$

where the $F \times K$ nonnegative matrix $\mathbf{B}$ and the $K \times N$ nonnegative matrix $\mathbf{H}$ are the K bases and activations, respectively (see Chapter 8). A significant shortcoming of this model is that the temporal structure of the input signal is not represented in any way. The basis matrix $\mathbf{B}$ will contain a set of spectra that can be used to compose an input sound, but there is no representation of their temporal relationships. In order to address this problem we consider an alternative decomposition that explicitly encodes sequences of spectra. We start using the following formulation:

$$\mathbf{v}_{(n)}(n, \ldots, n + N') = \mathbf{B}_{(n)}\mathbf{H} \tag{9.2}$$

$$\hat{\mathbf{V}} \approx \sum_n \mathbf{v}_{(n)} \tag{9.3}$$

where $\mathbf{v}_{(n)}(n, \ldots, n + N')$ represents an additive part of the nth to $(n + N')$th columns (time frames) of $\hat{\mathbf{V}}$. The constant N' is chosen so that for the maximum value of n we would have $n + N' = N - 1$, for reasons that should become clear momentarily. There are a couple of observations to make about this model. First of all, we now have a different representation of the bases. Instead of a single basis matrix $\mathbf{B}$, where each column is a basis, we have a set of basis matrices $\mathbf{B}_{(n)}$, where the parenthesized index relates to a time shift relating to its input. We note that each of these matrices is used to approximate a time-shifted version of the input and that all these matrices share the same activation patterns. By doing this we effectively force the bases of $\mathbf{B}_{(n+1)}$ to always get activated the same way as the bases of $\mathbf{B}_{(n)}$ have been in the previous time frame. In other words, we would always expect to see the kth basis of $\mathbf{B}_{(n)}$ to be followed in the next time frame by the kth basis of $\mathbf{B}_{(n+1)}$, etc. By doing this we create a set of bases that have a deterministic temporal evolution which spans as many time steps as we have $\mathbf{B}_{(n)}$ matrices.

To consider the effects of this new structure let us consider the toy input shown in Figure 9.1. The toy input in this case is shown in the top right panel and consists of a "spectrogram" that contains two types of repeating patterns. One pattern consists of two parallel components which over time drop by one frequency bin for a duration of two frames, and the other consists of two components which similarly rise in frequency.

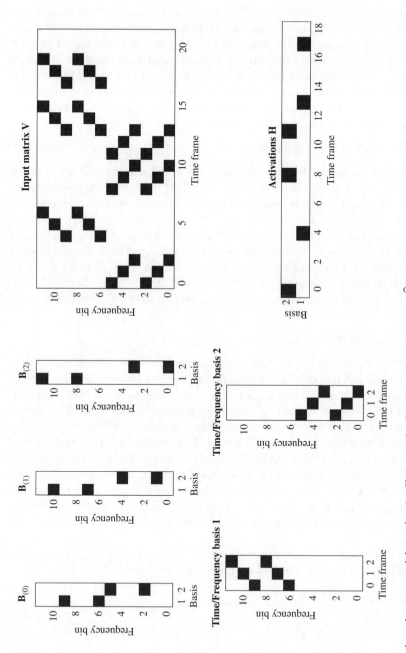

Figure 9.1 Learning temporal dependencies. The top right plot shows the input matrix $\hat{\mathbf{V}}$, which has a consistent left-right structure. The top left plot shows the learned matrices $\mathbf{B}_{(n)}$ and the bottom right plot shows the learned activations $\mathbf{H}$. The bottom left plot shows the bases again, only this time we concatenate the corresponding columns from each $\mathbf{B}_k$. We can clearly see that this sequence of columns learns bases that extend over time.

Although we could decompose this input using a regular NMF model, the resulting representation would not be very illuminating (it would be a set of six unordered bases, not revealing temporal structure very clearly). Instead we analyze this using the model above. We will ask for three $F \times 2$ matrices $\mathbf{B}_{(n)}$, $n \in \{0, 1, 2\}$, each made up of two bases. This will allow us to learn two time-frequency components which extend for three time frames each. Since this is exactly the structure in the input, we would expect to effectively learn the patterns therein. After estimating these parameters, we show the results in the remaining plots in the same figure. At the top left we see the three matrices $\mathbf{B}_{(n)}$. Note that the nth matrix will contain time frame n for all the bases. Below these plots we show the same information reordered so that each basis is grouped with its temporal components. In order to obtain the kth convolutive basis $\mathbf{B}_k$, we use:

$$\mathbf{B}_k = [\mathbf{B}_{(0)}(k), \mathbf{B}_{(1)}(k), \mathbf{B}_{(2)}(k), \ldots], \tag{9.4}$$

i.e. in the case of Figure 9.1 for the first basis we would concatenate the matrices $\mathbf{B}_{(0)}(1), \mathbf{B}_{(1)}(1), \mathbf{B}_{(2)}(1)$. Doing this, we now clearly see that the two learned bases have a temporal structure that reflects the input. The first one contains two rising components, and the second one contains two descending components. These two, when combined using the activations shown in the lower right plot, will compose the original input. From the activations we see that the second basis is activated at times 0, 8, and 11 (which are the start times of the descending components), and the other at times 4, 13, 17, which are the start times of the other component. The fact that some of the components overlap is not a notable complication since NMF is very good at resolving mixing.

9.1.2 Convolutive NMF as a Meta-Model

We now turn to the problem of parameter estimation for the above model NMF. There are of course many variants of NMF depending on how one likes to describe the cost function, the underlying noise model, any probabilistic aspects, etc. In order to not get caught up with these details we will introduce this model as a meta-model, so that we can easily use an existing NMF algorithm and adapt it to this process.

We note that in the above description we defined this model as a combination of multiple NMF models that are defined on time-shifted versions of the input and share their activations. The number of models that we use maps to the temporal extent of the estimates bases. Since we used this modular formulation, we will take advantage of it to define a training procedure that averages these models in order to perform estimation. The resulting meta-model will inherit the model specifications of the underlying NMF models.

We will start by observing that the factorization to solve is a set of tied factorizations over different lags of the input, which share the same activations:

$$\mathbf{v}_{(0)}(0, \ldots, N') = \mathbf{B}_{(0)}\mathbf{H} \tag{9.5}$$

$$\mathbf{v}_{(1)}(1, \ldots, 1 + N') = \mathbf{B}_{(1)}\mathbf{H} \tag{9.6}$$

$$\mathbf{v}_{(2)}(2, \ldots, 2 + N') = \mathbf{B}_{(2)}\mathbf{H} \tag{9.7}$$

$$\cdots \tag{9.8}$$

Each of these problems is easy to resolve independently using any NMF algorithm, but solving them together such that $\hat{\mathbf{V}} \approx \sum_n \mathbf{v}_{(n)}$ requires a slightly different approach.

To illustrate, let us show how this works with the KL-NMF model. In this model the objective is to minimize the KL divergence between the input $\hat{\mathbf{V}}$ and its approximation $\mathbf{BH}$. In this case we iterate over the following parameter updates:

$$\mathbf{R} = \frac{\hat{\mathbf{V}}}{\mathbf{BH}} \tag{9.9}$$

$$\mathbf{B} = \mathbf{B} \circ (\mathbf{RH}^T) \tag{9.10}$$

$$\mathbf{H} = \mathbf{H} \circ (\mathbf{B}^T \mathbf{R}) \tag{9.11}$$

where the $\circ$ operator is elementwise multiplication and the fraction is elementwise division. In order to resolve an ambiguity in this model, after every update we additionally have to normalize either $\mathbf{B}$ or $\mathbf{H}$ to a fixed ℓ_1 norm. Traditionally, we normalize the bases $\mathbf{B}$ to sum to 1. In order to estimate each $\mathbf{B}_{(n)}$ and $\mathbf{H}$ we will use the same form as with the equations above, but we need to modify the computation of $\mathbf{R}$ to account for the time shifting. We simply do so by

$$\mathbf{r}(n) = \frac{\hat{\mathbf{v}}(n)}{\sum_t \mathbf{B}_{(n)} \, \mathbf{h}(n - t - 1)} \tag{9.12}$$

$$\mathbf{B}_{(n)} = \mathbf{B}_{(n)} \circ (\mathbf{RH}^T) \tag{9.13}$$

$$\mathbf{H} = \mathbf{H} \circ (\mathbf{B}_{(n)}^T \mathbf{R}). \tag{9.14}$$

The only difference is that when we approximate the input using our model we need to add all the time-shifted reconstructions using each $\mathbf{B}_{(n)}$. This is done when computing the denominator of the expression to compute $\mathbf{R}$.

In a similar fashion we can adapt other NMF update algorithms to act the same way. We need to account for all time shifts and use the appropriate $\mathbf{B}_{(n)}$ in each, and at the end of each iteration we average the estimates of $\mathbf{H}$ for each corresponding $\mathbf{B}_{(n)}$ to obtain a single estimate for $\mathbf{H}$. Iterating in this manner can produce the model parameters by simply building on whichever basic NMF model we choose to start with. In the next section we will show a different approach, which results in the same updates, albeit via a direct derivation and not as a meta-model heuristic.

9.1.3 N-D Model

We can take the idea above and extend it to more dimensions. This will allow us to obtain components that cannot only be arbitrarily positioned left-right (corresponding to a shift along the time axis for spectrograms), but also up-down (corresponding to a shift along the frequency axis space), or in the case of higher-dimensional input over other dimensions as well. The general form will then include a whole set of tied factorization where each will not only approximate a different time lag of the input, but also a different frequency lag. In this case it will be easier to move to a more compact notation.

Although so far we have referred to this model as a convolutive model, we have encountered no convolutions. We will now show a more compact form that can be used to express the above operation, and also easily extend to more dimensions, albeit one that does not work as well as a meta-model and does not allow incorporating existing NMF models as easily. We start with the 1D version and we reformulate it as:

$$\hat{\mathbf{V}} \approx \sum_k \mathbf{B}_k \star \mathbf{h}_k \tag{9.15}$$

where the $\star$ operator denotes convolution over the left-right axis. In other terms, we have

$$\hat{\mathbf{v}}(n) \approx \sum_{k=1}^{K} \sum_{n'=0}^{N'-1} \mathbf{B}_k(n') h_k(n - n'). \tag{9.16}$$

In this version we have K matrix bases that extend over N' frames as self-contained $F \times N'$ matrices $\mathbf{B}_k$, and each one will be convolved with a $1 \times N$ vector h_k. The matrices $\mathbf{B}_k$ are the ones shown in the bottom left plots in Figure 9.1, and the vectors h_k are the rows of the matrix $\mathbf{H}$ shown in the bottom right of the same figure. Intuitively, what we do in this model is that we shift and scale each time-frequency basis using a convolution operation and then sum them all up.

Using this notation allows us to extend this model to employ shifting on other axes as well. For example, we can ask for components that shift not only left-right, but also up-down. That implies a model like:

$$\hat{\mathbf{V}} \approx \sum_{k=1}^{K} \mathbf{B}_k \star \mathbf{H}_k \tag{9.17}$$

where now the $\star$ operator performs 2D convolution between the matrices $\mathbf{B}_k$ and $\mathbf{H}_k$. In other terms:

$$\hat{v}(n,f) \approx \sum_{k} \sum_{n'f'} b_k(n',f') h_k(n - n', f - f'). \tag{9.18}$$

Depending on our preference we can crop the result of this convolution to return an output sized as $F \times N$, or restrict the size of $\mathbf{H}_k$ to be $(F - F' + 1) \times (N - N' + 1)$.

We can derive the estimation procedure as above using a meta-model formulation, or directly optimize the above formulation. For the case of using NMF with a KL divergence cost function, the updated equations for the 2D convolutive model above are:

$$r(n,f,n',f',k) = \frac{\sum_{n'f'} b_k(n',f') h_k(n - n', f - f')}{\sum_{k'} \sum_{n'f'} b_{k'}(n',f') h_{k'}(n - n', f - f')} \tag{9.19}$$

$$b_k(n',f') = \sum_{nf} b_k(n,f) r(n,f,n',f',k) \tag{9.20}$$

$$h_k(n,f) = \sum_{n'f'} h_k(n,f) r(n,f,n',f',k). \tag{9.21}$$

In order to avoid an oscillation assigning more energy to the components or the activations, we traditionally normalize the components to have their elements sum to a fixed value (usually 1). For more details and for the derivation of the general any-dimensional case see Smaragdis and Raj (2007).

9.1.4 Illustrative Examples

Finally, we turn to some applications that this model can be used for. In this section we will show three common applications: constructing time-frequency dictionaries, extracting coherent time-frequency objects from a mixture, and discovering shift-invariant structure from a recording. These applications serve as lower level steps on which one can build signal separators, pitch detectors, and content analysis systems.

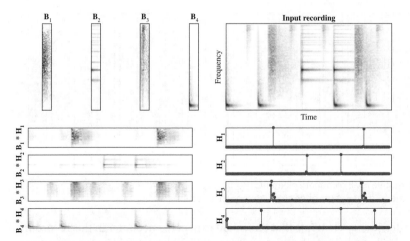

Figure 9.2 Extraction of time-frequency sources. The input to this case is shown in the top right plot. It is a drum pattern composed of four distinct drum sounds. The set of four top left plots shows the extracted time-frequency templates using 2D convolutive NMF. Their corresponding activations are shown in the step plots in the bottom right, and the individual convolutions of each template with its activation as the lower left set of plots. As one can see this model learns the time-frequency profile of the four drum sounds and correctly identifies where they are located.

9.1.4.1 Time-Frequency Component Extraction

Quite often what we would refer to as a component will have time-frequency structure. This is the case with many sounds that do not exhibit a sustained spectral structure. To illustrate such an example, consider the recording in Figure 9.2. It shows the spectrogram of a drum pattern which is composed of four different sounding drums. Since these drums do not have a static spectral profile it would be inappropriate to attempt to extract them using plain NMF methods. Doing so would potentially produce only an average spectrum for each sound, and in other cases not work at all. In this case an appropriate model for analysis is a 1D convolutive NMF with four components, one for each drum sound. The results of this analysis are shown in the same figure. On the left we see the four extracted templates $\mathbf{B}_k$ which, as we can easily see, have taken the shape of the four distinct drum spectra. Their corresponding activations, shown in the bottom right, show us where in time these templates are active. If we convolve $\mathbf{B}_k$ with $\mathbf{H}_k$ we obtain a reconstruction using only one of the drum sounds. These are shown in the bottom left, and upon inspection we note that they have successfully extracted each drum sound's contribution.

9.1.4.2 Time-Frequency Dictionaries

Another application of this method is that of extracting time-frequency dictionaries for a type of sound. Using plain NMF, this is a standard approach to building dictionaries of sounds that we can later use for various applications. Using this model we can learn slightly more rigid dictionaries that also learn some of the temporal aspects of each basis.

To do this we simply decide on the number of bases to use, and their temporal extent. As an example consider extracting such a dictionary out of a speech recording. For this example we will take a recording of a speaker and decompose it as a 1D convolutive NMF model of a few dozen bases. Compared to the previous section, this is a longer recording with more variation. Decomposing it with multiple components would extract

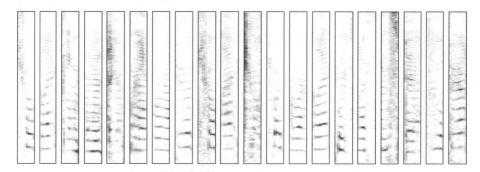

Figure 9.3 Convolutive NMF dictionary elements (**B**$_k$) for a speech recording. Note that each component has the form of a short phoneme-like speech inflection.

common time-frequency elements that one might expect to encounter in this recording. As one might guess, such a set of components would be phoneme-like elements at various pitches. We show a subset of such learned components from speech in Figure 9.3. Such a dictionary model works well as the basis for source separation using NMF for sounds with consistent temporal structure (usually music). Because of its rigid temporal constraints it does not always work as well for sounds exhibiting much more temporal variability, such as speech.

9.1.4.3 Shift-Invariant Transforms

Finally, we present an application for the 2D convolutive NMF model. In this case we make use of shifting over the frequency axis. To obtain meaningful results in this case we need to use a time-frequency transformation that exhibits a semantically relevant *shift invariance* along the frequency axis. One such case is the constant-Q transform (Brown, 1991), which exhibits a frequency axis such that a pitch shift would simply translate a spectrum along the frequency axis without changing its shape. This means that, unlike before, we can use a single component to represent all possible pitches of a specific sound, which can result in a significantly more compact dictionary. Obviously this approach has many applications to music signal analysis where pitch is a quantity that needs to be taken into account frequently.

To show how this model would be useful in such a case consider the recording in Figure 9.4. In this recording we have a violin playing a melody that often has two simultaneous notes. In the top right plot we see the constant-Q transform of this sound. Since we have only one instrument we decompose it using the model in (9.17), with one $F \times 1$ component **B**, which will be convolved with a $F \times N$ activation **H**. What this will result in is estimating a 1D function that will be shifted over both dimensions and replicated such that it approximates the input. Naturally, this function would be the constant-Q spectrum of a violin note, which will be shifted across frequency to represent different pitches, and shifted across time to represent all the played notes. The results of this analysis are shown in Figure 9.4. It is easy to see that the extracted template **B** looks like a harmonic series. A more interesting form is found for the activation matrix **H**. Since this is the function that will specify the pitch and time offset, it will effectively tell us what the pitch of the input was at every time step (corresponding to the peaks along the vertical dimension), and also encode the energy of the signal over time. This effectively becomes a pitch-time representation of the input that we can use to infer the notes being played.

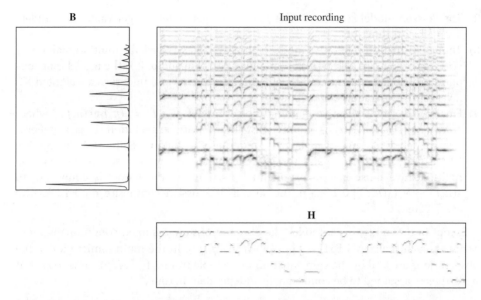

Figure 9.4 Convolutive NMF decomposition for a violin recording. Note how the one extracted basis **B** corresponds to a constant-Q spectrum that when 2D convolved with the activation **H** approximates the input. The peaks in **H** produce a pitch transcription of the recording by indicating energy at each pitch and time offset.

9.2 Overview of Dynamical Models

In the remainder of this chapter we investigate dynamical models that impose a temporal structure of the matrix **H**, whereas the previous section was about imposing some temporal structure of the dictionary **B**. The two scenarios are not mutually exclusive and can easily be combined. The dynamical NMF models that we will review are special cases of the general dynamic model given by

$$\widehat{\mathbf{v}}(n) \sim p(\widehat{\mathbf{v}}(n) \mid \mathbf{B}\mathbf{h}(n)) \tag{9.22}$$

$$\mathbf{h}(n) \sim p(\mathbf{h}(n) \mid \mathbf{h}(n-1)) \tag{9.23}$$

where (9.22) defines a probabilistic NMF observation model such that $\mathbb{E}\{\widehat{\mathbf{V}} \mid \mathbf{BH}\} = \mathbf{BH}$ and (9.23) introduces temporal dynamics by assuming a Markovian structure for the activation coefficients.

The variety of models proposed for the dynamical part (9.23) will be the topic of the next sections. Regarding the observation part (9.22), the literature concentrates on four models that we sketch here (see also Chapter 8 and see Section 9.9 for the definition of the random variables considered next).

1) The additive Gaussian noise model $\widehat{v}(n,f) = v(n,f) + \epsilon(n,f)$, with $\epsilon(n,f) \sim \mathcal{N}(\epsilon(n,f) \mid 0, \sigma^2)$, is a popular NMF model. It is not, however, a true generative model of nonnegative data because it can theoretically produce negative data values in low SNR regimes. It underlies the common quadratic cost function in the sense that $-\log p(\widehat{\mathbf{V}} \mid \mathbf{V}) = \frac{1}{\sigma^2}\sum_{nf}(\widehat{v}(n,f) - v(n,f))^2 + \mathrm{cst}$ (where the notation cst everywhere defines the terms independent of the parameters).

2) The Poisson model $\hat{v}(n,f) \sim \mathcal{P}(\hat{v}(n,f) \mid v(n,f))$ generates integer values and underlies a KL cost function.

3) Though formally a generative model of integer values as well, the multinomial model $\hat{\mathbf{v}}(n) \sim \mathcal{M}(\hat{\mathbf{v}}(n) \mid \|\hat{\mathbf{v}}(n)\|_1, \mathbf{v}(n))$ is a popular model in audio. It is the model that supports probabilistic latent component analysis (PLCA) and underlies a weighted KL divergence (Smaragdis *et al.*, 2006).

4) Finally, the multiplicative gamma noise model $\hat{v}(n,f) = v(n,f)\,\epsilon(n,f)$, where $\epsilon(n,f) \sim \mathcal{G}(\epsilon(n,f) \mid \alpha, \alpha)$ is gamma distributed with expectation 1, is a generative model of nonnegative data. It underlies the Itakura–Saito (IS) divergence. When $\hat{v}(n,f) = |x(n,f)|^2$ and the gamma shape parameter α equals one, i.e. the multiplicative noise has an exponential distribution, the model is equivalent to $x(n,f) \sim \mathcal{N}_c(x(n,f) \mid 0, \hat{v}(n,f))$, the so-called Gaussian composite model (Févotte *et al.*, 2009).

Except in the multinomial model, the observations are assumed conditionally independent, such that $p(\hat{\mathbf{V}} \mid \mathbf{BH}) = \prod_{nf} p(\hat{v}(n,f) \mid v(n,f))$. In the multinomial model, the observations are tied by the sum constraint (a sample from $\mathcal{M}(\cdot \mid N, \mathbf{p})$ sums to N) but nevertheless assumed to be conditionally independent in time.

9.3 Smooth NMF

9.3.1 Generalities

A straightforward approach to using *temporal continuity* is to apply some constraints that reduce fluctuations in each individual row of $\mathbf{H}$. This corresponds to the assumption that different rows of $\mathbf{H}$ are independent. Smoothing the rows of $\mathbf{H}$ is a way of capturing the temporal correlation of sound. Because this corresponds to a more physically realistic assumption, it can also improve the semantic relevance of the dictionary $\mathbf{B}$ and leads to more pleasant audio components in source separation scenarios. In this approach, the general equation (9.23) can be written as:

$$\mathbf{h}(n) \sim \prod_{k=1}^{K} p(h_k(n) \mid h_k(n-1)). \tag{9.24}$$

A natural choice for $p(h_k(n) \mid h_k(n-1))$ is a distribution that either takes its mode at $h_k(n-1)$ or is such that $\mathbb{E}\{h_k(n) \mid h_k(n-1)\} = h_k(n-1)$. A classical choice is the Gaussian random walk of the form

$$p(h_k(n) \mid h_k(n-1)) = \mathcal{N}(h_k(n) \mid h_k(n-1), \sigma^2), \tag{9.25}$$

which underlies the squared differences penalty

$$-\log p(\mathbf{H}) = \frac{1}{2\sigma^2} \sum_{kn} (h_k(n) - h_k(n-1))^2 + \text{cst.} \tag{9.26}$$

This choice of dynamical model has been used in a maximum a posteriori (MAP) setting with an additive Gaussian noise observation model by Chen *et al.* (2006) and with a Poisson observation model by Virtanen (2007) and Essid and Févotte (2013).

Like the additive Gaussian noise observation model, the Gaussian Markov chain does not comply with the nonnegative assumption of $\mathbf{H}$, from a generative perspective. As such, other works have considered other nonnegativity-preserving models based on gamma or *inverse-gamma* distribution. For instance, Févotte *et al.* (2009) proposes the use of Markov chains of the form

$$p(h_k(n) \mid h_k(n-1)) = \mathcal{IG}(h_k(n) \mid \alpha, (\alpha+1)h_k(n-1)) \tag{9.27}$$

or

$$p(h_k(n) \mid h_k(n-1)) = \mathcal{G}(h_k(n) \mid \alpha, (\alpha-1)/h_k(n-1)), \tag{9.28}$$

where $\mathcal{IG}$ refers to the inverse-gamma distribution defined in Section 9.7. Both priors are such that the mode of $p(h_k(n) \mid h_k(n-1))$ is $h_k(n-1)$. The shape parameter α controls the peakiness of the distribution and as such the correlation between the activations of adjacent frames. Févotte *et al.* (2009) describes an expectation-maximization (EM) algorithm for MAP estimation in the multiplicative noise observation model. Févotte (2011) describes a faster algorithm for a similar model where the dynamical model reduces to penalizing the IS data fitting term with the smoothing term $\sum_{kn} d^{\mathrm{IS}}(h_k(n) \mid h_k(n-1))$.

Finally, Virtanen *et al.* (2008) introduced the use of hierarchical gamma priors for smooth NMF. The construction of the chain involves a latent variable $z_k(n)$ such that

$$p(h_k(n) \mid z_k(n)) = \mathcal{G}(h_k(n) \mid \alpha_h, \alpha_h z_k(n)) \tag{9.29}$$

$$p(z_k(n) \mid h_k(n-1)) = \mathcal{G}(z_k(n) \mid \alpha_z + 1, \alpha_z h_k(n-1)). \tag{9.30}$$

The expectation of $h_k(n)$ given $h_k(n-1)$ is shown to be $h_k(n-1)$:

$$
\begin{aligned}
\mathbb{E}\{h_k(n) \mid h_k(n-1)\} &= \int h_k(n) p(h_k(n) \mid h_k(n-1)) \mathrm{d}h_k(n) \\
&= \int h_k(n) \left(\int p(h_k(n), z_k(n) \mid h_k(n-1)) \mathrm{d}z_k(n) \right) \mathrm{d}h_k(n) \\
&= \iint h_k(n) p(h_k(n) \mid z_k(n)) p(z_k(n) \mid h_k(n-1)) \mathrm{d}z_k(n) \mathrm{d}h_k(n) \\
&= \int \frac{1}{z_k(n)} p(z_k(n) \mid h_k(n-1)) \mathrm{d}h_k(n-1) = h_k(n-1).
\end{aligned}
\tag{9.31}
$$

As explained by Cemgil and Dikmen (2007), the hyperparameters α_h and α_z control the variance and skewness of $p(h_k(n) \mid h_k(n-1))$. The hierarchical gamma Markov chain offers a more flexible model than the plain gamma Markov chain, while offering computational advantages (conjugacy with the Poisson observation model). Cemgil and Dikmen (2007) and Dikmen and Cemgil (2010) have also investigated the hierarchical inverse-gamma Markov chain and mixed variants. Hierarchical gamma Markov chains have been considered in NMF under the Poisson observation model (Virtanen *et al.*, 2008; Nakano *et al.*, 2011; Yoshii and Goto, 2012).

9.3.2 A Special Case

For tutorial purposes, we now show how to derive a smooth NMF algorithm in a particular case. We will assume a multiplicative gamma noise model for $\hat{\mathbf{V}}$ and independent gamma Markov chains for $\mathbf{H}$. As explained earlier, the multiplicative gamma noise model is a truly generative model for nonnegative data that underlies a generative Gaussian variance model of complex-valued spectrograms when $\alpha = 1$. The gamma Markov chain is a simple model to work with – the proposed procedure can be generalized to more complex prior. The proposed procedure is a variant of that of Févotte (2011) and a special case of that of Févotte *et al.* (2013).

Our goal is to find a stationary point of the negative log-posterior:

$$C^{\text{MAP}}(\mathbf{B}, \mathbf{H}) = -\log p(\widehat{\mathbf{V}}, \mathbf{H} \mid \mathbf{B}) = -\log p(\widehat{\mathbf{V}} \mid \mathbf{BH}) - \log p(\mathbf{H}) \qquad (9.32)$$

where

$$p(\widehat{\mathbf{V}} \mid \mathbf{BH}) = \prod_{nf} \mathcal{G}(\widehat{v}(n,f) \mid \alpha, \alpha/v(n,f)) \qquad (9.33)$$

$$p(\mathbf{H}) = \prod_{k=1}^{K} \prod_{n=1}^{N-1} \mathcal{G}(h_k(n) \mid \alpha_h, \alpha_h/h_k(n-1)). \qquad (9.34)$$

These assumptions imply that $\mathbb{E}\{\widehat{\mathbf{V}} \mid \mathbf{BH}\} = \mathbf{BH}$ and $\mathbb{E}\{\mathbf{h}(n) \mid \mathbf{h}(n-1)\} = \mathbf{h}(n-1)$. In the following we will assume by convention (and for simplicity) that $h_k(-1) = h_k(N) = 1$. It is easily found that

$$-\log p(\widehat{\mathbf{V}} \mid \mathbf{BH}) = \alpha \sum_{nf} \frac{\widehat{v}(n,f)}{v(n,f)} + \log v(n,f) + \text{cst} \qquad (9.35)$$

and

$$-\log p(\mathbf{H})$$
$$= \sum_{kn} \left[\alpha_h \frac{h_k(n)}{h_k(n-1)} + \alpha_h \log h_k(n-1) + (1-\alpha_h) \log h_k(n) \right] + \text{cst}, \qquad (9.36)$$

where we recall that cst denotes the terms independent of the parameters **B** or **H**. Typical NMF algorithms proceed with alternate updates of **B** and **H**. The update of **B** given the current estimate of **H** boils downs to standard IS-NMF and can be performed with standard multiplicative rules (see Chapter 8). The norm of **B** should, however, be controlled (via normalization or penalization) so as to avoid degenerate solutions such that $\|\mathbf{B}\| \to +\infty$ and $\|\mathbf{H}\| \to 0$. Indeed, let $\mathbf{\Lambda}$ be a nonnegative diagonal matrix with coefficients $\{\lambda_k\}$. We have:

$$C^{\text{MAP}}(\mathbf{B}\mathbf{\Lambda}^{-1}, \mathbf{\Lambda}\mathbf{H}) = C^{\text{MAP}}(\mathbf{B}, \mathbf{H}) + N \sum_k \log \lambda_k \qquad (9.37)$$

which shows how degenerate solutions can be obtained by letting λ_k go to zero (Févotte, 2011; Févotte *et al.*, 2013). We now concentrate on the update of **H** given **B**. As can be seen from (9.35) and (9.36), adjacent columns of **H** are coupled in the optimization. We propose a left-to-right *block coordinate descent* approach that updates $\mathbf{h}(n)$ sequentially. As such, the optimization of (9.32) with respect to **H** involves the sequential optimization of

$$C^{\text{MAP}}(\mathbf{h}(n)) = \alpha \sum_f \left[\frac{\widehat{v}(n,f)}{\sum_k b_k(f) h_k(n)} + \log \sum_k b_k(f) h_k(n) \right]$$
$$+ \sum_k \left[\log h_k(n) + \alpha_h \left(\frac{h_k(n)}{\widetilde{h}_k(n-1)} + \frac{\widetilde{h}_k(n+1)}{h_k(n)} \right) \right] \qquad (9.38)$$

where $\tilde{h}_k(n-1)$ and $\tilde{h}_k(n+1)$ denote the values of $h_k(n-1)$ and $h_k(n+1)$ at the current and previous iteration, respectively. The minimum of $C^{\text{MAP}}(\mathbf{h}(n))$ does not have a closed-form expression and we need to resort to numerical optimization. A handy choice, very common in NMF, is to use *majorization-minimization* (MM). It consists of replacing the infeasible closed-form minimization of $C^{\text{MAP}}(\mathbf{h}(n))$ by the iterative minimization of an upper bound $Q(\mathbf{h}(n), \tilde{\mathbf{h}}(n))$ that is locally tight in the current parameter estimate $\tilde{\mathbf{h}}(n)$, see, for example, Févotte and Idier (2011) and Smaragdis *et al.* (2014). Denoting $\tilde{v}(n,f) = \sum_k b_k(f)\tilde{h}_k(n)$, the following inequalities apply, with equality for $\mathbf{h}(n) = \tilde{\mathbf{h}}(n)$. By convexity of $1/x$ and Jensen's inequality, we have:

$$\frac{1}{\sum_k b_k(f) h_k(n)} \leq \frac{1}{\tilde{v}^2(n,f)} \sum_k b_k(f) \frac{\tilde{h}_k^2(n)}{h_k(n)}. \tag{9.39}$$

By concavity of $\log x$ and the tangent inequality, we have

$$\log h_k(n) \leq (\log \tilde{h}_k(n) - 1) + \frac{h_k(n)}{\tilde{h}_k(n)}, \tag{9.40}$$

$$\log \sum_k b_k(f) h_k(n) \leq (\log \tilde{v}(n,f) - 1) + \frac{1}{\tilde{v}(n,f)} \sum_k b_k(f) h_k(n) \tag{9.41}$$

Plugging the latter inequalities in (9.38), we obtain

$$Q(\mathbf{h}(n), \tilde{\mathbf{h}}(n)) = \left(\alpha \tilde{q}_k(n) + \frac{1}{\tilde{h}_k(n)} + \frac{\alpha_h}{\tilde{h}_k(n-1)} \right) h_k(n)$$

$$+ (\alpha \tilde{p}_k(n) \tilde{h}_k^2(n) + \alpha_h \tilde{h}_k(n+1)) \frac{1}{h_k(n)} + \text{cst} \tag{9.42}$$

with

$$\tilde{p}_k(n) = \sum_f b_k(f) \frac{\hat{v}(n,f)}{\tilde{v}^2(n,f)}, \quad \tilde{q}_k(n) = \sum_f \frac{b_k(f)}{\tilde{v}(n,f)} \tag{9.43}$$

Minimization of $Q(\mathbf{h}(n), \tilde{\mathbf{h}}(n))$ leads to

$$h_k(n) = \left(\frac{\alpha \, \tilde{p}_k(n) \tilde{h}_k^2(n) + \alpha_h \tilde{h}_k(n+1)}{\alpha \, \tilde{q}_k(n) + \tilde{h}_k^{-1}(n) + \alpha_h \tilde{h}_k^{-1}(n-1)} \right)^{\frac{1}{2}}. \tag{9.44}$$

9.3.3 Illustrative Example

Following Févotte (2011), we consider for illustration the decomposition of a 108 s long music excerpt from *My Heart (Will Always Lead Me Back To You)* recorded by Louis Armstrong and His Hot Five in the 1920s. The band features a trumpet, a clarinet, a trombone, a piano, and a double bass. A STFT $\mathbf{X} = [x(n,f)]_{fn}$ of the original signal x (sampled at 11 kHz) was computed using a sine analysis window of length $L = 256$ (23 ms) with 50% overlap, leading to $F = 129$ frequency bins and $N = 9312$ frames. To illustrate the effect of smoothing of the rows of $\mathbf{H}$ we performed the following experiment. First we ran unpenalized NMF with the IS divergence (IS-NMF) with $K = 10$, retaining the solution with lowest final cost value among ten runs from 10 random initializations. Then we ran the smooth IS-NMF algorithm presented in Section 9.3.2 with

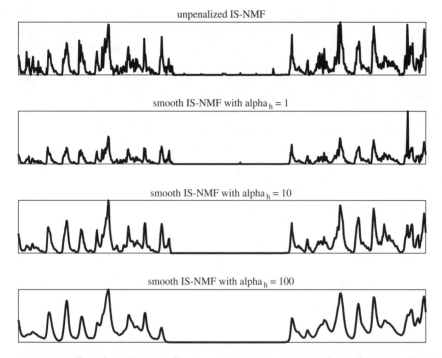

Figure 9.5 Effect of regularization for $\alpha_h = \{1, 10, 100\}$. A segment of one of the rows of **H** is displayed, corresponding to the activations of the accompaniment (piano and double bass). A trumpet solo occurs in the middle of the displayed time interval, where the accompaniment vanishes; the regularization smoothes out coefficients with small energies that remain in unpenalized IS-NMF.

B and **H**, respectively, fixed and initialized to the unpenalized solution. Figure 9.5 reports the results with $\alpha = 1$ and different values of α_h. It shows how the value of the hyperparameter α_h controls the degree of smoothness of **H**. Some works have addressed the estimation of this hyperparameter together with **B** and **H**, e.g. Dikmen and Cemgil (2010).

9.4 Nonnegative State-Space Models

9.4.1 Generalities

Smooth NMF does not capture the full extent of frame-to-frame dependencies in its input. In practice we will observe various temporal correlations between adjacent time frames which will be more nuanced than the continuity that smooth NMF implies. In other words, there is correlation both across (smoothness) and between (transitions) the coefficients of **H**. For real-valued time series, this type of structure can be handled with the classical linear dynamical system, using dynamics of the form $\mathbf{h}(n) = \mathbf{D}\mathbf{h}(n-1) + \epsilon(n)$, where $\epsilon(n)$ is a centered Gaussian innovation. This model is not natural in the NMF setting because it may not maintain nonnegativity in the activations. However, it is possible to design alternative dynamic models that maintain nonnegativity while preserving

$$\mathbb{E}\{\mathbf{h}(n) \mid \mathbf{D}\mathbf{h}(n-1)\} = \mathbf{D}\mathbf{h}(n-1). \tag{9.45}$$

A *nonnegative dynamical system* with multiplicative gamma innovations was proposed by Févotte *et al.* (2013) in conjunction with multiplicative gamma noise for the observation (IS-NMF model), similar to the model considered in Section 9.3.2. Note that in the case of the Gaussian linear dynamical system, integration of the activation coefficients from the joint likelihood $p(\widehat{\mathbf{V}}, \mathbf{H} \mid \mathbf{B})$ is feasible using the Kalman filter. Such computations are unfortunately intractable with a nonnegative dynamical system, and a MAP approach based on an MM algorithm, as in Section 9.3.2, is pursued by Févotte *et al.* (2013).

Dynamic filtering of the activation coefficients in the PLCA model has also been considered by Nam *et al.* (2012) and Mohammadiha *et al.* (2013), where the proposed algorithms use Kalman-like prediction strategies. Mohammadiha *et al.* (2013) consider a more general multistep predictor such that $\mathbf{h}(n) \approx \sum_{n'} \mathbf{D}(n')\mathbf{h}(n - n')$, and describe an approach for both the smoothing (which relies on both past and future data) and causal filtering (which relies only on past data) problems.

9.4.2 A Special Case

In this section we review the dynamic NMF model of Mohammadiha *et al.* (2015) which uses a continuous state-space approach to utilize the temporal dependencies in NMF. The model and underlying assumptions are described here and the main derivation steps are reviewed.

9.4.2.1 Statistical Model

We consider a dynamic NMF model in which the NMF coefficients $\mathbf{h}(n)$ are assumed to evolve over time according to the following *nonnegative vector autoregressive* (AR) model:

$$p(\widehat{\mathbf{v}}(n) \mid \mathbf{B}, \mathbf{h}(n)) = \mathcal{M}(\widehat{\mathbf{v}}(n) \mid \gamma(n), \mathbf{B}\mathbf{h}(n)) \tag{9.46}$$

$$p(\mathbf{h}(n) \mid \mathbf{D}, \mathbf{h}(n - 1), \dots, \mathbf{h}(n - N')) = \mathcal{E}\left(\mathbf{h}(n) \,\middle|\, \sum_{n'=1}^{N'} \mathbf{D}(n')\mathbf{h}(n - n') \right), \tag{9.47}$$

where $\gamma(n) = \sum_f \widehat{v}(n,f)$, N' is the order of the nonnegative vector AR model, $\mathbf{D}(n')$ is a $K \times K$ matrix, $\mathbf{D}$ denotes the union of $\mathbf{D}(n')$, $\forall n'$, and $\mathcal{E}(\cdot \mid \cdot)$ and $\mathcal{M}(\cdot \mid \cdot, \cdot)$ refer to the exponential and multinomial distributions defined in Section 9.9. Eq. (9.46) defines a PLCA observation model, in which the columns of $\mathbf{B}$ and $\mathbf{H}$ are assumed to sum to 1 so that $\mathbf{B}\mathbf{h}(n)$ defines a discrete probability distribution.

The conditional expected values of $\mathbf{h}(n)$ and $\widehat{\mathbf{v}}(n)$ under the model (9.46)–(9.47) are given by:

$$\mathbb{E}\{\mathbf{h}(n) \mid \mathbf{D}, \mathbf{h}(n - 1), \dots \mathbf{h}(n - N')\} = \sum_{n'=1}^{N'} \mathbf{D}(n')\mathbf{h}(n - n'), \tag{9.48}$$

$$\mathbb{E}\{\widehat{\mathbf{v}}(n) \mid \mathbf{B}, \mathbf{h}(n)\} = \left(\sum_f \widehat{v}(n,f) \right) \mathbf{B}\mathbf{h}(n), \tag{9.49}$$

which is used to obtain an NMF approximation of the input data as $\widehat{\mathbf{v}}(n) \approx \left(\sum_f \widehat{v}(n,f) \right) \mathbf{B}\mathbf{h}(n)$.

The distributions in (9.46)–(9.47) are chosen to be appropriate for nonnegative data. For example, it is well known that the conjugate prior for the multinomial likelihood is the Dirichlet distribution. However, it can be shown that the obtained state estimates in this case are no longer guaranteed to be nonnegative. Therefore, the exponential distribution is used in (9.46) for which, as will be shown later in Section 9.4.2.2, the obtained state estimates are always nonnegative.

As already mentioned, if we discard (9.47) we recover the basic PLCA model of Smaragdis *et al.* (2006). In this formulation, the observations $\hat{\mathbf{v}}(n)$ are assumed to be count data over F possible categories. Each vector $\mathbf{h}(n)$ is a probability vector that represents the contribution of each basis vector in explaining the observation, i.e. $h_k(n) = P(z(n) = k)$ where $z(n)$ is a latent variable used to index the basis vectors at time n. Moreover, each column of $\mathbf{B}$ is a probability vector that contains the underlying structure of the observations given the latent variable z and is referred to as a basis vector. More precisely, $b_k(f)$ is the probability that the fth element of $\hat{\mathbf{v}}(n)$ will be chosen in a single draw from the multinomial distribution in (9.46), i.e. $b_k(f) = P(\hat{\mathbf{v}}(n) = \mathbf{e}(f) \mid z(n) = k)$ with $\mathbf{e}(f)$ being an $F \times 1$ indicator vector whose fth element is equal to one (see Mohammadiha *et al.* (2013) for more explanation). Note that (by definition) $b_k(f)$ is time-invariant. In the following, this notation is abbreviated to $b_k(f) = P(f \mid z(n) = k)$.

It is worthwhile to compare (9.46)–(9.47) to the state-space model utilized in the Kalman filter and to highlight the main differences between the two. First, all the variables are constrained to be nonnegative in (9.46)–(9.47). Second, the process and observation noises are embedded into the specified distributions, which is different from the additive Gaussian noise utilized in the Kalman filtering. Finally, in the process equation, a multilag nonnegative vector AR model is used. It is also important to note that both state-space model parameters ($\mathbf{B}$ and $\mathbf{D}$) and state variables $\mathbf{H}$ should be estimated simultaneously.

In the following section, an EM algorithm is derived to compute maximum likelihood (ML) estimates of $\mathbf{D}$ and $\mathbf{B}$ and to compute a MAP estimate of the state variables $\mathbf{H}$. In the latter case, the estimation consists of prediction/propagation and update steps, similarly to the classical Kalman filter. However, a nonlinear update function is derived here in contrast to the linear additive update at Kalman filtering.

9.4.2.2 Estimation Algorithm

Let us denote the nonnegative parameters in (9.46)–(9.47) by $\theta = \{\mathbf{D}, \mathbf{H}, \mathbf{B}\}$. Given a nonnegative data matrix $\hat{\mathbf{V}}$, θ can be estimated by maximizing the MAP objective function for the model in (9.46)–(9.47), i.e. as

$$\mathcal{M}^{\mathrm{MAP}}(\theta) = \log p(\hat{\mathbf{V}}, \mathbf{H} \mid \mathbf{B}, \mathbf{D}) \tag{9.50}$$

$$= \log p(\hat{\mathbf{V}} \mid \mathbf{B}, \mathbf{H}) + \log p(\mathbf{H} \mid \mathbf{D}). \tag{9.51}$$

Maximizing $\mathcal{M}^{\mathrm{MAP}}(\theta)$ with respect to $\mathbf{B}, \mathbf{D}$, and $\mathbf{H}$ results in a MAP estimate of $\mathbf{H}$ and ML estimates of $\mathbf{B}$ and $\mathbf{D}$. For this maximization, an EM algorithm is derived by Mohammadiha *et al.* (2015) to iteratively update the parameters. EM is a commonly used approach to estimate the unknown parameters in the presence of latent variables, where a lower bound on $\mathcal{M}^{\mathrm{MAP}}(\theta)$ is maximized by iterating between an expectation (E) step and a maximization (M) step until convergence (Dempster *et al.*, 1977). It is a particular form of MM algorithm where the construction of the upper bound relies on the posterior of the latent variables and the complete likelihood. In our setting these are the variables $z(n)$ that index the basis vectors. In the E step, the posterior probabilities of these variables are obtained as:

$$P(z(n) = k \mid f, \tilde{\theta}) = \frac{P(f \mid z(n) = k)P(z(n) = k)}{\sum_{k=1}^{K} P(f \mid z(n) = k)P(z(n) = k)} \tag{9.52}$$

$$= \frac{\tilde{b}_k(f)\tilde{h}_k(n)}{\sum_k \tilde{b}_k(f)\tilde{h}_k(n)}, \tag{9.53}$$

where $\tilde{\theta}$ denotes the estimated parameters from the previous iteration of the EM algorithm. In the M step, the expected log-likelihood of the complete data:

$$Q(\theta, \tilde{\theta}) = \sum_{kn} P(z(n) = k \mid f, \tilde{\theta}) \log p(\hat{\mathbf{v}}(n), z(n) \mid \theta)$$
$$+ \sum_{n} \log p(\mathbf{h}(n) \mid \mathbf{D}, \mathbf{h}(n-1), \dots \mathbf{h}(n-N')) \tag{9.54}$$

is maximized with respect to θ to obtain a new set of estimates. As previously, we assume by convention $\tilde{h}_k(n - N') = 1$ for $n < N'$. $Q(\theta, \tilde{\theta})$ can be equivalently (up to a constant) written as:

$$Q(\theta, \tilde{\theta}) = \sum_{kfn} \hat{v}_{fn} P(z(n) = k \mid f, \tilde{\theta})(\log b_k(f) + \log h_k(n))$$
$$- \sum_{kn} \left(\log \eta_k(n) + \frac{h_k(n)}{\eta_k(n)} \right), \tag{9.55}$$

where $\eta(n) = \sum_{n'=1}^{N'} \mathbf{D}(n')\mathbf{h}(n - n')$. As mentioned in Section 9.4.2.1, $\mathbf{b}(k)$ and $\mathbf{h}(n)$ are probability vectors, and hence, to make sure that they sum to one, we need to impose two constraints $\sum_k h_k(n) = 1$ and $\sum_f b_k(f) = 1$. To solve the constrained optimization problem, we form the *Lagrangian* function $\mathcal{L}$ and maximize it:

$$\mathcal{L}(\theta, \alpha, \beta) = Q(\theta, \tilde{\theta}) + \sum_k \alpha_k \left(1 - \sum_f b_k(f) \right) + \sum_n \beta_n \left(1 - \sum_k h_k(n) \right), \tag{9.56}$$

where $\alpha = \{\alpha_k\}_k$ and $\beta = \{\beta_n\}_n$ are Lagrange multipliers. In the following, the maximization with respect to $\mathbf{B}$, $\mathbf{H}$, and $\mathbf{D}$ are successively presented.

Equation (9.56) can be easily maximized with respect to $\mathbf{B}$ to obtain:

$$b_k(f) = \frac{\sum_n \hat{v}(n, f) P(z(n) = k \mid f, \tilde{\theta})}{\alpha_k}, \tag{9.57}$$

where the Lagrange multiplier $\alpha_k = \sum_{nf} \hat{v}(n, f) P(z(n) = k \mid f, \tilde{\theta})$ ensures that $\mathbf{b}_k$ sums to one. Maximization with respect to $\mathbf{H}$ leads to a recursive algorithm, where $\mathbf{h}(1), \mathbf{h}(2), \dots$ are estimated sequentially, as in Section 9.3.2. The partial derivative of $\mathcal{L}$ with respect to $h_k(n)$ is set to zero to obtain

$$h_k(n) = \frac{\sum_f \hat{v}(n, f) P(z(n) = k \mid f, \tilde{\theta})}{\beta_n + 1/\eta_k(n)}, \tag{9.58}$$

where $\eta(n)$ is defined after (9.55). The Lagrange multiplier β_n has to be computed such that $\mathbf{h}(n)$ sums to one. This can be done using an iterative Newton's method (Mohammadiha *et al.*, 2013). Finally, we attend to the estimation of the nonnegative vector AR model parameters $\mathbf{D}$. Note that there are many approaches to estimate vector AR model parameters in the literature (Hamilton, 1994; Lütkepohl, 2005). However, since most of these approaches are based on least squares estimation, they are not suitable for a nonnegative framework. Moreover, they tend to be very time-consuming for high-dimensional data. First, note that $\mathbf{D}$, which is defined as $\mathbf{D} = [\mathbf{D}(1)\,\mathbf{D}(2)\dots\mathbf{D}(N')]$, is a $K \times KN'$ matrix. Let $KN' \times 1$ vector $\bar{\mathbf{h}}(n)$ represent the stacked state variables as $\bar{\mathbf{h}}(n) = [\mathbf{h}^T(n-1), \mathbf{h}^T(n-2), \dots, \mathbf{h}^T(n-N')]^T$. The parts of (9.56) that depend on $\mathbf{D}$ are equivalently written as:

$$\mathcal{L}(\mathbf{D}) = -\sum_{kn}\left(\log\left(\mathbf{D}\bar{\mathbf{h}}(n)\right)_k + \frac{h_k(n)}{(\mathbf{D}\bar{\mathbf{h}}(n))_k}\right) + \text{cst} \tag{9.59}$$

$$= -C^0(\mathbf{H} \mid \mathbf{D}\bar{\mathbf{H}}) + \text{cst}, \tag{9.60}$$

where $\bar{\mathbf{H}} = [\bar{\mathbf{h}}(0)\dots\bar{\mathbf{h}}(N-1)]$, $(\cdot)_k$ denotes the kth entry of its argument, and $C^0(\cdot \mid \cdot)$ is the IS divergence as defined in (8.6). Hence, the ML estimate of $\mathbf{D}$ can be obtained by performing IS-NMF in which the NMF coefficient matrix $\bar{\mathbf{H}}$ is held fixed and only the basis matrix $\mathbf{D}$ is optimized. This is done by executing

$$\mathbf{D} = \mathbf{D} \circ \frac{((\mathbf{D}\bar{\mathbf{H}})^{-2} \circ \mathbf{H})\bar{\mathbf{H}}^T}{(\mathbf{D}\bar{\mathbf{H}})^{-1}\bar{\mathbf{H}}^T}, \tag{9.61}$$

iteratively until convergence, with initial values $\mathbf{D} = \tilde{\mathbf{D}}$. Alternatively, (9.61) can be repeated only once, resulting in a generalized EM algorithm.

In the supervised source separation or speech enhancement, the presented dynamic NMF approach can be used to estimate all the model parameters simultaneously using the training data from individual sources. As convergence criterion, the stationarity of $\mathcal{M}^{\text{MAP}}$ or EM lower bound can be checked, or a fixed (sufficient) number of iterations can be simply used. In the testing step, $\mathbf{B}$ and $\mathbf{D}$ are held fixed and only the the state variables $\mathbf{H}$ are estimated from the mixture input.

9.5 Discrete Dynamical Models

9.5.1 Generalities

Time series data often have a hidden structure in which each time frame corresponds to a discrete hidden state $q(n)$. Moreover, there is typically a relationship between the hidden states at different time frames, in the form of temporal dynamics. For example, each time frame of a speech signal corresponds to a subunit of speech such as a phoneme, which can be modeled as a distinct state. The subunits evolve over time as governed by temporal dynamics. *Hidden Markov models* (HMMs) (Rabiner, 1989) have been used extensively to model such data. They model temporal dynamics with a transition matrix defined by the distribution $P(q(n) \mid q(n-1))$. There is a thread of literature (Ozerov *et al.*, 2009; Mysore *et al.*, 2010; Mysore and Smaragdis, 2011; Nakano *et al.*, 2010; Mohammadiha and Leijon, 2013) that combines these ideas with NMF to model nonnegative data with such a structure.

The notion of a state is incorporated in the NMF framework by associating distinct dictionary elements with each state. This is done by allowing each state to determine a different support of the activations, which we express with the distribution $p(\mathbf{h}(n) \mid q(n))$. This is to say that given a state, the model allows only certain dictionary elements to be active. Some techniques (Ozerov *et al.*, 2009; Nakano *et al.*, 2010) define the support of each state to be a single dictionary element, while other techniques (Mysore *et al.*, 2010; Mysore and Smaragdis, 2011; Mohammadiha and Leijon, 2013), called *nonnegative HMMs*, allow the support of each state to be a number of dictionary elements. Since only a subset of the dictionary elements are active at each time frame (as determined by the state at that time frame), we can interpret these models as imposing block sparsity on the dictionary elements (Mysore, 2012).

As in (9.23), there is a dependency between $\mathbf{h}(n)$ and $\mathbf{h}(n-1)$. However, unlike the continuous models, this dependency is only through the hidden states, which are in turn related through the temporal dynamics. Therefore $\mathbf{h}(n)$ is conditionally independent of $\mathbf{h}(n-1)$ given $q(n)$ or $q(n-1)$. In the case of discrete models, we can therefore replace (9.23) with

$$q(n) \sim P(q(n) \mid q(n-1)), \tag{9.62}$$

$$\mathbf{h}(n) \sim p(\mathbf{h}(n) \mid q(n)). \tag{9.63}$$

Since these models incorporate an HMM structure into an NMF framework, one can make use of the vast theory of Markov chains to extend these models in various ways. For example, one can incorporate high-level knowledge of a particular class of signals into the model, use higher order Markov chains, or use various natural language processing techniques. Language models were incorporated in this framework (Mysore and Smaragdis, 2012) as typically done in the speech recognition literature (Rabiner, 1989). Similarly, one can incorporate other types of temporal structure, such as music theory rules when modeling music signals.

The above techniques discuss how to model a single source using an HMM structure. However, in order to perform source separation, we need to model mixtures. This is typically done by combining the individual source models into a *nonnegative factorial HMM* (Ozerov *et al.*, 2009; Mysore *et al.*, 2010; Mysore and Smaragdis, 2011; Nakano *et al.*, 2011; Mohammadiha and Leijon, 2013), which allows each source to be governed by a distinct pattern of temporal dynamics. One issue with this strategy is that the computational complexity of inference is exponential in the number of sources. This can be circumvented using approximate inference techniques such as variational Bayesian (VB) inference (Mysore and Sahani, 2012), which makes the complexity linear in the number of sources.

9.5.2 A Special Case

We describe here the specific nonnegative HMM and nonnegative factorial HMM models of Mysore *et al.* (2010). For a detailed derivation, see Mysore (2010). In the nonnegative HMM, each state q corresponds to a distinct dictionary, which is to say that a different subset of the dictionary elements in the model are associated with each state and we in turn call these subsets, dictionaries. There is therefore a one-to-one correspondence between states and dictionaries.

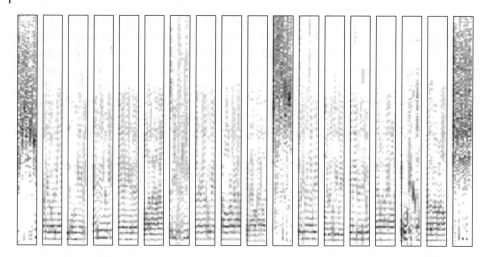

Figure 9.6 Dictionaries were learned from the speech data of a given speaker. Shown are the dictionaries learned for 18 of the 40 states. Each dictionary is composed of 10 elements that are stacked next to each other. Each of these dictionaries roughly corresponds to a subunit of speech, either a voiced or unvoiced phoneme.

An example of the dictionaries learned from a sample of speech is shown in Figure 9.6. Each dictionary in the figure is composed of dictionary elements that are stacked next to each other. Notice the visual similarity of these dictionaries with the dictionary elements learned from convolutive NMF shown in Figure 9.4. However, convolutive NMF dictionaries are defined over multiple time frames, so they can only model well data that is of the same fixed length as the dictionaries. For example, they can model a drum hit quite well, but have less flexibility to model data such as phonemes of speech with varying lengths. On the other hand, nonnegative HMMs model each time frame (with temporal dependencies between time frames) and therefore have more flexibility. They can model phonemes of speech of varying lengths quite well, but the increased flexibility comes with potential decreased accuracy for fixed length events such as a drum hit.

Similar to the dynamic NMF model in Section 9.4.2, the nonnegative HMM is a dynamic extension of PLCA. We now briefly describe the model and parameter estimation for the nonnegative HMM. The dictionary element k from dictionary (state) q is defined by a discrete distribution $P(f \mid k, q)$, which shows the relative magnitude of the frequency bins for that dictionary element. It is, by definition, time-invariant. At time n, the activation of dictionary element k from dictionary q is given by a discrete distribution $P(k(n) \mid q(n))$.

The complete set of distributions that form the nonnegative HMM are defined below, and include the two distributions mentioned above. Each of these distributions, except for the energy distributions, are discrete distributions. The energy distributions are Gaussian distributions.

1) Dictionary elements: $P(f \mid k, q)$ defines the dictionary element k of dictionary q. Unlike the previous models that were discussed in this chapter, in the nonnegative HMM there is a grouping of the dictionary elements (columns of **B**). The **B** matrix is essentially a concatenation of the individual dictionaries of the nonnegative HMM. Therefore, the dictionary q and dictionary element k together define a column of **B**.

2) Activations: $P(k(n) \mid q(n))$ defines the activations that correspond to dictionary q at time n. The concatenation of these activations for all dictionaries weighted by the relative weighting of the individual dictionaries corresponds to $\mathbf{h}(n)$ given by $P(\mathbf{h}(n) \mid q(n))$. This relative weighting at a given time frame is governed by the temporal dynamics mentioned below.

3) Transition matrix: $P(q(n) \mid q(n-1))$ defines a standard HMM transition matrix (Rabiner, 1989).

4) Prior probabilities: $P(q(1))$ defines a distribution over states at the first time frame.

5) Energy distributions: $p(g(n) \mid q(n))$ defines a distribution of the energies of state q, which intuitively corresponds to the range of observed loudness of each state.

Given the spectrogram $\widehat{\mathbf{V}}$ of a sound source, we use the EM algorithm to learn the model parameters of the nonnegative HMM. The E-step is computed as follows:

$$P(k(n), q(n) \mid f(n), \widehat{\mathbf{V}}, \boldsymbol{\gamma}) = \frac{\alpha(q(n))\beta(q(n))}{\sum_{q(n)} \alpha(q(n))\beta(q(n))} P(k(n) \mid f(n), q(n)) , \tag{9.64}$$

where

$$P(k(n) \mid f(n), q(n)) = \frac{P(k(n) \mid q(n))P(f(n) \mid k(n), q(n))}{\sum_{k(n)} P(k(n) \mid q(n))P(f(n) \mid k(n), q(n))} . \tag{9.65}$$

$P(k(n), q(n) \mid f(n), \widehat{\mathbf{V}}, \boldsymbol{\gamma})$ is the posterior distribution that is used to estimate the dictionary elements and activations. $\boldsymbol{\gamma}$ denotes the number of draws over each of the time frames in the spectrogram $(\gamma(1), \dots, \gamma(N))$. The number of draws in a given time frame intuitively corresponds to how loud the signal is at that time frame. The number of draws over all time frames is simply this information for the entire signal. Note that in spite of the dictionary elements $P(f \mid k, q)$ being time invariant, they are given a time index n in the above equation. This is simply done in order to correspond to the values of f, k, and q referenced at time n in the left-hand side of the equation, and is constant for all values of n.

The *forward-backward* variables $\alpha(q(n))$ and $\beta(q(n))$ are computed using the likelihoods of the data, $p(\widehat{v}(n), g(n) \mid q(n))$, for each state (as in classical HMMs (Rabiner, 1989)). The likelihoods are computed as follows:

$$p(\widehat{v}(n), g(n) \mid q(n))$$
$$= p(g(n) \mid q(n)) \prod_{f(n)} \left(\sum_{k(n)} P(f(n) \mid k(n), q(n))P(k(n) \mid q(n)) \right)^{\lambda v(n,f)} , \tag{9.66}$$

where λ is scaling factor.

The dictionary elements and their weights are estimated in the M-step as follows:

$$P(f \mid k, q) = \frac{\sum_n \widehat{v}(n, f)P(k(n), q(n) \mid f(n), \widehat{\mathbf{V}}, \boldsymbol{\gamma})}{\sum_{f(n)} \sum_n \widehat{v}(n, f)P(k(n), q(n) \mid f(n), \widehat{\mathbf{V}}, \boldsymbol{\gamma})} , \tag{9.67}$$

$$P(k(n) \mid q(n)) = \frac{\sum_{f(n)} \widehat{v}(n, f)P(k(n), q(n) \mid f(n), \widehat{\mathbf{V}}, \boldsymbol{\gamma})}{\sum_{k(n)} \sum_{f(n)} \widehat{v}(n, f)P(k(n), q(n) \mid f(n), \widehat{\mathbf{V}}, \boldsymbol{\gamma})} . \tag{9.68}$$

The transition matrix, $P(q(n) \mid q(n-1))$, and prior probability, $P(q(1))$, are computed exactly as in classical HMMs (Rabiner, 1989). The mean and variance of $p(g \mid q)$ are also learned from the data.

Nonnegative HMMs are learned from isolated training data of sounds sources. Once these models are learned, they can be combined into a nonnegative factorial HMM and used for source separation. If trained nonnegative HMMs are available for all sources (e.g., separation of speech from multiple speakers), then supervised source separation can be performed (Mysore *et al.*, 2010). This can be done efficiently using VB inference (Mysore and Sahani, 2012). If nonnegative HMMs are available for all sources except for one (e.g., separation of speech and noise), then source separation can be performed using semi-supervised separation (Mysore and Smaragdis, 2011).

9.6 The Use of Dynamic Models in Source Separation

In order to demonstrate the utility of dynamic models in context, we use a real-world source separation example. This time it will be an acoustic mixture of speech mixed with background noise from a factory (using the TIMIT and NOISEX-92 databases). The mixture is shown using a magnitude STFT representation in Figure 9.7. This particular case is interesting because of the statistics of speech. We note that human speech tends to have a smooth acoustic trajectory, which means that there is a strong temporal correlation between adjacent time frames. On the other hand, we also know that speech has a strong discrete hidden structure which is associated with the sequence of spoken phonemes. These properties make this example a good candidate for demonstrating the differences between the methods discussed so far and their effects on source separation.

We performed source separation using the three main approaches that we covered in this chapter. These include a static PLCA model (Smaragdis *et al.*, 2007), a dynamic PLCA model (Mohammadiha *et al.*, 2013), and a nonnegative HMM (Mysore *et al.*, 2010). In all three cases, we trained a model for speech and a model for background noise form training data. The dictionary size for the noise was fixed to 30 elements, whereas the speech model had 60 dictionary elements for PLCA and dynamic PLCA, and 40 states with 10 dictionary elements each for the nonnegative HMM. For the dynamic models, we learned the temporal statistics as well. In order to separate a mixture of test data of the sources, we fixed the learned **B** matrices for both the speech and noise models, and estimated their respective activations **H** using the context of each model. In Figure 9.7 we show the reconstruction of speech using each model. We also show objective metrics using SDR, SIR, and SAR (defined in Chapter 1) to evaluate the quality of separation in each case. These results are averaged over 20 different speakers to reduce biasing and initialization effects.

For the static PLCA model, we see that there is a detectable amount of visible suppression of the background noise, which amounts to a modest SIR of about 5dB. The dynamic PLCA model, on the other hand, by taking advantage of the temporal statistics of speech, does a much better job, resulting in more than double the SIR. Note, however, that in the process of adhering to the expected statistics, it introduces artifacts, which result in a lower SAR as compared to the static model. The nonnegative HMM results

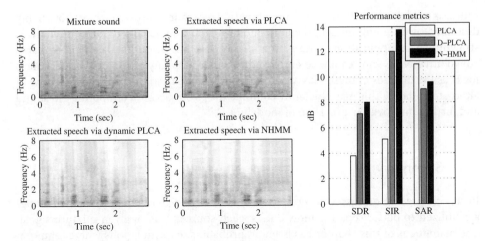

Figure 9.7 Example of dynamic models for source separation. The four spectrograms show the mixture and the extracted speech for three different approaches. D-PLCA denotes dynamic PLCA and N-HMM denotes nonnegative HMM. The bar plot shows a quantitative evaluation of the separation performance of each approach. Adapted from Smaragdis *et al.* (2014).

in an even higher SIR and a better SAR than the dynamic PLCA model. This is because the specific signal we are modeling has a temporal structure that is well described by a discrete dynamic model as we transition from phoneme to phoneme. By constraining our model to only use a small dictionary at each discrete state, we obtain a cleaner estimate of the source. An example of that can be seen when comparing the separation results in Figure 9.7, where unwanted artifacts between the harmonics of speech in the dynamic PLCA example are not present in the nonnegative HMM example since the dictionary elements within a state cannot produce such complex spectra.

9.7 Which Model to Use?

In addition to pondering on which cost function is the most appropriate to employ, we also have a decision to make on which model is best for a source separation approach. As always the answer depends on the nature of the sources in the mixture. In general the static model has found success in a variety of areas, but does not take advantage of temporal correlations. In domains where we do not expect a high degree of correlations across time (e.g., short burst-like sources) this model works well, but in cases where we expect a strong sense of continuity (e.g., a smooth source like a whale song), then a continuous dynamic model would work better. Furthermore, if we know that a source exhibits a behavior of switching through different states, each with its own unique character (e.g., speech), then a model like the nonnegative HMM is more appropriate since it will eliminate the concurrent use of elements that belong at different states and produce a more plausible reconstruction. Of course by using the generalized formulation we use in this article, there is nothing that limits us from employing different models concurrently. It is entirely plausible to design a source separation system where one source is modeled by a static model and other by a dynamic one, or even have both being described by different kinds of dynamic models. Doing so usually requires a relatively straightforward application of the estimation process that we outlined earlier. Similarly, convolutive NMF can readily be employed together with any of the proposed models for **H** as the updates of these two variables are independent in the considered setting of alternate updates. This may efficiently combine the two sources of temporality in sound and further improve the precision of modeling.

9.8   Summary

In this chapter we discussed several extensions of NMF where temporal dependencies are utilized to better separate individual speech sources from a given mixture signal. The main focus of this chapter has been on probabilistic formulations where temporal dependencies can be utilized to build informative prior distributions to be combined with appropriate probabilistic NMF formulations. The presented dynamic extensions are classified into continuous and discrete models, where both models are explained using a unified framework. The continuous models include smooth NMF and more recently proposed continuous state-space models. For the discrete models, we discussed the discrete state-space models based on HMM, where the output distributions are modeled using static NMF. Short simulation results and qualitative comparisons are also provided to give an insight into different models and their performance.

9.9   Standard Distributions

Gamma distribution:

$$\mathcal{G}(x \mid \alpha, \beta) = \frac{\beta^\alpha}{\Gamma(\alpha)} \, x^{\alpha-1} \, e^{-\beta\,x}, \quad x \geq 0 \tag{9.69}$$

$$\text{mode}(X) = \frac{\alpha-1}{\beta}, \quad \mathbb{E}\{X\} = \frac{\alpha}{\beta}, \quad \mathbb{E}\left\{\frac{1}{X}\right\} = \frac{\beta}{\alpha-1} \tag{9.70}$$

Exponential distribution:

$$\mathcal{E}(x \mid \lambda) = \lambda^{-1} e^{-\frac{x}{\lambda}}, \quad x \geq 0 \tag{9.71}$$

$$\mathbb{E}\{X\} = \lambda \tag{9.72}$$

Inverse-gamma distribution:

$$\mathcal{IG}(x \mid \alpha, \beta) = \frac{\beta^\alpha}{\Gamma(\alpha)} \, x^{-(\alpha+1)} \, e^{-\frac{\beta}{x}}, \quad x \geq 0 \tag{9.73}$$

$$\text{mode}(X) = \frac{\beta}{\alpha+1}, \quad \mathbb{E}\{X\} = \frac{\beta}{\alpha-1}, \quad \mathbb{E}\left\{\frac{1}{X}\right\} = \frac{\alpha}{\beta} \tag{9.74}$$

Multinomial:

$$\mathcal{M}(\mathbf{x} \mid N, \mathbf{p}) = \frac{N!}{x_1! \ldots x_K!} p_1^{x_1} \ldots p_K^{x_K}, \quad x_k \in \{0, \ldots, N\}, \sum_k x_k = N \tag{9.75}$$

Bibliography

Brown, J.C. (1991) Calculation of a constant q spectral transform. *Journal of the Acoustical Society of America*, **89** (1), 425–434.

Cemgil, A.T. and Dikmen, O. (2007) Conjugate gamma Markov random fields for modelling nonstationary sources, in *Proceedings of International Conference on Independent Component Analysis and Signal Separation*.

Chen, Z., Cichocki, A., and Rutkowski, T.M. (2006) Constrained non-negative matrix factorization method for EEG analysis in early detection of Alzheimer's disease, in *Proceedings of IEEE International Conference on Audio, Speech and Signal Processing*.

Dempster, A.P., Laird, N.M., and Rubin., D.B. (1977) Maximum likelihood from incomplete data via the EM algorithm. *Journal of the Royal Statistical Society: Series B*, **39**, 1–38.

Dikmen, O. and Cemgil, A.T. (2010) Gamma Markov random fields for audio source modeling. *IEEE Transactions on Audio, Speech, and Language Processing*, **18** (3), 589–601.

Essid, S. and Févotte (2013) Smooth nonnegative matrix factorization for unsupervised audiovisual document structuring. *IEEE Transactions on Multimedia*, **15** (2), 415–425.

Févotte, C. (2011) Majorization-minimization algorithm for smooth Itakura-Saito nonnegative matrix factorization, in *Proceedings of IEEE International Conference on Audio, Speech and Signal Processing*.

Févotte, C., Bertin, N., and Durrieu, J.L. (2009) Nonnegative matrix factorization with the Itakura-Saito divergence. With application to music analysis. *Neural Computation*, **21** (3), 793–830.

Févotte, C. and Idier, J. (2011) Algorithms for nonnegative matrix factorization with the beta-divergence. *Neural Computation*, **23** (9), 2421–2456.

Févotte, C., Le Roux, J., and Hershey, J.R. (2013) Non-negative dynamical system with application to speech and audio, in *Proceedings of IEEE International Conference on Audio, Speech and Signal Processing*.

Hamilton, J.D. (1994) *Time Series Analysis*, Princeton University Press, New Jersey.

Lütkepohl, H. (2005) *New Introduction to Multiple Time Series Analysis*, Springer.

Mohammadiha, N. and Leijon, A. (2013) Nonnegative HMM for babble noise derived from speech HMM: Application to speech enhancement. *IEEE Transactions on Audio, Speech, and Language Processing*, **21** (5), 998–1011.

Mohammadiha, N., Smaragdis, P., and Leijon, A. (2013) Prediction based filtering and smoothing to exploit temporal dependencies in NMF, in *Proceedings of IEEE International Conference on Audio, Speech and Signal Processing*.

Mohammadiha, N., Smaragdis, P., Panahandeh, G., and Doclo, S. (2015) A state-space approach to dynamic nonnegative matrix factorization. *IEEE Transactions on Signal Processing*, **63** (4), 949–959.

Mysore, G.J. (2010) *A Non-negative Framework for Joint Modeling of Spectral Structure and Temporal Dynamics in Sound Mixtures*, Ph.D. thesis, Stanford University.

Mysore, G.J. (2012) A block sparsity approach to multiple dictionary learning for audio modeling, in *Proceedings of International Conference on Machine Learning Workshop on Sparsity, Dictionaries, and Projections in Machine Learning and Signal Processing*.

Mysore, G.J. and Sahani, M. (2012) Variational inference in non-negative factorial hidden Markov models for efficient audio source separation, in *Proceedings of International Conference on Machine Learning*.

Mysore, G.J. and Smaragdis, P. (2011) A non-negative approach to semi-supervised separation of speech from noise with the use of temporal dynamics, in *Proceedings of IEEE International Conference on Audio, Speech and Signal Processing*.

Mysore, G.J. and Smaragdis, P. (2012) A non-negative approach to language informed speech separation, in *Proceedings of International Conference on Latent Variable Analysis and Signal Separation*.

Mysore, G.J., Smaragdis, P., and Raj, B. (2010) Non-negative hidden Markov modeling of audio with application to source separation, in *Proceedings of International Conference on Latent Variable Analysis and Signal Separation*.

Nakano, M., Le Roux, J., Kameoka, H., Kitano, Y., Ono, N., and Sagayama, S. (2010) Nonnegative matrix factorization with Markov-chained bases for modeling time-varying patterns in music spectrograms, in *Proceedings of International Conference on Latent Variable Analysis and Signal Separation*.

Nakano, M., Le Roux, J., Kameoka, H., Nakamura, T., Ono, N., and Sagayama, S. (2011) Bayesian nonparametric spectrogram modeling based on infinite factorial infinite hidden Markov model, in *Proceedings of IEEE Workshop on Applications of Signal Processing to Audio and Acoustics*.

Nam, J., Mysore, G.J., and Smaragdis, P. (2012) Sound recognition in mixtures, in *Proceedings of International Conference on Latent Variable Analysis and Signal Separation*.

O'Grady, P.D. and Pearlmutter, B.A. (2008) Discovering speech phones using convolutive non-negative matrix factorisation with a sparseness constraint. *Neurocomputing*, **72**, 88–101.

Ozerov, A., Févotte, C., and Charbit, M. (2009) Factorial scaled hidden Markov model for polyphonic audio representation and source separation, in *Proceedings of IEEE Workshop on Applications of Signal Processing to Audio and Acoustics*.

Rabiner, L.R. (1989) A tutorial on hidden Markov models and selected applications in speech recognition. *Proceedings of the IEEE*, **77** (2), 257–286.

Smaragdis, P. (2007) Convolutive speech bases and their application to supervised speech separation. *IEEE Transactions on Audio, Speech, and Language Processing*, **15** (1), 1–12.

Smaragdis, P., Févotte, C., Mysore, G., Mohammadiha, N., and Hoffman, M. (2014) Static and dynamic source separation using nonnegative factorizations: A unified view. *IEEE Signal Processing Magazine*, **31** (3), 66–75.

Smaragdis, P. and Raj, B. (2007) Shift-invariant probabilistic latent component analysis, *Tech. Rep. TR2007-009*, Mitsubishi Electric Research Laboratories.

Smaragdis, P., Raj, B., and Shashanka, M.V. (2006) A probabilistic latent variable model for acoustic modeling, in *Proceedings of Neural Information Processing Systems Workshop on Advances in Models for Acoustic Processing*.

Smaragdis, P., Raj, B., and Shashanka, M.V. (2007) Supervised and semi-supervised separation of sounds from single-channel mixtures, in *Proceedings of International Conference on Independent Component Analysis and Signal Separation*.

Virtanen, T. (2007) Monaural sound source separation by non-negative matrix factorization with temporal continuity and sparseness criteria. *IEEE Transactions on Audio, Speech, and Language Processing*, **15** (3), 1066–1074.

Virtanen, T., Cemgil, A.T., and Godsill, S. (2008) Bayesian extensions to non-negative matrix factorisation for audio signal modelling, in *Proceedings of IEEE International Conference on Audio, Speech and Signal Processing*.

Wang, W., Cichocki, A., and Chambers, J.A. (2009) A multiplicative algorithm for convolutive non-negative matrix factorization based on squared Euclidean distance. *IEEE Transactions on Signal Processing*, **57** (7), 2858–2864.

Yoshii, K. and Goto, M. (2012) Infinite composite autoregressive models for music signal analysis, in *Proceedings of International Society for Music Information Retrieval Conference*.

Part III

Multichannel Separation and Enhancement

10

Spatial Filtering

Shmulik Markovich-Golan, Walter Kellermann, and Sharon Gannot

This chapter is dedicated to spatial filtering with applications to noise reduction and interference cancellation. Spatial filtering should be understood here as the application of linear, often time-varying, filters to a multichannel signal acquired by microphones which are spatially distributed in the physical space (reverberant enclosures in the context of audio processing). The spatial diversity of the microphone arrangement allows for a spatially selective processing of the sound field. The term "beamformer" was historically used for a special class of spatial filtering system which directs a beam of high sensitivity to a certain region in space (or towards a certain direction) relative to the microphone arrangement. With the introduction of more complex propagation regimes that take reverberation into account, the term became more general. In the following, we use the term "beamformer" for all multiple-input single-output systems that produce a single output signal by combining a set of filtered microphone signals. Equivalently, the spatial filtering effect (not the optimization criteria for its design) by linear multiple-input multiple-output systems can be understood as the superposition of multiple beamformers. The output of the beamformer comprises an enhanced desired signal (or a combination of enhanced sources, e.g. a desired conversation) with (hopefully negligible) residual noise and competing sources.

In the following, we first introduce some fundamentals of array processing in Section 10.1, i.e. the signal model and the characterization of spatial selectivity by beampatterns, directivity and sensitivity. We then discuss array topologies in Section 10.2 followed by data-independent beamforming in Section 10.3 and data-dependent beamforming in Section 10.4. For a general treatment of the latter, we review the use of relative transfer functions (RTFs) and various design criteria. Data-dependent beamforming is then related to binary masking and also to the spatial filtering properties of blind source separation (BSS) algorithms, and the use of BSS algorithms as a component within data-dependent beamformers is delineated. We present a widely used structure for realizing data-dependent beamformers, the *generalized sidelobe canceler* (GSC), in Section 10.5. Finally, we address the special role of single-channel postfiltering as a complement to multichannel spatial filtering in Section 10.6. We provide a summary in Section 10.7.

Audio Source Separation and Speech Enhancement, First Edition.
Edited by Emmanuel Vincent, Tuomas Virtanen and Sharon Gannot.
© 2018 John Wiley & Sons Ltd. Published 2018 by John Wiley & Sons Ltd.
Companion Website: https://project.inria.fr/ssse/

10.1 Fundamentals of Array Processing

In this section, we explore some fundamental concepts of array processing. Unless otherwise stated, these definitions are applicable to all arrays (not necessarily microphone arrays). For a comprehensive review on array types (not specifically for speech applications), see Van Trees (2002).

We focus on the single target source scenario, namely $J = 1$, and will henceforth omit the source index j. Since audio is a wideband signal, a convolution operation is applied to each microphone signal, rather than a multiplication by scalar. See discussion by Frost (1972) on the application of finite impulse response (FIR) filtering for dealing with wideband signals. As explained in Chapter 2, if the frame length of the short-time Fourier transform (STFT) is sufficiently large, the time-domain filtering translates to frequency-domain multiplication. Cases for which this assumption is violated will only be briefly addressed.

A *beamformer* is determined by a, possibly time-varying, frequency-dependent vector $\mathbf{w}(n,f) = [w_1(n,f), \dots, w_I(n,f)]^T$ comprising one complex-valued weight per microphone per frequency. The beamformer is applied to the array input signals in the STFT domain, namely $\mathbf{x}(n,f)$. The output of the beamformer in the STFT domain is given by

$$\hat{s}(n,f) = \mathbf{w}^H(n,f)\,\mathbf{x}(n,f). \tag{10.1}$$

Finally, the time-domain output signal $\hat{s}(t)$ is reconstructed by applying the inverse STFT. If the frequency-dependent weighting $w_i(n,f)$ should be implemented in the time domain, then FIR filters are typically used whose frequency response approximate $w_i(n,f)$. A schematic block diagram of a beamformer is depicted in Figure 10.1.

The source spatial image is given by $\mathbf{c}(n,f) = \mathbf{a}(n,f)s(n,f)$, where $s(n,f)$ is a point source located at position $\mathbf{p}$ and the acoustic transfer function $\mathbf{a}(n,f)$ should be understood as location-dependent. The vector of received microphone signals is then given by $\mathbf{x}(n,f) = \mathbf{c}(n,f) + \mathbf{u}(n,f)$, where $\mathbf{u}(n,f)$ is the noise spatial image as received by the

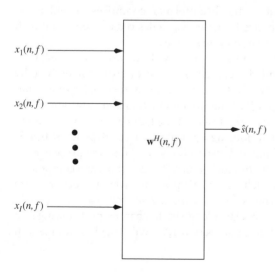

Figure 10.1 General block diagram of a beamformer with filters
$\mathbf{w}(n,f) = [w_1(n,f), \dots, w_I(n,f)]^T.$

$x_1(n,f)$

$x_2(n,f)$

$\mathbf{w}^H(n,f)$

$\hat{s}(n,f)$

$x_I(n,f)$

microphones. Finally, the beamformer output is given by

$$\hat{s}(n,f) = \mathbf{w}^H(n,f)\mathbf{x}(n,f) = \mathbf{w}^H(n,f)\mathbf{a}(n,f)s(n,f) + \mathbf{w}^H(n,f)\mathbf{u}(n,f), \tag{10.2}$$

where $\mathbf{w}^H(n,f)\mathbf{u}(n,f)$ is the residual interference at the beamformer output.

10.1.1 Beampattern

For the sake of simplicity, we focus on the special case of a narrowband source and a *uniform linear array*, in which the microphones are spatially aligned with each other with equal inter-microphone distance ℓ. We further assume that the source position is static, located in the far field and that the emitted wave propagates in an anechoic environment, i.e. the time-invariant impulse response relating the source and the microphones is dominated by the direct path of the impulse response and hence solely determined by the direction of arrival (DOA) of the source signal. Correspondingly, the beamformer is also assumed time-invariant, and its array response to different directions is often referred to as the *beampattern*.

Recall the spherical coordinate system, with θ the azimuth angle, φ the elevation angle, and $\mathbf{k}$ a unit vector pointing from the axis origin towards θ, φ

$$\mathbf{k} = \begin{bmatrix} \cos\theta \ \cos\varphi \\ \sin\theta \ \cos\varphi \\ \sin\varphi \end{bmatrix}. \tag{10.3}$$

For details, the reader is referred to Figure 3.6. The radius is irrelevant for defining the classical far-field beampattern.

The time-invariant acoustic transfer function $\mathbf{a}(f)$ can be substituted by the simpler steering vector $\mathbf{d}(\alpha, \ell/\lambda)$, solely determined by the angle-of-arrival α with respect to the microphone axis and the ratio of the microphone spacing and signal wavelength $\frac{\ell}{\lambda}$, where $\lambda = c/v_f$ is the center wavelength of the narrowband signal, with v_f the center frequency. In that case, the time difference of arrival (TDOA) between the first microphone and microphone i is given by $(i-1)\ell \cos(\alpha)/c$. The far-field steering vector in (3.9) then reads

$$\mathbf{d}\left(\alpha, \frac{\ell}{\lambda}\right) = \begin{bmatrix} 1 \\ e^{-2j\pi\frac{\ell}{\lambda}\cos\alpha} \\ \vdots \\ e^{-2j\pi(I-1)\frac{\ell}{\lambda}\cos\alpha} \end{bmatrix}. \tag{10.4}$$

The (complex-valued) array response is called the beampattern and in the case of a uniform linear array is given by

$$b\left(\alpha, \frac{\ell}{\lambda}\right) = \mathbf{w}^H \mathbf{d}\left(\alpha, \frac{\ell}{\lambda}\right) = \sum_{i=1}^{I} w_i e^{-2j\pi(i-1)\frac{\ell}{\lambda}\cos\alpha}. \tag{10.5}$$

Since we discuss the narrowband case, the frequency index is also omitted and hence $w_i(n,f)$ is simply denoted w_i. The beampattern plays an important role in analyzing the array performance.

Define the *delay-and-sum* (DS) beamformer, steered towards α_0, as the beamformer with weights $w_i = \frac{1}{I} e^{-2j\pi(i-1)\frac{\ell}{\lambda}\cos\alpha_0}$. The DS beamformer first compensates the relative delays between the target signal images and then averages the delayed microphone signals. Consequently, the target signal images are coherently summed. Denote the absolute squared beampattern as the *beampower*. In the special DS case, it is given by

$$\left| b\left(\alpha, \frac{\ell}{\lambda}\right) \right|^2 = \left| \frac{\sin\left(I\pi\frac{\ell}{\lambda}(\cos\alpha - \cos\alpha_0)\right)}{I\sin\left(\pi\frac{\ell}{\lambda}(\cos\alpha - \cos\alpha_0)\right)} \right|^2. \tag{10.6}$$

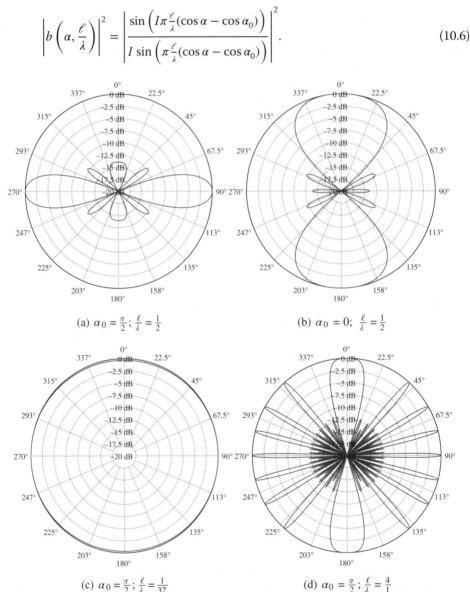

(a) $\alpha_0 = \frac{\pi}{2}$; $\frac{\ell}{\lambda} = \frac{1}{2}$

(b) $\alpha_0 = 0$; $\frac{\ell}{\lambda} = \frac{1}{2}$

(c) $\alpha_0 = \frac{\pi}{2}$; $\frac{\ell}{\lambda} = \frac{1}{32}$

(d) $\alpha_0 = \frac{\pi}{2}$; $\frac{\ell}{\lambda} = \frac{4}{1}$

Figure 10.2 Beampower (10.6) in polar coordinates of the DS beamformer for a uniform linear array. The beampower is normalized to unity at the look direction. The different behavior of the beampower as a function of the array look direction and the signal wavelength is clearly demonstrated.

Typical beampowers as functions of the steering angle and of the ratio $\frac{\ell}{\lambda}$ are depicted in Figure 10.2 in polar coordinates.

In Figures 10.2a and 10.2b we set $\frac{\ell}{\lambda} = \frac{1}{2}$. A uniform linear array with $\frac{\ell}{\lambda} = \frac{1}{2}$ is usually referred to as a *standard linear array* (Van Trees, 2002). In Figure 10.2a the steering direction is perpendicular to the array axis and in Figure 10.2b it is parallel to it. The former look direction is called *broadside* and the latter *endfire*. Note that the shapes of the beampower or respectively the (absolute value of) the beampatterns are very distinct for these two look directions. The consequences of setting the wavelength to a very high value (or, correspondingly, very low frequency), i.e. $\frac{\ell}{\lambda} \ll 1$, can be deduced from Figure 10.2c, where the beampower is almost omnidirectional, and setting it to a very low value, i.e. $\frac{\ell}{\lambda} \gg 1$, can be deduced from Figure 10.2d, where the beampower exhibits *grating lobes*, which are the result of spatial aliasing.

10.1.2 Directivity

An important attribute of a beamformer is its *directivity*, defined as the response towards the look direction divided by the integral over all other possible directions. The directivity in dB scale is called the directivity index. In its most general form (Levin *et al.*, 2013a), applicable to any propagation regime (e.g., reverberant environment), the directivity at a given frequency can be defined as

$$\text{Dir}_{\text{gen}}(\mathbf{w}, \mathbf{p}_0) = \frac{|\mathbf{w}^H \mathbf{a}(\mathbf{p}_0)|^2}{\int |\mathbf{w}^H \mathbf{a}(\mathbf{p})|^2 d\mathbf{p}} \tag{10.7}$$

where, as elaborated in Section 3.2.2, $\mathbf{a}(\mathbf{p})$ is the vector of acoustic transfer functions explicitly parameterizing the 3D source position $\mathbf{p}$, here assumed time-independent and frequency-dependent (the frequency dependency is omitted for brevity), and $\mathbf{p}_0$ is the array look direction.

The most common far-field definition of directivity is assumed, with the abstract acoustic transfer function parameterization substituted by the wave propagation vector in (10.3). The directivity in spherical coordinates is then given by

$$\text{Dir}_{\text{sph}}(\mathbf{w}, \theta_0, \varphi_0) = \frac{|\mathbf{w}^H \mathbf{d}(\mathbf{k}_0)|^2}{\frac{1}{4\pi} \int_0^{2\pi} \int_{-\pi/2}^{\pi/2} |b(\theta, \varphi)|^2 \sin \varphi \, d\theta d\varphi} \tag{10.8}$$

with $\mathbf{k}_0 = [\cos \theta_0 \, \cos \varphi_0, \sin \theta_0 \, \cos \varphi_0, \sin \varphi_0]^T$ the look direction of the array. Assuming that the response in the look direction is equal to 1, this expression simplifies to (Van Trees, 2002)

$$\text{Dir}_{\text{sph}}(\mathbf{w}, \theta, \varphi) = (\mathbf{w}^H \mathbf{\Omega} \mathbf{w})^{-1} \tag{10.9}$$

where $\mathbf{\Omega}$ is the covariance matrix of a diffuse sound field whose entries $\omega_{ii'}$ are given in (3.6).

Maximizing the directivity with respect to the array weights results in[1]

$$\text{Dir}_{\text{max}}(\theta_0, \varphi_0) = \mathbf{d}^H(\theta_0, \varphi_0) \mathbf{\Omega}^{-1} \mathbf{d}(\theta_0, \varphi_0). \tag{10.10}$$

---

[1] The array weights that maximize the directivity are given by the MVDR beamformer (10.32) with $\mathbf{\Sigma}_u = \mathbf{\Omega}$. This relation will be discussed in Section 10.4.4. See Parsons (1987) for more details.

As evident from this expression, the directivity may depend on the steering direction. It can be shown that the maximum directivity attained by the standard linear array ($\frac{\ell}{\lambda} = \frac{1}{2}$) is equal to the number of microphones I, which is independent of the steering angle. The array weights in this case are given by $w_i = \frac{1}{I}$, $i \in \{1, \ldots, I\}$ assuming broadside look direction. Any beamformer that attains a directivity significantly higher than I is called a *superdirective* beamformer. Parsons (1987) also proved that the directivity of an array with endfire look direction and with vanishing inter-microphone distance, i.e. $\frac{\ell}{\lambda} \to 0$, approaches I^2. It was also claimed, without a proof, that "it is most unlikely" that another beamformer can attain higher directivity.

10.1.3 Sensitivity

Another attribute of a beamformer is its *sensitivity* to array imperfections, i.e., its sensitivity to perturbations of the microphone positions and beamformer weights.

The signal-to-noise ratio (SNR) at the output of the microphone array for a source impinging the array from $\mathbf{k}_0$ is given by

$$\text{SNR}_{\text{out}} = \frac{\sigma_s^2 |\mathbf{w}^H \mathbf{d}(\mathbf{k}_0)|^2}{\mathbf{w}^H \boldsymbol{\Sigma}_\mathbf{u} \mathbf{w}}. \tag{10.11}$$

If the noise is spatially white, i.e. $\boldsymbol{\Sigma}_\mathbf{u} = \sigma_u^2 \mathbf{I}_I$ with $\mathbf{I}_I$ the $I \times I$ identity matrix, then

$$\text{SNR}_{\text{out}} = \frac{\sigma_s^2}{\sigma_u^2} \frac{|\mathbf{w}^H \mathbf{d}(\mathbf{k}_0)|^2}{\mathbf{w}^H \mathbf{w}} = \text{SNR}_{\text{in}} \frac{|\mathbf{w}^H \mathbf{d}(\mathbf{k}_0)|^2}{\mathbf{w}^H \mathbf{w}} \tag{10.12}$$

with $\text{SNR}_{\text{in}} = \frac{\sigma_s^2}{\sigma_u^2}$, the SNR at the input.

Further assuming unit gain in the look direction, the SNR improvement, called *white noise gain*, is given by

$$\frac{1}{\mathbf{w}^H \mathbf{w}} = \|\mathbf{w}\|_2^{-2} \tag{10.13}$$

where $\|\cdot\|_2$ stands for the ℓ_2 norm of a vector. The sensitivity of an array to perturbations of the microphone positions and to the beamformer weights is inversely proportional to its white noise gain (Cox *et al.*, 1987)[2]

$$\varsigma = \|\mathbf{w}\|_2^2. \tag{10.14}$$

It was further shown that there is a tradeoff between the array directivity and its sensitivity and that the superdirective beamformer with directivity I^2 suffers from infinite sensitivity to mismatch between the nominal design parameters and the actual parameters (Cox *et al.*, 1987). It was therefore proposed to constrain the norm of $\mathbf{w}$ in order to obtain a more robust design. It should be noted that if the microphone position perturbations are coupled (e.g., if all microphones share the same packaging) a modified norm constraint should be applied to guarantee low sensitivity (Levin *et al.*, 2013b).

---

2 This sensitivity should not be confused with the sensitivity to erroneous steering direction, which is closely related to the array *beamwidth*, as defined by, for example, Er and Cantoni (1983). See also the discussion by Anderson *et al.* (2015).

10.2 Array Topologies

In the following, we introduce generic topologies of microphone arrays by generalizing the uniform linear array along various parameters.

The uniform linear array as assumed in the previous sections represents the most commonly used array topology. Corresponding data-independent beamformers are relatively easy to analyze and design (see Section 10.3). If for a narrowband signal with temporal frequency $v_f = c/\lambda$ a spacing of $\ell = \lambda/2$ is used (standard linear array), then this choice is optimum insofar as both the white noise gain (10.13) and the directivity in a diffuse sound field are maximized if the array is combined with a DS beamformer (see Figure 10.2).

Uniform spatial selectivity. As a first obvious limitation of the uniform linear array, we can overcome the dependency of its spatial selectivity on the steering angle (see Figure 10.2) by using *uniform circular arrays*, where all microphones are placed equidistantly on a circular contour and thus provide approximately uniform spatial selectivity in any plane parallel to the array plane, due to the array's approximate rotational invariance (Johnson and Dudgeon, 1993; Teutsch and Kellermann, 2006; Mabande *et al.*, 2012). They allow efficient beamforming design and processing in the cylindrical harmonics domain (Teutsch, 2007; Rafaely, 2015) and are commonly placed close to or in the center of the sound field of interest.

2D spatial selectivity. Another limitation of the uniform linear array is its limitation to a single angular variable for spatial discrimination of sources in the far field, as is reflected by the rotational invariance (independence of φ) of the beampattern (10.5). While a uniform circular array exhibits such a dependency on the second angular variable, this cannot be independently controlled by spatial filtering algorithms. For independent spatial filtering of far-field sources with respect to azimuth and elevation, 2D or 3D array topologies are required. Among those, uniform planar arrays are an obvious extension of uniform linear arrays (Flanagan *et al.*, 1985; Silverman *et al.*, 1997), while uniform spherical arrays extend uniform circular arrays to controlling two independent spatial dimensions. A uniform placement of microphones on a sphere corresponds to constructing a Platonic solid and thus can only be approximated for more than $I = 20$ microphones. Analysis and design are typically based on wavefield representations in the spherical harmonics domain (Meyer and Elko, 2002; Abhayapala and Ward, 2002; Rafaely, 2005; Li and Duraiswami, 2007; Elko and Meyer, 2012; Rafaely, 2015).

Broadband signals. Ensuring the desirable properties of the uniform linear array for broadband signals at each frequency would require $\ell = \lambda/2$ for any λ along the wavelength axis, which is obviously infeasible for arrays with discrete microphones. As a common approximation, 1D arrays with logarithmic microphone spacing (Van der Wal *et al.*, 1996), and, even more practically relevant, so-called *nested arrays* with 1D (linear) and 2D (planar) realizations (Flanagan *et al.*, 1985; Kellermann, 1991) are widely used. Harmonically nested arrays use spacings which are powers of two relative to a minimum spacing, which is chosen according to the highest frequency of interest, so that each spacing can be associated with an octave of the frequency range. With the minimum spacing in the center of the array and I being even, half of the microphones dedicated to a given octave band can be used for the next lower octave band as well, so that the number of microphones with its associated hardware can be kept

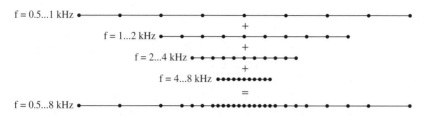

Figure 10.3 Harmonically nested linear arrays with $l = 25$ for $f = 0.5$ to 8 kHz.

minimum for a given frequency range of operation (see Figure 10.3 and Kellermann (1991)). The concept of nesting arrays for broadband signals can also be applied to 2D planar arrays (Flanagan *et al.*, 1985), circular arrays (Johnson and Dudgeon, 1993), and spherical arrays (Jin *et al.*, 2014), although for the latter two the usability of a given microphone for multiple frequency ranges is not straightforward.

Embodied arrays. The above array topologies are typically conceived and evaluated by assuming point-like microphones operating in a free soundfield. In reality, the free-field assumption is violated due to the physical size of the microphones and the necessary mounting, which is most relevant for high frequencies when wavelengths are on the order of the physical size of the surrounding solid bodies. Then, the design of spatial filtering involves typically knowledge of the acoustic properties (absorptive/reflective/scattering/diffractive/diffusive) of the surrounding materials and the geometric shape of the bodies. For the design of microphone arrays as individual devices one will aim at either free-field approximations (Rafaely, 2005; Jin *et al.*, 2014) or assume analytically tractable properties for rigid or highly absorptive bodies for the embodiment (Meyer and Elko, 2002; Teutsch, 2007). The reflection properties of scatterers can also be exploited, e.g. to reduce the number of microphones (Li and Duraiswami, 2005).

If microphone arrays need to be incorporated into smart devices (e.g., smart phones, hearing aids, car interiors, multimodal human/machine interfaces, robot heads, etc.) other constraints add to or even supersede the design criteria for the microphone arrays. Very often the total array size is the main constraint, which requires spatial filtering methods using microphone pairs where $\ell/\lambda \ll 1/2$ at the lower end of the frequency range. This leads to the importance of differential beamforming as further discussed in Section 10.3.

Designing spatial filters for embodied microphone arrays requires the measurement or modeling of the specific transfer functions from the positions of sources of interest to the microphones. With respect to far-field sources, such object-related transfer functions are usually measured for a grid of points on a sphere with a radius assuring sufficient distance from the embodiment, e.g. 1 m for head-related transfer functions for designing hearing aids on a human head.

Combination of generic array types. Obviously, the above generic arrays can be combined. For instance, several 1D arrays can be combined to form a 2D array, e.g. two linear arrays forming an "L" or a cross (Johnson and Dudgeon, 1993), and planar arrays with orthogonal normal vectors can be combined to allow volumetric beamforming

(Flanagan *et al.*, 1985; Silverman *et al.*, 1997). Transparent circular or spherical arrays, i.e. when the array approximates microphones in the free field, are often complemented by additional microphones in the array center. Sometimes, it is useful to interpret and analyze such configurations as sparse versions of originally fully populated uniform 2D or 3D arrays.

The range of possible array topologies increases further once microphone networks are considered (see Chapter 19).

10.3 Data-Independent Beamforming

In this section, we consider some design principles and structures for beamformers which do not explicitly account for the statistics of the received signals. However, prior knowledge of the positions of the sources of interest must be available. The spatial selectivity of the time-invariant beamformer $\mathbf{w}(f)$ then depends (besides the frequency only) on the positions of the sources which lead to the observation vector $\mathbf{x}(n,f)$, and not on the signal statistics.

For the most common approach, it is assumed that the sources are in the far field of the array (Johnson and Dudgeon, 1993). Then, the design problem can be formulated as the problem of determining $\mathbf{w}(f)$ such that the beamformer response $b(\theta, \varphi, f) = \mathbf{w}^H(f)\mathbf{d}(\theta, \varphi, f)$ for a sound wave impinging from a given DOA (specified by azimuth θ and elevation φ) minimizes the error relative to a predefined target response $b^{\text{tgt}}(\theta, \varphi, f)$ with respect to a suitable norm, e.g. the ℓ_2 norm:

$$\widehat{\mathbf{w}}(f) = \underset{\mathbf{w}}{\operatorname{argmin}} \frac{1}{4\pi} \int_0^{2\pi} \int_{-\pi/2}^{\pi/2} \|\mathbf{w}^H(f)\mathbf{d}(\theta, \varphi, f) - b^{\text{tgt}}(\theta, \varphi, f)\|_2^2 \sin \varphi \; d\theta d\varphi$$

$$(10.15)$$

where

$$\mathbf{d}(\theta, \varphi, f) = \begin{bmatrix} e^{-2j\pi\mathbf{k}^T(\theta,\varphi,f)\mathbf{m}_1/\lambda} \\ \vdots \\ e^{-2j\pi\mathbf{k}^T(\theta,\varphi,f)\mathbf{m}_I/\lambda} \end{bmatrix} \tag{10.16}$$

represents the 2D steering vector analogous to (10.4), $\mathbf{k}$ is defined in (10.3), and $\mathbf{m}_i$ represent the microphone locations in Cartesian coordinates. For the uniform linear array with spacing ℓ (c.f. Section 10.1.1), (10.15) simplifies to

$$\widehat{\mathbf{w}}(f) = \underset{\mathbf{w}}{\operatorname{argmin}} \frac{1}{2} \int_0^{\pi} \left\| \sum_{i=1}^{I} w_i(f) e^{-2j\pi(i-1)\frac{\ell}{\lambda}\cos\alpha} - b^{\text{tgt}}(\alpha, f) \right\|_2^2 \sin \alpha \; d\alpha. \tag{10.17}$$

For a given frequency f, this can be directly linked to the FIR filter design problem for frequency-selective filtering of discrete time-domain signals (Oppenheim and Schafer, 1975; Johnson and Dudgeon, 1993) by identifying the normalized spatial sampling interval $\frac{\ell v_f}{c} = \frac{\ell}{\lambda}$ with the temporal sampling interval $1/f_s$ and identifying the angular variable $\cos \alpha$ with the normalized temporal frequency v_f/f_s. As a consequence, all the windowing functions designed for spectral analysis (see, for example, Harris (1978)) can be used for spatial filtering by uniform linear arrays with the spatial selectivity along $\cos\alpha$ corresponding to the spectral selectivity along the v axis. This is especially useful for classical beamforming tasks, such as directing a narrow beam (angular range of high sensitivity) towards a target direction, while suppressing all other directions, which corresponds to highlighting a narrow frequency range in spectral analysis. Aside from designing single narrow beams, other desired functions b^{tgt} are common and more are conceivable (Johnson and Dudgeon, 1993): DOA ranges could be specified, e.g., to extract multiple desired sources or suppress undesired sources, depending on available prior knowledge on source locations, with design methods similar to bandpass and bandstop FIR filters in conventional filter design. Obviously, the beamformer responses do not always need to be specified for all DOAs, as, e.g. when using only two microphones, a null-steering beamformer is only specified by a single null ($b^{\text{tgt}}(\theta_0, \varphi_0, f) = 0$) for a given DOA ($\theta_0, \varphi_0$).

Given a beamformer $\mathbf{w}$ according to (10.15), the beamformer response $b(\theta, \varphi, f)$ can be rotated by introducing an additional phase shift for each microphone signal, i.e. replacing $\mathbf{k}(\theta, \varphi, f)$ by $\mathbf{k}(\theta, \varphi, f) - \mathbf{k}(\theta_0, \varphi_0, f)$. If $\mathbf{k}(\pi/2, 0, f)$ (see (10.3)) is considered as the original look direction, then the new look direction is given by $\mathbf{k}(\theta_0, \varphi_0)$ with the according changes of the steering vector $\mathbf{d}(\theta, \varphi, f)$ and the beamformer response $\mathbf{w}^H(f)\mathbf{d}(\theta, \varphi, f)$. Note that, unless the microphone array is spherically or – for beamforming in a 2D plane – circularly invariant, the beampattern will not be identical for different steering angles due to the nonlinear mapping of the angles via trigonometric functions (see (10.6)). Moreover, for beamforming in the time domain, whenever the phase shifts in the frequency domain correspond to noninteger time delays, interpolation of the discrete-time signals requires dedicated interpolation filters (Laakso *et al.*, 1996), so that for practical reasons only a finite set of phase shifts will be available (Johnson and Dudgeon, 1993; Van Veen and Buckley, 1988). For circularly symmetric arrays, polynomial beamformers (Kajala and Hämäläinen, 2001) are attractive for avoiding the steering angle-specific phase shifts of the individual microphone signals. Instead, a single parameter controls the steering direction of the beamformer. If such an array has S axes of symmetries in a given plane, then only a frequency range of π/S needs to be specified and all degrees of freedom for the beamformer design can be used for the design of the reduced angular range (Mabande *et al.*, 2012).

Aside from prescribing beampatterns, data-independent beamformers can also be designed on the basis of assumptions on the involved sound fields. These designs are typically based on the spatial covariance matrix of the microphone signals. As the designs for assumed covariance matrices are identical to those for corresponding measured covariance matrices, we refer to the treatment of data-dependent beamforming in Section 10.4 for details.

For broadband acoustic signals like speech or music, beamformers must typically be designed for a continuous frequency range that often spans multiple octaves. Considering a beamformer design for the full audio range from 20 Hz to 20 kHz and a uniform linear array according to (10.17), the normalized spatial sampling interval $\frac{\ell v_f}{c} = \frac{\ell}{\lambda}$ at the lowest and the highest frequency vary by a factor of 1000, with the according effects on spatial resolution at the low end and aliasing at the high end of the frequency range (see Figure 10.2 for illustration). Wideband beamformers may be implemented in the STFT domain as indicated in (10.1), or in the time domain (Johnson and Dudgeon, 1993). The time domain is especially attractive if the delay required for the STFT should be avoided and the frequency-dependent channel weights $w_i(f)$ should be realized by low-order FIR filters. Then, the FIR filter design for realizing the frequency response $w_i(f)$ for microphone i involves another approximation problem.

A typical design in wideband acoustic beamforming should assure a distortionless response for the look direction over the entire frequency range, or at least the magnitude should be constant, $|b^{tgt}(\theta_0, \varphi_0, f)| = 1$. In order to avoid magnitude distortions if the desired source is slightly off the look direction, constant beamwidth designs (Goodwin and Elko, 1993; Ward *et al.*, 1995) are highly desirable and also motivate the use of nested arrays (see Section 10.2), which can cover multiple octaves by several arrays with the same beamformer applied to each individual octave.

Given a uniform linear array with spacing ℓ, the ratio of ℓ to the wavelength λ decreases for decreasing frequency, and considering (10.17), the interchannel phase difference (IPD) between adjacent microphones $-2\pi\frac{\ell v_f}{c}\cos\alpha$ changes only very little over α if $\frac{\ell}{\lambda} = \frac{\ell v_f}{c}$ is very small, so that the spatial selectivity becomes small if, e.g. $w_i(f) = 1, \forall i$ (DS beamformer steered to the broadside, c.f. Figure 10.2c). For still achieving high spatial selectivity, the small IPDs between adjacent microphones can be amplified by choosing weights $w_i(f)$ with large absolute values. These *differential* beamformers (Elko, 1996, 2004) approximate the spatial derivative of the sound pressure field and achieve a higher directivity (10.8) than the so-called additive arrays (Benesty and Chen, 2013) in diffuse noise fields, which also motivates the term superdirectional (Elko, 2000) or superdirective (Bitzer and Simmer, 2001) beamforming (see also Section 10.1.2).

While differential arrays can maximize the array gain relative to diffuse noise fields, the array gain relative to spatially white noise (e.g. caused by sensor noise or microphone mismatch, and expressed by the white noise gain or sensitivity, see above) is known to be a major challenge for its deployment. Therefore, beamformer design methods are desirable, which maximize directivity in diffuse sound fields while at the same time guaranteeing a minimum array gain over spatially white noise. While early iterative approaches designed the beamformer and then verified fulfillment of the white noise gain constraint (Cox, 1973; Cox *et al.*, 1986), the general frequency-invariant wideband design with distortionless response and white noise gain constraint has recently been formulated as a convex optimization problem which can be numerically solved in one step (Mabande *et al.*, 2009), see Figure 10.4. This generic approach also allows to optimize beamformers for arrays attached to or embodied into scatterers, as desirable for hearing aids or humanoid robots. The free-field model in the steering vector $\mathbf{d}(\theta, \varphi, f)$ (10.16) is then substituted by the object-related acoustic transfer functions from a certain DOA to the microphones, which provides significant benefits in real-world applications (Barfuss *et al.*, 2015).

Finally, it should be mentioned that the beamformer design methods discussed for sources in the far field can also be applied to sources in the near field after an according transform (Kennedy *et al.*, 1998).

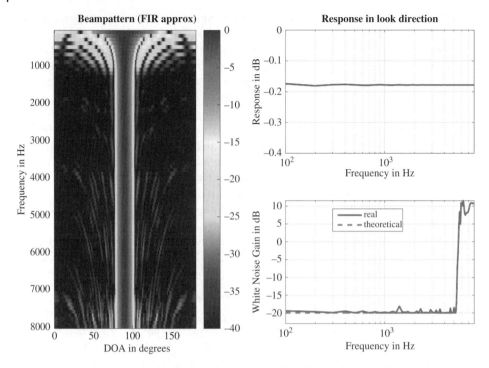

Figure 10.4 Constant-beamwidth distortionless response beamformer design with desired response towards broadside and constrained white noise gain after FIR approximation of $w_i(f)$ (Mabande *et al.*, 2009) for the array of Figure 10.3.

10.4 Data-Dependent Spatial Filters: Design Criteria

This section is dedicated to data-dependent beamformers, which depend on the input signal statistics. Compared with data-independent beamforming explored in Section 10.3, data-dependent designs can attain higher performance due to their ability to adapt to the actual acoustic propagation of the sound source and to the statistics of the target and interfering sources. In many cases, these beamformers are also *adaptive*, i.e. time-varying. The higher performance is often traded off for substantially higher computational complexity and sometimes for degraded robustness. In this section we explore many popular data-dependent beamforming criteria.

10.4.1 The Relative Transfer Function

Designing data-dependent spatial filters requires knowledge or estimates of the acoustic transfer functions relating the sources and the microphones. The RTF, which is a generalization of the TDOA-based steering vector (which assumes free-field propagation) to relative responses (represented by complex numbers in the STFT domain) suitable for representing propagation in reverberant enclosures. The RTF is defined and discussed in Chapter 3. When the reverberation phenomenon is not severe, the desired signal at the output of the spatial filter can be defined as the desired source image at a reference microphone. The problem of estimating the RTF is not as cumbersome as estimating the acoustic transfer function since it is a nonblind problem). In Chapter 11, a plethora of methods will be explored. To that end, the data-dependent spatial filters that will be presented in the following sections can use the source RTFs rather than their corresponding acoustic transfer functions. Yet, for the sake of simplicity and readability, the spatial filters will be defined using acoustic transfer functions.

10.4.2 General Criterion for the Narrowband Model

Assume the narrowband approximation in the STFT domain holds. Further assume that the received microphone signals comprise J_p point sources of interest and $J - J_p$ noise sources with arbitrary spatial characteristics. The microphone signals can be written as

$$\mathbf{x}(n,f) = \mathbf{A}(n,f)\mathbf{s}(n,f) + \mathbf{u}(n,f) \tag{10.18}$$

where $\mathbf{A}(n,f) = [\mathbf{a}_1(n,f), \dots, \mathbf{a}_{J_p}(n,f)]$, $\mathbf{s}(n,f) = [s_1(n,f), \dots, s_{J_p}(n,f)]^T$, and $\mathbf{u}(n,f) = \sum_{j=J_p+1}^{J} \mathbf{c}_j(n,f)$ is the contribution of all noise sources. The time-frequency indices (n,f) will be omitted in the following for brevity.

In the most general form, define $\mathbf{d} = \mathbf{Q}^H\mathbf{s}$ as the desired output vector, where $\mathbf{Q}$ is a $J_p \times D$ matrix of weights controlling the contributions of the signals of interest at the output, and $\hat{\mathbf{d}} = \mathbf{W}^H\mathbf{x}$ the actual $D \times 1$ vector output of the filtering matrix $\mathbf{W}$. The filtering matrix $\mathbf{W}$ is set to satisfy the following minimum mean square error (MMSE) criterion:

$$\mathbf{W}_{\text{MO-MWF}} = \underset{\mathbf{W}}{\arg\min} \, \mathbb{E}\{\text{tr}((\hat{\mathbf{d}} - \mathbf{d})(\hat{\mathbf{d}} - \mathbf{d})^H)\} \tag{10.19}$$

$$= \underset{\mathbf{W}}{\arg\min} \, (\mathbf{Q} - \mathbf{A}^H\mathbf{W})^H\mathbf{\Sigma}_{\mathbf{s}}(\mathbf{Q} - \mathbf{A}^H\mathbf{W}) + \mathbf{W}^H\mathbf{\Sigma}_{\mathbf{u}}\mathbf{W} \tag{10.20}$$

where $\mathbf{\Sigma}_{\mathbf{s}} = \text{Diag}(\sigma_1^2, \dots, \sigma_{J_p}^2)$ and $\mathbf{\Sigma}_{\mathbf{u}}$ are the covariance matrices of $\mathbf{s}$ and $\mathbf{u}$, respectively. The result of the optimization is $\mathbf{W}_{\text{MO-MWF}}$, the multiple-output *multichannel Wiener filter* (MWF) matrix, which is given by

$$\mathbf{W}_{\text{MO-MWF}} = \mathbf{\Sigma}_{\mathbf{x}}^{-1}\mathbf{A}\mathbf{\Sigma}_{\mathbf{s}}\mathbf{Q} = (\mathbf{A}\mathbf{\Sigma}_{\mathbf{s}}\mathbf{A}^H + \mathbf{\Sigma}_{\mathbf{u}})^{-1}\mathbf{A}\mathbf{\Sigma}_{\mathbf{s}}\mathbf{Q} \tag{10.21}$$

with $\mathbf{\Sigma}_{\mathbf{x}} = \mathbf{A}\mathbf{\Sigma}_{\mathbf{s}}\mathbf{A}^H + \mathbf{\Sigma}_{\mathbf{u}}$ the covariance matrix of $\mathbf{x}$. This expression of the MMSE estimator is valid for any distribution of the variables, provided that the filtering process is linear. Similarly to Section 5.2.3 for single-channel filtering, it can also be interpreted as the maximum a posteriori (MAP) estimator and the MMSE estimator for zero-mean Gaussian distributed variables, without requiring the linearity assumption. Indeed, assuming that

$$\begin{bmatrix} \mathbf{x} \\ \mathbf{d} \end{bmatrix} \sim \mathcal{N}_c\left(\begin{bmatrix} \mathbf{x} \\ \mathbf{d} \end{bmatrix} \,\middle|\, \begin{bmatrix} \mathbf{0}_I \\ \mathbf{0}_D \end{bmatrix}, \begin{bmatrix} \mathbf{\Sigma}_{\mathbf{x}} & \mathbf{\Sigma}_{\mathbf{xd}} \\ \mathbf{\Sigma}_{\mathbf{dx}} & \mathbf{\Sigma}_{\mathbf{d}} \end{bmatrix}\right), \tag{10.22}$$

the MAP estimate $\hat{\mathbf{d}}$ and the posterior mean $\mathbb{E}\{\mathbf{d} \mid \mathbf{x}\}$ are given by

$$\hat{\mathbf{d}} = \mathbb{E}\{\mathbf{d} \mid \mathbf{x}\} = \mathbf{\Sigma}_{\mathbf{dx}}\mathbf{\Sigma}_{\mathbf{x}}^{-1}\mathbf{x} \tag{10.23}$$

with $\mathbf{\Sigma}_{\mathbf{dx}} = \mathbf{Q}^H\mathbf{\Sigma}_{\mathbf{s}}\mathbf{A}^H$ (Bishop, 2006).

In the more widely-used scenario, which is the focus of this chapter, a single desired combination of the signals of interest $d = \mathbf{q}^H\mathbf{s}$ is considered, where the desired response vector $\mathbf{q}$ is a vector of weights controlling the contribution of the signals at the desired output, and $\hat{d} = \mathbf{w}^H \mathbf{x}$ is the actual output of the beamformer $\mathbf{w}$. The beamformer weights are set to satisfy the following weighted MMSE criterion (Markovich-Golan *et al.*, 2012b), referred to as the multiple speech distortion weighted MWF criterion:

$$\mathbf{w}_{\text{MSDW-MWF}} = \underset{\mathbf{w}}{\text{argmin}} \, (\mathbf{q} - \mathbf{A}^H \mathbf{w})^H \mathbf{\Lambda} \mathbf{\Sigma}_s (\mathbf{q} - \mathbf{A}^H \mathbf{w}) + \mathbf{w}^H \mathbf{\Sigma}_u \mathbf{w} \tag{10.24}$$

where $\mathbf{\Lambda} = \text{Diag}(\lambda_1, \ldots, \lambda_{J_p})$ is a diagonal weight matrix with tradeoff factors on its diagonal, controlling the deviation of the individual contributions of the signals of interest at the output from their respective desired contribution. The multiple speech distortion weighted MWF optimizing the criterion in (10.24) is given by

$$\mathbf{w}_{\text{MSDW-MWF}} = (\mathbf{A}\mathbf{\Lambda}\mathbf{\Sigma}_s \mathbf{A}^H + \mathbf{\Sigma}_u)^{-1} \mathbf{A}\mathbf{\Lambda}\mathbf{\Sigma}_s \mathbf{q}. \tag{10.25}$$

By setting $\mathbf{\Lambda} = \mathbf{I}_{J_p}$ we get the MWF for estimating a desired combination of all signals of interest $d = \mathbf{q}^H \mathbf{s}$:

$$\mathbf{w}_{\text{M-MWF}} = (\mathbf{A}\mathbf{\Sigma}_s \mathbf{A}^H + \mathbf{\Sigma}_u)^{-1} \mathbf{A}\mathbf{\Sigma}_s \mathbf{q} = \mathbf{\Sigma}_x^{-1} \mathbf{A}\mathbf{\Sigma}_s \mathbf{q}. \tag{10.26}$$

Various widely-used criteria can be derived from the multiple speech distortion weighted MWF criterion (10.24) by setting the values of the weight matrix $\mathbf{\Lambda}$, as explained in the following section. See also Van Trees (2002), Van Veen and Buckley (1988), and Cox *et al.* (1987) for more information on beamforming techniques.

10.4.3   MWF and SDW-MWF

Starting from the single desired source case $J_p = 1$, i.e. $d = q^* s_1$, the MWF can be derived by rewriting the MMSE criterion as

$$\mathbf{w}_{\text{MWF}} = \underset{\mathbf{w}}{\text{argmin}} |q - \mathbf{a}_1^H \mathbf{w}|^2 \sigma_{s_1}^2 + \mathbf{w}^H \mathbf{\Sigma}_u \mathbf{w}. \tag{10.27}$$

The minimizer of the cost function in (10.27) is the MWF

$$\mathbf{w}_{\text{MWF}} = (\sigma_{s_1}^2 \mathbf{a}_1 \mathbf{a}_1^H + \mathbf{\Sigma}_u)^{-1} \sigma_{s_1}^2 \mathbf{a}_1 q = \frac{\sigma_{s_1}^2 \mathbf{\Sigma}_u^{-1} \mathbf{a}_1}{1 + \sigma_{s_1}^2 \mathbf{a}_1^H \mathbf{\Sigma}_u^{-1} \mathbf{a}_1} q. \tag{10.28}$$

The MWF cost function comprises two terms. The first term, $|q - \mathbf{a}_1^H \mathbf{w}|^2 \sigma_{s_1}^2$, is the power of the speech distortion induced by spatial filtering, while the second term, $\mathbf{w}^H \mathbf{\Sigma}_u \mathbf{w}$, is the noise power at the output of the beamformer. These two terms are also known as artifacts and interference, respectively, in the source separation literature. To gain further control on the cost function, a tradeoff factor may be introduced, resulting in the *speech distortion weighted MWF* cost function (Doclo *et al.*, 2005)

$$\mathbf{w}_{\text{SDW-MWF}} = \underset{\mathbf{w}}{\text{argmin}} |q - \mathbf{a}_1^H \mathbf{w}|^2 \sigma_{s_1}^2 + \mu \mathbf{w}^H \mathbf{\Sigma}_u \mathbf{w} \tag{10.29}$$

where μ is a tradeoff factor between speech distortion and noise reduction. The relation between μ and the matrix $\mathbf{\Lambda}$ in (10.24) will be shortly discussed. Minimizing the criterion in (10.29) yields

$$\mathbf{w}_{\text{SDW-MWF}} = \frac{\sigma_{s_1}^2 \mathbf{\Sigma}_u^{-1} \mathbf{a}_1}{\mu + \sigma_{s_1}^2 \mathbf{a}_1^H \mathbf{\Sigma}_u^{-1} \mathbf{a}_1} q. \tag{10.30}$$

Note that by selecting $\mu \neq 1$ an increase in the mean square error (MSE) is expected, however, this additional degree of freedom allows the system designer to reduce one of the error terms in (10.27) (either distortion or noise power) at the expense of increasing the other error term.

Interestingly, the speech distortion weighted MWF can be obtained from the general multiple speech distortion weighted MWF in (10.25) by setting $\Lambda = \mu^{-1}I_{J_p}$ and $J_p = 1$.

Time-domain implementation of the (single-source) MWF-based speech enhancement is proposed by Doclo and Moonen (2005) and Doclo *et al.* (2005). The covariance matrix of the received microphone signals comprises speech and noise components. Using the generalized singular value decomposition, the mixture and noise covariance matrices can be jointly diagonalized (Doclo and Moonen, 2002). Utilizing the low-rank structure of the speech component, a time-recursive and reduced-complexity implementation is proposed. The complexity can be further reduced by shortening the length of the generalized singular value decomposition-based filters.

In later work, Benesty *et al.* (2008) derived a similar solution to the speech distortion weighted MWF from a different perspective. They proposed to minimize the noise variance at the output of the beamformer while constraining the maximal distortion incurred to the speech signal, denoted $\sigma_{d,\max}^2$. The beamformer which optimizes the latter criterion is called the *parametric MWF*

$$\mathbf{w}_{\text{PMWF}} = \underset{\mathbf{w}}{\arg\min} \ \mathbf{w}^H \mathbf{\Sigma_u w} \quad \text{s.t.} \quad \mathbb{E}\{|d - \widehat{d}|^2\} \leq \sigma_{d,\max}^2. \tag{10.31}$$

The expression of the parametric MWF is similar to that of the speech distortion weighted MWF in (10.30), where $\sigma_{d,\max}^2$ is used for controlling the aforementioned tradeoff between the error terms. The relation between the parameters $\sigma_{d,\max}^2$ of the parametric MWF and μ of the speech distortion weighted MWF does not have a closed-form representation in the general case.

10.4.4 MVDR, Maximum SNR, and LCMV

By tuning μ in the range $(0, +\infty)$, the speech distortion level can be traded off for the residual noise level. For $\mu \rightarrow +\infty$, maximum noise reduction but maximum speech distortion are obtained. Setting $\mu = 1$, the speech distortion weighted MWF identifies with the MWF. For $\mu \rightarrow 0$, the speech distortion weighted MWF identifies with the *minimum variance distortionless response* (MVDR) beamformer, with a strict distortionless response $\mathbf{w}^H \mathbf{a}_1 = q$,

$$\mathbf{w}_{\text{MVDR}} = \frac{\mathbf{\Sigma_u}^{-1} \mathbf{a}_1}{\mathbf{a}_1^H \mathbf{\Sigma_u}^{-1} \mathbf{a}_1} q, \tag{10.32}$$

which optimizes the following constrained minimization:

$$\mathbf{w}_{\text{MVDR}} = \underset{\mathbf{w}}{\arg\min} \ \mathbf{w}^H \mathbf{\Sigma_u w} \quad \text{s.t.} \quad \mathbf{a}_1^H \mathbf{w} = q. \tag{10.33}$$

The MVDR beamformer was applied to speech signals by, for example, Affes and Grenier (1997) and Gannot *et al.* (2001). For more information regarding the speech distortion weighted MWF and MVDR beamformers and their relations, see Doclo *et al.* (2010). In Section 10.6 we will discuss in detail the decomposition of the speech distortion weighted MWF into an MVDR beamformer and a subsequent postfiltering stage.

It is also easy to establish (Van Trees, 2002) that the MVDR and the following *minimum power distortionless response* criterion are equivalent, provided that the steering vector is perfectly known:

$$\mathbf{w}_{\text{MPDR}} = \underset{\mathbf{w}}{\arg\min}\ \mathbf{w}^H \mathbf{\Sigma}_x \mathbf{w} \quad \text{s.t.} \quad \mathbf{a}_1^H \mathbf{w} = q. \tag{10.34}$$

The resulting beamformer

$$\mathbf{w}_{\text{MPDR}} = \frac{\mathbf{\Sigma}_x^{-1} \mathbf{a}_1}{\mathbf{a}_1^H \mathbf{\Sigma}_x^{-1} \mathbf{a}_1} q \tag{10.35}$$

exhibits, however, higher sensitivity to perturbations in microphone positions and rounding errors in the beamformer weights than the MVDR beamformer (Cox, 1973).

Finally, it can be shown (Cox *et al.*, 1987) that the *maximum SNR* beamformer that maximizes the output SNR

$$\mathbf{w}_{\text{MSNR}} = \underset{\mathbf{w}}{\arg\max} \frac{|\mathbf{a}_1^H \mathbf{w}|^2}{\mathbf{w}^H \mathbf{\Sigma}_u \mathbf{w}} \tag{10.36}$$

is given by

$$\mathbf{w}_{\text{MSNR}} = \zeta \mathbf{\Sigma}_u^{-1} \mathbf{a}_1 \tag{10.37}$$

with ζ an arbitrary scalar. The maximum SNR beamformer is equal to the MVDR and the minimum power distortionless response beamformers if they satisfy the same constraint $\mathbf{a}_1^H \mathbf{w} = q$. The maximum SNR beamformer was applied to speech enhancement by Araki *et al.* (2007) and Warsitz and Haeb-Umbach (2007).

Selecting $\mathbf{\Lambda} = \mu^{-1} \mathbf{\Sigma}_s^{-1}$ in (10.25), we obtain at the limit

$$\lim_{\mu \to 0} \mathbf{w}_{\text{MSDW-MWF}} = \mathbf{\Sigma}_u^{-1} \mathbf{A} (\mathbf{A}^H \mathbf{\Sigma}_u^{-1} \mathbf{A})^{-1} \mathbf{q}, \tag{10.38}$$

which is called the *linearly constrained minimum variance* (LCMV) beamformer. It is easily verified that the LCMV beamformer is equivalent to the *linearly constrained minimum power* beamformer (Van Trees, 2002)

$$\mathbf{w}_{\text{LCMP}} = \mathbf{\Sigma}_x^{-1} \mathbf{A} (\mathbf{A}^H \mathbf{\Sigma}_x^{-1} \mathbf{A})^{-1} \mathbf{q}. \tag{10.39}$$

The LCMV beamformer optimizes the following criterion:

$$\mathbf{w}_{\text{LCMV}} = \underset{\mathbf{w}}{\arg\min}\ \mathbf{w}^H \mathbf{\Sigma}_u \mathbf{w} \quad \text{s.t.} \quad \mathbf{A}^H \mathbf{w} = \mathbf{q}, \tag{10.40}$$

while the linearly constrained minimum power criterion is obtained by substituting $\mathbf{\Sigma}_u$ by $\mathbf{\Sigma}_x$ in (10.40). The linearly constrained minimum power beamformer is known to be much more sensitive to microphone position perturbations and numerical errors than the LCMV beamformer (Cox, 1973). Note that while an interfering source can be perfectly nulled out by adding a proper constraint to the LCMV criterion, its power will only be suppressed by the minimization operation of the MVDR criterion. Interestingly, the MVDR beamformer can also direct an almost perfect null towards an interfering source, provided that the respective spatial covariance matrix $\mathbf{\Sigma}_u$ is a rank-1 matrix. Similar relations, with the proper modifications, apply to the linearly constrained minimum power and minimum power distortionless response beamformers. An LCMV beamformer was applied to speech signals by Markovich-Golan *et al.* (2012b).

10.4.5 Criteria for Full-Rank Covariance Models

The beamformers we have seen so far rely on the narrowband approximation (10.18). As elaborated in Chapter 3, this approximation does not hold for highly reverberant environments, especially if the STFT is implemented with short frames. To circumvent this issue, interframe and interband information can be utilized. Beamformers then involve STFT coefficients from multiple frames or frequency bands as inputs. Consequently, the dimension of the beamformer increases. See Section 19.2.2 for details about this approach.

The underlying MMSE criterion can also be used when this approximation does not hold, e.g. with full-rank covariance models (Duong *et al.*, 2010). Using the models in Chapter 3 (see (3.1) and its extensions), we define the target signal to be estimated as the vector $\mathbf{c}_j(n,f)$ of STFT coefficients of the spatial image of source j. Beamforming can then be achieved using a matrix of weights $\mathbf{W}(n,f)$ as $\hat{\mathbf{c}}_j(n,f) = \mathbf{W}^H(n,f)\mathbf{x}(n,f)$. The MMSE criterion is expressed as

$$\underset{\mathbf{W}}{\operatorname{argmin}} \ \mathbb{E}\{\|\mathbf{W}^H(n,f)\mathbf{x}(n,f) - \mathbf{c}_j(n,f)\|_2^2\}, \tag{10.41}$$

which is optimized by the MWF

$$\mathbf{W}_j(n,f) = \mathbf{\Sigma}_{\mathbf{x}}^{-1}(n,f)\mathbf{\Sigma}_{\mathbf{c}_j}(n,f) \tag{10.42}$$

with $\mathbf{\Sigma}_{\mathbf{x}}(n,f) = \sum_{j=1}^{J} \mathbf{\Sigma}_{\mathbf{c}_j}(n,f)$. This estimator can also be derived in a MAP perspective assuming that the source spatial images $\mathbf{c}_j(n,f)$ are Gaussian distributed.

Finally, Schwartz *et al.* (2015) combined the two perspectives (narrowband and full-rank models) in a nested MVDR beamforming structure.

10.4.6 Binary Masking and Beamforming

The enhancement and separation criteria explored thus far are obtained by (constrained) power minimization and hence can be expressed in terms of the second-order statistics of the signals. These criteria assume that the signals are characterized by a Gaussian distribution and that they may be simultaneously active. However, as we have seen in Chapter 5, speech signals are assumed sparse in the STFT domain. In the single-channel context, a speech signal can therefore be segregated from a mixture by applying a binary mask. In the multichannel context, spatial filters that utilize sparsity can be designed. This is the focus of the current section.

The simplest model assumes that a single source indexed by $j = z(n,f)$ is active in each time-frequency bin (Roman *et al.*, 2003; Yılmaz and Rickard, 2004; Izumi *et al.*, 2007). If we further assume that $z(n,f)$ is uniformly distributed in $\{1, \dots, J\}$ and that the noise $\mathbf{u}(n,f)$ is Gaussian with covariance $\mathbf{\Sigma}_{\mathbf{u}}(f)$, the sources $s_j(n,f)$ and the model parameters $\theta = \{\mathbf{a}_j(f), \mathbf{\Sigma}_{\mathbf{u}}(f)\}$ can be jointly estimated by maximizing the log-likelihood

$$\underset{s,\theta}{\operatorname{argmax}} \sum_{nf} - \log \det(\pi\mathbf{\Sigma}_{\mathbf{u}}(f)) - (\mathbf{x}(n,f) - \mathbf{a}_{z(n,f)}(f) \, s_{z(n,f)}(n,f))^H$$

$$\mathbf{\Sigma}_{\mathbf{u}}^{-1}(f)(\mathbf{x}(n,f) - \mathbf{a}_{z(n,f)}(f) \, s_{z(n,f)}(n,f)). \tag{10.43}$$

Given $z(n,f)$ and θ, it turns out that the optimal value of the predominant source is obtained by the MVDR beamformer $s_{z(n,f)}(n,f) = \mathbf{w}_{\text{MVDR}}^H(f)\mathbf{x}(n,f)$, where $\mathbf{w}_{\text{MVDR}}(f)$

is given by (10.32) by identifying $\mathbf{a}_1$ with $\mathbf{a}_{z(n,f)}(f)$ and setting $q = 1$. The other sources $s_j(n,f)$, $j \neq z(n,f)$, are set to zero. This can be interpreted as a conventional MVDR beamformer followed by a binary postfilter equal to 1 for the predominant source and 0 for the other sources (see Section 10.6).

A variant of this approach assumes that a subset of sources is active in each time-frequency bin where the number of active sources is smaller than the number of microphones (Gribonval, 2003; Rosca *et al.*, 2004; Togami *et al.*, 2006; Aissa-El-Bey *et al.*, 2007; Thiergart *et al.*, 2014).

10.4.7 Blind Source Separation and Beamforming

Just as beamforming, BSS aims at extracting desired signals from a mixture of source signals or from background noise. If multiple microphones and linear filters are used for blindly separating J sources, this corresponds to a spatial filtering as with beamforming by a filter matrix $\mathbf{W}$, see (10.21). The main difference between BSS and beamforming lies in the criteria for determination of the filter matrices $\mathbf{W}$, which in turn results from the fact that for BSS no knowledge of the array topology nor the source positions is assumed, so that $\mathbf{A}(n,f)$ in the model (10.18) remains unknown. While traditionally beamformers utilized the array topology and the source positions, modern beamformers often do not require such information (e.g., Markovich *et al.* (2009) requires information on the activity pattern of the sources of interest during the beamformer learning stage instead).

The determination of the filters $\mathbf{W}$ in BSS algorithms is thus based on the assumption that the sources produce statistically independent, or at least uncorrelated, signals.

Still closely related to the beamforming criteria above are blind spatial filtering techniques based on eigenanalysis of the spatial covariance matrix of the microphone signals $\mathbf{\Sigma}_\mathbf{x}$, such as principal component analysis, which allows dominant eigenvectors to be interpreted as steering vectors (for more, see Chapter 13).

While the uncorrelated outputs of the principal component analysis-based eigenanalysis are not necessarily identical to the original statistically independent inputs (Hyvärinen *et al.*, 2004), independent component analysis (ICA) based on instantaneous observation vectors for a given frequency bin f and time frame n, $\mathbf{x}(n,f)$, aims at identifying a filter matrix $\mathbf{W}$ which approximates statistical independence between the J output signals of interest. Then, similarly as for narrowband subspace methods, additional mechanisms are required to align the separated outputs pertaining to the same signal source across all frequencies to circumvent the so-called internal permutation problem and effectively provide a spatial filtering effect comparable to a beamformer (see Chapter 13). To avoid the internal permutation problem of frequency-domain ICA (FD-ICA), convolutive BSS algorithms can be derived directly from time-domain criteria to capture the broadband nature of the involved signals. This also allows the direct exploitation of the temporal statistics of the observations, i.e. nonwhiteness and nonstationarity, in addition to nongaussianity (Buchner *et al.*, 2004, 2005). Second-order statistics-based algorithms exploiting spatial and temporal correlation proved to be particularly robust and effective, also for real-time applications (Smaragdis, 1998; Parra and Spence, 2000; Buchner *et al.*, 2005), and can conveniently be implemented in the STFT domain for computational efficiency (Aichner *et al.*, 2007). Performance can be further improved if all three properties of the source signals are exploited as, for example, by the TRINICON cost function (Buchner *et al.*,

2004, 2007). In the context of a discussion of design criteria for spatial filtering, it is stressed here that although ICA algorithms suggest a separation based only on statistical criteria, for multiple microphones the resulting filtering matrix $\mathbf{W}$ reflects physically plausible spatial filtering properties: in free sound fields it will point beams of increased sensitivity to the individual sources (Parra and Spence, 2000), and in reflective environments it will also include correlated reflections in the respective output signal (Zheng *et al.*, 2014). If additional knowledge on array topologies or even source locations is available, BSS methods can advantageously be combined with data-dependent beamforming schemes (Saruwatari *et al.*, 2001; Zheng *et al.*, 2009), e.g. for estimating RTFs for data-dependent beamforming (see Chapter 11) or to provide reference information for noise, interference, and reverberation in underdetermined scenarios (Takahashi *et al.*, 2009; Zheng *et al.*, 2009; Schwarz *et al.*, 2012; Reindl *et al.*, 2014) without requiring source activity information.

10.5 Generalized Sidelobe Canceler Implementation

Griffiths and Jim (1982) proposed a decomposition of the MVDR beamformer into two orthogonal beamformers called GSC, where one beamformer is responsible for satisfying the distortionless response constraint and the other for noise power minimization. An elegant proof that any LCMV beamformer can be formulated and implemented as a GSC is given by Breed and Strauss (2002) and an according LCMV beamformer is depicted in Figure 10.5.

The main benefit in using the GSC structure is the resulting decomposition of a usually intricate constrained optimization into (1) a *fixed beamformer* $\mathbf{w}_0(n,f)$, (2) a so-called *blocking matrix* $\mathbf{B}(n,f)$, and (3) an unconstrained optimization problem for determining a noise canceler $\mathbf{g}(n,f) = [g_1(n,f), \ldots, g_{I-J_p}(n,f)]^T$. The fixed beamformer $\mathbf{w}_0(n,f)$ is responsible for implementing J_p linear constraints, one of which

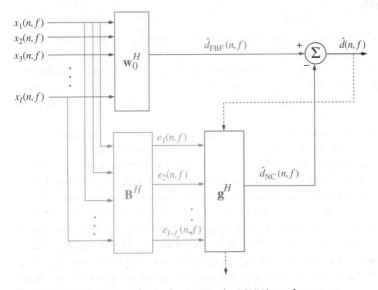

Figure 10.5 GSC structure for implementing the LCMV beamformer.

will typically enforce a distortionless response for the target source (corresponding to the MVDR beamformer constraint) and, in the LCMV case, the remaining $J_p - 1$ constraints usually try to suppress interfering point sources at known locations. The blocking matrix $\mathbf{B}(n,f)$ is orthogonal to $\mathbf{w}_0(n,f)$ and thus should suppress the target(s) which should be preserved by $\mathbf{w}_0$. It can be implemented as a projection matrix to the null space of the set of constrained sources and its structure can be simplified to avoid excess computational load (Markovich-Golan *et al.*, 2012a). The outputs of the blocking matrix, $e_1(n,f)$, ..., $e_{I-J_p}(n,f)$ represent estimates for noise and interference, which the noise canceler $\mathbf{g}(n,f)$ linearly combines to produce an estimate for the remaining noise in the output $\widehat{d}_{\mathrm{FBF}}(n,f)$ of the fixed beamformer. Optimizing $\mathbf{g}(n,f)$ corresponds to an unconstrained supervised system identification problem, which can be solved by least squares methods or iteratively by adaptive filters, hence allowing the entire beamformer to track changes in the noise statistics (Affes and Grenier, 1997; Gannot *et al.*, 2001). It has also been shown that for nongaussian signals a minimum mutual information criterion can further improve the performance (Reindl *et al.*, 2014).

As long as the blocking matrix $\mathbf{B}(n,f)$ is time-invariant and just suppresses a fixed steering direction for the target, it must be expected that already slight changes in the target source position and all significant reflections of the target signal lead to target signal leakage into $e_i(n,f)$ and consequently to the highly undesired signal cancellation effect in $\widehat{d}(n,f)$. In practical implementations for acoustic human/machine interfaces with some uncertainty regarding the exact target position, adaptive blocking matrices are necessary (Hoshuyama *et al.*, 1999; Herbordt and Kellermann, 2003), which must be adapted while, ideally, only the target source is active, whereas the noise canceler $\mathbf{g}(n,f)$ should then only be adapted if the target is inactive (Herbordt *et al.*, 2005). More recent BSS-based methods, however, allow the blocking matrix to be adapted continuously (Zheng *et al.*, 2009; Reindl *et al.*, 2014).

10.6 Postfilters

The performance of the beamformers discussed in this chapter may be limited by the presence of spatially distributed undesired sources or when the number of interfering sources exceed the number of available microphones. Single-channel spectral enhancement methods, which do not take spatial characteristics into account, can therefore serve as *postfilters* at the output of the beamformer to further reduce the residual noise and interference. Another benefit of the single-microphone methods is their ability to adapt to fast changes in the noise statistics, making them particularly suitable to nonstationary noise environments. The latter property will be discussed in Chapter 11 together with the estimation procedures for the beamformer and postfilter parameters.

Balan and Rosca (2002) gave a mathematical justification for applying nonlinear postfiltering, provided that it can be stated as an MMSE estimator of a nonlinear function of the desired signal (see also the discussion by Van Trees (2002) for general arrays). Assuming that the desired source and the noise signals are jointly complex Gaussian, it can be shown that a sufficient statistics (in the Bayesian sense) for the MMSE estimator of any nonlinear function of the desired signal d, namely $\mathbb{E}\{|\rho(d) - \rho(\widehat{d})|^2\}$, is the output of the MVDR beamformer, where $\rho(\cdot)$ stands for an arbitrary nonlinear function. This result is only applicable for a single desired speaker. Schwartz *et al.* (2017)

recently extended it to the multispeaker case. It was shown that the LCMV beamformer is the sufficient statistic in this case, and a multispeaker postfilter was derived. In this section and in Chapter 11 we will only focus on postfilters for the case of a single desired source. This relation states that the MMSE estimator of $\rho(d)$ given the microphone signals can be evaluated by first applying the MVDR beamformer to the microphone signals and subsequently applying a single-channel postfilter to its output. This concatenated processing structure reduces complexity, as the design of the MVDR beamformer as the first step considers only spatial properties and is independent of the nonstationary speech source. By setting $\rho(d) = d$ we simply get the result that the MWF can be decomposed as an MVDR followed by a single-channel Wiener filter. By setting $\rho(d) = |d|$ or $\rho(d) = \log|d|$ we obtain the Ephraim and Malah short-time spectral amplitude estimator (Ephraim and Malah, 1984) or log-spectral amplitude estimator (Ephraim and Malah, 1985), respectively.

Similarly, the speech distortion weighted MWF in (10.30) can be interpreted as a multichannel MVDR beamformer, which is independent of the speech temporal variations, followed by a speech distortion weighted single-channel Wiener postfilter which accounts for the instantaneous SNR (Doclo *et al.*, 2010). Let the desired source be $d = s_1$, the first source, with the respective acoustic transfer function $\mathbf{a}_1$. The speech distortion weighted MWF can then be decomposed as follows:

$$
\mathbf{w}_{\text{SDW-MWF}} = \underbrace{\frac{\boldsymbol{\Sigma}_{\mathbf{u}}^{-1}\mathbf{a}_1}{\mathbf{a}_1^H \boldsymbol{\Sigma}_{\mathbf{u}}^{-1}\mathbf{a}_1}}_{\mathbf{w}_{\text{MVDR}}} \times \underbrace{\frac{\sigma_{s_1}^2}{\sigma_{s_1}^2 + \mu \sigma_u^{\text{out } 2}}}_{w_{\text{SDW-SWF}}}
\tag{10.44}
$$

where $\mathbf{w}_{\text{MVDR}}$ is the MVDR beamformer, $\sigma_u^{\text{out } 2}$ is the noise power at its output, and $w_{\text{SDW-SWF}}$ is a single-channel Wiener filter.

A plethora of postfilters can be found in the literature, differing in the procedures for estimating the speech and noise statistics (Zelinski, 1988; Meyer and Simmer, 1997; Marro *et al.*, 1998; McCowan and Bourlard, 2003; Kolossa and Orglmeister, 2004; Leukimmiatis *et al.*, 2006; Hoffmann *et al.*, 2009; Reindl *et al.*, 2013).

10.7 Summary

Beamforming has become a widely used speech enhancement tasks. Many current commercial products, e.g. teleconferencing systems and cellular phones, are utilizing beamforming techniques to achieve enhanced speech quality and intelligibility. In this chapter, we have explored some of the widely used criteria and techniques for designing beamformers. In Chapter 11, estimation procedures that are necessary for constructing the beamformers will be discussed. Other techniques for spatial filtering and the associated parameter estimation can be found in Chapters 12, 13, and 14. Chapter 18 discusses the special case of designing binaural beamformers.

Enhancing the robustness to mismatch between the design of the beamformer in nominal conditions and the actual application conditions is still an open challenge. New optimal and suboptimal criteria for distributed beamforming for ad hoc microphone arrays is another current research topic. Distributed algorithms will be briefly explored in Chapter 19.

Bibliography

Abhayapala, T.D. and Ward, D.B. (2002) Theory and design of high order sound field microphones using spherical microphone array, in *Proceedings of IEEE International Conference on Audio, Speech and Signal Processing*, vol. 2, pp. 1949–1952.

Affes, S. and Grenier, Y. (1997) A signal subspace tracking algorithm for microphone array processing of speech. *IEEE Transactions on Speech and Audio Processing*, **5** (5), 425–437.

Aichner, R., Buchner, H., and Kellermann, W. (2007) Exploiting narrowband efficiency for broadband convolutive blind source separation. *EURASIP Journal on Advances in Signal Processing*, **2007**, 1–9.

Aissa-El-Bey, A., Abed-Meraim, K., and Grenier, Y. (2007) Blind separation of underdetermined convolutive mixtures using their time-frequency representation. *IEEE Transactions on Audio, Speech, and Language Processing*, **15** (5), 1540–1550.

Anderson, C.A., Teal, P.D., and Poletti, M.A. (2015) Spatially robust far-field beamforming using the von Mises (-Fisher) distribution. *IEEE/ACM Transactions on Audio, Speech, and Language Processing*, **23** (12), 2189–2197.

Araki, S., Sawada, H., and Makino, S. (2007) Blind speech separation in a meeting situation with maximum SNR beamformers, in *Proceedings of IEEE International Conference on Audio, Speech and Signal Processing*, vol. 1, pp. 41–44.

Balan, R. and Rosca, J. (2002) Microphone array speech enhancement by Bayesian estimation of spectral amplitude and phase, in *Proceedings of IEEE Workshop on Sensor Array and Multichannel Processing*, pp. 209–213.

Barfuss, H., Hümmer, C., Schwarz, A., and Kellermann, W. (2015) HRTF-based robust least-squares frequency-invariant beamforming, in *Proceedings of IEEE Workshop on Applications of Signal Processing to Audio and Acoustics*, pp. 1–5.

Benesty, J. and Chen, J. (2013) *Study and Design of Differential Microphone Arrays*, Springer.

Benesty, J., Chen, J., and Huang, Y. (2008) *Microphone array signal processing*, vol. 1, Springer.

Bishop, C. (2006) *Pattern Recognition and Machine Learning*, Springer.

Bitzer, J. and Simmer, K.U. (2001) Superdirective microphone arrays, in *Microphone Arrays: Signal Processing Techniques and Applications*, Springer, chap. 2, pp. 19–38.

Breed, B.R. and Strauss, J. (2002) A short proof of the equivalence of LCMV and GSC beamforming. *IEEE Signal Processing Letters*, **9** (6), 168–169.

Buchner, H., Aichner, R., and Kellermann, W. (2004) TRINICON: A versatile framework for multichannel blind signal processing, in *Proceedings of IEEE International Conference on Audio, Speech and Signal Processing*, vol. 3, pp. 889–892.

Buchner, H., Aichner, R., and Kellermann, W. (2005) A generalization of blind source separation algorithms for convolutive mixtures based on second-order statistics. *IEEE Transactions on Speech and Audio Processing*, **13** (1), 120–134.

Buchner, H., Aichner, R., and Kellermann, W. (2007) TRINICON-based blind system identification with application to multiple-source localization and separation, in *Blind Speech Separation*, Springer, pp. 101–147.

Cox, H. (1973) Resolving power and sensitivity to mismatch of optimum array processors. *Journal of the Acoustical Society of America*, **54** (3), 771–785.

Cox, H., Zeskind, R., and Kooij, T. (1986) Practical supergain. *IEEE Transactions on Acoustics, Speech, and Signal Processing*, **34** (3), 393–398.

Cox, H., Zeskind, R., and Owen, M. (1987) Robust adaptive beamforming. *IEEE Transactions on Acoustics, Speech, and Signal Processing*, **35** (10), 1365–1376.

Doclo, S., Gannot, S., Moonen, M., and Spriet, A. (2010) Acoustic beamforming for hearing aid applications, in *Handbook on Array Processing and Sensor Networks*, Wiley-IEEE Press.

Doclo, S. and Moonen, M. (2002) GSVD-based optimal filtering for single and multimicrophone speech enhancement. *IEEE Transactions on Signal Processing*, **50** (9), 2230–2244.

Doclo, S. and Moonen, M. (2005) Multimicrophone noise reduction using recursive GSVD-based optimal filtering with ANC postprocessing stage. *IEEE Transactions on Speech and Audio Processing*, **13** (1), 53–69.

Doclo, S., Spriet, A., Wouters, J., and Moonen, M. (2005) Speech distortion weighted multichannel Wiener filtering techniques for noise reduction, in *Speech Enhancement*, Springer, pp. 199–228.

Duong, N.Q.K., Vincent, E., and Gribonval, R. (2010) Under-determined reverberant audio source separation using a full-rank spatial covariance model. *IEEE Transactions on Audio, Speech, and Language Processing*, **18** (7), 1830–1840.

Elko, G. (2000) Superdirectional microphone arrays, in *Acoustic Signal Processing for Telecommunication*, Kluwer, chap. 10, pp. 181–237.

Elko, G. (2004) Differential microphone arrays, in *Audio Signal Processing for Next-Generation Multimedia Communication Systems*, Kluwer, pp. 2–65.

Elko, G.W. (1996) Microphone array systems for hands-free telecommunication. *Speech Communication*, **20** (3), 229–240.

Elko, G.W. and Meyer, J.M. (2012) Using a higher-order spherical microphone array to assess spatial and temporal distribution of sound in rooms. *Journal of the Acoustical Society of America*, **132** (3), 1912–1912.

Ephraim, Y. and Malah, D. (1984) Speech enhancement using a minimum-mean square error short-time spectral amplitude estimator. *IEEE Transactions on Acoustics, Speech, and Signal Processing*, **32** (6), 1109–1121.

Ephraim, Y. and Malah, D. (1985) Speech enhancement using a minimum mean square error log-spectral amplitude estimator. *IEEE Transactions on Acoustics, Speech, and Signal Processing*, **33** (2), 443–445.

Er, M. and Cantoni, A. (1983) Derivative constraints for broad-band element space antenna array processors. *IEEE Transactions on Acoustics, Speech, and Signal Processing*, **31** (6), 1378–1393.

Flanagan, J., Johnston, J., Zahn, R., and Elko, G. (1985) Computer-steered microphone arrays for sound transduction in large rooms. *Journal of the Acoustical Society of America*, **78** (5), 1508–1518.

Frost, O.L. (1972) An algorithm for linearly constrained adaptive array processing. *Proceedings of the IEEE*, **60** (8), 926–935.

Gannot, S., Burshtein, D., and Weinstein, E. (2001) Signal enhancement using beamforming and nonstationarity with applications to speech. *IEEE Transactions on Signal Processing*, **49** (8), 1614–1626.

Goodwin, M.M. and Elko, G.W. (1993) Constant beamwidth beamforming, in *Proceedings of IEEE International Conference on Audio, Speech and Signal Processing*, vol. 1, pp. 169–172.

Gribonval, R. (2003) Piecewise linear source separation, in *Proceedings of SPIE Wavelets: Applications in Signal and Image Processing*, pp. 297–310.

Griffiths, L.J. and Jim, C.W. (1982) An alternative approach to linearly constrained adaptive beamforming. *IEEE Transactions on Antennas and Propagation*, **30** (1), 27–34.

Harris, F.J. (1978) Use of windows for harmonic analysis. *Proceedings of the IEEE*, **66** (1), 51–83.

Herbordt, W., Buchner, H., Nakamura, S., and Kellermann, W. (2005) Outlier-robust DFT-domain adaptive filtering for bin-wise stepsize controls, and its application to a generalized sidelobe canceller, in *Proceedings of International Workshop on Acoustic Echo and Noise Control*.

Herbordt, W. and Kellermann, W. (2003) Adaptive beamforming for audio signal acquisition, in *Adaptive Signal Processing*, Springer, chap. 6, pp. 155–194.

Hoffmann, E., Kolossa, D., and Orglmeister, R. (2009) Time frequency masking strategy for blind source separation of acoustic signals based on optimally-modified log-spectral amplitude estimator, in *Proceedings of International Conference on Independent Component Analysis and Signal Separation*, pp. 581–588.

Hoshuyama, O., Sugiyama, A., and Hirano, A. (1999) A robust adaptive beamformer for microphone arrays with a blocking matrix using constrained adaptive filters. *IEEE Transactions on Signal Processing*, **47** (10), 2677–2684.

Hyvärinen, A., Karhunen, J., and Oja, E. (2004) *Independent Component Analysis*, Wiley.

Izumi, Y., Ono, N., and Sagayama, S. (2007) Sparseness-based 2ch BSS using the EM algorithm in reverberant environment, in *Proceedings of IEEE Workshop on Applications of Signal Processing to Audio and Acoustics*, pp. 147–150.

Jin, C.T., Epain, N., and Parthy, A. (2014) Design, optimization and evaluation of a dual-radius spherical microphone array. *IEEE/ACM Transactions on Audio, Speech, and Language Processing*, **22** (1), 193–204.

Johnson, D.H. and Dudgeon, D.E. (1993) *Array Signal Processing: Concepts and Techniques*, Prentice Hall.

Kajala, M. and Hämäläinen, M. (2001) Filter-and-sum beamformer with adjustable filter characteristics, in *Proceedings of IEEE International Conference on Audio, Speech and Signal Processing*, vol. 5, pp. 2917–2920.

Kellermann, W. (1991) A self-steering digital microphone array, in *Proceedings of IEEE International Conference on Audio, Speech and Signal Processing*, p. 3581–3584.

Kennedy, R.A., Abhayapala, T.D., and Ward, D.B. (1998) Broadband nearfield beamforming using a radial beampattern transformation. *IEEE Transactions on Signal Processing*, **46** (8), 2147–2156.

Kolossa, D. and Orglmeister, R. (2004) Nonlinear postprocessing for blind speech separation, in *Proceedings of International Conference on Independent Component Analysis and Signal Separation*, pp. 832–839.

Laakso, T., Välimäki, V., Karjalainen, M., and Laine, U. (1996) Splitting the unit delay. *IEEE Signal Processing Magazine*, **13** (1), 34–60.

Leukimmiatis, S., Dimitriadis, D., and Maragos, P. (2006) An optimum microphone array post-filter for speech applications, in *Proceedings of Interspeech*, pp. 2142–2145.

Levin, D., Habets, E.A.P., and Gannot, S. (2013a) A generalized theorem on the average array directivity factor. *IEEE Signal Processing Letters*, **20** (9), 877–880.

Levin, D., Habets, E.A.P., and Gannot, S. (2013b) Robust beamforming using sensors with nonidentical directivity patterns, in *Proceedings of IEEE International Conference on Audio, Speech and Signal Processing*.

Li, Z. and Duraiswami, R. (2005) Hemispherical microphone arrays for sound capture and beamforming, in *Proceedings of IEEE Workshop on Applications of Signal Processing to Audio and Acoustics*, pp. 106–109.

Li, Z. and Duraiswami, R. (2007) Flexible and optimal design of spherical microphone arrays for beamforming. *IEEE Transactions on Audio, Speech, and Language Processing*, **15** (2), 702–714.

Mabande, E., Buerger, M., and Kellermann, W. (2012) Design of robust polynomial beamformers for symmetric arrays, in *Proceedings of IEEE International Conference on Audio, Speech and Signal Processing*, pp. 1–4.

Mabande, E., Schad, A., and Kellermann, W. (2009) Design of robust superdirective beamformers as a convex optimization problem, in *Proceedings of IEEE International Conference on Audio, Speech and Signal Processing*, pp. 77–80.

Markovich, S., Gannot, S., and Cohen, I. (2009) Multichannel eigenspace beamforming in a reverberant noisy environment with multiple interfering speech signals. *IEEE Transactions on Audio, Speech, and Language Processing*, **17** (6), 1071–1086.

Markovich-Golan, S., Gannot, S., and Cohen, I. (2012a) A sparse blocking matrix for multiple constraints GSC beamformer, in *Proceedings of IEEE International Conference on Audio, Speech and Signal Processing*, pp. 197–200.

Markovich-Golan, S., Gannot, S., and Cohen, I. (2012b) A weighted multichannel Wiener filter for multiple sources scenarios, in *Proceedings of IEEE Convention of Electrical and Electronics Engineers in Israel*.

Marro, C., Mahieux, Y., and Simmer, K. (1998) Analysis of noise reduction and dereverberation techniques based on microphone arrays with postfiltering. *IEEE Transactions on Speech and Audio Processing*, **6** (3), 240–259.

McCowan, I. and Bourlard, H. (2003) Microphone array post-filter based on noise field coherence. *IEEE Transactions on Speech and Audio Processing*, **11** (6), 709–716.

Meyer, J. and Elko, G. (2002) A highly scalable spherical microphone array based on an orthonormal decomposition of the soundfield, in *Proceedings of IEEE International Conference on Audio, Speech and Signal Processing*, vol. 2, pp. 1781–1784.

Meyer, J. and Simmer, K.U. (1997) Multi-channel speech enhancement in a car environment using Wiener filtering and spectral subtraction, in *Proceedings of IEEE International Conference on Audio, Speech and Signal Processing*, pp. 21–24.

Oppenheim, A. and Schafer, R.W. (1975) *Digital Signal Processing*, Prentice Hall.

Parra, L. and Spence, C. (2000) Convolutive blind separation of non-stationary sources. *IEEE Transactions on Speech and Audio Processing*, **8** (3), 320–327.

Parsons, A.T. (1987) Maximum directivity proof for three-dimensional arrays. *Journal of the Acoustical Society of America*, **82** (1), 179–182.

Rafaely, B. (2005) Analysis and design of spherical microphone arrays. *IEEE Transactions on Speech and Audio Processing*, **13** (1), 135–143.

Rafaely, B. (2015) *Fundamentals of Spherical Array Processing*, Springer.

Reindl, K., Meier, S., Barfuss, H., and Kellermann, W. (2014) Minimum mutual information-based linearly constrained broadband signal extraction. *IEEE/ACM Transactions on Audio, Speech, and Language Processing*, **22** (6), 1096–1108.

Reindl, K., Zheng, Y., Schwarz, A., Meier, S., Maas, R., Sehr, A., and Kellermann, W. (2013) A stereophonic acoustic signal extraction scheme for noisy and reverberant environments. *Computer Speech and Language*, **27** (3), 726–745.

Roman, N., Wang, D., and Brown, G. (2003) Speech segregation based on sound localization. *Journal of the Acoustical Society of America*, **114** (4), 2236–2252.

Rosca, J.P., Borss, C., and Balan, R.V. (2004) Generalized sparse signal mixing model and application to noisy blind source separation, in *Proceedings of IEEE International Conference on Audio, Speech and Signal Processing*, pp. III–877–III–880.

Saruwatari, H., Kurita, S., and Takeda, K. (2001) Blind source separation combining frequency-domain ICA and beamforming, in *Proceedings of IEEE International Conference on Audio, Speech and Signal Processing*, vol. 5, pp. 2733–2736.

Schwartz, O., Gannot, S., and Habets, E.A.P. (2017) Multispeaker LCMV beamformer and postfilter for source separation and noise reduction. *IEEE/ACM Transactions on Audio, Speech, and Language Processing*, **25** (5), 940–951.

Schwartz, O., Habets, E.A.P., and Gannot, S. (2015) Nested generalized sidelobe canceller for joint dereverberation and noise reduction, in *Proceedings of IEEE International Conference on Audio, Speech and Signal Processing*.

Schwarz, A., Reindl, K., and Kellermann, W. (2012) A two-channel reverberation suppression scheme based on blind signal separation and Wiener filtering, in *Proceedings of IEEE International Conference on Audio, Speech and Signal Processing*, pp. 1–4.

Silverman, H.F., Patterson, W.R.I., Flanagan, J.L., and Rabinkin, D. (1997) A digital processing system for source location and sound capture by large microphone arrays, in *Proceedings of IEEE International Conference on Audio, Speech and Signal Processing*, vol. 1, pp. 251–254.

Smaragdis, P. (1998) Blind separation of convolved mixtures in the frequency domain. *Neurocomputing*, **22** (1), 21–34.

Takahashi, Y., Takatani, T., Osako, K., Saruwatari, H., and Shikano, K. (2009) Blind spatial subtraction array for speech enhancement in noisy environment. *IEEE Transactions on Audio, Speech, and Language Processing*, **17** (4), 650–664.

Teutsch, H. (2007) *Modal Array Signal Processing: Principles and Applications of Acoustic Wavefield Decomposition*, Springer.

Teutsch, H. and Kellermann, W. (2006) Acoustic source detection and localization based on wavefield decomposition using circular microphone arrays. *Journal of the Acoustical Society of America*, **5** (120), 2724–2736.

Thiergart, O., Taseska, M., and Habets, E.A.P. (2014) An informed parametric spatial filter based on instantaneous direction-of-arrival estimates. *IEEE/ACM Transactions on Audio, Speech, and Language Processing*, **22** (12), 2182–2196.

Togami, M., Sumiyoshi, T., and Amano, A. (2006) Sound source separation of overcomplete convolutive mixtures using generalized sparseness, in *Proceedings of International Workshop on Acoustic Echo and Noise Control*.

Van der Wal, M., Start, E., and de Vries, D. (1996) Design of logarithmically spaced constant-directivity transducer arrays. *Journal of the Acoustical Society of America*, **44** (6), 497–507.

Van Trees, H.L. (2002) *Detection, Estimation, and Modulation Theory*, vol. IV, Wiley.

Van Veen, B.D. and Buckley, K.M. (1988) Beamforming: A versatile approach to spatial filtering. *IEEE ASSP Magazine*, pp. 4–24.

Ward, D.B., Kennedy, R.A., and Williamson, R.C. (1995) Theory and design of broadband sensor arrays with frequency invariant far-field beam patterns. *Journal of the Acoustical Society of America*, **97** (2), 91–95.

Warsitz, E. and Haeb-Umbach, R. (2007) Blind acoustic beamforming based on generalized eigenvalue decomposition. *IEEE Transactions on Audio, Speech, and Language Processing*, **15** (5), 1529–1539.

Yılmaz, Ö. and Rickard, S.T. (2004) Blind separation of speech mixtures via time-frequency masking. *IEEE Transactions on Signal Processing*, **52** (7), 1830–1847.

Zelinski, R. (1988) A microphone array with adaptive post-filtering for noise reduction in reverberant rooms, in *Proceedings of IEEE International Conference on Audio, Speech and Signal Processing*, pp. 2578–2581.

Zheng, Y., Reindl, K., and Kellermann, W. (2009) BSS for improved interference estimation for blind speech signal extraction with two microphones, in *Proceedings of IEEE International Workshop on Computational Advances in Multi-Sensor Adaptive Processing*.

Zheng, Y., Reindl, K., and Kellermann, W. (2014) Analysis of dual-channel ICA-based blocking matrix for improved noise estimation. *EURASIP Journal on Advances in Signal Processing*, p. 26.

11

Multichannel Parameter Estimation

Shmulik Markovich-Golan, Walter Kellermann, and Sharon Gannot

In this chapter we will explore some widely used structures and estimation procedures for tracking the parameters that are required for constructing the data-dependent spatial filters for audio signals that are discussed in Chapter 10. As before, the spatial filters are designed to extract a desired source contaminated by background noise, in the case of a single speaker, or by interfering speakers and background noise, in the multiple speakers case.

The spatial filters explored in Chapter 10 (mainly, those referred to as beamformers) assume that certain parameters are available for their computation, namely the relative transfer functions (RTFs) of the speakers, the covariance matrices of the background noise and the speakers, and/or the cross-covariance between the mixture signals and the desired signal.

The variety of optimization criteria in Chapter 10 leads to different data-dependent spatial filters, yet most of them rely on similar parameters and therefore a common estimation framework can be derived. In general, estimates of speech presence probability (SPP) are used to govern the estimation of noise and speech spatial covariance matrices. These estimates are then utilized to estimate source RTF vectors. Finally, the data-dependent spatial filters are designed based on the latter estimates. A high-level block diagram of the common estimation framework is depicted in Figure 11.1.

Methods utilizing multichannel inputs for estimating SPP, which may govern the adaptation of spatial filters, are presented in Section 11.1. Tracking the spatial covariance matrices of speech and noise is discussed in Section 11.2. Weakly and strongly guided methods for estimating the RTF in the case of a single speaker or multiple speakers case are surveyed in Section 11.3. Section 11.4 summarizes the chapter.

11.1 Multichannel Speech Presence Probability Estimators

The SPP is a fundamental and crucial component of many speech enhancement algorithms, including both single-channel and multichannel methods. SPP governs the adaptation of various components which contribute to the calculation of the spatial filters described in Chapter 10. More specifically, it can be used to govern

Audio Source Separation and Speech Enhancement, First Edition.
Edited by Emmanuel Vincent, Tuomas Virtanen and Sharon Gannot.
© 2018 John Wiley & Sons Ltd. Published 2018 by John Wiley & Sons Ltd.
Companion Website: https://project.inria.fr/ssse/

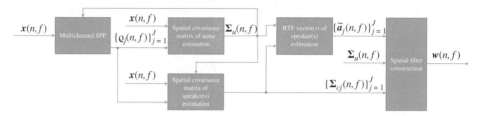

Figure 11.1 High-level block diagram of the common estimation framework for constructing data-dependent spatial filters.

the estimation of noise and speech covariance matrices (see Section 11.2) and of RTFs (see Section 11.3).

The problem of estimating the SPP is derived from the classic detection problem, also known as the radar problem (Van Trees, 2002), and its goal is to identify the spectro-temporal activity pattern of speech contaminated by noise, e.g. determining if a time-frequency bin contains a noisy speech component or just noise. While voice activity detection (VAD) typically provides a binary decision in the time domain, SPP can be seen as solving a regression problem and produces a probability estimate, rather than a binary decision, per time-frequency bin. Speech enhancement using an SPP model was first presented by McAulay and Malpass (1980).

When used to govern the adaptation of spatial filters, misdetection of speech and its false classification as noise might lead to major distortion, also known as the self-cancellation phenomenon. On the other hand, false alarms, i.e. time-frequency bins containing noise which are mistakenly classified as desired speech, result in increased noise level at the output of the spatial filter, since it is designed to pass through them. Depending on the usage of the SPP, estimating the power spectrum of either speech or noise, and on the desired compromise between misdetection and false alarm rates, hard or soft decision rules can be defined.

Most single-channel SPP estimators rely on exploiting the nonstationarity of speech to discriminate against noise which is assumed to be stationary during the considered observation interval. However, with low signal-to-noise ratio (SNR) and nonstationary background noise and interference signals, the accuracy of the estimation is degraded. For more details on single-channel SPP, please refer to Chapter 6.

Alternative single-channel methods utilize the harmonic structure of voiced speech phonemes, also known as the harmonicity property. Quatieri and McAulay (1990) modeled voiced speech as harmonics of a fundamental frequency for the purpose of noise reduction in the presence of white noise. Given an estimate of the fundamental frequency, the corresponding phase and amplitude parameters and those of its harmonics, along with goodness of fit, are obtained by solving a least squares problem or by nearest neighbor match to training sequences.

Several contributions extend SPP estimation to utilize spatial information when using an array of microphones, also known as *multichannel SPP*. The spatial dimension can be used for (1) discriminating speech- and noise-dominated time-frequency bins based on the observed interchannel coherence (IC) (speech is coherent whereas noise in many cases tends to be incoherent or diffuse), (2) improving the SNR of the microphone signals by spatial filtering, and (3) distinguishing between different speakers in the multiple speaker case.

Here we present some of these methods. In Section 11.1.1, the single-channel Gaussian signal model of both speech and noise is extended to multichannel input, yielding a multichannel SPP. In Section 11.1.2, the coherence property of the speech source, assuming diffuse noise, that is manifested in the *direct-to-diffuse ratio* (DDR) is used to estimate the prior SPP. In Section 11.1.3, the relation between the variances of the upper and lower branches of the generalized sidelobe canceler (GSC) is incorporated for estimating the SPP. The case of multiple speakers is considered in Section 11.1.4, and by incorporating the latter DDR-based prior SPP with estimated speaker positions, the SPPs of individual speakers can be obtained.

An extension of the harmonicity-based SPP (see Quatieri and McAulay (1990)) to the spatial domain, where directions of arrival (DOAs) and harmonicity parameters (pitch, amplitudes and phases) are estimated jointly, is proposed by Pessentheiner *et al.* (2016). The multidimensional state vector is tracked by optimizing a combined cost function and seeking a sparse representation in this multidimensional space.

11.1.1 Multichannel Gaussian Model-Based SPP

All derivations in this section refer to a specific time-frequency bin (n, f) and are replicated for all time-frequency bins. For brevity, the time and frequency indices are omitted in the rest of this section. Considering the case of a single speech source contaminated by noise, the received microphone signals

$$\mathbf{x} = \mathbf{c} + \mathbf{u} \tag{11.1}$$

and the speech and noise components thereof are modeled as complex Gaussian random variables

$$\mathbf{c} \sim \mathcal{N}_c(\mathbf{c} \mid \mathbf{0}_I, \mathbf{\Sigma}_c) \tag{11.2}$$

$$\mathbf{u} \sim \mathcal{N}_c(\mathbf{u} \mid \mathbf{0}_I, \mathbf{\Sigma}_u). \tag{11.3}$$

The vector of received speech components in the short-time Fourier transform (STFT) domain can be expressed as a multiplication of the speech source and a vector of acoustic transfer functions

$$\mathbf{c} = \mathbf{a}s \tag{11.4}$$

where $\mathbf{\Sigma}_c = \sigma_s^2 \mathbf{a}\mathbf{a}^H$ is the spatial covariance matrix of the speech component at the microphone signals. The set of (i, i')th elements of the spatial covariance matrices of all frequencies corresponds to the cross-spectrum between microphones i and i'. Consequently, a multichannel Gaussian model is adopted for the noise-only and for the noisy speech hypotheses, denoted by $\mathcal{H}_0$ and $\mathcal{H}_1$, respectively

$$\mathbf{x} \mid \mathcal{H}_0 \sim \mathcal{N}_c(\mathbf{x} \mid \mathbf{0}_I, \mathbf{\Sigma}_u) \tag{11.5}$$

$$\mathbf{x} \mid \mathcal{H}_1 \sim \mathcal{N}_c(\mathbf{x} \mid \mathbf{0}_I, \mathbf{\Sigma}_c + \mathbf{\Sigma}_u). \tag{11.6}$$

The single-channel SPP was extended to the multichannel case by Souden *et al.* (2010b) and is defined as:

$$\varrho = P(\mathcal{H}_1 \mid \mathbf{x}). \tag{11.7}$$

Using Bayes rule, the multichannel SPP can be computed as

$$\varrho = \frac{\rho}{1+\rho},\tag{11.8}$$

where ρ is the *generalized likelihood ratio*, which for our case equals

$$\rho = \frac{P(\mathcal{H}_1)}{P(\mathcal{H}_0)} \cdot \frac{1}{1 + \text{tr}(\mathbf{\Sigma}_\mathbf{u}^{-1}\mathbf{\Sigma}_\mathbf{c})} \cdot \exp\left(\frac{\mathbf{x}^H\mathbf{\Sigma}_\mathbf{u}^{-1}\mathbf{\Sigma}_\mathbf{c}\mathbf{\Sigma}_\mathbf{u}^{-1}\mathbf{x}}{1 + \text{tr}(\mathbf{\Sigma}_\mathbf{u}^{-1}\mathbf{\Sigma}_\mathbf{c})}\right),\tag{11.9}$$

with $P(\mathcal{H}_1)$ and $P(\mathcal{H}_0) = 1 - P(\mathcal{H}_1)$ the prior probabilities of speech presence or absence, respectively. The generalized likelihood ratio is an expression comprising a division of the measurement likelihoods according to two hypothesis. By examining if the ratio is higher or lower than 1, we can conjecture which hypothesis is more probable. In order to simplify the expression in (11.9) we define the multichannel *a priori SNR* as

$$\xi = \text{tr}(\mathbf{\Sigma}_\mathbf{u}^{-1}\mathbf{\Sigma}_\mathbf{c})\tag{11.10}$$

and also define

$$\beta = \gamma\xi = \mathbf{x}^H\mathbf{\Sigma}_\mathbf{u}^{-1}\mathbf{\Sigma}_\mathbf{c}\mathbf{\Sigma}_\mathbf{u}^{-1}\mathbf{x}\tag{11.11}$$

where

$$\gamma = \mathbf{x}^H\mathbf{\Sigma}_\mathbf{u}^{-1}\mathbf{\Sigma}_\mathbf{c}\mathbf{\Sigma}_\mathbf{u}^{-1}\mathbf{x}(\sigma_{c_1}^2\mathbf{a}^H\mathbf{\Sigma}_\mathbf{u}^{-1}\mathbf{a}^H)^{-1}\tag{11.12}$$

is the multichannel *a posteriori SNR*. Then, substituting (11.9), (11.10) and (11.11) in (11.8) yields the multichannel SPP

$$\varrho = \left(1 + \frac{P(\mathcal{H}_0)}{P(\mathcal{H}_1)} \cdot (1+\xi) \cdot e^{-\frac{\beta}{1+\xi}}\right)^{-1}.\tag{11.13}$$

Note that the single-channel SPP (of the first microphone) can be derived as a special case of the multichannel SPP by substituting

$$\xi_1 = \frac{\sigma_{c_1}^2}{\sigma_{u_1}^2}\tag{11.14}$$

$$\beta_1 = \gamma_1 \cdot \xi_1,\tag{11.15}$$

where $\gamma_1 = |x_1|^2/\sigma_{u_1}^2$ is the single-channel posterior SNR and $\sigma_{c_1}^2, \sigma_{u_1}^2$ denote the speech and noise variances at the first microphone, respectively.

In the following we will show the equivalence between the multichannel SPP and a single-channel SPP estimated at the output of the minimum variance distortionless response (MVDR) beamformer $\mathbf{w}_{\text{MVDR}} = \mathbf{\Sigma}_\mathbf{u}^{-1}\mathbf{a}/(\mathbf{a}^H\mathbf{\Sigma}_\mathbf{u}^{-1}\mathbf{a})$ (see Chapter 10). The signal at the output of the MVDR is $\mathbf{w}_{\text{MVDR}}^H\mathbf{x}$. Correspondingly, the noise power and SNR at its output can be shown to be equal to $(\mathbf{a}^H\mathbf{\Sigma}_\mathbf{u}^{-1}\mathbf{a})^{-1}$ and $\sigma_{c_1}^2\mathbf{a}^H\mathbf{\Sigma}_\mathbf{u}^{-1}\mathbf{a}$, respectively. Note that the single-channel SNR and posterior SNR (see (11.14) and (11.15), respectively) at the output of the MVDR are respectively given by $\xi_1 = \sigma_{c_1}^2\mathbf{a}^H\mathbf{\Sigma}_\mathbf{u}^{-1}\mathbf{a}$ and $\beta_1 = \sigma_{c_1}^2|\mathbf{a}^H\mathbf{\Sigma}_\mathbf{u}^{-1}\mathbf{x}|^2$, and are equal to (11.10) and (11.11), respectively. Hence, the multichannel SPP can be interpreted as a single-channel SPP applied to the output of an MVDR spatial filter.

An example for estimated SPP is depicted in Figure 11.2. A desired speech signal contaminated by noise containing both coherent and spatially-white components at an SNR of 0 dB is received by a linear array consisting of four microphones with 8 cm spacing.

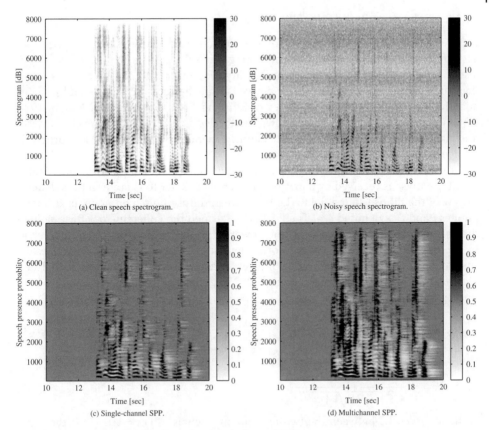

(a) Clean speech spectrogram.

(b) Noisy speech spectrogram.

(c) Single-channel SPP.

(d) Multichannel SPP.

Figure 11.2 Example of single-channel and multichannel SPP in a noisy scenario.

The superiority of the multichannel SPP over the single-channel SPP is evident from this example. The improvement of using the multichannel SPP over the single-channel SPP depends on the spatial properties of the noise and of the speech, and equals the gain of the MVDR spatial filter for the same case (Souden *et al.*, 2010b). Two interesting special cases are the spatially white noise case and the coherent noise case. In the first case of spatially white noise, the spatial covariance matrix equals $\Sigma_{\mathbf{u}} = \sigma_u^2 \mathbf{I}_I$ where $\mathbf{I}_I$ is the $I \times I$ identity matrix. For this case the multichannel SNR equals $I \cdot \xi_1$ and is higher than the single-channel SNR by a factor of the number of microphones (assuming that the SNRs at all microphones are equal). In the second case of coherent noise, the spatial covariance matrix equals $\Sigma_{\mathbf{u}} = \sigma_{u,\text{coh}}^2 \mathbf{a}_{u,\text{coh}} \mathbf{a}_{u,\text{coh}}^H + \sigma_{u,\text{sw}}^2 \mathbf{I}_I$, where $\mathbf{a}_{u,\text{coh}}$ and $\sigma_{u,\text{coh}}^2$ are the vector of acoustic transfer functions relating the coherent interference and the microphone signals and the corresponding variance. It is further assumed that the microphones also contain spatially white noise components with variance $\sigma_{u,\text{sw}}^2$. In this case, a perfect speech detection is obtained, i.e. $\varrho \to 1$ and $\varrho \to 0$ during speech active and speech inactive time-frequency bins, respectively, regardless of the coherent noise power (assuming that vectors of acoustic transfer functions of speech and coherent noise are not parallel and that the spatially white sensor noise power $\sigma_{u,\text{sw}}^2$ is sufficiently low).

11.1.2 Coherence-Based Prior SPP

As presented before in Chapter 6 for the single-channel case and in the previous section for the multichannel case, computing the SPP requires the *prior SPP* $P(\mathcal{H}_1)$, also known as the speech a priori probability. The prior SPP can be either set to a constant (Gerkmann *et al.*, 2008) or derived from the received signals and updated adaptively according to past estimates of SPP and SNR (Ephraim and Malah, 1984; Cohen and Berdugo, 2001) (also known as the decision-directed approach). Taseska and Habets (2013b) adopted the multichannel generalization of the SPP and proposed to incorporate IC information in the prior SPP.

Let us consider a scenario where a single desired speech component contaminated by diffuse noise is received by a pair of microphones. Refer to Chapter 3 for a detailed review of the diffuse sound field. The IC between diffuse noise components at microphones i and i' is denoted $\omega_{ii'}$. The DDR, also termed the coherent-to-diffuse ratio, is defined as the SNR in this case, i.e. the power ratio of the directional speech received by the microphone and the diffuse noise. Heuristically, high and low DDR values are transformed into low and high prior SPP, respectively. For more details refer to Thiergart *et al.* (2012) and Taseska and Habets (2013b). The DDR is estimated for any pair of microphones i and i' by

$$\text{DDR}_{ii'} = \Re\left(\frac{\omega_{ii'} - \text{IC}_{ii'}}{\text{IC}_{ii'} - e^{j\text{IPD}_{ii'}}}\right) \tag{11.16}$$

where

$$\text{IC}_{ii'} = \frac{\mathbb{E}\{x_i x_{i'}^*\}}{\sqrt{\mathbb{E}\{|x_i|^2\}}\sqrt{\mathbb{E}\{|x_{i'}|^2\}}} \tag{11.17}$$

is the observed IC between the two microphone signals, $\text{IPD}_{ii'} = \angle(c_i c_{i'}^*)$ is the interchannel phase difference (IPD), $\Re(\cdot)$ denotes the real component of a complex number and $\mathbb{E}\{\cdot\}$ denotes the expectation operator. The IC is computed from estimates of the power spectra and cross power spectrum of microphones i and i' (see Section 11.2), and the IPD is approximated by the phase of the corresponding cross power spectrum according to

$$\text{IPD}_{ii'} = \angle\,\mathbb{E}\{x_i x_{i'}^*\}, \tag{11.18}$$

assuming that both the SNR and the DDR are high. Schwarz and Kellermann (2015) provide a comprehensive comparison of these and other DDR estimation methods. Taseska and Habets (2013b) suggested relating the prior SPP to the estimated DDR using

$$P(\mathcal{H}_0) = P_{\min}(\mathcal{H}_0) + (P_{\max}(\mathcal{H}_0) - P_{\min}(\mathcal{H}_0))\frac{10^{\varsigma\tau/10}}{10^{\varsigma\tau/10} + \text{DDR}_{ii'}^\tau} \tag{11.19}$$

where $P_{\min}(\mathcal{H}_0)$ and $P_{\max}(\mathcal{H}_0)$ determine the minimum and maximum values of $P(\mathcal{H}_0)$, ς (in dB) controls the offset along the DDR axis, and τ controls the steepness of the transition region.

An example for estimating the DDR using a dual microphone array with spacing of 8 cm and applying the DOA-independent estimator (see Schwarz and Kellermann (2015) for further details) is depicted in Figure 11.3. Evidently, high DDR values correspond to early speech spectral components. Since reverberations tend to be spatially diffuse, the estimated DDR can also be used for dereverberation.

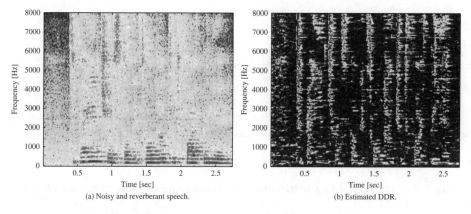

(a) Noisy and reverberant speech.

(b) Estimated DDR.

Figure 11.3 Example of a DDR estimator in a noisy and reverberant scenario.

11.1.3 Multichannel SPP Within GSC Structures

For multichannel speech enhancement based on adaptive beamforming, GSC structures (see Figure 10.5) play a dominant role due to their characteristic feature of turning the constrained MVDR or linearly constrained minimum variance (LCMV) optimization problems into unconstrained ones. Thereby, GSC structures allow for the use of conventional and unconstrained adaptive filtering algorithms. The adaptation, however, typically requires knowledge about dominance of the desired source or background noise, i.e. the estimation of SPP for speech signal extraction. In particular, when both the blocking matrix and the noise canceler should be adaptive, the blocking matrix should be adapted during target speech dominance and the noise canceler should be adapted during target speech inactivity. For speech enhancement, an SPP was first realized as a broadband SNR estimator based on the power ratio of the outputs of the fixed beamformer and the blocking matrix, respectively (Hoshuyama *et al.*, 1998). Herbordt *et al.* (2003) refined this idea for individual frequency bins and later extended it to robust statistics (Herbordt *et al.*, 2005).

Cohen *et al.* (2003) and Gannot and Cohen (2004) presented a related method which comprises an MVDR implemented as a GSC structure (Gannot *et al.*, 2001) with a subsequent modified version of the log-spectral amplitude estimator (Cohen and Berdugo, 2001). As depicted in Figure 11.4, the required SPP is estimated by incorporating spatial information through the GSC structure. In brief, deviation from stationarity is calculated for both the fixed beamformer and the blocking matrix outputs. A larger change in the lower branch power, i.e. the blocking matrix output, indicates a change in noise statistics, while a larger measure of nonstationarity at the upper branch, i.e. the fixed beamformer output, indicates speech presence. This information can be utilized to enhance the performance of the single-channel speech enhancement algorithm that now incorporates the more reliable spatial information. Similar to the above, the SPP decisions can be fed back to the beamformer to better control its adaptation. Hence, if nonspeech segments are detected, the noise canceler can be updated and if speech is detected, the RTF estimator can be updated (see Section 11.3), allowing for source tracking and updating the blocking matrix.

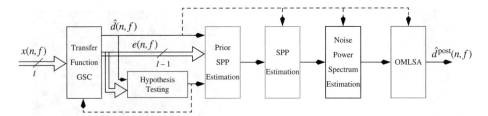

Figure 11.4 Postfilter incorporating spatial information (Cohen *et al.*, 2003; Gannot and Cohen, 2004).

11.1.4 Multiple Speakers Position-Based SPP

Consider the J speakers scenario in which the microphone signals can be formulated as

$$\mathbf{x} = \sum_{j=1}^{J} \mathbf{c}_j + \mathbf{u}. \tag{11.20}$$

For designing spatial filters able to extract a desired speaker while suppressing all other speakers and noise, knowledge of speaker activity patterns (or a soft version thereof, i.e. SPPs) is required. We discuss a few representative concepts here.

Taseska and Habets (2013a) proposed using a multichannel Wiener filter (MWF) for extracting a desired source from a multichannel convolutive mixture of sources. By incorporating position estimates into the SPP and classifying the dominant speaker per time-frequency bin, the "interference" components' covariance matrix, comprising noise and interfering speakers, and the desired speaker components' covariance matrix are estimated and utilized for constructing a spatial filter. Speaker positions are derived by triangulation of DOA estimates obtained from distributed subarrays of microphones with known positions.

The SPPs of individual sources are defined as

$$\varrho_j = P(\mathcal{H}_1^j \mid \mathbf{x}) = P(\mathcal{H}_1^j \mid \mathbf{x}, \mathcal{H}_1) \cdot \varrho, \tag{11.21}$$

where $\mathcal{H}_1^j$ denotes the hypothesis that speaker j is active (per time-frequency bin), and ϱ is the previously defined SPP (for any speaker activity).

The conditional SPPs given the microphone signals are replaced by conditional SPPs given the estimated 3D position of the dominant active speaker, denoted $\hat{\mathbf{p}}$, i.e. it is assumed that

$$P(\mathcal{H}_1^j \mid \hat{\mathbf{p}}, \mathcal{H}_1) \approx P(\mathcal{H}_1^j \mid \mathbf{x}, \mathcal{H}_1). \tag{11.22}$$

The estimated position, given that few speakers are active, is modeled as a mixture of Gaussian variables centered at the source positions

$$P(\hat{\mathbf{p}} \mid \mathcal{H}_1) = \sum_{j=1}^{J} \pi_j \mathcal{N}(\hat{\mathbf{p}} \mid \mathbf{p}_j, \mathbf{\Sigma}_{\mathbf{p}_j}) \tag{11.23}$$

where $\mathbf{p}_j$, $\mathbf{\Sigma}_{\mathbf{p}_j}$, and π_j are the mean, covariance, and mixing coefficient of the Gaussian vector distribution which corresponds to the estimated position of source j. The parameters of the distribution of $\hat{\mathbf{p}}$ are estimated by an expectation-maximization (EM) procedure given a batch of estimated positions. For a detailed explanation, see Taseska and Habets (2013a).

Taseska and Habets (2013b, 2014) further extended this work by designing an MWF to extract sources arriving from a predefined "spot", i.e. a bounded area, while suppressing all other sources outside of the spot. The authors call this method "spotforming".

More recently, speech activity detection in multisource scenarios has also been tackled using artificial neural networks (Meier and Kellermann, 2016). For wideband signal activity detection and SPP, cross-correlation sequences estimates between microphone pairs were found to be the most useful input features of the neural networks. Even with networks of small size – much smaller than for blind source separation (BSS) – and very small amounts of training data, traditional approaches exploiting the same parameters can be outperformed.

11.2 Covariance Matrix Estimators Exploiting SPP

The spatial covariance matrix of the noise can be estimated by recursively averaging instantaneous covariance matrices weighted according to the SPP

$$\hat{\Sigma}_{\mathbf{u}}(n,f) = \lambda'_u(n,f)\hat{\Sigma}_{\mathbf{u}}(n-1,f) + (1 - \lambda'_u(n,f))\mathbf{x}(n,f)\mathbf{x}^H(n,f), \tag{11.24}$$

where

$$\lambda'_u(n,f) = (1 - \varrho(n,f))\lambda_u + \varrho(n,f) \tag{11.25}$$

is a time-varying recursive averaging factor and λ_u is selected such that its corresponding estimation period ($\frac{1}{1-\lambda_u}$ frames) is shorter than the stationarity time of the background noise. Alternatively, a hard binary weighting, obtained by applying a threshold to the SPP, can be used instead of the soft weighting. Note that when the time-frequency bin is dominated by noise, i.e. for $\varrho(n,f) \to 0$, the regression factor reaches its lower bound $\lambda'_u(n,f) \to \lambda_u$. On the other hand, when the time-frequency bin is dominated by speech, i.e. $\varrho(n,f) \to 1$, the regression factor reaches its upper bound $\lambda'_u(n,f) \to 1$, i.e. the previous estimate of the spatial covariance matrix of the noise is maintained.

The hypothesis that speaker j is present and the corresponding SPP are denoted in Section 11.1.4 as $\mathcal{H}^j_1(n,f)$ and $\varrho_j(n,f)$, respectively. Similarly to (11.24), the spatial covariance matrix of the spatial image of source j, denoted $\Sigma_{\mathbf{c}_j}(n,f) = \sigma^2_{s_j}(n,f)\mathbf{a}_j(f)\mathbf{a}^H_j(f)$, can be estimated by

$$\hat{\Sigma}_{\mathbf{c}_j}(n,f) = \lambda'_{s_j}(n,f)\hat{\Sigma}_{\mathbf{c}_j}(n-1,f) + (1 - \lambda'_{s_j}(n,f))(\mathbf{x}(n,f)\mathbf{x}^H(n,f) - \hat{\Sigma}_{\mathbf{u}}(n-1,f)) \tag{11.26}$$

where

$$\lambda'_{s_j}(n,f) = \varrho_j(n,f) \cdot \lambda_s + (1 - \varrho_j(n,f)) \tag{11.27}$$

is a time-varying recursive averaging factor. The SPP-dependent regression factor reaches its lower and upper bounds, i.e. λ_s and 1, when the SPP reaches its upper and lower bounds, respectively. Note that the time-varying spatial covariance matrix of source j has two multiplicative components: the usually slowly time-varying coherence matrix, which is determined by the characteristics of the acoustic environment, $\mathbf{a}_j(f)\mathbf{a}^H_j(f)$, and the power spectrum of the source, $\sigma^2_{s_j}(n,f)$, which is typically changing much faster (depending on the nonstationarity of the described source). The parameter

λ_s is selected such that its corresponding estimation period ($\frac{1}{1-\lambda_s}$ frames) is shorter than the coherence time of the acoustic impulse response of speaker j, i.e. the time period over which the impulse response is assumed to be time-invariant. The impulse response may vary due to movement of the speaker or changes in the acoustic environment. Note that (1) usually the estimation period is longer than the stationarity interval of speech, therefore, although the spatial structure of $\mathbf{\Sigma}_{c_j}(n,f)$ is maintained, the estimated variance is an average of the speech variances over multiple stationarity intervals, and (2) the estimate $\hat{\mathbf{\Sigma}}_{c_j}(n,f)$ keeps its past value when speaker j is absent. Some typical values for slowly time-varying environments are $\lambda_s = 0.9$ and $\lambda_u = 0.99$ which, for STFT window length $K = 1024$ with 75% overlap at a sampling rate of 16 kHz, correspond to integration times of 208 ms and 1648 ms for estimating speech and noise covariance matrices, respectively.

In the case of static nonstationary sources, the different stationarity times of the spatial covariance matrix and the power spectrum of the signals can be exploited. The spatial covariance matrix, reflecting the spatial properties of the soundfield, can be estimated from longer observation intervals, thereby leading to reliable estimates. The nonstationary covariance matrices can then be computed by combining instantaneous power spectrum estimates with the spatial covariance matrix estimate (Reindl *et al.*, 2012).

11.3 Methods for Weakly Guided and Strongly Guided RTF Estimation

In this section, several methods for estimating RTFs are presented. Single speaker scenarios and multiple speakers scenarios are considered in Section 11.3.1 and Section 11.3.2, respectively. In the single speaker case, whereas in the more intricate case of multiple speakers both weakly guided and strongly guided methods are surveyed. As explained in Chapter 1, weakly guided methods utilize some general properties of the speech signal, e.g. nonstationarity, or the mixing process, whereas strongly guided methods utilize scenario-specific information, e.g. speaker activity pattern or its DOA. RTF estimation is an essential component in the implementation of the blocking matrix and the fixed beamformer of GSC structures. The blocking matrix is responsible for extracting noise reference signals. The RTFs are also used to construct the fixed beamformer, such that the desired signal components in all microphones are phase-aligned with the desired signal component at the reference microphone and then combined with scalar positive gains which correspond to the desired signal powers (in essence equivalent to a spatial matched filter).

11.3.1 Single-Speaker Case

Two common approaches for weakly guided RTF estimation for the single speaker case are the covariance subtraction (Doclo and Moonen, 2005; Cohen, 2004; Souden *et al.*, 2010a; Serizel *et al.*, 2014) and the covariance whitening (Markovich *et al.*, 2009; Bertrand and Moonen, 2012; Serizel *et al.*, 2014) methods. Both of these approaches rely on the estimated noisy speech and noise-only covariance matrices, i.e. $\hat{\mathbf{\Sigma}}_x$ and

$\widehat{\boldsymbol{\Sigma}}_{\mathbf{u}}$. Given the estimated covariance matrices, the *covariance subtraction* approach estimates the speaker RTF by

$$\tilde{\mathbf{a}}_{CS} = \frac{1}{\mathbf{i}_1^H (\widehat{\boldsymbol{\Sigma}}_{\mathbf{x}} - \widehat{\boldsymbol{\Sigma}}_{\mathbf{u}}) \mathbf{i}_1} (\widehat{\boldsymbol{\Sigma}}_{\mathbf{x}} - \widehat{\boldsymbol{\Sigma}}_{\mathbf{u}}) \mathbf{i}_1 \tag{11.28}$$

where a rank-1 model is assumed for the estimated spatial covariance matrix of speech, which is given by $\widehat{\boldsymbol{\Sigma}}_{\mathbf{x}} - \widehat{\boldsymbol{\Sigma}}_{\mathbf{u}}$, and $\mathbf{i}_1 = [1, 0, \ldots, 0]^T$ is a $I \times 1$ selection vector for extracting the component of the reference microphone, here assumed to be the first microphone. Theoretically, $\widehat{\boldsymbol{\Sigma}}_{\mathbf{x}} - \widehat{\boldsymbol{\Sigma}}_{\mathbf{u}}$ should be positive definite, however in practice, due to estimation errors, minimum value constraints are applied.

The *covariance whitening* approach estimates the RTF by applying the generalized eigenvalue decomposition to $\widehat{\boldsymbol{\Sigma}}_{\mathbf{x}}$ with $\widehat{\boldsymbol{\Sigma}}_{\mathbf{u}}$ as the whitening matrix. The principal generalized eigenvector $\tilde{\mathbf{a}}_{\mathbf{u}}$ is defined as the normalized vector which satisfies

$$\widehat{\boldsymbol{\Sigma}}_{\mathbf{x}} \tilde{\mathbf{a}}_{\mathbf{u}} = \lambda \widehat{\boldsymbol{\Sigma}}_{\mathbf{u}} \tilde{\mathbf{a}}_{\mathbf{u}} \tag{11.29}$$

with the largest λ, called the principal generalized eigenvalue. Finally, an estimate for the RTF is obtained by normalizing the dewhitened eigenvector, namely $\widehat{\boldsymbol{\Sigma}}_{\mathbf{u}} \tilde{\mathbf{a}}_{\mathbf{u}}$, by the reference microphone component

$$\tilde{\mathbf{a}}_{CW} = (\mathbf{i}_1^H \widehat{\boldsymbol{\Sigma}}_{\mathbf{u}} \tilde{\mathbf{a}}_{\mathbf{u}})^{-1} \widehat{\boldsymbol{\Sigma}}_{\mathbf{u}} \tilde{\mathbf{a}}_{\mathbf{u}}. \tag{11.30}$$

Markovich-Golan and Gannot (2015) reported a preliminary analysis and comparison of the covariance subtraction and covariance whitening methods. Affes and Grenier (1997) proposed an early *subspace tracking* method for estimating (a variant of) the RTF, which is specifically tailored to small speaker movements around its nominal position.

Alternative methods utilize the speech nonstationarity property, assuming that the noise has slowly time-varying statistics. Gannot *et al.* (2001) formulated the problem of estimating the RTF of microphone i as a least squares problem by concatenating several time segments (each time segment consisting of multiple time frames). The equation associated with the lth time segment utilizes $\widehat{\sigma}_{x_i x_1}^l$, the estimated cross power spectrum of microphone i and the reference microphone in the lth time segment. This cross power spectrum satisfies

$$\widehat{\sigma}_{x_i x_1}^l = \tilde{a}_i (\widehat{\sigma}_{x_1}^l)^2 + \widehat{\sigma}_{u_i x_1}^l + \epsilon_i^l \tag{11.31}$$

where we use the relation $\mathbf{x} = \tilde{\mathbf{a}} x_1 + \mathbf{u}$. The unknowns are $\tilde{a}_i$, i.e. the required RTF, and $\widehat{\sigma}_{u_i x_1}^l$, which is a nuisance parameter. The error term of the lth equation is denoted by ϵ_i^l. Multiple least squares problems, one for each microphone, are solved for estimating the RTF vector. Note that the latter method, also known as the nonstationarity-based RTF estimation, does not require a prior estimate of the noise covariance, since it simultaneously solves for the RTF and the noise statistics. Similarly, a weighted least squares problem with exponential weighting can be defined and implemented using a recursive least squares algorithm (Dvorkind and Gannot, 2005). Considering speech sparsity in the STFT domain, Cohen (2004) incorporated the SPPs into the weights of the weighted least squares problem, resulting in a more accurate solution. Sparsity of the speech signals in the frequency domain increases the convergence time of RTF estimation methods until sufficient signal energy is collected in the entire band. Koldovský *et al.* (2015) utilized time-domain sparsity of the RTF for reducing convergence time by interpolating missing spectral components.

The equivalent of the RTF in the time domain is a filter which transforms the desired signal component at the reference microphone to a corresponding component at another microphone. The relative impulse response between a pair of microphones can be estimated in the time domain using least squares fit (Chen *et al.*, 2008).

11.3.2 The Multiple-Speaker Case

In the multiple-speaker case, the problem of estimating the RTFs becomes more intricate. In the first group of methods, it is assumed that time intervals can be identified where only one source is active and for this case numerous clustering methods have been proposed (see Chapter 12). Rather than clustering and classification, (Taseska and Habets, 2015) estimated RTFs based on instantaneous observation vectors projected onto the signals subspace constructed by smoothing past observation vectors.

As an alternative to clustering methods, and allowing immediate tracking of sources, Gürelli and Nikias (1995) utilized blind system identification methods, which are based on the analysis of the null space of the spatial covariance matrix of the microphone signals, to estimate transfer functions, e.g. as a key component of the transfer function GSC (Gannot *et al.*, 2001), but also for identifying the blocking matrix for general GSC-based implementations of the general MVDR and LCMV. For practical applications in audio, a normalized least mean squares based method (Huang *et al.*, 1999) allowed a real-time identification of the acoustic transfer functions for a single source from microphone pairs for low background noise levels and relatively short filters. Their method can be viewed as an adaptive variant of the method proposed by Gürelli and Nikias (1995).

For two simultaneously active sources, the RTFs can be estimated by TRINICON, a generic independent component analysis (ICA)-based concept for broadband blind multiple-input multiple-output signal processing, which was proposed by Buchner *et al.* (2005) and later extended to more sources (Buchner *et al.*, 2007). Thereby the internal permutation problem characteristic for narrowband ICA algorithms is avoided and efficient online adaptation using only second-order statistics is possible. Meier and Kellermann (2015) presented an analysis and evaluation of ICA methods for RTF estimation. Markovich-Golan *et al.* (2010) presented an alternative method, where the recursive estimator of the RTF based on subspace tracking (Affes and Grenier, 1997) was extended to multiple moving sources.

If the number of simultaneously active point sources and microphones becomes large, then fully blind ICA-based methods, which try to identify RTFs by separating all sources simultaneously, become computationally very complex and the performance must be expected to degrade rapidly even with moderate reverberation. In most cases, however, it suffices to determine the RTFs for a finite number of sources, especially for signal extraction by MVDR or LCMV. Then, there is typically some DOA information for the source of interest available (from separate DOA estimation algorithms, if necessary), which can be used to add a geometric constraint to the ICA cost function (Zheng *et al.*, 2009). Thereby, two-channel ICA-based source separation provides one output in which all components, which are correlated to the signal from the given DOA, are suppressed. If the DOA points to the direction of a desired speaker, as typical in MVDR, this output contains only information on all other sources, and thus corresponds to one output of a blocking matrix in a GSC structure. Then, the filters of the ICA demixing system represent the RTFs and correspond to the according filters of a blocking matrix in GSC structure. As opposed to noise reference information, which is only computed during target speaker inactivity or during low SPP, the ICA-based noise reference is computed continuously and thus provides an instantaneous noise estimate. This is especially desirable for nonstationary, e.g. speech-like or burst-like noise and interference (Zheng *et al.*, 2014) and can also beneficially be used to provide noise power spectrum estimates for single-channel postfilters.

More generally, given I microphones, $I - 1$ independent microphone pairs can then produce $I - 1$ outputs of the blocking matrix for further use in the noise canceler of the GSC (Reindl *et al.*, 2013), which renders adaptation control for the blocking matrix unnecessary. By definition, ICA is designed for exploiting higher-order statistics, so that by using the constrained ICA criteria to the blocking matrix and a minimum mutual information criterion for the noise canceler, the second-order statistics-based LCMV can be generalized to a linearly constrained minimum mutual information beamformer and still be implemented in a GSC structure (Reindl *et al.*, 2014). In a further refinement, significantly improved convergence behavior could be demonstrated for the extraction of multiple sources (Markovich-Golan *et al.*, 2017).

11.4 Summary

In this chapter we have considered some practical aspects for estimating the crucial parameters of the data-dependent spatial filters that were presented in Chapter 10.

Several extensions of the single-channel SPP to the multichannel case were presented in Section 11.1. These extensions offer higher accuracy by utilizing spatial information as well as the ability to distinguish between different sources in the multiple speakers case.

Considering both the single speaker case and the multiple speaker case, several methods for estimating the spatial covariance matrices of the background noise and of the speakers, as well as weakly guided and strongly guided methods for estimating the corresponding RTF vectors, were presented in Sections 11.2 and 11.3, respectively. The methods explored in this chapter utilize many sources of information, i.e. prior knowledge of the speakers' activity patterns, level of stationarity, the DOAs of the sources, SPP decisions, as well as multiple inference techniques, e.g. statistical independence and clustering procedures.

Bibliography

Affes, S. and Grenier, Y. (1997) A signal subspace tracking algorithm for microphone array processing of speech. *IEEE Transactions on Speech and Audio Processing*, **5** (5), 425–437.

Bertrand, A. and Moonen, M. (2012) Distributed node-specific LCMV beamforming in wireless sensor networks. *IEEE Transactions on Signal Processing*, **60**, 233–246.

Buchner, H., Aichner, R., and Kellermann, W. (2007) TRINICON-based blind system identification with application to multiple-source localization and separation, in *Blind Speech Separation*, Springer, pp. 101–147.

Buchner, H., Aichner, R., Stenglein, J., Teutsch, H., and Kellermann, W. (2005) Simultaneous localization of multiple sound sources using blind adaptive MIMO filtering, in *Proceedings of IEEE International Conference on Audio, Speech and Signal Processing*, pp. III–97–100.

Chen, J., Benesty, J., and Huang, Y. (2008) A minimum distortion noise reduction algorithm with multiple microphones. *IEEE Transactions on Audio, Speech, and Language Processing*, **16** (3), 481–493.

Cohen, I. (2004) Relative transfer function identification using speech signals. *IEEE Transactions on Speech and Audio Processing*, **12** (5), 451–459.

Cohen, I. and Berdugo, B. (2001) Speech enhancement for non-stationary noise environments. *Signal Processing*, **81** (11), 2403–2418.

Cohen, I., Gannot, S., and Berdugo, B. (2003) An integrated real-time beamforming and postfiltering system for nonstationary noise environments. *EURASIP Journal on Advances in Signal Processing*, **2003**, 1064–1073.

Doclo, S. and Moonen, M. (2005) Multimicrophone noise reduction using recursive GSVD-based optimal filtering with ANC postprocessing stage. *IEEE Transactions on Speech and Audio Processing*, **13** (1), 53–69.

Dvorkind, T.G. and Gannot, S. (2005) Time difference of arrival estimation of speech source in a noisy and reverberant environment. *Signal Processing*, **85** (1), 177–204.

Ephraim, Y. and Malah, D. (1984) Speech enhancement using a minimum-mean square error short-time spectral amplitude estimator. *IEEE Transactions on Acoustics, Speech, and Signal Processing*, **32** (6), 1109–1121.

Gannot, S., Burshtein, D., and Weinstein, E. (2001) Signal enhancement using beamforming and nonstationarity with applications to speech. *IEEE Transactions on Signal Processing*, **49** (8), 1614–1626.

Gannot, S. and Cohen, I. (2004) Speech enhancement based on the general transfer function GSC and postfiltering. *IEEE Transactions on Speech and Audio Processing*, **12** (6), 561–571.

Gerkmann, T., Breithaupt, C., and Martin, R. (2008) Improved a posteriori speech presence probability estimation based on a likelihood ratio with fixed priors. *IEEE Transactions on Audio, Speech, and Language Processing*, **16** (5), 910–919.

Gürelli, M.I. and Nikias, C.L. (1995) EVAM: an eigenvector-based algorithm for multichannel blind deconvolution of input colored signals. *IEEE Transactions on Signal Processing*, **43** (1), 134–149.

Herbordt, W., Buchner, H., Nakamura, S., and Kellermann, W. (2005) Outlier-robust DFT-domain adaptive filtering for bin-wise stepsize controls, and its application to a generalized sidelobe canceller, in *Proceedings of International Workshop on Acoustic Echo and Noise Control*.

Herbordt, W., Trini, T., and Kellermann, W. (2003) Robust spatial estimation of the signal-to-interference ratio for non-stationary mixtures, in *Proceedings of International Workshop on Acoustic Echo and Noise Control*, pp. 247–250.

Hoshuyama, O., Begasse, B., Sugiyama, A., and Hirano, A. (1998) A real time robust adaptive microphone array controlled by an SNR estimate, in *Proceedings of IEEE International Conference on Audio, Speech and Signal Processing*, vol. 6, pp. 3605–3608.

Huang, Y., Benesty, J., and Elko, G.W. (1999) Adaptive eigenvalue decomposition algorithm for real time acoustic source localization system, in *Proceedings of IEEE International Conference on Audio, Speech and Signal Processing*, vol. 2, pp. 937–940.

Koldovský, Z., Málek, J., and Gannot, S. (2015) Spatial source subtraction based on incomplete measurements of relative transfer function. *IEEE/ACM Transactions on Audio, Speech, and Language Processing*, **23** (8), 1335–1347.

Markovich, S., Gannot, S., and Cohen, I. (2009) Multichannel eigenspace beamforming in a reverberant noisy environment with multiple interfering speech signals. *IEEE Transactions on Audio, Speech, and Language Processing*, **17** (6), 1071–1086.

Markovich-Golan, S. and Gannot, S. (2015) Performance analysis of the covariance subtraction method for relative transfer function estimation and comparison to the covariance whitening method, in *Proceedings of IEEE International Conference on Audio, Speech and Signal Processing*, pp. 544–548.

Markovich-Golan, S., Gannot, S., and Cohen, I. (2010) Subspace tracking of multiple sources and its application to speakers extraction, in *Proceedings of IEEE International Conference on Audio, Speech and Signal Processing*, pp. 201–204.

Markovich-Golan, S., Gannot, S., and Kellermann, W. (2017) Combined LCMV-TRINICON beamforming for separating multiple speech sources in noisy and reverberant environments. *IEEE/ACM Transactions on Audio, Speech, and Language Processing*, **25** (2), 320–332.

McAulay, R. and Malpass, M. (1980) Speech enhancement using a soft-decision noise suppression filter. *IEEE Transactions on Acoustics, Speech, and Signal Processing*, **28** (2), 137–145.

Meier, S. and Kellermann, W. (2015) Analysis of the performance and limitations of ICA-based relative impulse response identification, in *Proceedings of European Signal Processing Conference*, pp. 414–418.

Meier, S. and Kellermann, W. (2016) Artificial neural network-based feature combination for spatial voice activity detection, in *Proceedings of Interspeech*, pp. 2987–2991.

Pessentheiner, H., Hagmüller, M., and Kubin, G. (2016) Localization and characterization of multiple harmonic sources. *IEEE/ACM Transactions on Audio, Speech, and Language Processing*, **24** (8), 1348–1363.

Quatieri, T.F. and McAulay, R.J. (1990) Noise reduction using a soft-decision sine-wave vector quantizer, in *Proceedings of IEEE International Conference on Audio, Speech and Signal Processing*, pp. 821–824.

Reindl, K., Markovich-Golan, S., Barfuss, H., Gannot, S., and Kellermann, W. (2013) Geometrically constrained TRINICON-based relative transfer function estimation in underdetermined scenarios, in *Proceedings of IEEE Workshop on Applications of Signal Processing to Audio and Acoustics*, pp. 1–4.

Reindl, K., Meier, S., Barfuss, H., and Kellermann, W. (2014) Minimum mutual information-based linearly constrained broadband signal extraction. *IEEE/ACM Transactions on Audio, Speech, and Language Processing*, **22** (6), 1096–1108.

Reindl, K., Zheng, Y., Schwarz, A., Meier, S., Maas, R., Sehr, A., and Kellermann, W. (2012) A stereophonic acoustic signal extraction scheme for noisy and reverberant environments. *Computer Speech and Language*, **27** (3), 726–745.

Schwarz, A. and Kellermann, W. (2015) Coherent-to-diffuse power ratio estimation for dereverberation. *IEEE/ACM Transactions on Audio, Speech, and Language Processing*, **23** (6), 1006–1018.

Serizel, R., Moonen, M., Van Dijk, B., and Wouters, J. (2014) Low-rank approximation based multichannel Wiener filter algorithms for noise reduction with application in cochlear implants. *IEEE/ACM Transactions on Audio, Speech, and Language Processing*, **22** (4), 785–799.

Souden, M., Benesty, J., and Affes, S. (2010a) A study of the LCMV and MVDR noise reduction filters. *IEEE Transactions on Signal Processing*, **58** (9), 4925–4935.

Souden, M., Chen, J., Benesty, J., and Affes, S. (2010b) Gaussian model-based multichannel speech presence probability. *IEEE Transactions on Audio, Speech, and Language Processing*, **18** (5), 1072–1077.

Taseska, M. and Habets, E.A.P. (2013a) MMSE-based source extraction using position-based posterior probabilities, in *Proceedings of IEEE International Conference on Audio, Speech and Signal Processing*, pp. 664–668.

Taseska, M. and Habets, E.A.P. (2013b) Spotforming using distributed microphone arrays, in *Proceedings of IEEE Workshop on Applications of Signal Processing to Audio and Acoustics*, pp. 1–4.

Taseska, M. and Habets, E.A.P. (2014) Informed spatial filtering for sound extraction using distributed microphone arrays. *IEEE/ACM Transactions on Audio, Speech, and Language Processing*, **22** (7), 1195–1207.

Taseska, M. and Habets, E.A.P. (2015) Relative transfer function estimation exploiting instantaneous signals and the signal subspace, in *Proceedings of European Signal Processing Conference*, pp. 404–408.

Thiergart, O., Galdo, G.D., and Habets, E.A.P. (2012) Signal-to-reverberant ratio estimation based on the complex spatial coherence between omnidirectional microphones, in *Proceedings of IEEE International Conference on Audio, Speech and Signal Processing*, pp. 309–312.

Van Trees, H.L. (2002) *Detection, Estimation, and Modulation Theory*, vol. IV, Wiley.

Zheng, Y., Reindl, K., and Kellermann, W. (2009) BSS for improved interference estimation for blind speech signal extraction with two microphones, in *Proceedings of IEEE International Workshop on Computational Advances in Multi-Sensor Adaptive Processing*, pp. 253–256.

Zheng, Y., Reindl, K., and Kellermann, W. (2014) Analysis of dual-channel ICA-based blocking matrix for improved noise estimation. *EURASIP Journal on Advances in Signal Processing*, **2014**, 26.

12

Multichannel Clustering and Classification Approaches

Michael I. Mandel, Shoko Araki, and Tomohiro Nakatani

This chapter describes methods for estimating time-frequency masks of source activity from multichannel observations using clustering and classification techniques. Such methods are similar to the speech presence probability (SPP) estimates in Chapter 11, but can be applied to any signal, not just speech, and can be applied in the presence of nonstationary noise, not just stationary noise. Clustering algorithms estimate time-frequency masks by grouping together time-frequency bins with similar characteristics. Classification algorithms estimate these masks based on a comparison of time-frequency bins in the signal under analysis to those of previously seen training data. Because clustering algorithms only compare parts of the test signal to one another, they typically do not require training data. Classification algorithms, in contrast, are extremely dependent on the characteristics and quality of their training data. In the notation of Section 1.3.3, clustering is generally a learning-free method, while classification is a separation-based training method.

This chapter is also related to Chapter 14, which describes a complete generative model of the joint spatial and time-frequency characteristics of multichannel signals that can be used to separate or enhance target signals of interest. The methods described in the current chapter, in contrast, focus on estimating only the spatial properties of the signals and then use these estimates to perform separation. Some of these methods can incorporate models of the time-frequency characteristics of the signals involved, but most operate in a signal-agnostic manner, basing separations on spatial information alone. This less complete modeling involves fewer parameters and hidden variables, and thus may be able to operate on shorter excerpts or with more rapidly changing scenes than more complicated models and may require less training data for classification models.

The structure of this chapter is as follows. Section 12.1 introduces the two-channel time-frequency clustering problem, which is easiest to visualize, and describes several approaches to source separation in this setting. Section 12.2 describes approaches to the more general problem of multichannel clustering, including generalizations of two-channel methods and native multichannel methods. It also describes several approaches to estimating the number of sound sources present in a mixture. Section 12.3 describes multichannel classification approaches, including those

Audio Source Separation and Speech Enhancement, First Edition.
Edited by Emmanuel Vincent, Tuomas Virtanen and Sharon Gannot.
© 2018 John Wiley & Sons Ltd. Published 2018 by John Wiley & Sons Ltd.
Companion Website: https://project.inria.fr/ssse/

developed for two-channel recordings and multichannel recordings, along with a discussion of generalization in these systems with regards to data dependence. Section 12.4 describes the use of the masks estimated by any of these methods to perform spatial filtering. Section 12.5 summarizes the chapter.

12.1 Two-Channel Clustering

The general idea of spatial clustering is to use differences in spatial characteristics of broadband sound sources to determine which time-frequency bins belong to each source. It can be thought of as simultaneously localizing sound sources in space and in regions of a time-frequency representation like a spectrogram. These two processes are mutually beneficial and better localization in one domain can lead to better localization in the other domain. We will first illustrate the concepts of spatial clustering in the two-microphone setting to build intuition as well as mirroring to some extent the historical development of spatial clustering approaches.

Two-channel spatial clustering is based on the use of interchannel level, phase, and time differences (ILD, IPD, and ITD, respectively) to discriminate between sources. Recall from Chapter 3 that the relative transfer function (RTF) between all microphones and reference microphone 1 for source j is defined as

$$\tilde{\mathbf{a}}_j(f) = \frac{1}{a_{1j}(f)} \mathbf{a}_j(f), \tag{12.1}$$

which has a one-to-one mapping to IPD and ILD,

$$\mathrm{ILD}_{ij}(f) = 20 \log_{10} |\tilde{a}_{ij}(f)| \tag{12.2}$$

$$\mathrm{IPD}_{ij}(f) = \angle \tilde{a}_{ij}(f). \tag{12.3}$$

Note that these quantities were defined in Chapter 3 according to the narrowband approximation as a single value at each frequency for each source. Thus for two-channel observations these parameters consist of a single scalar at each frequency for source j, $\mathrm{ILD}_{2j}(f)$ and $\mathrm{IPD}_{2j}(f)$. As parameters of source models, these quantities are generally assumed to change slowly over time and are time-independent for sources that are not moving. Many of the models discussed in this chapter make the assumption that the sources are not moving.

If, further, the RTF is assumed to be dominated by the direct path signal, it can be well approximated by the relative steering vector, which in the case of two microphones is

$$\tilde{\mathbf{d}}_j(f) = [1, e^{-2j\pi v_f (r_{2j} - r_{1j})/c}]^T \tag{12.4}$$

where v_f is the frequency of index f, c is the speed of sound in air, and r_{2j} and r_{1j} are the distances between source j and microphones 2 and 1, respectively. In this case, the IPD (i.e., the phase of $\tilde{d}_{2j}(f)$) is entirely determined by the time difference of arrival (TDOA)

$$\Delta_{2j} = \frac{r_{2j} - r_{1j}}{c}. \tag{12.5}$$

As discussed in Chapter 3, below the spatial aliasing frequency for a given microphone spacing, an observed IPD can be converted directly back to the underlying ITD via

$$\mathrm{ITD}_{ij}(f) = \frac{\angle \tilde{a}_{ij}(f)}{2\pi v_f} \tag{12.6}$$

but above the spatial aliasing frequency it cannot because of phase wrapping issues. Instead, there are various techniques for estimating it.

An observed mixture of J sources can also be characterized by ILD and IPD, but as time-varying quantities,

$$\text{ILD}\,(n,f) = 20\log_{10}\left|\frac{x_2(n,f)}{x_1(n,f)}\right| \tag{12.7}$$

$$\text{IPD}\,(n,f) = \angle\frac{x_2(n,f)}{x_1(n,f)}. \tag{12.8}$$

In subsequent discussion, it should be clear whether these quantities refer to a time-varying observation or time-invariant source model parameters.

Two different approaches to spatial clustering have been developed in order to deal with the problem of ambiguity in the IPD-to-ITD mapping caused by spatial aliasing: narrowband and wideband clustering. *Narrowband clustering* performs spatial clustering independently in each frequency band and then matches the estimated sources across frequencies. *Wideband clustering* performs clustering of observations at all frequencies simultaneously in a space that is frequency-independent, typically based on an estimated ITD. Note that both approaches perform clustering at all frequencies, but it is the manner of coordinating estimated sources and parameters across frequency that differs between them. Narrowband clustering thus requires more sophisticated time-based grouping procedures, while wideband clustering requires more sophisticated frequency-based grouping, including models of spatial recording processes.

12.1.1   Wideband Clustering with Simple IPD to ITD Mapping

One of the first spatial clustering approaches was called DUET (Jourjine *et al.*, 2000; Yılmaz and Rickard, 2004). It clusters all time-frequency bins together in a joint ILD-ITD space, thus representing each source as a single, frequency-invariant centroid in this space. The ITD coordinate is computed using the naive mapping from IPD in (12.6), and so is not robust to spatial aliasing. The use of a single centroid across frequency is compatible with the assumption of an anechoic point source, but does not accurately represent sources in reverberant environments or with diffuse spatial characteristics.

Figure 12.2 shows this ITD-ILD histogram for the portions of the IPD and ILD observations shown in Figure 12.1 that are under the spatial aliasing frequency. Because the microphones used were spaced 6 cm from one another, this frequency is approximately 2.8 kHz. At a sampling rate of 44.1 kHz and a frame size of 32 ms, this corresponds to the first 87 rows of the spectrogram out of 707 rows in total. Peaks in this histogram were then manually identified, and are plotted as white circles in Figure 12.2. Various procedures have been recommended for identifying these peaks automatically, but none has been shown to be sufficiently robust to be used in all situations. These peaks were then used to derive the binary masks shown in the figure by assigning each point in the spectrogram to the peak that yields the smallest error between the left channel and the right channel modified by the peak's ITD and ILD parameters. Even though only frequencies below the spatial aliasing threshold can be used to construct the histogram, once an ITD is identified, it predicts a unique IPD across all frequencies and can be used to construct

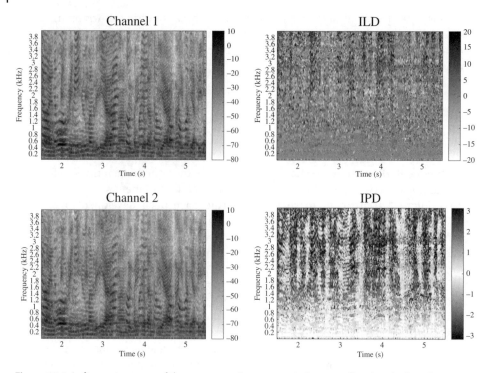

Figure 12.1 Left: spectrograms of three-source mixture recorded on two directional microphones spaced approximately 6 cm apart (Sawada *et al.*, 2011). Right: ILD and IPD of this recording.

a mask for the entire spectrogram. The reliability of this mask, however, may suffer at higher frequencies, as inaccuracies accumulate from ITD estimation at low frequencies.

12.1.2 Wideband Clustering with Latent ITD Variable

In order to overcome this issue, Mandel *et al.* (2007) introduced a binaural spatial clustering system known as MESSL that includes a latent variable representing ITD. This system models the straightforward ITD-to-IPD mapping of (12.4), which is unambiguous, instead of the ambiguous IPD-to-ITD mapping of (12.6). It is then possible to marginalize over this latent variable to compute the posterior probability of each time-frequency bin coming from each source, i.e. a time-frequency mask for each source. MESSL models the likelihood of a time-frequency bin coming from a particular source at a particular delay as a Gaussian distribution, meaning that its likelihood across all delays is a *Gaussian mixture model* (GMM). Each delay candidate leads to a Gaussian, and the marginalization across delay candidates leads to the GMM. The parameters of these Gaussians can then be estimated in alternation with the masks using the *expectation-maximization* (EM) algorithm.

We will briefly review the GMM here followed by its specific application to source separation as the MESSL algorithm. See Chapter 7 for a related discussion of the use of GMMs as general classifiers, and Chapter 17 for their use in automatic speech

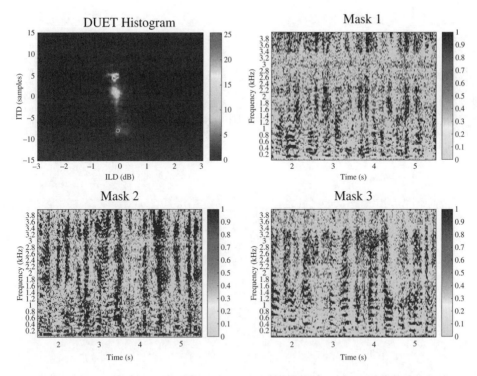

Figure 12.2 Histogram of ITD and ILD features extracted by DUET from the recording in Figure 12.1 along with separation masks estimated using manually selected parameters.

recognition (ASR). The GMM is a generative model of a $D \times 1$ vector observation, $\mathbf{o}$. It has the form

$$p(\mathbf{o}) = \sum_{k=1}^{K} \pi_k \, \mathcal{N}(\mathbf{o} \mid \boldsymbol{\mu}_k, \boldsymbol{\Sigma}_k). \tag{12.9}$$

The parameters of the model are $\boldsymbol{\theta} = \{\pi_1, \dots, \pi_K, \boldsymbol{\mu}_1, \dots, \boldsymbol{\mu}_K, \boldsymbol{\Sigma}_1, \dots, \boldsymbol{\Sigma}_K\}$, which correspond to the mixture weights, means, and covariances of K Gaussians. The weights are normalized so that $\pi_k \geq 0$ and $\sum_k \pi_k = 1$. The means are each $D \times 1$ vectors, representing the center of each Gaussian. The covariances are in general $D \times D$ matrices, representing the scale of each dimension and the degree of correlation between dimensions. The simplifying assumption is often made that data dimensions are uncorrelated with one another, leading to a diagonal covariance matrix, which has D free parameters along the diagonal. All of these parameters are typically learned in an unsupervised manner by maximizing the log-likelihood of a sequence of observations, $\mathbf{O} = [\mathbf{o}(0), \dots, \mathbf{o}(N-1)]$, under the model

$$\mathcal{M}^{\mathrm{ML}}(\boldsymbol{\theta}) = \sum_{n} \log \left(\sum_{k=1}^{K} \pi_k \, \mathcal{N}(\mathbf{o}(n) \mid \boldsymbol{\mu}_k, \boldsymbol{\Sigma}_k) \right). \tag{12.10}$$

This expression is nonconvex and multimodal, so has no unique solution, nor is there a closed-form solution for one of its maxima. Instead, the EM algorithm (Dempster

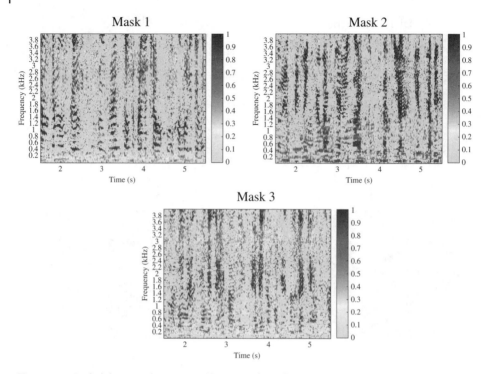

Figure 12.3 Probabilistic masks estimated by MESSL from the mixture shown in Figure 12.1 using Markov random field mask smoothing (Mandel and Roman, 2015).

et al., 1977) introduces a set of latent random variables, $q(n)$, that represent the event of the data point $\mathbf{o}(n)$ coming from a particular Gaussian. These variables allow the definition of two alternating updates, the expectation step (E-step) for the expected memberships $\gamma_k(n) = \mathbb{E}\{q(n) = k \mid \mathbf{O}, \theta\}$ and the maximization step (M-step) for the model parameters θ, which are guaranteed not to decrease the total log-likelihood of the data, (12.10). The E-step update is

$$\gamma_k(n) = \mathbb{E}\{q(n) = k \mid \mathbf{O}, \theta^{(m)}\} = P(q(n) = k \mid \mathbf{O}, \theta^{(m)}) \tag{12.11}$$

$$= \frac{\pi_k^{(m)} \, \mathcal{N}(\mathbf{o}(n) \mid \boldsymbol{\mu}_k^{(m)}, \boldsymbol{\Sigma}_k^{(m)})}{\sum_{k'=1}^{K} \pi_{k'}^{(m)} \, \mathcal{N}(\mathbf{o}(n) \mid \boldsymbol{\mu}_{k'}^{(m)}, \boldsymbol{\Sigma}_{k'}^{(m)})} \tag{12.12}$$

where $(\cdot)^{(m)}$ indicates a parameter estimate at step m. The M-step updates are

$$\pi_k^{(m+1)} = \frac{1}{N} \sum_n \gamma_k(n) \tag{12.13}$$

$$\boldsymbol{\mu}_k^{(m+1)} = \frac{\sum_n \gamma_k(n) \mathbf{o}(n)}{\sum_n \gamma_k(n)} \tag{12.14}$$

$$\boldsymbol{\Sigma}_k^{(m+1)} = \frac{\sum_n \gamma_k(n)(\mathbf{o}(n) - \boldsymbol{\mu}_k^{(m+1)})(\mathbf{o}(n) - \boldsymbol{\mu}_k^{(m+1)})^T}{\sum_n \gamma_k(n)} \tag{12.15}$$

By iterating these updates, a set of model parameters can typically be found so that the data are well represented by the model.

Note that because these updates are unsupervised, they are entirely guided by the goal of describing the data. No effort is made to enforce any structure on the $q(n)$ variables or any relationship between the K Gaussians, making interpretation of these quantities difficult. If the observations, $\mathbf{O}$, fall into clear clusters, then GMM modeling can discover that natural clustering, especially if initialized well, as it typically provides a good fit between model and data. If, however, the observations are relatively homogeneous, there could be many sets of very different GMM parameters that yield very similar likelihoods.

In the case of MESSL, each source is represented as a GMM over L discrete values of ITD on a predefined grid, $\{\Delta_1, \ldots, \Delta_\tau, \ldots, \Delta_L\}$, where τ will be used as the general index into this set. In order to accommodate all possible angles of arrival, the maximum and minimum values of Δ_τ should be greater than the maximum and minimum possible delays between the two microphones. The grid can be uniform or concentrated around expected source positions. The spacing between grid elements can be anything from a fraction of a sample to many samples. These J source models are then themselves mixed together, creating a larger combined model that is still a GMM, but with $K = JL$. The observed data to be modeled are IPD (n,f) and ILD (n,f) of an acoustic mixture. For the purposes of GMM modeling, they are assumed to be statistically independent and each modeled by a separate Gaussian. The means, $\boldsymbol{\mu}_k^{\text{IPD}}$ and $\boldsymbol{\mu}_k^{\text{ILD}}$, are $F \times 1$ vectors representing the IPD and ILD of each source at each frequency. The $F \times F$ covariance matrices are assumed to be diagonal. Because MESSL clusters individual time-frequency bins, it uses the variables $q_{j\tau}(n,f)$ to represent cluster assignments. This is in contrast to the standard GMM, which would only use $q_{j\tau}(n)$ to represent the assignment of an entire spectrogram frame to a Gaussian. In reverberation, the distribution of the ILD of each source, measured in decibels, is close to a simple frequency-dependent Gaussian, so the ILD of the mixture can similarly be modeled as a mixture of J Gaussians.

Thus, following (12.10), the total log-likelihood for the MESSL GMM is

$$\mathcal{M}^{\text{ML}}(\theta) = \sum_{nf} \log p(\text{ IPD }(n,f), \text{ILD }(n,f) \mid \theta) \tag{12.16}$$

$$= \sum_{nf} \log \left(\sum_{j\tau} P(q_{j\tau}(n,f) \mid \theta) \cdot \right.$$

$$\left. p(\text{ IPD }(n,f), \text{ILD }(n,f) \mid q_{j\tau}(n,f), \theta) \right) \tag{12.17}$$

$$= \sum_{nf} \log \left(\sum_{j\tau} P(q_{j\tau}(n,f) \mid \theta) \cdot \mathcal{N}(\text{ IPD }(n,f) \mid \mu_{j\tau}^{\text{IPD}}(f), \sigma_{j\tau}^{\text{IPD}\ 2}(f)) \cdot \right.$$

$$\left. \mathcal{N}(\text{ ILD }(n,f) \mid \mu_j^{\text{ILD}}(f), \sigma_j^{\text{ILD}\ 2}(f)) \right). \tag{12.18}$$

In order to flexibly enforce the relationship between ITD and IPD, MESSL initializes its IPD means to

$$\mu_{j\tau}^{\text{IPD}}(f) = \angle e^{-2j\pi v_f \Delta_\tau}, \tag{12.19}$$

the IPD that would be observed for an anechoic signal with an ITD of Δ_τ s. Because this is just the initialization of the parameters, they are free to adapt to the observations, for example due to early echoes (see Chapter 3). Mandel *et al.* (2010) further enforce a particular ITD for each source by initializing $\pi_{j\tau}$ (as defined in (12.13)) using a source localization technique (see Chapter 4). Specifically, $\pi_{j\tau}$ is initialized to be high for the ITD that source j is identified as most likely originating from, falling off for ITDs farther away.

The EM algorithm attempts to find the source spatial parameters, θ, and settings of hidden variables $q_{j\tau}(n,f)$ that maximize (12.16), the total log-likelihood. The estimated parameters can be used to produce an estimate of the source location by, for example, computing the expected ITD or the mode of the ITD distribution. The estimated hidden variables can by used to produce a mask for each source representing the time-frequency bins that that source dominates,

$$w_j(n,f) = \sum_\tau \gamma_{j\tau}(n,f). \tag{12.20}$$

The application of the mask to each channel, as in Chapter 5, yields an estimate of the spatial image of the source, $\hat{c}_j(n,f)$,

$$\hat{\mathbf{c}}_j(n,f) = w_j(n,f)\mathbf{x}(n,f). \tag{12.21}$$

An improved estimate of these spatial images or even an estimate of the original clean signal can be made by deriving a spatial filter from the estimated mask, as described in Chapter 10 and Section 12.4.

Historically, MESSL was first developed only for IPD observations with the latent ITD variables by Mandel *et al.* (2007). The ILD observations were added by Mandel and Ellis (2007). Mandel *et al.* (2010) presented the entire model and thoroughly evaluated the separation performance of various modeling choices, such as the enforcement of frequency-independence in the model parameters, a prior model connecting ITD parameters to ILD parameters based on anechoic head-related transfer functions, and a warping of the $\gamma_k(n,f)$ probabilities into more effective separation masks. More recently, Mandel and Roman (2015) incorporated MESSL into a *Markov random field* framework to smooth the estimated masks in a probabilistically sound manner (see Fig. 12.3).

When applying MESSL to the separation of moving sources, its assumption of a fixed set of spatial model parameters across all observation frames is not realistic. Schwartz and Gannot (2014) addressed this shortcoming by using the recursive EM algorithm. The online nature of this procedure allows the model parameters to follow moving sources by smoothly adapting the spatial parameters of each source model. Schwartz and Gannot (2014) further generalized MESSL to calibrated arrays of more than two microphones, in particular using ITD observations from an array made up of many pairs of microphones distributed around a room. One Gaussian is then associated with each point in 3D space where a source may be active, and the mean of that Gaussian is set to the IPDs that would be observed at all microphone pairs if the source was located at that point. They show that using an appropriate amount of smoothing allows accurate tracking of two speech sources moving at 1.25 m/s in a simulated recording with a high reverberation time (RT60) of 700 ms. Dorfan and Gannot (2015) further generalized this approach to a distributed recursive EM algorithm so that it can be applied in sensor networks, which have severe limitations on processing and communication resources and Dorfan *et al.* (2015) extended it further to perform separation in addition to localization, showing very promising results.

12.1.3 Incorporating Pitch into Localization-Based Clustering

As described in Chapter 7, the process by which humans understand the complicated acoustic world around them is known as auditory scene analysis. This process consists of two stages. In the first, known as simultaneous segregation, energy observed at different frequencies at approximately the same time is grouped into segments that come from the same source. In the second, known as sequential grouping, segments are associated with those from the same source over time, creating auditory streams. These two processes use different auditory cues, including spatial and pitch cues as two important elements. The contribution of these cues to each process is still an active area of research (Schwartz *et al.*, 2012; Bremen *et al.*, 2013). The field of computational auditory scene analysis (CASA) has developed various computational systems to model this process, with the goals of better understanding of human hearing and better separation of sound mixtures. CASA systems typically follow this two-stage approach, but differ on the cues used in each stage. Because of their basis in human hearing, they typically utilize two-channel recordings.

Shamsoddini and Denbigh (2001) used wideband location information to classify spectrogram frames as target (assumed to be the source directly ahead of the listener) or noise. Using a series of carefully designed rules, each type of frame is processed differently, incorporating pitch information for voiced speech after the location-based processing. Human listening tests and ASR experiments show a substantial improvement in intelligibility of target isolated monosyllabic words when mixed with synchronized monosyllabic words from different locations.

Christensen *et al.* (2007) reversed the use of these two features. Specifically, they used pitch as the primary cue for simultaneous segregation, producing contiguous time-frequency "fragments". Location cues were then averaged over each of these fragments, leading to 35% more accurate localization performance in the presence of reverberation in their experiments. Christensen *et al.* (2009) extended this work to weight the localization cues within source fragments based on various measures of reliability. Several of the measures were found to perform similarly, increasing localization accuracy by an additional 24%, including those based on various models of the *precedence effect*, the human ability to accurately localize sound sources even in the presence of loud reflections from other directions (Litovsky *et al.*, 1999).

Woodruff and Wang (2010) also used pitch as the primary cue for simultaneous segregation and location for sequential grouping. Importantly, in addition to evaluating localization accuracy, they also evaluated the separation performance of their system in terms of the proportion of energy from the target source captured by the estimated time-frequency mask. They showed that their approach outperforms baseline separation systems using either pitch or location cues by themselves. Woodruff and Wang (2013) extended this work to use both pitch and location cues for simultaneous grouping, and location cues for sequential grouping. Through the development of a sophisticated tracking system based on a hidden Markov model (HMM), this system is able to estimate and separate a changing number of sound sources over time. Source number estimation is further discussed in Section 12.2.3. This system also utilizes a training corpus to learn the parameters of a multilayer perceptron that predicts the HMM state from a set of observations. In this way, this system is not just performing spatial clustering, but some amount of classification as well. In experiments on simulated two-source binaural mixtures, this system outperforms several others, including the implementation of MESSL by Mandel *et al.* (2010).

12.2 Multichannel Clustering

While human listeners are limited to processing two input audio channels, computational systems for performing source separation and enhancement can utilize as many microphones as are available. As the number of microphones in an array increases, the complexity of modeling their relationships grows quadratically. As detailed in Chapters 3 and 11, these relationships can be summarized in frequency-dependent spatial covariance matrices. This section discusses two approaches to multichannel spatial clustering, one that applies two-channel models to all pairs of microphones, and another that natively models multichannel sources.

12.2.1 Generalization of Wideband Clustering to more than Two Channels

As discussed in Section 12.1, wideband clustering approaches must include some model of the relationship between the spatial parameters of a source at different frequencies. This becomes more difficult in the multichannel case, where this relationship depends on both the spatial characteristics of the array (the location of the microphones) and the sources (diffuse or point sources). Assuming that the microphone locations are known, as Schwartz and Gannot (2014) do, is reasonable when building a source separation system for a particular microphone array, even a single array in mass production. The same algorithm can be used across several different arrays if it goes through a calibration procedure to identify the locations of the microphones in each array. This is more difficult, however, for recordings coming from unknown arrays, such as those that might contribute to online multimedia repositories of user-generated content. In such situations, a procedure that is blind to source and microphone positions, as described by Bagchi *et al.* (2015), is desirable, and will be described here. Both of these approaches make no explicit accommodation for diffuse sources, except for the uncertainty permitted by the variances of the Gaussians in the model.

A multichannel spatial clustering approach that is blind to source and microphone positions cannot translate location information between microphones. It can, however, still assume that the same source dominates a time-frequency bin across all microphones. With this assumption in place, two-channel spatial clustering methods can be applied to all pairs of microphones and coordinated through the mask estimated by each pair. Bagchi *et al.* (2015) do this using MESSL as the two-channel method, which has a clean statistical interpretation. Building on (12.16) by combining pairs of microphones as independent observations, the multichannel MESSL total log-likelihood is

$$\mathcal{M}^{\mathrm{ML}}(\theta) = \frac{2}{I} \sum_{i=1}^{I} \sum_{i'=1}^{i-1} \mathcal{M}^{\mathrm{ML}}(\theta_{ii'}) \tag{12.22}$$

$$= \frac{2}{I} \sum_{i=1}^{I} \sum_{i'=1}^{i-1} \sum_{nf} \log \left(\sum_{j\tau_{ii'}} P(q_{j\tau_{ii'}}(n,f) \mid \theta_{ii'}) \cdot \right.$$

$$\left. p(\mathrm{IPD}_{ii'}(n,f), \mathrm{ILD}_{ii'}(n,f) \mid q_{j\tau_{ii'}}(n,f), \theta_{ii'}) \right) \tag{12.23}$$

where $\mathrm{IPD}_{ii'}(n,f)$ and $\mathrm{ILD}_{ii'}(n,f)$ are the interchannel observations and $\theta_{ii'}$ is the set of parameters for the MESSL model between microphones i and i'. The latent variables $q_{j\tau_{ii'}}(n,f)$ are specific to each microphone pair as well (as indicated by the ii' index to τ), because without calibration it is not possible to map a delay at one pair to a delay at another. The variables marginalized across delays as $z_j(n,f) = \sum_{\tau_{ii'}} q_{j\tau_{ii'}}(n,f)$ are, however, shared across all microphone pairs. The $\frac{2}{I}$ term counteracts overcounting of observations, as there are $\frac{I(I-1)}{2}$ microphone pairs, but only $I - 1$ independent microphone pair observations. We have found that averaging together all pairs with this normalization term is more robust than selecting a single reference microphone to be paired with each of the other microphones. Figure 12.4 shows how this approach is implemented procedurally.

Note that this approach suffers to some extent from the source permutation problem, in that each parallel MESSL model has its own ordering for the sources, and these orderings must be coordinated somehow. In practice, this alignment can be achieved by initializing all of the models from the same set of masks estimated by another algorithm, or by comparing masks for the sources across pairs. Matching source masks across

---

function MultichannelMessl
  Input: IPD(n,f): IPDs for all pairs of mics, $F \times N \times I(I-1)/2$
    ILD(n,f): ILDs for all pairs of mics, $F \times N \times I(I-1)/2$
    θ: initial parameters for all pairwise MESSL models
  Output: $w_j(n,f)$: Masks for each source, $F \times N \times J$
    θ: parameters for all pairwise MESSL models
  for $c \leftarrow 1$ **to** MaxIter **do**
    for $i, i' \in$ AllPairs (I) **do**
      $w_{ii'j\tau}(n,f) \leftarrow$ MesslEStep $(\theta_{ii'}, \mathrm{ILD}_{ii'}(n,f), \mathrm{IPD}_{ii'}(n,f))$
    end
    $w_j(n,f) \leftarrow \left(\prod_{ii'} \sum_{\tau} w_{ii'j\tau}(n,f) \right)^{2/I}$
    for $i, i' \in$ AllPairs (I) **do**
      $\tilde{w}_{ii'j\tau}(n,f) \leftarrow w_{ii'j\tau}(n,f) \frac{w_j(n,f)}{\sum_{\tau} w_{ii'j\tau}(n,f)}$
      $\theta_{ii'} \leftarrow$ MesslMStep $(\tilde{w}_{ii'j\tau}(n,f), \mathrm{ILD}_{ii'}(n,f), \mathrm{IPD}_{ii'}(n,f))$
    end
  end

Figure 12.4 Multichannel MESSL algorithm. Computes masks for each pair of microphones in the E-step, then combines masks across pairs, then re-estimates parameters for each pair from the global masks.

microphone pairs is typically easier than matching source masks across frequencies, as described in the next section, because of a greater degree of consistency.

12.2.2 Narrowband Clustering Followed by Permutation Alignment

One possible approach to multichannel clustering is to combine the interchannel features of all the possible microphone pairs (Bagchi *et al.*, 2015), as detailed in the previous subsection. Instead of taking such a strategy, this section describes clustering on the observation vector $\mathbf{x}(n,f)$ which consists of all the channel observations. Using this feature, clustering still relies on spatial features of sources. If we can assume an anechoic condition and no spatial aliasing, with frequency normalization in phase, we can easily apply a wideband clustering (Araki *et al.*, 2007). However, in highly reverberant conditions, source spatial information is affected differently in different frequency bins. Moreover, if spatial aliasing is present due to the wide microphone spacing or high-frequency components, we face phase-wrapping problems. These issues make wideband clustering difficult.

To circumvent such issues, this section explains the two-stage narrowband approach to spatial clustering (Sawada *et al.*, 2011), which is illustrated in Figure 12.5. After transforming the signal into the time-frequency domain with the STFT

$$\mathbf{x}(n,f) = \sum_{k=1}^{J} \mathbf{a}_k(f) s_k(n,f) + \mathbf{u}(n,f), \tag{12.24}$$

and extracting features from this representation, the two-stage approach performs "narrowband clustering" at each frequency bin. While this clustering is robust, the order of the output signals is different at each frequency. This problem is referred to as the "permutation problem" and solved by the "permutation alignment" step so that separated signals can be reconstructed properly in the time domain. Through this two-stage approach, masks $w_j(n,f)$ for separating signals are estimated. The separated signals are estimated by (12.21) and their waveforms are reconstructed by using the inverse STFT. Figure 12.6 illustrates example spectra of (a) the original sources, (b) the mixtures, the masks (c) before and (d) after the permutation alignment, and (e) the estimated source images.

The remainder of this subsection explains each stage of the two-stage approach in detail.

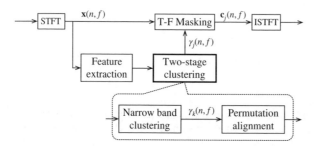

Figure 12.5 Narrowband clustering followed by permutation alignment.

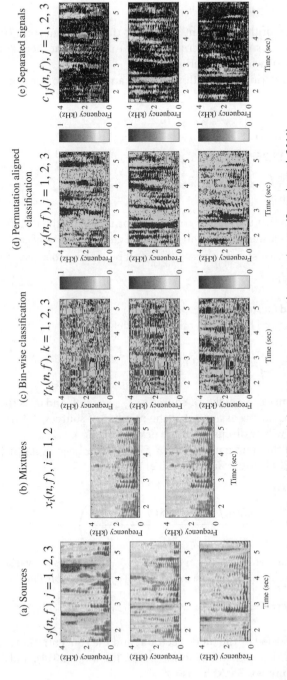

Figure 12.6 Example spectra and masks via a multichannel clustering system on reverberant mixtures (Sawada *et al.*, 2011).

12.2.2.1 Feature Extraction

When the source signals are assumed to be sparse (see Chapter 5), (12.24) becomes

$$\mathbf{x}(n,f) \approx \mathbf{a}_{k'}(f)s_{k'}(n,f) + \mathbf{u}(n,f), \tag{12.25}$$

where k' is the index of the dominant source for time-frequency bin (n,f). Here, we use k instead of j for the source subscript in order to clarify that there are permutation ambiguities in the narrowband clustering.

As a feature for clustering, this section employs an observation vector normalized to unit norm

$$\bar{\mathbf{x}}(n,f) = \frac{\mathbf{x}(n,f)}{\|\mathbf{x}(n,f)\|_2}. \tag{12.26}$$

The substitution of (12.25) into (12.26) gives

$$\bar{\mathbf{x}}(n,f) = \frac{s_{k'}(n,f)}{|s_{k'}(n,f)|} \frac{\mathbf{a}_{k'}(f)}{\|\mathbf{a}_{k'}(f)\|_2}. \tag{12.27}$$

Hence, it can be seen that the feature contains information on the RTF (see Chapter 3), and that subsequent clustering relies on the spatial characteristics of each source image. The remaining unknown source phase $\frac{s_{k'}(n,f)}{|s_{k'}(n,f)|}$ will be taken into account in the following step.

12.2.2.2 Narrowband Clustering

The features are clustered by fitting a mixture model,

$$p(\bar{\mathbf{x}}(n,f) \mid \theta) = \sum_{k=1}^{J} \pi_k(f)p(\bar{\mathbf{x}}(n,f) \mid \theta_k(f)), \tag{12.28}$$

where $\pi_k \geq 0$ is the mixture weight for source k, $\sum_{k=1}^{J} \pi_k = 1$, and θ_k is the set of model parameters for source k. We assume that the number of sources, J, is given. The case where J is unknown will be discussed in the next subsection.

Several variants of the source model $p(\bar{\mathbf{x}}(n,f) \mid \theta_k)$ have been described in the literature. Sawada *et al.* (2011) use a Gaussian(-like) model

$$p(\bar{\mathbf{x}}(n,f) \mid \theta_k(f))$$
$$= \frac{1}{(\pi\sigma_k^2)^{I-1}} \exp\left(-\frac{\|\bar{\mathbf{x}}(n,f) - (\bar{\mathbf{a}}_k^H(f)\bar{\mathbf{x}}(n,f))\bar{\mathbf{a}}_k(f)\|_2}{\sigma_k^2}\right), \tag{12.29}$$

where $\bar{\mathbf{a}}_k$ is the centroid with unit-norm (i.e., $\|\bar{\mathbf{a}}_k\|_2 = 1$) and σ_k^2 is the variance for source k. Tran and Haeb-Umbach (2010) use a complex *Watson distribution*

$$p(\bar{\mathbf{x}}(n,f) \mid \theta_k(f)) = \frac{(I-1)!}{2\pi^I \mathcal{K}(1,I;\kappa_k)} \exp(\kappa_k |\bar{\mathbf{a}}_k^H(f)\bar{\mathbf{x}}(n,f)|^2), \tag{12.30}$$

where source k is modeled by the mode vector $\bar{\mathbf{a}}_k$ and the concentration κ_k, and $\mathcal{K}$ is the Kummer function. Thiemann and Vincent (2013), Ito *et al.* (2014), and Higuchi *et al.* (2016) employ the complex GMM by using

$$p(\mathbf{x}(n,f) \mid \theta_k(f)) = \mathcal{N}_c(\mathbf{x}(n,f) \mid \mathbf{0}_I, \sigma_k^2(n,f)\mathbf{R}_k(f)) \tag{12.31}$$

where source k is modeled by the power spectrum $\sigma_k^2(n,f)$ and the spatial covariance matrix $\mathbf{R}_k(f)$. Note that this method operates on the unnormalized observation vectors $\mathbf{x}(n,f)$ directly. Ito *et al.* (2016) use a complex angular central Gaussian model.

Narrowband clustering is then realized by estimating the parameter sets θ that maximize the log-likelihood

$$\mathcal{M}^{\text{ML}}(\theta) = \log p(\theta \mid \bar{\mathcal{X}}) = \sum_{nf} \log p(\bar{\mathbf{x}}(n,f) \mid \theta) \qquad (12.32)$$

with $\bar{\mathcal{X}} = \{\bar{\mathbf{x}}(n,f)\}_{nf}$. This optimization is efficiently performed using an EM algorithm. Letting $\theta^{(m)}$ be the parameter sets for the mth iteration, the Q function is given by

$$Q(\theta, \theta^{(m)}) = \sum_{nf} \sum_{k=1}^{J} \gamma_k(n,f) \log(\pi_k(f) p(\bar{\mathbf{x}}(n,f) \mid \theta)) \qquad (12.33)$$

where $\gamma_k(n,f) = P(z(n,f) = k \mid \bar{\mathbf{x}}(n,f), \theta^{(m)})$ is the posterior probability that source k is dominant at time-frequency bin (n,f) (i.e., $z(n,f)$ is the dominant source index). The Q function is maximized by iterating the following E- and M-steps. The E-step calculates the posterior probability

$$\gamma_k(n,f) = P(z(n,f) = k \mid \bar{\mathbf{x}}(n,f), \theta) \qquad (12.34)$$

$$= \frac{\pi_k(f) p(\bar{\mathbf{x}}(n,f) \mid \theta_k^{(m)})}{\sum_{k=1}^{J} \pi_k(f) p(\bar{\mathbf{x}}(n,f) \mid \theta_k^{(m)})}. \qquad (12.35)$$

The M-step updates the parameters

$$\theta^{(m+1)} \leftarrow \underset{\theta}{\text{argmax}}\, Q(\theta, \theta^{(m)}). \qquad (12.36)$$

The parameter update rules with a Gaussian(-like) model, a complex Watson distribution, and a complex Gaussian distribution can be found in the papers by Sawada *et al.* (2011), Tran and Haeb-Umbach (2010), and Ito *et al.* (2014), respectively.

12.2.2.3 Permutation Alignment

In the posterior probability $\gamma_k(n,f)$ given by (12.35), the order of the sources may be different in different frequency bins, as can be seen in Figure 12.6c. Thus the source indices must be reordered so that the same index corresponds to the same source at all frequencies, as in Figure 12.6d. To do this, we wish to find the optimal permutations $\Pi(f)$ of $\{1, \dots, J\}$ which match k with j, that is $\Pi(f)(k) = j$. Various approaches have been proposed to identify these permutations. For example, the source could be aligned using spatial information obtained from the estimated cluster centroids, but these estimates tend to be unstable in the presence of reverberation and noise.

One promising way to match speech signals across frequency even in a reverberant and noisy condition is to utilize their common amplitude modulation (AM) structure. As can be seen from Figure 12.6a, the frequency components of a speech signal tend to have common patterns of silence, onsets, and offsets. Furthermore, these patterns tend to be different for different sources. We can therefore solve the permutation problem by choosing a permutation that synchronizes the source activities at all frequencies.

One measure of source activity that works well in practice is the posterior probability sequence

$$\boldsymbol{\gamma}_k(f) = \begin{bmatrix} \gamma_k(0,f) \\ \vdots \\ \gamma_k(N-1,f) \end{bmatrix}. \tag{12.37}$$

The permutations can then be determined so that the activity features in all frequency bins are highly correlated. Specifically, the permutation is selected according to

$$\underset{\Pi(f)}{\mathrm{argmax}} \sum_{ff'} \left(\rho(\boldsymbol{\gamma}_j(f), \boldsymbol{\gamma}_j(f')) - \sum_{j' \neq j} \rho(\boldsymbol{\gamma}_{j'}(f), \boldsymbol{\gamma}_j(f')) \right) \tag{12.38}$$

where $\rho(\cdot, \cdot)$ is the correlation coefficient between two vectors and $j = \Pi(f)(k)$. This optimization can be realized through a clustering approach where centroid vector estimation and permutation alignment are performed in an iterative fashion. For more details, see Sawada *et al.* (2011).

12.2.2.4 Time-Frequency Masking

A time-frequency mask $w_j(n,f)$ can then be derived from this clustering for each source either as a soft mask

$$w_j(n,f) = \gamma_j(n,f) \tag{12.39}$$

or as a binary mask

$$w_j(n,f) = \begin{cases} 1 & \text{if } \gamma_j(n,f) > \gamma_{j'}(n,f), \forall j' \neq j \\ 0 & \text{otherwise.} \end{cases} \tag{12.40}$$

Either way, the separated signals can be calculated by using (12.21).

12.2.3 Source Number Estimation

The discussion so far has assumed that the number of sources J is known. In many practical situations, however, J is unknown and must be blindly estimated before spatial clustering. Source number estimation is thus an important issue in the area of clustering-based source separation research. In this context, this section explores the source number estimation problem.

There have been several methods for estimating the source number based on information criteria (Balan, 2008) or *nonparametric Bayesian* approaches (Otsuka *et al.*, 2014), however, this section explains source number estimation methods that are highly related to the clustering approach. For example, with the DUET approach the number of sources can be estimated by counting the number of peaks in the power-weighted histogram of the relative attenuation and delay (see Figure 12.2 and Rickard (2007)).

Another approach is based on maximum a posteriori (MAP) estimation, where a sparse prior is applied to the mixture weights (Araki *et al.*, 2009). In this case, clustering is performed with a large number of clusters, $J' > J$, and the sparse prior on the mixture weights encourages many of the clusters to have small weights. For example, in the case of a GMM, the sparse prior for the set of mixture weights $\pi = \{\pi_k\}_k$ is given by a *Dirichlet distribution*:

$$p(\pi) = \frac{1}{B(\phi)} \prod_{k=1}^{J'} \pi_k^{\phi-1} \tag{12.41}$$

with a small hyperparameter $0 < \phi < 1$, where $\pi_k \geq 0$, $\sum_{k=1}^{J'} \pi_k = 1$, and $B(\phi)$ is the beta distribution. Then, clustering is performed by maximizing $Q(\theta, \theta^{(m)}) + \log p(\theta)$, where Q is given by (12.33) and $p(\theta) = p(\pi)$. This clustering results in large weights for active source clusters and small weights for noise clusters. Finally, the number of mixture weights which are larger than a threshold gives the number of sources. Thresholding of variance or concentration parameters can also be combined with this in order to ignore large diffuse noise clusters.

Drude *et al.* (2014) counted the dominant cluster in a "deflational" way. First, the single most dominant source component is estimated via a variational Bayesian (VB) algorithm and then subtracted from the observation. Then, the second most dominant source component is estimated and subtracted, and the same procedure is continued until the maximum number of possible sources is reached. The number of sources is given by thresholding the mixture weights and variance or concentration parameters.

Ito *et al.* (2016) estimated the number of sources using the common amplitude modulation structure of speech signals. Here again, clustering is performed using a large number of clusters ($J' > J$). Instead of using frequency-dependent mixture weights $\pi_k(f)$, as in (12.28), they employed time-dependent but frequency-independent mixture weights $\pi_k(n)$, in order to model the activity of each source k. By using such mixture weights and appropriate permutation alignment during the parameter optimization, some of the J' clusters will emerge with high posterior probabilities and others with low posterior probabilities. The number of sources can be estimated by counting the number of clusters that have high posterior probabilities.

Another important problem is the situation where the number of sources is changing over time, that is, each source is sometimes active or inactive. To tackle this issue, each target source should be detected, localized, and tracked. For example, Chong *et al.* (2014) describe a system that can track a multitarget posterior distribution with a changing number of sources by modeling each target state with a Bernoulli random finite state.

12.3 Multichannel Classification

This section describes supervised classification approaches to multichannel source separation. Figure 12.7 shows a system diagram for a typical multichannel classification system where multichannel features are utilized for source classification. In contrast to clustering approaches, classification approaches use a dataset for training a classifier. Once a classifier is trained, its parameters are fixed and it is used to estimate time-frequency masks for separating test mixtures. This section first describes two-channel classification approaches and then several approaches to generalizing them to multichannel recordings.

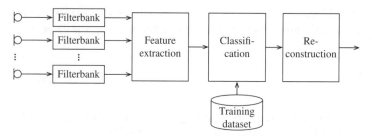

Figure 12.7 System diagram of a multichannel classification system.

12.3.1 Two-Channel Classification

One of the first multichannel classification systems was that of Roman *et al.* (2003). Like DUET, it characterizes individual time-frequency bins using ITD and ILD. Whereas DUET clusters these bins into one group per frequency, Roman *et al.* (2003) classify these bins as being either target dominated or interference dominated, using pretrained classifiers. The ground truth, introduced by this paper, is known as the ideal binary mask, and has been used by many subsequent systems performing source separation and enhancement by classification. Roman *et al.* (2003) assumed two sources were present in every mixture and performed classification conditioned on the frequency band and the azimuth of the two sources, i.e. a separate classifier was trained for every pair of source azimuths at each frequency. At test time, an initial localization step estimated the azimuths, which were then used to select the appropriate set of per-frequency classifiers. Classification was performed using kernel-based estimation of the likelihood ratio between the target-dominant and interference-dominant cases. These classifiers, which only operated on spatial cues, were trained exhaustively on white noise filtered by anechoic head-related transfer functions. The system was able to successfully separate anechoic mixtures of two speech sources with an additional nonspeech noise, improving both SNRs and intelligibility for listeners and ASR. Harding *et al.* (2006) extended this approach to work in reverberant mixtures, achieving word error rates in reverberation that were comparable to those of Roman *et al.* (2003) on anechoic mixtures.

Recently, deep neural networks (DNNs) have been applied to two-channel classification. For a full introduction to DNNs, see Chapter 7. We review them here briefly for completeness. DNNs are networks composed of computational units that are typically organized into layers. The hth layer of neurons takes inputs $\mathbf{x}_h$ and derives outputs $\mathbf{y}_h$ as

$$\mathbf{y}_h = g_h(\mathbf{Z}_h \mathbf{x}_h), \tag{12.42}$$

where $g_h(\cdot)$ is a nonlinear function that is differentiable almost everywhere. Popular nonlinearities include the rectified linear unit $g_{\mathrm{ReLU}}(x) = \max(0, x)$ and the hyperbolic tangent $g_{\mathrm{tanh}}(x) = \frac{e^x - e^{-x}}{e^x + e^{-x}}$. A multilayer perceptron consists of two such layers, and a DNN typically consists of three or more. Using the chain rule of differentiation, the weights of the network, $\mathbf{Z}_h$, can be optimized so that the network minimizes a differentiable loss function of interest, such as the mean square error (MSE) for continuous outputs or the cross-entropy for binary outputs. Using the backpropagation algorithm (Rumelhart *et al.*, 1986), this differentiation can be performed efficiently.

Jiang *et al.* (2014) described a DNN used for two-channel classification. Their system employed spatial features computed from two-channel observations as the input to the DNN, which outputted an estimate of the ideal binary mask. Because the ideal binary mask is a binary output, the cost function being minimized was the cross-entropy between the true ideal binary mask and the predictions of the network. The network's input features consisted of per-frequency ILD, ITD, and cross-correlation functions between the two channel signals. The target source was fixed at 0°, eliminating the need to identify which of several sources is the target. By training the network on simulated spatial recordings containing a variety of interferer configurations, the network was shown to be able to generalize to new interferer angles for up to five sources. The addition of single-channel features to the network improves performance when the interferer is collocated with the target, but not when spatial features are discriminative.

12.3.2 Generalization to More than Two Channels

Observations with more than two channels contain more spatial information about the targets and noise, and therefore should permit improved classification performance. In order to leverage this information, several multichannel classification approaches have been described in the literature that generalize two-channel approaches.

One generalization is to employ multichannel features as DNN inputs. This is a simple extension of the two-channel approach in the previous subsection. Various spatial features from multichannel observations may be employed, including ILD, IPD and ITD between all the microphone pairs, direction of arrival (DOA), time-frequency masks, and pre-enhanced signals, which can be calculated using all of the channels. For example, Pertilä and Nikunen (2015) employed an average of the differences between the measured IPD and theoretical IPD over all the microphone pairs, and its average over nonlook directions.

Instead of estimating classification results as time-frequency masks, it is also possible to estimate the magnitude spectra of target sources. In this case, magnitude STFT coefficients $|s_j(n,f)|$ (or log Mel-filterbank spectra) of the target source are estimated as outputs of a DNN. This method is sometimes referred to as a *denoising autoencoder*. The denoising autoencoder was originally a single-channel technique, however, its multichannel extension has recently been explored. Araki *et al.* (2015), for example, showed that denoising autoencoder performance can be improved by employing auxiliary features from the multichannel observations. As the auxiliary features, they used pre-enhanced target and noise signals, which are estimated with spatial clustering-based masks. These auxiliary features were then concatenated with the original input features to the DNN. As the dimension of these auxiliary features is independent of the number of microphones, this approach can easily be extended to systems with three channels or more without changing the configuration of the DNN.

Another approach to multichannel generalization is to first estimate a mask for each channel with a single-channel classifier, and then merge the masks for all the channels. For example, Heymann *et al.* (2015, 2016) took this approach, where single-channel classification for each channel was performed with a bidirectional long short-term memory (LSTM) neural network, and then the obtained masks were merged into a single mask with a median filter.

Another type of generalization is integrating single-channel classification approaches, which are discussed in Chapter 7, with multichannel clustering. This approach estimates a soft time-frequency mask by simultaneously clustering the spectral and spatial features of the signal. Nakatani *et al.* (2013) provide a typical example, which integrates a factorial HMM-based single-channel classifier (Section 7.2) with the narrowband multichannel spatial clustering of Section 12.2.2.

Related to multichannel classification, it should be noted that joint training of multichannel classification systems and ASR is now an active area of research (Sainath *et al.*, 2016; Xiao *et al.*, 2016) for speech recognition purposes. With this approach, spatial filtering is embedded into the feature extraction process in DNN-based acoustic modeling, and is optimized jointly with the acoustic model based on speech recognition criteria. This topic is discussed in Section 17.4.2.

12.3.3   Generalization in Classification Systems

Because classification systems are trained by example, the composition of the training dataset is an integral component of the final performance of the system. In particular, it is difficult to train a system by example to generalize to new conditions and therefore performance deterioration is expected if there is a mismatch between the training and test datasets. A classic solution to this situation in noise robust ASR is *multicondition training* (Lippmann *et al.*, 1987), in which recognition performance in noisy test environments is significantly improved by training on data that contains the same or similar noise conditions. This can be considered to be a generalization to new noise instances and new noise types. Similar approaches have been shown to explicitly assist in speech enhancement with DNNs (Chen *et al.*, 2016; Vincent *et al.*, 2017), even including the synthesis of new noises or the modification of existing noises to increase the diversity of the training noises. See Chapter 7 for an expanded discussion of this type of generalization in the context of single-channel separation.

In the context of multichannel source separation and enhancement, there are several additional dimensions of generalization that must be considered, in particular those related to spatial configurations of target and noise. Jiang *et al.* (2014) show that a DNN performing enhancement on two-channel inputs can generalize to new noise positions, provided that the target is in a fixed position. They also report small-scale experiments showing that the system can generalize to nearby target positions.

The spatial stationarity of the target source is dictated by the problem to be solved, which includes interfering talkers speaking simultaneously with the target talker. For speech enhancement, when there is only a single talker in the presence of nonspeech noise, it is clear which source should be enhanced. Deciding which of several talkers is the target is more difficult, generally requiring input from a human user, and is known as the *source selection problem*. Xu *et al.* (2015) and Weng *et al.* (2014) solve this problem for single-channel separation and recognition, respectively, by producing two outputs from their systems, one for the louder talker and one for the quieter talker. Thus the task of source selection is delayed to subsequent processing by assuming that the number of speech sources involved in a mixture is known. This solution is also insufficient when the two talkers are received at equal loudness, but this symmetry could be broken using other speaker characteristics like gender, age, etc.

Sainath *et al.* (2015) and Xiao *et al.* (2016) describe multichannel ASR systems where speech is mixed with nonspeech noise, solving the source selection problem and enabling the systems to generalize to new target positions. They are also able to generalize to recordings made or simulated in new rooms that were not used for training the network. This is also the case for the system described by Mandel and Barker (2016), which is able to generalize to new microphone array geometries for ASR through the use of spatial clustering. Creating a spatial classification system that can generalize in this way is an active area of research at the time of this book's publication. The problem is challenging because typical DNNs require their inputs to have a fixed dimensionality, but microphone arrays can differ in the number of microphones they utilize. A related problem to generalizing across array geometries is generalizing to different physical arrays with ostensibly the same geometry, as is encountered in mass-produced devices. Automatic self-calibration algorithms (Tashev, 2005) can be used to avoid the need to carefully select microphones with matching responses and mount them precisely in the same positions and enclosures.

12.4 Spatial Filtering Based on Masks

This chapter has introduced time-frequency masking, which can be conducted as in (12.21) and has been shown to be effective for separation and enhancement in terms of interference and noise reduction. Because it is time-varying, however, it often introduces nonlinear spectral distortion into the enhanced signal. Such distortion causes adverse effects, not only for human listening, known as musical noise artifacts, but also for ASR. Also, it does not benefit from the superior potential of multichannel filtering as opposed to single-channel masking.

In contrast, spatial filtering, discussed in Chapter 10, enhances a signal by

$$\hat{s}_j(n,f) = \mathbf{w}_j^H(f)\mathbf{x}(n,f) \tag{12.43}$$

with filter $\mathbf{w}_j(f)$ and avoids such artifacts because it is time-invariant linear processing. Conventional approaches to designing such spatial filters require spatial information (e.g., array geometry and target direction) in advance (Barker *et al.*, 2015). For example, minimum variance distortionless response (MVDR) beamforming (Section 10.4.4) requires steering vector for the target source. The estimation of such spatial information tends to have errors in practice.

In order to realize advanced spatial filtering, which is robust against the errors in microphone and target position estimation and can be designed in a blind manner, certain attempts have been studied to combine the clustering and classification approaches with spatial filtering. Figure 12.8 illustrates one way to combine these two types of approaches, known as *mask-based beamforming*. Clustering or classification-based masks are utilized to estimate the parameters for designing the spatial filter coefficients. Although there are several methods to combine masks and spatial filters, e.g. Madhu and Martin (2011), the rest of this subsection describes several typical examples that have a scheme in Figure 12.8. It should be noted that this subsection is closely related

Figure 12.8 Block diagram of spatial filtering based on masks.

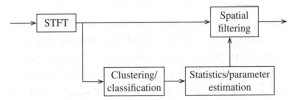

to Chapter 11. Note also that (12.43) is taken as yielding $\hat{c}_{ij}$ instead of $\hat{s}_j$ in a blind signal processing scenario, assuming that the gain of the reference microphone i is one for coping with the scale ambiguity problem.

12.4.1 Mask-Based Beamforming using Covariance Subtraction

As the spatial filter, Higuchi *et al.* (2016) employed a minimum power distortionless response beamformer (see Section 10.4.4)

$$\mathbf{w}_j(f) = \frac{\mathbf{\Sigma}_\mathbf{x}(f)\mathbf{a}_j(f)}{\mathbf{a}_j^H(f)\mathbf{\Sigma}_\mathbf{x}(f)\mathbf{a}_j(f)}. \tag{12.44}$$

To obtain an accurate minimum power distortionless response beamformer $\mathbf{w}_j(f)$ for target j, accurately estimating the steering vector $\mathbf{a}_j(f)$ is essential. The steering vector for target j is given by the principal eigenvector of the covariance matrix $\mathbf{\Sigma}_{\mathbf{c}_j}(f)$ of the observations when only target source j is active. In practice, however, $\mathbf{\Sigma}_{\mathbf{c}_j}(f)$ cannot be obtained directly and must be estimated. Using masks, it can be estimated as

$$\mathbf{\Sigma}_{\mathbf{c}_j}(f) = \mathbf{\Sigma}_{\mathbf{c}_j+\mathbf{u}}(f) - \mathbf{\Sigma}_\mathbf{u}(f), \tag{12.45}$$

$$\mathbf{\Sigma}_{\mathbf{c}_j+\mathbf{u}}(f) = \frac{1}{\sum_n w_j(n,f)} \sum_n w_j(n,f)\mathbf{x}(n,f)\mathbf{x}^H(n,f) \tag{12.46}$$

$$\mathbf{\Sigma}_\mathbf{u}(f) = \frac{1}{\sum_n w_u(n,f)} \sum_n w_u(n,f)\mathbf{x}(n,f)\mathbf{x}^H(n,f). \tag{12.47}$$

where $\mathbf{\Sigma}_{\mathbf{c}_j+\mathbf{u}}(f)$ is the covariance matrix of the target j with noise without other interference signals, and $\mathbf{\Sigma}_\mathbf{u}(f)$ is the covariance matrix of noise. The masks for target j and noise, $w_j(n,f)$ and $w_u(n,f)$ can be estimated using the clustering or classification approaches discussed in Section 12.2. For example, Higuchi *et al.* (2016) employed the complex GMM-based clustering approach for mask estimation, and obtained promising results. Here it should be noted that the signal extracted with $w_j(n,f)$ still includes the noise $u(n,f)$ component, therefore we have to subtract the noise influence by using (12.45).

12.4.2 Mask-Based Multichannel Wiener Filtering

Covariance subtraction used in Section 12.4.1 for estimating covariance matrices of target signals can also be employed to compute a multichannel Wiener filter (MWF) (see Section 10.4.3 and Souden *et al.* (2013))

$$\mathbf{w}_j(f) = \frac{\mathbf{\Sigma}_{\mathbf{x}-\mathbf{c}_j}^{-1}(f)\mathbf{\Sigma}_{\mathbf{c}_j}(f)}{\mu + \mathrm{tr}\,(\mathbf{\Sigma}_{\mathbf{x}-\mathbf{c}_j}^{-1}(f)\mathbf{\Sigma}_{\mathbf{c}_j}(f))}\mathbf{e}_1, \tag{12.48}$$

where $\mathbf{e}_1 = [1, 0, \ldots, 0]^T$, $\mu = 1$ for the MWF, and $\mathbf{\Sigma}_{\mathbf{c}_j}(f)$ and $\mathbf{\Sigma}_{\mathbf{x}-\mathbf{c}_j}(f)$ are the covariance matrices of target source j and interference plus noise, respectively. For accurate MWF estimation, the accurate estimation of these matrices is essential. By using estimated masks, Souden *et al.* (2013) estimated $\mathbf{\Sigma}_{\mathbf{c}_j}(f)$ in a way that is similar to Section 12.4.1, and $\mathbf{\Sigma}_{\mathbf{x}-\mathbf{c}_j}(f)$ can be approximated by $\mathbf{\Sigma}_{\mathbf{x}-\mathbf{c}_j}(f) = \mathbf{\Sigma}_\mathbf{x}(f) - \mathbf{\Sigma}_{\mathbf{c}_j}(f)$.

12.4.3 Mask-Based Maximum SNR Beamforming

Heymann *et al.* (2015, 2016) use estimated masks to derive a maximum signal-to-noise ratio (SNR) beamformer (Section 10.4.4). Here, the beamformer coefficients that maximize the SNR of the observation are obtained from the principal eigenvector of the following generalized eigenvalue problem:

$$\Sigma_{c_j}(f)\mathbf{w}_j(f) = \lambda_j(f)\Sigma_{\mathbf{u}}(f)\mathbf{w}_j(f).$$ (12.49)

In this context, providing accurate covariance matrices for target Σ_{c_j} and noise $\Sigma_{\mathbf{u}}$ is essential. These matrices can be estimated using the masks $w_j(n,f)$ produced by a clustering or classification approach. Heymann *et al.* (2015, 2016) estimated these matrices by

$$\Sigma_{c_j}(f) = \sum_n w_j(n,f)\mathbf{x}(n,f)\mathbf{x}^H(n,f)$$ (12.50)

$$\Sigma_{\mathbf{u}}(f) = \sum_n w_u(n,f)\mathbf{x}(n,f)\mathbf{x}^H(n,f).$$ (12.51)

Heymann *et al.* (2015, 2016) examined two different types of mask estimators: Watson mixture model clustering and DNN-based classification, where the latter includes both feedforward and bidirectional LSTM networks. They concluded that DNN-based classification outperforms Watson mixture model-based clustering for this purpose.

12.4.4 Classification-Based Multichannel Wiener Filtering

Sivasankaran *et al.* (2015) and Nugraha *et al.* (2016) combined classification-based spectral estimation with a time-varying MWF. By modeling the source image with a multivariate complex Gaussian distribution (see Section 14.1.1) $c_j(n,f) \sim \mathcal{N}_c(c_j(n,f) \mid 0_I, \sigma_j^2(n,f)\mathbf{R}_j(f))$, where $\sigma_j^2(n,f)$ and $\mathbf{R}_j(f)$ are the power spectrum and the spatial covariance matrix of source j, respectively, the MWF is given by

$$\mathbf{W}_j^H(n,f) = \sigma_j^2(n,f)\mathbf{R}_j(f)\left(\sum_{j'=1}^J \sigma_{j'}^2(n,f)\mathbf{R}_{j'}(f)\right)^{-1}.$$ (12.52)

For estimating this MWF, Nugraha *et al.* (2016) have proposed estimating the magnitude spectrum $\sigma_j(n,f)$ in (12.52) with a denoising autoencoder-based scheme. $\mathbf{R}_j(f)$ was calculated by iterating

$$\hat{c}_j(n,f) = \mathbf{W}^H(n,f)\mathbf{x}(n,f)$$ (12.53)

$$\mathbf{R}_j(f) = \left(\sum_n \sigma_j^2(n,f)\right)^{-1} \sum_n \hat{c}_j(n,f)\hat{c}_j^H(n,f).$$ (12.54)

12.5 Summary

This chapter described source separation and enhancement methods using spatial clustering or classification. The use of multiple microphones enables the use of spatial information to separate sources in addition to time-frequency information used by single-channel methods described in Chapter 7. In terms of the general taxonomy of methods described in Chapter 1, spatial clustering is learning-free, while spatial classification is a separation-based training method. Because both types of methods are based on time-frequency masking, as discussed in Chapter 7, they can be used to separate sources from underdetermined mixtures, i.e. where there are more sources than channels. As the introduced algorithms utilize more information from mixtures, such as spatial information, care must be taken to ensure that the datasets used for training and testing fully cover the necessary dimensions of diversity, either of sources or of spatial configurations. Such diversity is necessary to ensure that the algorithms will properly generalize to new conditions.

Bibliography

Araki, S., Hayashi, T., Delcroix, M., Fujimoto, M., Takeda, K., and Nakatani, T. (2015) Exploring multi-channel features for denoising-autoencoder-based speech enhancement, in *Proceedings of IEEE International Conference on Audio, Speech and Signal Processing*, pp. 116–120.

Araki, S., Sawada, H., and Makino, S. (2009) Blind sparse source separation for unknown number of sources using Gaussian mixture model fitting with Dirichlet prior, in *Proceedings of IEEE International Conference on Audio, Speech and Signal Processing*, pp. 33–36.

Araki, S., Sawada, H., Mukai, R., and Makino, S. (2007) Underdetermined blind sparse source separation for arbitrarily arranged multiple sensors. *Signal Processing*, **87** (8), 1833–1847.

Bagchi, D., Mandel, M.I., Wang, Z., He, Y., Plummer, A., and Fosler-Lussier, E. (2015) Combining spectral feature mapping and multi-channel model-based source separation for noise-robust automatic speech recognition, in *Proceedings of IEEE Workshop on Automatic Speech Recognition and Understanding*, pp. 496–503.

Balan, R. (2008) Information theory based estimator of the number of sources in a sparse linear mixing model, in *Proceedings of Annual Conference on Information Sciences and Systems*, pp. 269–273.

Barker, J., Marxer, R., Vincent, E., and Watanabe, S. (2015) The third 'CHiME' speech separation and recognition challenge: Dataset, task and baselines, in *Proceedings of IEEE Workshop on Automatic Speech Recognition and Understanding*, pp. 504–511.

Bremen, P., Middlebrooks, J.C., Middlebrooks, J., Onsan, Z., Gutschalk, A., Micheyl, C., and Oxenham, A. (2013) Weighting of spatial and spectro-temporal cues for auditory scene analysis by human listeners. *PLoS ONE*, **8** (3), 1–12.

Chen, J., Wang, Y., and Wang, D. (2016) Noise perturbation for supervised speech separation. *Speech Communication*, **78**, 1–10.

Chong, N., Wong, S., Vo, B.T., Sven, N., and Murray, I. (2014) Multiple moving speaker tracking via degenerate unmixing estimation technique and cardinality balanced multi-target multi-Bernoulli filter (DUET-CBMeMBer), in *Proceedings of IEEE International Conference on Intelligent Sensors, Sensor Networks and Information Processing*, pp. 1–6.

Christensen, H., Ma, N., Wrigley, S.N., and Barker, J. (2007) Integrating pitch and localisation cues at a speech fragment level, in *Proceedings of Interspeech*, pp. 2769–2772.

Christensen, H., Ma, N., Wrigley, S.N., and Barker, J. (2009) A speech fragment approach to localising multiple speakers in reverberant environments, in *Proceedings of IEEE International Conference on Audio, Speech and Signal Processing*, pp. 4593–4596.

Dempster, A., Laird, N.M., and Rubin, D.B. (1977) Maximum likelihood from incomplete data via the EM algorithm. *Journal of the Royal Statistical Society: Series B*, **39** (1), 1–38.

Dorfan, Y., Cherkassky, D., and Gannot, S. (2015) Speaker localization and separation using incremental distributed expectation-maximization, in *Proceedings of European Signal Processing Conference*, pp. 1256–1260.

Dorfan, Y. and Gannot, S. (2015) Tree-based recursive expectation-maximization algorithm for localization of acoustic sources. *IEEE/ACM Transactions on Audio, Speech, and Language Processing*, **23** (10), 1692–1703.

Drude, L., Chinaev, A., Tran, D.H.V., and Haeb-Umbach, R. (2014) Source counting in speech mixtures using a variational EM approach for complex Watson mixture models, in *Proceedings of IEEE International Conference on Audio, Speech and Signal Processing*, pp. 6834–6838.

Harding, S., Barker, J., and Brown, G. (2006) Mask estimation for missing data speech recognition based on statistics of binaural interaction. *IEEE Transactions on Audio, Speech, and Language Processing*, **14** (1), 58–67.

Heymann, J., Drude, L., Chinaev, A., and Haeb-Umbach, R. (2015) BLSTM supported GEV beamformer front-end for the 3rd CHiME challenge, in *Proceedings of IEEE Workshop on Automatic Speech Recognition and Understanding*, pp. 444–451.

Heymann, J., Drude, L., and Haeb-Umbach, R. (2016) Neural network based spectral mask estimation for acoustic beamforming, in *Proceedings of IEEE International Conference on Audio, Speech and Signal Processing*, pp. 196–200.

Higuchi, T., Ito, N., Yoshioka, T., and Nakatani, T. (2016) Robust MVDR beamforming using time-frequency masks for online/offline ASR in noise, in *Proceedings of IEEE International Conference on Audio, Speech and Signal Processing*, pp. 5210–5214.

Ito, N., Araki, S., and Nakatani, T. (2016) Complex angular central Gaussian mixture model for directional statistics in mask-based microphone array signal processing, in *Proceedings of European Signal Processing Conference*, pp. 409–413.

Ito, N., Araki, S., Yoshioka, T., and Nakatani, T. (2014) Relaxed disjointness based clustering for joint blind source separation and dereverberation, in *Proceedings of International Workshop on Acoustic Echo and Noise Control*, pp. 268–272.

Jiang, Y., Wang, D., Liu, R., and Feng, Z. (2014) Binaural classification for reverberant speech segregation using deep neural networks. *IEEE/ACM Transactions on Audio, Speech, and Language Processing*, **22** (12), 2112–2121.

Jourjine, A., Rickard, S., and Yılmaz, Ö. (2000) Blind separation of disjoint orthogonal signals: demixing n sources from 2 mixtures, in *Proceedings of IEEE International Conference on Audio, Speech and Signal Processing*, vol. 5, pp. 2985–2988.

Lippmann, R., Martin, E., and Paul, D. (1987) Multi-style training for robust isolated-word speech recognition, in *Proceedings of IEEE International Conference on Audio, Speech and Signal Processing*, vol. 12, pp. 705–708.

Litovsky, R., Colburn, S., Yost, W., and Guzman, S. (1999) The precedence effect. *Journal of the Acoustical Society of America*, **106** (4), 1633–1654.

Madhu, N. and Martin, R. (2011) A versatile framework for speaker separation using a model-based speaker localization approach. *IEEE Transactions on Audio, Speech, and Language Processing*, **19** (7), 1900–1912.

Mandel, M.I. and Barker, J.P. (2016) Multichannel spatial clustering for robust far-field automatic speech recognition in mismatched conditions, in *Proceedings of Interspeech*, pp. 1991–1995.

Mandel, M.I. and Ellis, D.P.W. (2007) EM localization and separation using interaural level and phase cues, in *Proceedings of IEEE Workshop on Applications of Signal Processing to Audio and Acoustics*, pp. 275–278.

Mandel, M.I., Ellis, D.P.W., and Jebara, T. (2007) An EM algorithm for localizing multiple sound sources in reverberant environments, in *Proceedings of Neural Information Processing Systems*, pp. 953–960.

Mandel, M.I. and Roman, N. (2015) Enforcing consistency in spectral masks using Markov random fields, in *Proceedings of European Signal Processing Conference*, pp. 2028–2032.

Mandel, M.I., Weiss, R.J., and Ellis, D.P.W. (2010) Model-based expectation maximization source separation and localization. *IEEE Transactions on Audio, Speech, and Language Processing*, **18** (2), 382–394.

Nakatani, T., Araki, S., Yoshioka, T., Delcroix, M., and Fujimoto, M. (2013) Dominance based integration of spatial and spectral features for speech enhancement. *IEEE Transactions on Audio, Speech, and Language Processing*, **21** (12), 2516–2531.

Nugraha, A.A., Liutkus, A., and Vincent, E. (2016) Multichannel audio source separation with deep neural networks. *IEEE/ACM Transactions on Audio, Speech, and Language Processing*, **24** (9), 1652–1664.

Otsuka, T., Ishiguro, K., Sawada, H., and G.Okuno, H. (2014) Bayesian nonparametrics for microphone array processing. *IEEE Transactions on Audio, Speech, and Language Processing*, **22** (2), 493–504.

Pertilä, P. and Nikunen, J. (2015) Distant speech separation using predicted time-frequency masks from spatial features. *Speech Communication*, **68**, 97–106.

Rickard, S. (2007) The DUET blind source separation algorithm, in *Blind Speech Separation*, Springer, pp. 217–241.

Roman, N., Wang, D., and Brown, G.J. (2003) Speech segregation based on sound localization. *Journal of the Acoustical Society of America*, **114** (4), 2236–2252.

Rumelhart, D.E., Hinton, G.E., and Williams, R.J. (1986) Learning representations by back-propagating errors. *Nature*, **323** (6088), 533–536.

Sainath, T., Weiss, R., Wilson, K., Narayanan, A., and Bacchiani, M. (2016) Factored spatial and spectral multichannel raw waveform CLDNNs, in *Proceedings of IEEE International Conference on Audio, Speech and Signal Processing*, pp. 5075–5079.

Sainath, T.N., Weiss, R.J., Senior, A., Wilson, K.W., and Vinyals, O. (2015) Learning the speech front-end with raw waveform CLDNNs, in *Proceedings of Interspeech*, pp. 1–5.

Sawada, H., Araki, S., and Makino, S. (2011) Underdetermined convolutive blind source separation via frequency bin-wise clustering and permutation alignment. *IEEE Transactions on Audio, Speech, and Language Processing*, **19** (3), 516–527.

Schwartz, A., McDermott, J.H., and Shinn-Cunningham, B. (2012) Spatial cues alone produce inaccurate sound segregation: The effect of interaural time differences. *Journal of the Acoustical Society of America*, **132** (1), 357–368.

Schwartz, O. and Gannot, S. (2014) Speaker tracking using recursive EM algorithms. *IEEE/ACM Transactions on Audio, Speech, and Language Processing*, **22** (2), 392–402.

Shamsoddini, A. and Denbigh, P. (2001) A sound segregation algorithm for reverberant conditions. *Speech Communication*, **33** (3), 179–196.

Sivasankaran, S., Nugraha, A.A., Vincent, E., Cordovilla, J.A.M., Dalmia, S., Illina, I., and Liutkus, A. (2015) Robust ASR using neural network based speech enhancement and feature simulation, in *Proceedings of IEEE Workshop on Automatic Speech Recognition and Understanding*, pp. 482–489.

Souden, M., Araki, S., Kinoshita, K., Nakatani, T., and Sawada, H. (2013) A multichannel MMSE-based framework for speech source separation and noise reduction. *IEEE Transactions on Audio, Speech, and Language Processing*, **21** (9), 1913–1928.

Tashev, I. (2005) Beamformer sensitivity to microphone manufacturing tolerances, in *Proceedings of International Conference on Systems for Automation of Engineering and Research*, pp. 1–5.

Thiemann, J. and Vincent, E. (2013) A fast EM algorithm for Gaussian model-based source separation, in *Proceedings of European Signal Processing Conference*, pp. 1–5.

Tran, D.H.V. and Haeb-Umbach, R. (2010) Blind speech separation employing directional statistics in an expectation maximization framework, in *Proceedings of IEEE International Conference on Audio, Speech and Signal Processing*, pp. 241–244.

Vincent, E., Watanabe, S., Nugraha, A.A., Barker, J., and Marxer, R. (2017) An analysis of environment, microphone and data simulation mismatches in robust speech recognition. *Computer Speech and Language*, **46**, 535–557.

Weng, C., Yu, D., Seltzer, M.L., and Droppo, J. (2014) Single-channel mixed speech recognition using deep neural networks, in *Proceedings of IEEE International Conference on Audio, Speech and Signal Processing*, pp. 5632–5636.

Woodruff, J. and Wang, D. (2010) Sequential organization of speech in reverberant environments by integrating monaural grouping and binaural localization. *IEEE Transactions on Audio, Speech, and Language Processing*, **18** (7), 1856–1866.

Woodruff, J. and Wang, D. (2013) Binaural detection, localization, and segregation in reverberant environments based on joint pitch and azimuth cues. *IEEE Transactions on Audio, Speech, and Language Processing*, **21** (4), 806–815.

Xiao, X., Watanabe, S., Erdogan, H., Lu, L., Hershey, J., Seltzer, M.L., Chen, G., Zhang, Y., Mandel, M., and Yu, D. (2016) Deep beamforming networks for multi-channel speech recognition, in *Proceedings of IEEE International Conference on Audio, Speech and Signal Processing*, pp. 5745–5749.

Xu, Y., Du, J., Dai, L.R., and Lee, C.H. (2015) A regression approach to speech enhancement based on deep neural networks. *IEEE/ACM Transactions on Audio, Speech, and Language Processing*, **23** (1), 7–19.

Yılmaz, Ö. and Rickard, S. (2004) Blind separation of speech mixtures via time-frequency masking. *IEEE Transactions on Signal Processing*, **52** (7), 1830–1847.

13

Independent Component and Vector Analysis

Hiroshi Sawada and Zbyněk Koldovský

The concept of blind source separation (BSS) has been introduced in previous chapters. The term "blind" means that no a priori information is used for separation and that all parameters are estimated from observed signals based on assumed (general) properties of the unknown original signals. For example, in Chapter 8, nonnegative matrix factorization (NMF) was introduced as a blind method which relies only on nonnegativity. This chapter is devoted to methods that rely on signal independence, a condition that is often encountered in real-world situations where signals originate from different processes that have no mutual connection between each other, e.g. speech and noise.

Efficient mathematical models of the independence come from probability theory. The signals to be separated can be modeled as stochastically independent random processes. Then, objective functions that quantify the independence can be derived based on the model and used to find independent signals. This gives rise to *independent component analysis* (ICA), a tool popular in BSS.

In principle, ICA assumes instantaneous mixing while audio mixtures are convolutive. This chapter is mainly focused on a solution that is called *frequency-domain ICA* (FD-ICA). The convolutive mixture is transformed by short-time Fourier transform (STFT) into a set of instantaneous mixtures, one mixture per frequency bin. Each frequency is separated using ICA independently of the others. Because of the indeterminacy of scaling and of the order of the components separated by ICA, so-called permutation and scaling problems arise. They must be resolved to separate the signals in the time domain as well. We further introduce a more advanced solution, which is based on *independent vector analysis* (IVA). Here, the separation and the permutation problems are solved simultaneously.

This chapter is organized as follows. Section 13.1 recalls basic facts about convolutive audio mixtures and their time-frequency representations. Section 13.2 is devoted to the FD-ICA solution. In particular, the fundamentals of ICA, basic models, and algorithms are introduced. Several solutions to the scaling ambiguity and permutation problem are described. Section 13.3 defines IVA and describes the most popular algorithms. Section 13.4 is devoted to experiments with real-world mixtures of speech signals and, finally, Section 13.5 summarizes the chapter.

Audio Source Separation and Speech Enhancement, First Edition.
Edited by Emmanuel Vincent, Tuomas Virtanen and Sharon Gannot.
© 2018 John Wiley & Sons Ltd. Published 2018 by John Wiley & Sons Ltd.
Companion Website: https://project.inria.fr/ssse/

13.1 Convolutive Mixtures and their Time-Frequency Representations

The mixing process in acoustic environments is linear and convolutive. In the time domain, a mixture of J signals observed through I microphones is given by (3.4), or as follows.

Let $s_1(t), \ldots, s_J(t)$ be J source signals and $x_1(t), \ldots, x_I(t)$ be I microphone observations. The convolutive mixture model is formulated as

$$x_i(t) = \sum_{j=1}^{J} \sum_{\tau=-\infty}^{+\infty} a_{ij}(\tau)s_j(t-\tau) \tag{13.1}$$

where t represents time and $a_{ij}(\tau)$ is the impulse response of source j to microphone i.

The fundamentals of ICA are based on instantaneous mixtures, i.e. when there are no delays or reverberation. Therefore, before applying ICA to a convolutive mixture, the original problem has to be transformed into one or more instantaneous ones. Basically, there are two approaches used in practice: the time-domain and the STFT domain formulations. We will focus on the latter solution. For readers interested in the time-domain approaches, see Buchner *et al.* (2005), Bourgeois and Minker (2009), and Koldovský and Tichavský (2011).

STFT applied to the microphone observations can be compactly represented by an $F \times N \times I$ complex-valued tensor $\mathcal{X} = [x_i(n,f)]_{fni}$ with elements $x_i(n,f)$; see the left-hand side in Figure 13.1. Here, F denotes the number of frequency bins, N is the number of time frames in STFT, and n and f denote the time frame and the frequency bin, respectively. The STFT has to be applied with a sufficiently long window to cover the main part of the impulse responses so that the convolutive mixture model (13.1) can be approximated[1] by

$$x_i(n,f) = \sum_{j=1}^{J} a_{ij}(f)s_j(n,f) \tag{13.2}$$

where $n \in \{0, \ldots, N-1\}$ and $f \in \{0, \ldots, F-1\}$ (Smaragdis, 1998; Murata *et al.*, 2001).

The approximation (13.2) constitutes an instantaneous mixture for frequency bin f. It can be written in a compact matrix form

$$\mathbf{X}(f) = \mathbf{A}(f)\mathbf{S}(f), \tag{13.3}$$

where $\mathbf{X}(f) = [x_i(n,f)]_{in}$, $\mathbf{S}(f) = [s_j(n,f)]_{jn}$, and $\mathbf{A}(f) = [a_{ij}(f)]_{ij}$, or in a vector form for frame n

$$\mathbf{x}(n,f) = \mathbf{A}(f)\mathbf{s}(n,f) \tag{13.4}$$

with $\mathbf{x}(n,f) = [x_1(n,f), \ldots, x_I(n,f)]^T$ and $\mathbf{s}(n,f) = [s_1(n,f), \ldots, s_J(n,f)]^T$.

$\mathbf{X}(f)$ corresponds to the fth horizontal slice of $\mathcal{X}$; see the upper right part in Figure 13.1.

---

1 The conditions under which the approximation is exact are specified in Section 2.3.2.

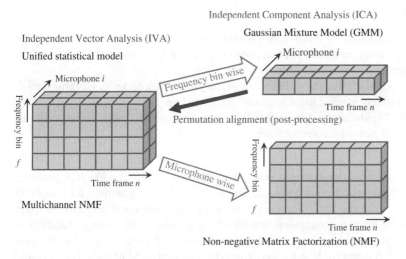

Figure 13.1 Multichannel time-frequency representation of an observed signal (left) and its slices: frequency-wise and microphone-wise (right). Methods discussed in this chapter are shown in red.

13.2 Frequency-Domain Independent Component Analysis

FD-ICA is a source separation method that employs ICA to separate the convolutive audio mixtures in the STFT domain (Smaragdis, 1998; Parra and Spence, 2000; Anemüller and Kollmeier, 2000; Murata *et al.*, 2001; Schobben and Sommen, 2002; Sawada *et al.*, 2003; Asano *et al.*, 2003; Saruwatari *et al.*, 2003; Mitianoudis and Davies, 2003; Nesta *et al.*, 2011). The basic principle is that ICA is applied separately to each $\mathbf{X}(f)$, where the goal is to obtain matrices $\widehat{\mathbf{S}}(f) = [\hat{s}_j(n,f)]_{jn}, f \in \{0, \ldots, F-1\}$ such that their rows contain frequency components of individual signals $\hat{s}_j(n,f)$. The separation proceeds by multiplying $\mathbf{X}(f)$ by a $J \times I$ separating matrix $\mathbf{W}^H(f)$, so

$$\widehat{\mathbf{S}}(f) = \mathbf{W}^H(f) \, \mathbf{X}(f), \qquad \text{or} \qquad \hat{\mathbf{s}}(n,f) = \mathbf{W}^H(f)\mathbf{x}(n,f). \tag{13.5}$$

The matrix $\mathbf{W}^H(f)$ is said to be separating whenever $\widehat{\mathbf{S}}(f)$ is equal to $\mathbf{S}(f)$ up to the order and scales of its rows, or, equivalently, the separated signals $\hat{\mathbf{s}}(n,f)$ are equal to the source signals $\mathbf{s}(n,f)$ up to their order and scales.

The fact that the rows of $\widehat{\mathbf{S}}(f)$ can have an order different from those of $\mathbf{S}(f)$, and that they can be multiplied by ordinary nonzero scaling factors, gives rise to indeterminacies that are inherent to the FD-ICA. To obtain the complete separated signal, the corresponding rows of $\widehat{\mathbf{S}}(0), \ldots, \widehat{\mathbf{S}}(F-1)$ have to be collected to form the STFT matrix, which is then transformed back to the time domain (see Figure 13.6). This is called the permutation problem and is addressed in Section 13.2.9. The scaling ambiguity means that the spectra of the separated signals are modified by a random filter (each frequency is multiplied by a random number). Methods to cope with this problem are described in Section 13.2.8.

13.2.1 ICA Principle

ICA is a core mechanism of many BSS methods, one of which is FD-ICA. During the following three subsections, we put the audio separation problem and FD-ICA aside, and focus on the basic principles of ICA.

ICA became popular after the publication of the pioneering paper of Comon (1994). It has been intensively studied for at least two decades (Lee, 1998; Haykin, 2000; Hyvärinen *et al.*, 2001; Cichocki and Amari, 2002; Comon and Jutten, 2010). To begin, let us consider a general instantaneous mixture

$$\mathbf{X} = \mathbf{AS}. \tag{13.6}$$

This mixture represents, for example, one of the mixtures within the FD-ICA problem (13.3). In the following and up to and including Section 13.2.6, we assume for simplicity that this mixture is in the time domain and we adopt the corresponding time-domain notation. We will limit our considerations to the case when the mixing matrix $\mathbf{A}$ is square and nonsingular, which means that the determined problem $I = J$ is studied. There are many extensions of ICA for underdetermined mixtures but these issues would go beyond the focus of this chapter (see, for example, Ferreol *et al.* (2005)).

The fundamental assumption in ICA is that the source signals represented by the rows of $\mathbf{S}$ are mutually independent. The problem to separate $\mathbf{S}$ from $\mathbf{X}$ is therefore formulated as finding an $I \times I$ matrix $\mathbf{W}$ such that $\hat{\mathbf{S}} = \mathbf{W}^H \mathbf{X}$ are independent. The rows of $\hat{\mathbf{S}}$ are referred to as independent components of $\mathbf{X}$. If $\mathbf{W}$ is such that

$$\hat{\mathbf{S}} = \mathbf{W}^H \mathbf{X} = \mathbf{\Pi} \mathbf{\Lambda} \mathbf{S}, \tag{13.7}$$

where $\mathbf{\Pi}$ and $\mathbf{\Lambda}$ denote, respectively, a permutation and a diagonal matrix with nonzero entries on its diagonal, then $\mathbf{W}^H$ is a *demixing* matrix. $\mathbf{\Pi}$ and $\mathbf{\Lambda}$ represent indeterminacies: the order and scales of rows of $\hat{\mathbf{S}}$ can be arbitrary because these circumstances have no influence on their independence.

The key question here is whether ICA always separates the signals. If $\mathbf{W}^H$ yields independent components, does it immediately follow that $\mathbf{W}^H$ is demixing? The strong property of ICA is that many theories and models give the positive answer to this question, which makes ICA highly attractive. The models of signal independence lead to concrete solutions of ICA. The following section describes the most popular model.

13.2.2 Nongaussianity-Based Separation

Assume that each source signal $s_j(t)$ is a sequence of independently and identically distributed (i.i.d.) samples drawn from a distribution $p(s_j)$ with zero mean and finite variance. We assume that the distributions are *nongaussian* or at most one of them is Gaussian. The purpose of this condition will become clear later.

Now it is easy to put the assumption of mutual independence of the source signals in concrete terms by building the joint probability distribution of the source signals as

$$p(\mathbf{s}) = \prod_{j=1}^{J} p(s_j), \tag{13.8}$$

with $\mathbf{s} = [s_1, \ldots, s_J]^T$. This actually means that the mathematical model of signal independence is, here, the stochastic independence of the corresponding random processes (Papoulis and Pillai, 2002).

Using (13.8) and standard rules for transforming probability distributions, the joint probability distribution of the mixed signals $\mathbf{x} = \mathbf{As}$ is

$$p(\mathbf{x}) = |\det \mathbf{A}|^{-1} \prod_{j=1}^{J} p(s_j), \tag{13.9}$$

Similarly, when $\hat{\mathbf{s}} = \mathbf{W}^H \mathbf{x}$ where $\mathbf{W}^H$ is a regular transform matrix, the probability distribution of the transformed signals is

$$p(\hat{\mathbf{s}}) = |\det \mathbf{W}^H \mathbf{A}|^{-1} \prod_{j=1}^{J} p((\mathbf{A}^{-1}(\mathbf{W}^H)^{-1}\hat{\mathbf{s}})_j). \tag{13.10}$$

Let the empirical marginal distribution of the jth transformed signal $\hat{s}_j$ be denoted by $\hat{p}(\hat{s}_j)$. To quantify the level of independence of the transformed signals, we can define an objective function that measures the discrepancy between the joint distribution $\hat{p}(\hat{\mathbf{s}})$ and the product of marginal distributions $\prod_{j=1}^{J} \hat{p}(\hat{s}_j)$.

An appropriate criterion borrowed from information theory (Cover and Thomas, 2006) is the Kullback–Leibler (KL) divergence between the probability distributions, which, in this special case, coincides with the so-called *mutual information* of the transformed signals $\hat{s}_1, \dots, \hat{s}_j$. It is defined as

$$\mathbb{I}\{\hat{s}_1, \dots, \hat{s}_j\} = \int \hat{p}(\hat{\mathbf{s}}) \log \frac{\hat{p}(\hat{\mathbf{s}})}{\prod_{j=1}^{J} \hat{p}(\hat{s}_j)} \, d\hat{\mathbf{s}}. \tag{13.11}$$

The key property is that $\mathbb{I}\{\hat{s}_1, \dots, \hat{s}_j\} \geq 0$ and that $\mathbb{I}\{\hat{s}_1, \dots, \hat{s}_j\} = 0$ if and only if $\hat{s}_1, \dots, \hat{s}_j$ are independent.

Mutual information has nice algebraic properties. It can be written as

$$\mathbb{I}\{\hat{s}_1, \dots, \hat{s}_j\} = \sum_{j=1}^{J} \mathbb{H}\{\hat{s}_j\} - \mathbb{H}\{\hat{s}_1, \dots, \hat{s}_j\}, \tag{13.12}$$

where

$$\mathbb{H}\{\hat{s}_j\} = -\int \hat{p}(\hat{s}_j) \log \hat{p}(\hat{s}_j) \, d\hat{s}_j = -\mathbb{E}_{\hat{p}(\hat{s}_j)}\{\log \hat{p}(\hat{s}_j)\} \tag{13.13}$$

is the *entropy* of $\hat{s}_j$. $\mathbb{H}\{\hat{s}_1, \dots, \hat{s}_j\}$ denotes the joint entropy of $\hat{s}_1, \dots, \hat{s}_j$, which is defined analogously to (13.13) (Cover and Thomas, 2006). Under the so-called orthogonal constraint, the joint entropy is independent with respect to the transform matrix $\mathbf{W}^H$, and then mutual information gets the form

$$\mathbb{I}\{\hat{s}_1, \dots, \hat{s}_j\} = \sum_{j=1}^{J} \mathbb{H}\{\hat{s}_j\} + \text{cst.} \tag{13.14}$$

The *orthogonal constraint* requires that the observed signals $\mathbf{x}$ are uncorrelated and normalized (all unit variance) and that the transform matrix $\mathbf{W}^H$ is orthogonal, i.e. $\mathbf{W}^H \mathbf{W} = \mathbf{I}_J$ with $\mathbf{I}_J$ the $J \times J$ identity matrix. This does not mean any practical restriction because the observed signals $\mathbf{x}$ can be transformed by a regular matrix so that they become uncorrelated and normalized. For example, *principal component analysis* can be used to perform this step (see Section 13.2.5.2 below).

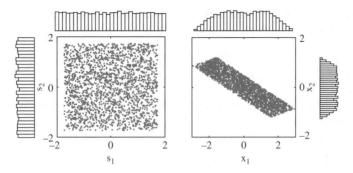

Figure 13.2 The scatter plot on the left-hand side illustrates the joint probability distribution of two normalized and uniformly distributed signals. The histograms correspond to marginal distributions of the two variables. The left-hand side corresponds to the same signals when mixed by matrix $\mathbf{A} = \begin{pmatrix} 1.3 & 0.4 \\ -0.6 & 0.1 \end{pmatrix}$. In the latter (mixed) case, the histograms are closer to a Gaussian.

Let $\mathbf{y}$ denote the decorrelated signals. These could be considered newly observed ones because the mixing matrix is unknown anyway. Using the fact that independent signals are also uncorrelated, it follows that the demixing matrix to be applied to $\mathbf{y}$ for yielding its independent components must be orthogonal. ICA can thus be solved through finding $\mathbf{W}^H$ that minimizes $\sum_{j=1}^{J} \mathbb{H}\{\hat{s}_j\}$ under the orthogonal constraint.

The latter formulation of ICA has nice interpretation through the physical meaning of entropy as a measure of "uncertainty". The Gaussian probability distribution has the highest entropy among all distributions under a unit variance constraint (Cover and Thomas, 2006). The goal to minimize the entropy of each output signal thus means to make them all as nongaussian as possible. While the mixing of random variables increases their "uncertainty", the separation should do the very opposite. This principle can also be interpreted through a generalized formulation of the central limit theorem, which, roughly speaking, says that the distribution of the average of J random variables tends to be Gaussian when J approaches infinity. This phenomenon is illustrated in Figure 13.2, where two uniformly distributed signals are mixed together, which turns their marginal distributions towards Gaussian (Cardoso, 1998).

It is worth recalling here the assumption from the beginning of this subsection: at most one source signal can be Gaussian distributed. Consider the case when there are two mixed Gaussian signals. These signals are independent if and only if they are uncorrelated. Only Gaussian signals have such a property. Uncorrelated Gaussian signals can be mixed by an arbitrary orthogonal matrix while they remain independent. The assumption that at most one of the signals is allowed to be Gaussian thus ensures the separability of the mixture (Eriksson and Koivunen, 2004).

13.2.3 Modeling the Signal Probability Distributions

Entropy is a function of the probability distribution of each source, which is unknown in the blind scenario. One way is to estimate the distribution from available samples of the signal in a parametric or a nonparametric way. It is also possible to assume a nongaussian distribution, though significantly different from the true one. The conditions under which such approaches still converge to the correct solution have been studied in

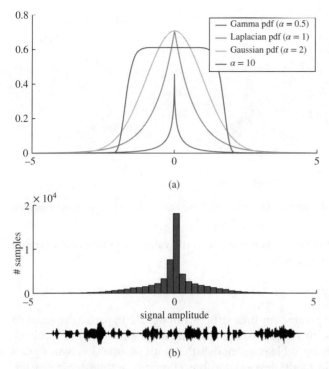

(a)

(b)

Figure 13.3 (a) Examples of generalized Gaussian distributions for $\alpha = 0.5$ (gamma), $\alpha = 1$ (Laplacian), $\alpha = 2$ (Gaussian), and $\alpha = 10$. The latter case demonstrates the fact that for $\alpha \to +\infty$ the distribution is uniform. (b) Histogram of a female utterance, which can be modeled as the Laplacian or gamma distribution.

many works, e.g. Pham and Garat (1997) and Koldovský *et al.* (2006). Unless the selected distribution is inappropriate for the given signal, it influences more the separation accuracy than the convergence itself. For example, time-domain samples of speech are well modeled as Laplacian random variables (Gazor and Zhang, 2003), see Figure 13.3.

Figure 13.3a shows four examples of distributions that belong to the *generalized Gaussian* family parameterized by one parameter $\alpha > 0$ (Varanasi and Aazhang, 1989; Kokkinakis and Nandi, 2005). For zero mean and variance one, the distribution is defined as

$$p(s_j) = \frac{\alpha \beta_\alpha}{2\Gamma(1/\alpha)}\, e^{-(\beta|s_j|)^\alpha}, \qquad \beta = \sqrt{\frac{\Gamma(3/\alpha)}{\Gamma(1/\alpha)}}, \tag{13.15}$$

where α controls the rate of decay, and $\Gamma(\cdot)$ is the gamma function. This class encompasses the ordinary Gaussian distribution for $\alpha = 2$. For $\alpha \neq 2$, the distributions are nongaussian, in particular the Laplacian distribution for $\alpha = 1$ and the uniform distribution in the limit $\alpha \to +\infty$.

An example of a complex-valued distribution that is far from Gaussian is

$$p(s_j) \propto \exp\left(-\sqrt{|s_j|^2 + \gamma}\right) \tag{13.16}$$

where γ is a parameter that is useful for obtaining the second-order derivative of $\log p(s_j)$ if $\gamma > 0$. Figure 13.4 shows that the source model (13.16) is more peaky than Gaussian

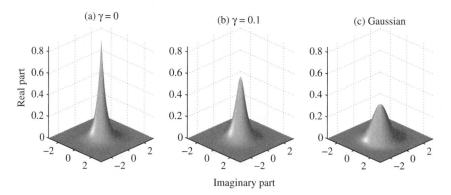

Figure 13.4 Complex-valued source models. (a) and (b) are based on (13.16). (c) is a complex Gaussian distribution.

at the origin, meaning that the source is more sparse (its values are more concentrated around the origin) than Gaussian noise.

13.2.4 Alternative Models

The previous model does not assume any time structure of signals. In particular, possible dependencies between adjacent samples of signals or their dynamics are not involved. In this subsection, we briefly introduce two models that both assume Gaussian signals but, compared to the previous model, involve the nonstationarity or time dependencies. More advanced hybrid models are mentioned afterwards.

13.2.4.1 Nonstationarity

The first model embodies the *nonstationarity*. Each signal is assumed to be a sequence of independent Gaussian random variables whose variances depend on time (Matsuoka *et al.*, 1995; Pham and Cardoso, 2001). For practical reasons, piecewise stationarity is assumed (the variance is constant over blocks of signals).

Specifically, assume that the signal matrices of samples $\mathbf{S}$ and $\mathbf{X}$ are partitioned into N nonoverlapping blocks of the same length T:

$$\mathbf{S} = [\mathbf{S}(0), \dots, \mathbf{S}(N-1)], \qquad \mathbf{X} = [\mathbf{X}(0), \dots, \mathbf{X}(N-1)]. \tag{13.17}$$

The model assumes that the jth row of $\mathbf{S}(n)$, $n \in \{0, \dots, N-1\}$, $j \in \{1, \dots, J\}$, which is the nth block of the jth source signal, is a Gaussian distributed i.i.d. sequence with zero mean and variance $\sigma_{s_j}^2(n)$.

The mixing model $\mathbf{X} = \mathbf{A}\,\mathbf{S}$ holds within each block, i.e. $\mathbf{X}(n) = \mathbf{A}\,\mathbf{S}(n)$, which points to the important fact that $\mathbf{A}$ is constant over the blocks. The covariance matrix in the nth block of the observed data is

$$\Sigma_{\mathbf{x}}(n) = \frac{1}{T} \sum_{t=0}^{T-1} \mathbf{x}(t + nT)\mathbf{x}^H(t + nT) \tag{13.18}$$

$$= \mathbf{A}\left(\frac{1}{T} \sum_{t=0}^{T-1} \mathbf{s}(t + nT)\mathbf{s}^H(t + nT)\right)\mathbf{A}^H \tag{13.19}$$

$$= \mathbf{A}\,\mathrm{Diag}(\sigma_{s_1}^2(n), \dots, \sigma_{s_J}^2(n))\mathbf{A}^H. \tag{13.20}$$

The last equality follows from the fact that the source signals are independent, so their covariance matrix is diagonal. It also means that the mixing matrix jointly diagonalizes the covariance matrices $\mathbf{\Sigma}_{\mathbf{x}}(0), \dots, \mathbf{\Sigma}_{\mathbf{x}}(N-1)$.

Hence, the demixing matrix $\mathbf{W}^H = \mathbf{A}^{-1}$ can be found as a matrix that provides an approximate joint diagonalization of the sample covariance matrices $\hat{\mathbf{\Sigma}}_{\mathbf{x}}(n)$, $n \in \{0, \dots, N-1\}$. In other words, $\mathbf{W}$ should have the property that the matrices $\mathbf{W}^H\hat{\mathbf{\Sigma}}_{\mathbf{x}}(n)\mathbf{W}$, $n \in \{0, \dots, N-1\}$, are all approximately diagonal. The approximate joint diagonalization of a set of matrices is a well-studied problem that has to be solved numerically by optimizing an appropriate criterion (see, for example, Cardoso and Souloumiac (1996), Theis and Inouye (2006), and Tichavský and Yeredor (2009)).

13.2.4.2 Nonwhiteness

The second alternative model assumes that each source signal is a weak stationary Gaussian process. It involves possible *nonwhiteness* of the signal spectrum as well as dependencies of adjacent samples, which was not considered by the previous models.

The *time-lagged covariance matrices* of the observed signals are defined as

$$\mathbf{\Sigma}_{\mathbf{x}}(\tau) = \frac{1}{T - |\tau|} \sum_t \mathbf{x}(t)\mathbf{x}^H(t - \tau) \tag{13.21}$$

$$= \mathbf{A} \left(\frac{1}{T - |\tau|} \sum_t \mathbf{s}(t)\mathbf{s}^H(t - \tau) \right) \mathbf{A}^H \tag{13.22}$$

$$= \mathbf{A}\mathbf{\Sigma}_{\mathbf{s}}(\tau)\mathbf{A}^H, \tag{13.23}$$

where τ is the time lag and $\mathbf{\Sigma}_{\mathbf{s}}(\tau)$ is time-lagged covariance matrix of the source signals. $\mathbf{\Sigma}_{\mathbf{s}}(\tau)$ is diagonal due to the signals' independence and, in general, it is not zero for $\tau \neq 0$ due to the possible sample dependencies (Molgedey and Schuster, 1994).

Similarly to the previous model, the problem to estimate the demixing matrix $\mathbf{W}^H$ here leads to the approximate joint diagonalization of the sample-based estimates of (13.21). A popular algorithm based on this model, especially in biomedical applications, is known under the name of SOBI (Belouchrani *et al.*, 1997). A statistically optimum algorithm derived from autoregressive (AR) modeling of the source signals is WASOBI (Tichavský and Yeredor, 2009).

13.2.4.3 Hybrid Models

Real-world signals, such as speech, embody various features, therefore more advanced methods aim to capture several signal diversities by a joint model (Adalı *et al.*, 2014). Hybrid models combine two or three of the aforementioned signal diversities: the nongaussianity, nonstationarity, and nonwhiteness. Since speech signals exhibit all of these diversities, hybrid models can be exploited for separating them in a more flexible way.

A common approach is to assume that the signals are piecewise stationary. This enables us to involve the nonstationarity and to combine it either with the nonwhiteness or with the nongaussianity. In the former case, the signal is assumed to be Gaussian and stationary within each block. For example, the BARBI algorithm (Tichavský *et al.*, 2009) assumes block-wise stationary AR processes. The second combination assumes that each signal is a nongaussian i.i.d. sequence whose distribution is block-dependent (see, for example, Koldovský *et al.* (2009)). MULTICOMBI runs two different BSS algorithms EFICA and WASOBI and combines their results in order to separate signals that are simultaneously nongaussian and nonwhite (Tichavský and Koldovský, 2011).

TRINICON, already mentioned in Chapter 10, is a framework based on which several BSS algorithms were proposed for separating audio signals (Buchner *et al.*, 2005). These methods exploit all three signal diversities by using a general information-theoretic approach (Kellermann *et al.*, 2006).

13.2.5 ICA Algorithms

13.2.5.1 Natural Gradient

The *natural gradient* algorithm is a nongaussianity model-based ICA method (Bell and Sejnowski, 1995; Amari *et al.*, 1996; Anemüller *et al.*, 2003) for maximum likelihood (ML) estimation of the demixing matrix $\mathbf{W}^H$. Assuming that the separated signals $\hat{s}_j(t)$ are independent and distributed as $p(s_j)$, the normalized log-likelihood is (Papoulis and Pillai, 2002)

$$\mathcal{M}^{\mathrm{ML}}(\mathbf{W}) = \frac{1}{T} \sum_t \log p(\mathbf{x}(t)) \tag{13.24}$$

$$= 2 \log |\det \mathbf{W}^H| + \sum_{j=1}^{J} \frac{1}{T} \sum_t \log p(\hat{s}_j(t)). \tag{13.25}$$

For $T \to +\infty$, the sample mean may be replaced by the expectation operator over the empirical sample distribution:

$$\mathcal{M}^{\mathrm{ML}}(\mathbf{W}) = 2 \log |\det \mathbf{W}^H| + \sum_{j=1}^{J} \mathbb{E}_{\hat{p}(\hat{s}_j)} \{ \log p(\hat{s}_j) \}. \tag{13.26}$$

It is worth comparing $\mathcal{M}^{\mathrm{ML}}(\mathbf{W})$ with (13.14), which points to the relation between the theory of Section 13.2.2 and the ML estimate that is considered here. Under the orthogonal constraint, the first term in (13.26) is zero. The second term is equal to $-\mathbb{H}\{\hat{s}_j\}$ provided that the distributions $p(s_j)$ and $\hat{p}(\hat{s}_j)$ are equal. This is satisfied when $\mathbf{W}^H$ is the exact separating matrix because then $\hat{s}_j = s_j$. The objective functions thus share the same optimum point.

The maximization of (13.26) through the classical gradient approach is performed by iterative updates with a small stepsize parameter η, $\mathbf{W}^H \leftarrow \mathbf{W}^H + \eta \cdot \frac{\partial \mathcal{M}^{\mathrm{ML}}}{\partial (\mathbf{W}^H)^*}$, where the gradient of (13.26) is

$$\frac{\partial \mathcal{M}^{\mathrm{ML}}}{\partial (\mathbf{W}^H)^*} = \mathbf{W}^{-1} - \mathbb{E}\{\Phi(\hat{s})\mathbf{x}^H\} \tag{13.27}$$

with so-called *score functions*

$$\Phi(\hat{s}) = [\Phi(\hat{s}_1), \dots, \Phi(\hat{s}_J)]^T, \quad \Phi(\hat{s}_j) = -\frac{\partial \log p(\hat{s}_j)}{\partial \hat{s}_j^*}. \tag{13.28}$$

Since the distribution $p(s_j)$ is not known, it is replaced by an appropriate model. For example, using the distribution in (13.16) yields the choice

$$\Phi(\hat{s}_j) = \frac{\hat{s}_j}{2\sqrt{|\hat{s}_j|^2 + \gamma}}. \tag{13.29}$$

The drawback of the gradient (13.27) is that it involves matrix inversion, which is computationally demanding. Therefore, the natural gradient proposed by Amari *et al.* (1996) and Cichocki and Amari (2002)

$$\frac{\partial \mathcal{M}^{\mathrm{ML}}}{\partial (\mathbf{W}^H)^*} \mathbf{W}\mathbf{W}^H = (\mathbf{I}_J - \mathbb{E}\{\Phi(\hat{\mathbf{s}})\hat{\mathbf{s}}^H\})\mathbf{W}^H \tag{13.30}$$

is commonly used instead. Finally, the natural gradient algorithm iterates as

$$\mathbf{W}^H \leftarrow \mathbf{W}^H + \eta(\mathbf{I}_J - \mathbb{E}\{\Phi(\hat{\mathbf{s}})\hat{\mathbf{s}}^H\})\mathbf{W}^H. \tag{13.31}$$

This algorithm is popular for its simplicity. It can be modified for adaptive (real-time) processing by removing the expectation operator from (13.31), by which the stochastic gradient method is obtained (Cardoso and Laheld, 1996).

13.2.5.2 FastICA

FastICA is a popular nongaussianity-based fixed-point algorithm first proposed by Hyvärinen (1999). It is based on the optimization of the objective function (13.14) under the orthogonal constraint, which is ensured by a preprocessing step. It consists of decorrelating and normalizing the observed signals so that their sample covariance is the identity matrix. Specifically, the preprocessed signals are

$$\mathbf{Y} = \hat{\boldsymbol{\Sigma}}_{\mathbf{x}}^{-1/2} \, \mathbf{X} \tag{13.32}$$

where $\hat{\boldsymbol{\Sigma}}_{\mathbf{x}}^{1/2}$ is a matrix that satisfies $\hat{\boldsymbol{\Sigma}}_{\mathbf{x}}^{1/2}\hat{\boldsymbol{\Sigma}}_{\mathbf{x}}^{1/2} = \hat{\boldsymbol{\Sigma}}_{\mathbf{x}}$. The way to compute such a matrix is by applying the eigenvalue decomposition to the empirical covariance $\hat{\boldsymbol{\Sigma}}_{\mathbf{x}} = \frac{1}{T}\mathbf{X}\mathbf{X}^H$, which is symmetric and positive definite, so its eigenvalues are positive and its eigenvectors are orthogonal. The eigenvalue decomposition gives $\hat{\boldsymbol{\Sigma}}_{\mathbf{x}} = \mathbf{V}\mathrm{Diag}(\lambda_1, \dots, \lambda_J)\mathbf{V}^H$ where $\mathbf{V}\mathbf{V}^H = \mathbf{I}_J$ and $\lambda_1, \dots, \lambda_J$ are the eigenvalues of $\hat{\boldsymbol{\Sigma}}_{\mathbf{x}}$. Then,

$$\hat{\boldsymbol{\Sigma}}_{\mathbf{x}}^{1/2} = \mathbf{V}\,\mathrm{Diag}(\lambda_1^{1/2}, \dots, \lambda_J^{1/2})\mathbf{V}^H. \tag{13.33}$$

Note that $\mathbf{Y}$ could be multiplied by an arbitrary unitary matrix while still being orthogonal and normalized. The separating transform can therefore be searched through finding an appropriate unitary matrix $\mathbf{U}$ such that $\mathbf{U}\mathbf{Y}$ are as independent as possible. The separating matrix $\mathbf{W}^H$ is then obtained as $\mathbf{W}^H = \mathbf{U}\hat{\boldsymbol{\Sigma}}_{\mathbf{x}}^{-1/2}$.

When the unknown signal distributions are replaced by their modeled counterparts and the expectation in (13.14) is replaced by the sample average, the following contrast function is obtained

$$C^{\mathrm{MI}}(\mathbf{U}) = \sum_{j=1}^{J} \frac{1}{T} \sum_{t} g(\mathbf{u}_j \mathbf{y}(t)), \tag{13.34}$$

where $\mathbf{u}_j$ is the jth row of $\mathbf{U}$, and $g(\cdot)$ is a suitable nonlinear and nonquadratic function, which is ideally equal to $-\log p(s_j)$ (Cardoso, 1998).

One-Unit FastICA The rows of $\mathbf{U}$ are, as variables, separated within $C^{\mathrm{MI}}(\mathbf{U})$, so they can be optimized separately. The one-unit algorithm estimates one row of $\mathbf{U}$ as a vector $\mathbf{u}$ under the constraint $\|\mathbf{u}\|_2 = 1$. Starting with an initial guess, the algorithm iterates

$$\mathbf{u} \leftarrow \mathbf{Y}g'(\mathbf{Y}^H\mathbf{u}) - \mathbf{u}\,g''(\mathbf{u}^H\mathbf{Y})\mathbf{1}_T \tag{13.35}$$

$$\mathbf{u} \leftarrow \mathbf{u}/\|\mathbf{u}\|_2 \tag{13.36}$$

until convergence is achieved. Here, $g'(\cdot)$ and $g''(\cdot)$ denote the first and second derivatives of $g(\cdot)$ and are applied elementwise; $\mathbf{1}_T$ is a vector of ones of size $T \times 1$. In general, it is not known in advance which row of $\mathbf{U}$ is estimated (corresponds to $\mathbf{u}$); this is influenced by the initialization.

Deflation and Symmetric FastICA In order to separate all signals, there are two variants of FastICA: deflation and symmetric. Both algorithms enforce the orthogonality constraint. The *deflation* approach estimates $\mathbf{U}$ row by row where every estimated row must be orthogonal to the previous ones. This is achieved by subtracting the projection of the current row onto the subspace that is spanned by the previous rows after each one-unit iteration. Once convergence for the given row is achieved, the next row is estimated.

The symmetric algorithm estimates all rows of $\mathbf{U}$ in parallel. One-unit iterations that are performed with each row are followed by the symmetric orthonormalization

$$\mathbf{U} \leftarrow (\mathbf{U}\mathbf{U}^H)^{-1/2}\mathbf{U}. \tag{13.37}$$

The process is repeated until convergence is achieved. For a more advanced version of FastICA see, for example, Koldovský *et al.* (2006) and Koldovský and Tichavský (2015).

13.2.5.3 JADE

Joint cumulants of random variables are higher-order statistics with appealing properties (Papoulis and Pillai, 2002). In particular, the cumulants are multilinear. A cumulant of a set of random variables is equal to zero if these random variables can be divided into two independent nonempty sets. By fixing all but two parameters of a cumulant, a cumulant matrix is defined. It is possible to select cumulant matrices such that the demixing transform can be sought through their joint approximate diagonalization.

JADE was proposed by Cardoso (1993) and is based on the approximate joint diagonalization of cumulant matrices involving all cumulants of order two and four. The joint diagonalization algorithm utilizes Jacobi rotations under the orthogonal constraint.

13.2.6 A Comparative Experiment

The accompanying web page contains a simple experiment where 15 different signals are artificially mixed and separated. The source signals are artificial random sequences obeying the aforementioned models and speech signals. There are five i.i.d. signals generated from the generalized Gaussian distribution with the shape parameter α, respectively, equal to 0.1, 0.5 (gamma), 1 (Laplacian), 2 (Gaussian), and $+\infty$ (uniform). Three other sources are nonwhite Gaussian AR stationary sources. These signals are generated as Gaussian i.i.d. sequences filtered by all-pole filters, respectively, with coefficients [1, 0.5], [1, −0.5], and [1, 0.9]. Next, there are three piecewise Gaussian i.i.d. sources, each consisting of 10 blocks. The variances of these signals on the blocks are randomly generated from the uniform distribution on [0, 1].

The signals are normalized to zero mean and unit variance. Then they are mixed by a random matrix and separated by selected algorithms. The separation is evaluated in terms of the signal-to-interference ratio (SIR) after the original order of signals is restored.

Figure 13.5 shows SIR by two different algorithms, EFICA (Koldovský *et al.*, 2006) and BARBI (Tichavský *et al.*, 2009), achieved in a typical trial. While EFICA assumes

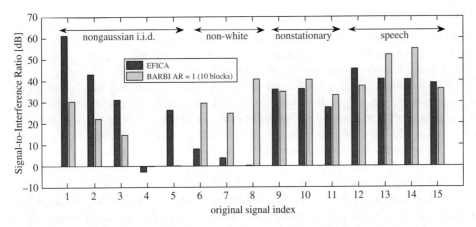

Figure 13.5 A single-trial SIR achieved by EFICA and BARBI when separating 15 different signals (five i.i.d. sequences, three nonwhite AR Gaussian processes, three piecewise Gaussian i.i.d. processes (nonstationary) and four speech signals).

the nongaussianity-based model, BARBI is based on the hybrid model of signals that assumes the AR Gaussian piecewise stationary processes (AR of order 1, 10 blocks). The resulting SIRs of the separated signals correspond with the assumed signal properties. EFICA cannot separate the fourth signal because it is Gaussian as well as the signals 6 through 8. The other signals are nongaussian or behave like nongaussians, so EFICA separates them with good SIRs. The best SIR is achieved with the nongaussian i.i.d. signals (1 through 3 and 5). By contrast, BARBI exploits nonwhiteness and nonstationarity. It thus achieves high SIRs on signals 6 through 15 while low SIRs with the i.i.d. sequences 1 through 5.

13.2.7 Required Post-Processing

We now go back to ICA performed in the frequency-domain for separating speech and audio convolutive mixtures. In FD-ICA, ICA has been performed in a frequency binwise manner, so we need to perform some post-processing to construct proper separated signals in the time domain. As shown in Figure 13.6, the post-processing includes permutation and scaling alignments. The next two subsections discuss these two steps. Since scale-aligned separated signals are used to perform permutation alignment, we first discuss the scaling alignment.

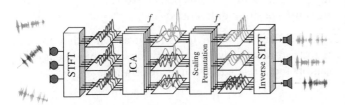

Figure 13.6 Flow of FD-ICA for separating convolutive mixtures.

13.2.8 Scaling Ambiguity

There is *scaling ambiguity* in an ICA solution, as we have described in Section 13.2.1. For each frequency f, a separated source must be multiplied by a frequency-dependent scaling factor in order to reconstruct the spectrum of the separated source.

When no additional knowledge of the signal spectrum is available, the only properly defined spectrum is that of the spatial image of the source on microphones. Therefore, the scaling ambiguity is resolved by reconstructing the separated signals as they are observed at microphones under the constraint that the sum of all rescaled separated signals is equal to the microphone signal (Murata *et al.*, 2001; Matsuoka and Nakashima, 2001; Takatani *et al.*, 2004).

The rescaling proceeds as follows. For the separation matrix $\widehat{\mathbf{W}}^H(f)$, which was obtained by applying ICA, we calculate the inverse transpose matrix $\widehat{\mathbf{A}}(f) = [\widehat{\mathbf{a}}_1(f), \dots, \widehat{\mathbf{a}}_J(f)] = (\widehat{\mathbf{W}}^H(f))^{-1}$ or, alternatively, its Moore–Penrose pseudo-inverse. Then, according to (13.5), it holds that

$$\widehat{\mathbf{A}}(f)\widehat{\mathbf{s}}(n,f) = \sum_{j=1}^{J} \widehat{\mathbf{a}}_j(f)\widehat{s}_j(n,f) = \mathbf{x}(n,f). \tag{13.38}$$

Hence, the complete spatial image of $\widehat{s}_j(n,f)$ on the microphones is the vector output

$$\widehat{\mathbf{c}}_j(n,f) = \widehat{\mathbf{a}}_j(f)\widehat{s}_j(n,f). \tag{13.39}$$

If we focus on a specific reference microphone i, we have a scaled separated signal $\widehat{a}_{ij}(f)\widehat{s}_j(n,f)$ on the reference microphone.

(13.38) is also important for the interpretation of the mixing process. $\widehat{\mathbf{A}}(f)$ can be regarded as an estimated mixing matrix and thus the vectors $\widehat{\mathbf{a}}_j(f), j \in \{1, \dots, J\}$, can be regarded as estimated mixing vectors. They are used for estimating time differences of arrival (TDOAs), as will be explained in Section 13.2.9.2.

13.2.9 Permutation Problem

The permutation correction must be done so that each separated signal in the time domain contains frequency components from the same source signal (Sawada *et al.*, 2004). Numerous approaches to this *permutation problem* have been proposed (Smaragdis, 1998; Parra and Spence, 2000; Schobben and Sommen, 2002; Buchner *et al.*, 2004; Kurita *et al.*, 2000; Saruwatari *et al.*, 2003; Ikram and Morgan, 2002; Sawada *et al.*, 2004, 2007b; Mukai *et al.*, 2004b,a; Soon *et al.*, 1993; Murata *et al.*, 2001; Anemüller and Kollmeier, 2000). Here, we discuss two methods:

- activity sequence clustering (utilizing source signal characteristics);
- TDOA clustering (utilizing source position differences).

The former method is effective for sources whose time structures (when they are loud or silent) are clearly different and is robust to reverberations. The latter method is effective in a low reverberant situation and is related to source localization.

13.2.9.1 Activity Sequence Clustering

This method utilizes source signal characteristics such as whether the given source is active or silent. The simplest measure for such activities is the magnitude spectrum $|\hat{s}_j(n,f)|$ of a separated signal. Considering that the time frame index n varies, $|\hat{s}_j(f)| = [|\hat{s}_j(n,f)|]_n$ represents the signal activity of the jth separated signal at frequency bin f. Magnitude spectra exhibit high correlation coefficients between neighboring frequencies. Thus, the correlation coefficient $\rho(|\hat{s}_j(f)|, |\hat{s}_{j'}(f+1)|)$ is high if the jth separation at bin f and the j'th separation at bin $f+1$ belong to the same source. However, if two frequencies f and f' are far apart, the correlation coefficient $\rho(|\hat{s}_j(f)|, |\hat{s}_j(f')|)$ is usually not high even for the same source. The left-hand side in Figure 13.7 shows such a situation.

Another measure for source activities is the power ratio (Sawada *et al.*, 2007a)

$$\varrho_j(n,f) = \frac{\|\hat{c}_j(n,f)\|_2^2}{\sum_{j=1}^{J} \|\hat{c}_j(n,f)\|_2^2} \tag{13.40}$$

of the rescaled separated signals using (13.39). As shown in the right-hand side of Figure 13.7, power ratios exhibit a high correlation coefficient for the same source even if the two frequencies are far apart. More notably, power ratios exhibit a negative correlation coefficient for different sources. This is caused by the exclusiveness of the power ratio: if one source is active and dominates a time-frequency slot (n,f), then the power ratios of the other sources are close to zero. The concept of power ratio has been

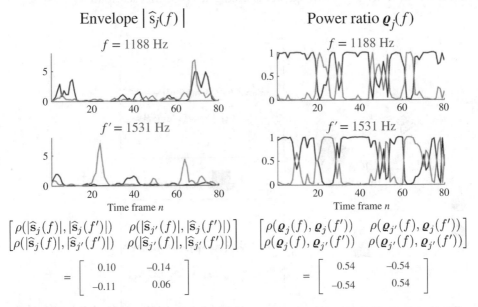

Figure 13.7 Comparison of two criteria for permutation alignment: magnitude spectrum and power ratio. Permutations are aligned as each color (blue or green) corresponds to the same source. Power ratios generally exhibit a higher correlation coefficient for the same source and a more negative correlation coefficient for different sources.

generalized to posterior probability sequences (Sawada *et al.*, 2011), which are utilized for the depermutation in clustering-based underdetermined BSS.

After calculating the power ratio, we basically interchange the indices j of separated signals so that the correlation coefficient $\rho(\varrho_j(f), \varrho_j(f'))$ between the power ratios $\varrho_j(f) = [\varrho_j(n,f)]_n$ at different frequency bins f and f' is maximized for the same source. The optimization procedure has been described in detail by Sawada *et al.* (2007b, 2011).

13.2.9.2 TDOA Clustering

This method utilizes source position differences. If the sources originate from different positions, their TDOAs subject to a given microphone pair differ in general. This can be used for permutation alignment.

Recall that we have estimated the mixing vector $\hat{\mathbf{a}}_j(f) = [\hat{a}_{1j}(f), \dots, \hat{a}_{Ij}(f)]^T$ of source j with (13.38). Assuming the anechoic model, TDOA for source j between microphones 1 and 2 at frequency f is estimated as

$$\Delta_{12j}(f) = \frac{\angle(\hat{a}_{2j}(f)/\hat{a}_{1j}(f))}{2\pi v_f} . \tag{13.41}$$

Figure 13.8 shows an example of the TDOA-based permutation alignment. In a two-source two-microphone situation, TDOAs are estimated by (13.41) for $f \in \{0, \dots, F-1\}$ and every separated signal. By clustering the TDOAs, the permutation ambiguities are aligned based on the averaged TDOA of each source.

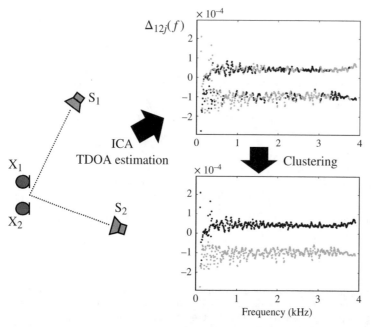

Figure 13.8 TDOA estimation and permutation alignment. For a two-microphone two-source situation (left), ICA is applied in each frequency bin and TDOAs for two sources between the two microphones are estimated (right upper). Each color (navy or orange) corresponds to the same source. Clustering for TDOAs aligns the permutation ambiguities (right lower).

13.3 Independent Vector Analysis

So far we have seen methods to solve the permutation problem by post-processing after ICA. This section, taking a different path, introduces IVA (Hiroe, 2006; Kim *et al.*, 2007; Lee *et al.*, 2007; Ono, 2012), in which the permutation problem is solved simultaneously with the separation.

13.3.1 Formulation

Compared to FD-ICA, the input to IVA is the whole observation tensor $\mathcal{X}$ introduced in Section 13.1. It means that all frequency channels are processed simultaneously or, in other words, the demixing matrices $\mathcal{W} = \{\mathbf{W}(f)\}_f$ are sought simultaneously. The key difference compared to FD-ICA is that IVA takes into account possible dependencies between frequency components that correspond to the same source. In FD-ICA, these dependencies are exploited only in the post-processing stage when solving the permutation problem.

Therefore, for the jth source signal, there is a joint distribution of the frequency components of $\mathbf{s}_j(n) = [s_j(n, 0), \dots, s_j(n, F-1)]^T$, which is, in general, not decomposable into a product of marginal distributions because they are dependent. Let us denote the joint distribution of source j by $\bar{p}(\mathbf{s}_j)$. The log-likelihood of the joint model reads

$$\mathcal{M}^{\text{ML}}(\mathcal{W}) = \sum_f 2 \log |\det \mathbf{W}^H(f)| + \sum_{j=1}^{J} \frac{1}{N} \sum_n \log \bar{p}(\widehat{\mathbf{s}}_j(n)). \tag{13.42}$$

For $N \to +\infty$, the corresponding objective function is

$$\mathcal{M}^{\text{ML}}(\mathcal{W}) = \sum_f 2 \log |\det \mathbf{W}^H(f)| + \sum_{j=1}^{J} \mathbb{E}_{\widehat{p}(\widehat{\mathbf{s}}_j)} \{\log \bar{p}(\widehat{\mathbf{s}}_j)\}. \tag{13.43}$$

It is even more difficult to estimate $\bar{p}(\mathbf{s}_j)$ than the marginal distributions of individual frequency components as required in FD-ICA. Therefore, the typical approach here is to select a nongaussian distribution that is not decomposable as a product of marginals. An example of such a distribution is

$$\bar{p}(\mathbf{s}_j) \propto \exp\left(-\sqrt{\sum_f |s_j(f)|^2 + \gamma}\right), \tag{13.44}$$

which is an ad hoc extension of (13.16).

13.3.2 Algorithms

13.3.2.1 Natural Gradient

Similarly to Section 13.2.5.1, the natural gradient algorithm is derived for IVA (Hiroe, 2006; Kim *et al.*, 2007) to optimize (13.43) as

$$\mathbf{W}^H(f) \leftarrow \mathbf{W}^H(f) + \eta \left(\mathbf{I}_J - \mathbb{E}\{\Phi_f(\widehat{\mathbf{S}})\widehat{\mathbf{s}}^H(f)\}\right)\mathbf{W}^H(f), \tag{13.45}$$

where $\hat{\mathbf{S}} = [\hat{\mathbf{s}}_1, \ldots, \hat{\mathbf{s}}_j]^T$. The vector of score functions is defined as

$$\Phi_f(\hat{\mathbf{S}}) = [\Phi_f(\hat{\mathbf{s}}_1), \ldots, \Phi_f(\hat{\mathbf{s}}_j)]^T, \quad \Phi_f(\hat{\mathbf{s}}_j) = -\frac{\partial \log \bar{p}(\hat{\mathbf{s}}_j)}{\partial \hat{s}_j^*(f)}. \tag{13.46}$$

If we assume the distribution of the form defined in (13.44), the score function gets the form

$$\Phi_f(\hat{\mathbf{s}}_j) = \frac{\hat{s}_j(f)}{2\sqrt{\sum_f |\hat{s}_j(f)|^2 + \gamma}}. \tag{13.47}$$

13.3.2.2 FastIVA

Extending FastICA from Section 13.2.5.2 to IVA is also possible, which gives rise to *FastIVA* (Lee *et al.*, 2007). This is done by changing the nonlinear function $g(\cdot)$ in (13.34) to

$$g(\mathbf{s}_j) = \sqrt{\sum_f |s_j(f)|^2 + \gamma}. \tag{13.48}$$

The first- and second-order partial derivatives of $g(\cdot)$ with respect to $s_j^*(f)$ are, respectively,

$$g_f'(\mathbf{s}_j) = \frac{\partial g(\mathbf{s}_j)}{\partial s_j^*(f)} = \frac{s_j(f)}{2\sqrt{\sum_f |s_j(f)|^2 + \gamma}}, \tag{13.49}$$

and

$$g_f''(\mathbf{s}_j) = \frac{1}{2\sqrt{\sum_f |s_j(f)|^2 + \gamma}} \left[1 - \frac{|s_j(f)|^2}{2\left(\sum_f |s_j(f)|^2 + \gamma\right)} \right]. \tag{13.50}$$

The above two algorithms inherit the characteristics of the original natural gradient and FastICA. Finally, another method for IVA is worth mentioning which has fast and stable convergence. It is known under the acronym of AuxIVA. It maximizes (13.42) using an auxiliary function-based optimization algorithm (see Ono (2011, 2012)). The relationship between IVA and multichannel NMF (Ozerov and Févotte, 2010; Sawada *et al.*, 2013) has been clarified by Ozerov *et al.* (2012) and Kitamura *et al.* (2016).

13.4 Example

We performed experiments to examine how the described FD-ICA and IVA methods separate real-world sound mixtures. The experimental setup is illustrated in Figure 13.9. Four loudspeakers and microphones were located in a room with a reverberation time (RT60) of 130 ms. The distance of the loudspeakers from the microphones was 1.2 m, and each loudspeaker was situated at a different angle. Two male and two female utterances of 7 s in length were each played by the respective loudspeaker; two datasets (A and B) of the signals were considered. The spatial images of the signals were recorded by the microphones and signal mixtures were created by summing selected recordings.

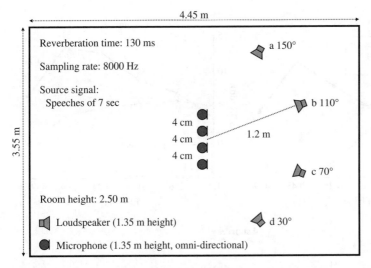

Figure 13.9 Experimental setups: source positions a and b for a two-source setup, a and b and c for a three-source setup, and a through d for a four-source setup. The number of microphones used is always the same as the number of sources.

The soundproof room basically shut down the background noise, but low frequency noise (around 60 Hz) came into the microphone signals. The sampling frequency was 8 kHz.

Three setups were considered with two (a and b), three (a through c) and four (a through d) sources, respectively. The number of microphones used was the same as that of the sources in each setup. The mixtures were separated by the three methods described in this chapter: FD-ICA (FastICA followed by natural gradient) with activity sequence (power ratio) clustering from Section 13.2.9.1, FD-ICA with TDOA estimations from Section 13.2.9.2, and FastIVA from Section 13.3.2.

Figure 13.10 shows the results achieved by the three methods with dataset A when the length of the STFT window was varied from 128 through 4096. To evaluate, the signal-to-distortion ratio (SDR) computed by means of the BSS_EVAL toolbox (Vincent *et al.*, 2006) was used as the main criterion. It was evaluated over the separated spatial images of the sources and the average was taken.

In some cases, permutation alignments were not successful. Figure 13.11 shows such an example. The SDR achieved by the methods mostly grows with the length of STFT but only to a certain limit, which depends on the method and the number of sources. The performance of each method drops when the length of STFT achieves the critical value; these phenomena are caused by the fundamental limitations of the frequency-domain BSS, which were described by Araki *et al.* (2003).

Figure 13.12 shows the results for mixtures of less active sources from dataset B, which have longer and less overlapping silent periods than the sources from dataset A. In this case, IVA works better than with dataset A and competes with FD-ICA (power ratio). These results are indicative of the fact that the performances of BSS algorithms depend on many aspects, and that there is no BSS method that achieves superior separation in every situation.

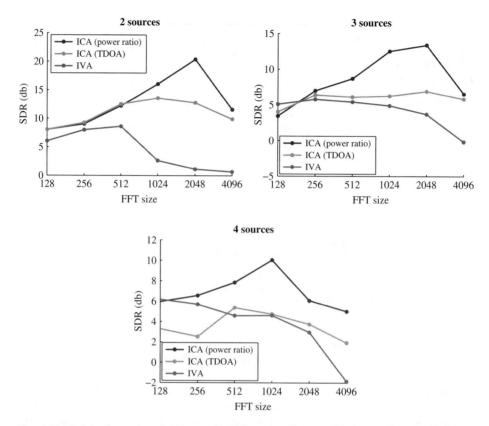

Figure 13.10 Experimental results (dataset A): SDR averaged over separated spatial images of the sources.

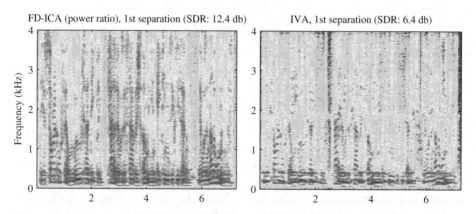

Figure 13.11 Spectrograms of the separated signals (dataset A, two sources, STFT window size of 1024). Here, IVA failed to solve the permutation alignment. These two sources are difficult to separate because the signals share the same silent period at around 2.5 s.

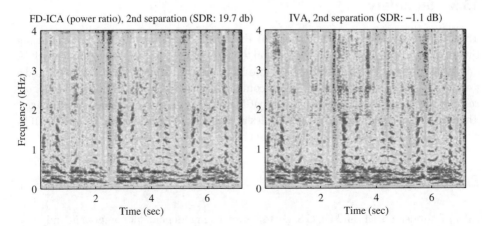

FD-ICA (power ratio), 2nd separation (SDR: 19.7 db)

IVA, 2nd separation (SDR: −1.1 dB)

Figure 13.11 (*Continued*)

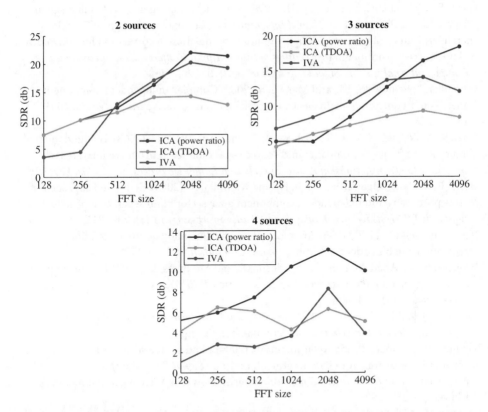

Figure 13.12 Experimental results with less active sources (dataset B): SDR averaged over separated spatial images of the sources.

13.5 Summary

This chapter described popular tools for the blind separation of audio signals using their independence. We mainly focused on the frequency-domain separation using ICA and IVA. Fundamental principles, basic models, and algorithms were introduced, and a practical example was presented. The accompanying web page provides audio files corresponding to the source signals, their spatial images, and the separated signal images. The web page also contains other examples and codes to test and compare various ICA and BSS algorithms utilizing signal independence.

Bibliography

Adalı, T., Anderson, M., and Fu, G.S. (2014) Diversity in independent component and vector analyses: Identifiability, algorithms, and applications in medical imaging. *IEEE Signal Processing Magazine*, **31** (3), 18–33.

Amari, S., Cichocki, A., and Yang, H.H. (1996) A new learning algorithm for blind signal separation, in *Proceedings of Neural Information Processing Systems*, pp. 757–763.

Anemüller, J. and Kollmeier, B. (2000) Amplitude modulation decorrelation for convolutive blind source separation, in *Proceedings of International Conference on Independent Component Analysis and Signal Separation*, pp. 215–220.

Anemüller, J., Sejnowski, T.J., and Makeig, S. (2003) Complex independent component analysis of frequency-domain electroencephalographic data. *Neural Networks*, **16** (9), 1311–1323.

Araki, S., Mukai, R., Makino, S., Nishikawa, T., and Saruwatari, H. (2003) The fundamental limitation of frequency domain blind source separation for convolutive mixtures of speech. *IEEE Transactions on Audio, Speech, and Language Processing*, **11** (2), 109–116.

Asano, F., Ikeda, S., Ogawa, M., Asoh, H., and Kitawaki, N. (2003) Combined approach of array processing and independent component analysis for blind separation of acoustic signals. *IEEE Transactions on Speech and Audio Processing*, **11** (3), 204–215.

Bell, A. and Sejnowski, T. (1995) An information-maximization approach to blind separation and blind deconvolution. *Neural Computation*, 7 (6), 1129–1159.

Belouchrani, A., Abed-Meraim, K., Cardoso, J.F., and Moulines, E. (1997) A blind source separation technique using second-order statistics. *IEEE Transactions on Signal Processing*, **45** (2), 434–444.

Bourgeois, J. and Minker, W. (2009) *Time-Domain Beamforming and Blind Source Separation: Speech Input in the Car Environment*, Springer.

Buchner, H., Aichner, R., and Kellermann, W. (2004) Blind source separation for convolutive mixtures: A unified treatment, in *Audio Signal Processing for Next-Generation Multimedia Communication Systems* (eds Y. Huang and J. Benesty), Kluwer, pp. 255–293.

Buchner, H., Aichner, R., and Kellermann, W. (2005) A generalization of blind source separation algorithms for convolutive mixtures based on second-order statistics. *IEEE Transactions on Speech and Audio Processing*, **13** (1), 120–134.

Cardoso, J.F. (1993) Blind beamforming for non-Gaussian signals. *IEE Proceedings-F*, pp. 362–370.

Cardoso, J.F. (1998) Blind signal separation: statistical principles. *Proceedings of the IEEE*, **86** (10), 2009–2025.

Cardoso, J.F. and Laheld, B.H. (1996) Equivariant adaptive source separation. *IEEE Transactions on Signal Processing*, **44** (12), 3017–3030.

Cardoso, J.F. and Souloumiac, A. (1996) Jacobi angles for simultaneous diagonalization. *SIAM Journal on Matrix Analysis and Applications*, **17** (1), 161–164.

Cichocki, A. and Amari, S. (2002) *Adaptive Blind Signal and Image Processing*, Wiley.

Comon, P. (1994) Independent component analysis, a new concept? *Signal Processing*, **36**, 287–314.

Comon, P. and Jutten, C. (2010) *Handbook of Blind Source Separation: Independent Component Analysis and Applications*, Elsevier.

Cover, T. and Thomas, J. (2006) *Elements of Information Theory*, Wiley.

Eriksson, J. and Koivunen, V. (2004) Identifiability, separability, and uniqueness of linear ICA models. *IEEE Signal Processing Letters*, **11** (7), 601–604.

Ferreol, A., Albera, L., and Chevalier, P. (2005) Fourth-order blind identification of underdetermined mixtures of sources (FOBIUM). *IEEE Transactions on Signal Processing*, **53** (5), 1640–1653.

Gazor, S. and Zhang, W. (2003) Speech probability distribution. *IEEE Signal Processing Letters*, **10** (7), 204–207.

Haykin, S. (ed.) (2000) *Unsupervised Adaptive Filtering (Volume I: Blind Source Separation)*, Wiley.

Hiroe, A. (2006) Solution of permutation problem in frequency domain ICA using multivariate probability density functions, in *Proceedings of International Conference on Independent Component Analysis and Signal Separation*, pp. 601–608.

Hyvärinen, A. (1999) Fast and robust fixed-point algorithm for independent component analysis. *IEEE Transactions on Neural Networks*, **10** (3), 626–634.

Hyvärinen, A., Karhunen, J., and Oja, E. (2001) *Independent Component Analysis*, Wiley.

Ikram, M.Z. and Morgan, D.R. (2002) A beamforming approach to permutation alignment for multichannel frequency-domain blind speech separation, in *Proceedings of IEEE International Conference on Audio, Speech and Signal Processing*, pp. 881–884.

Kellermann, W., Buchner, H., and Aichner, R. (2006) Separating convolutive mixtures with TRINICON, in *Proceedings of IEEE International Conference on Audio, Speech and Signal Processing*, vol. V, pp. 961–964.

Kim, T., Attias, H.T., Lee, S.Y., and Lee, T.W. (2007) Blind source separation exploiting higher-order frequency dependencies. *IEEE Transactions on Audio, Speech, and Language Processing*, pp. 70–79.

Kitamura, D., Ono, N., Sawada, H., Kameoka, H., and Saruwatari, H. (2016) Determined blind source separation unifying independent vector analysis and nonnegative matrix factorization. *IEEE/ACM Transactions on Audio, Speech, and Language Processing*, **24** (9), 1626–1641.

Kokkinakis, K. and Nandi, A.K. (2005) Exponent parameter estimation for generalized gaussian probability density functions with application to speech modeling. *Signal Processing*, **85** (9), 1852–1858.

Koldovský, Z., Málek, J., Tichavský, P., Deville, Y., and Hosseini, S. (2009) Blind separation of piecewise stationary non-gaussian sources. *Signal Processing*, **89** (12), 2570–2584.

Koldovský, Z. and Tichavský, P. (2011) Time-domain blind separation of audio sources on the basis of a complete ICA decomposition of an observation space. *IEEE Transactions on Audio, Speech, and Language Processing*, **19** (2), 406–416.

Koldovský, Z. and Tichavský, P. (2015) Improved variant of the FastICA algorithm, in *Advances in Independent Component Analysis and Learning Machines* (eds E. Bingham, S. Kaski, J. Laaksonen, and J. Lampinen), Elsevier, pp. 53–74.

Koldovský, Z., Tichavský, P., and Oja, E. (2006) Efficient variant of algorithm FastICA for independent component analysis attaining the Cramér-Rao lower bound. *IEEE Transactions on Neural Networks*, **17** (5), 1265–1277.

Kurita, S., Saruwatari, H., Kajita, S., Takeda, K., and Itakura, F. (2000) Evaluation of blind signal separation method using directivity pattern under reverberant conditions, in *Proceedings of IEEE International Conference on Audio, Speech and Signal Processing*, pp. 3140–3143.

Lee, I., Kim, T., and Lee, T.W. (2007) Fast fixed-point independent vector analysis algorithms for convolutive blind source separation. *Signal Processing*, **87** (8), 1859–1871.

Lee, T.W. (1998) *Independent Component Analysis - Theory and Applications*, Kluwer.

Matsuoka, K. and Nakashima, S. (2001) Minimal distortion principle for blind source separation, in *Proceedings of International Conference on Independent Component Analysis and Signal Separation*, pp. 722–727.

Matsuoka, K., Ohya, M., and Kawamoto, M. (1995) A neural net for blind separation of nonstationary signals. *Neural Networks*, **8** (3), 411–419.

Mitianoudis, N. and Davies, M.E. (2003) Audio source separation of convolutive mixtures. *IEEE Transactions on Speech and Audio Processing*, **11** (5), 489–497.

Molgedey, L. and Schuster, H.G. (1994) Separation of a mixture of independent signals using time delayed correlations. *Physical Review Letters*, **72** (23), 3634–3637.

Mukai, R., Sawada, H., Araki, S., and Makino, S. (2004a) Frequency domain blind source separation for many speech signals, in *Proceedings of International Conference on Independent Component Analysis and Signal Separation*, Springer, pp. 461–469.

Mukai, R., Sawada, H., Araki, S., and Makino, S. (2004b) Frequency domain blind source separation using small and large spacing sensor pairs, in *Proceedings of IEEE International Symposium on Circuits and Systems*, vol. V, pp. 1–4.

Murata, N., Ikeda, S., and Ziehe, A. (2001) An approach to blind source separation based on temporal structure of speech signals. *Neurocomputing*, **41**, 1–24.

Nesta, F., Svaizer, P., and Omologo, M. (2011) Convolutive BSS of short mixtures by ICA recursively regularized across frequencies. *IEEE Transactions on Audio, Speech, and Language Processing*, **19** (3), 624–639.

Ono, N. (2011) Stable and fast update rules for independent vector analysis based on auxiliary function technique, in *Proceedings of IEEE Workshop on Applications of Signal Processing to Audio and Acoustics*, pp. 189–192.

Ono, N. (2012) Fast stereo independent vector analysis and its implementation on mobile phone, in *Proceedings of International Workshop on Acoustic Signal Enhancement*, pp. 1–4.

Ozerov, A. and Févotte, C. (2010) Multichannel nonnegative matrix factorization in convolutive mixtures for audio source separation. *IEEE Transactions on Audio, Speech, and Language Processing*, **18** (3), 550–563.

Ozerov, A., Vincent, E., and Bimbot, F. (2012) A general flexible framework for the handling of prior information in audio source separation. *IEEE Transactions on Audio, Speech, and Language Processing*, **20** (4), 1118–1133.

Papoulis, A. and Pillai, S.U. (2002) *Probability, Random Variables and Stochastic Processes*, McGraw-Hill.

Parra, L. and Spence, C. (2000) Convolutive blind separation of non-stationary sources. *IEEE Transactions on Speech and Audio Processing*, **8** (3), 320–327.

Pham, D.T. and Cardoso, J.F. (2001) Blind separation of instantaneous mixtures of nonstationary sources. *IEEE Transactions on Signal Processing*, **49** (9), 1837–1848.

Pham, D.T. and Garat, P. (1997) Blind separation of mixture of independent sources through a quasi-maximum likelihood approach. *IEEE Transactions on Signal Processing*, **45** (7), 1712–1725.

Saruwatari, H., Kurita, S., Takeda, K., Itakura, F., Nishikawa, T., and Shikano, K. (2003) Blind source separation combining independent component analysis and beamforming. *EURASIP Journal on Applied Signal Processing*, **2003** (11), 1135–1146.

Sawada, H., Araki, S., and Makino, S. (2007a) Measuring dependence of bin-wise separated signals for permutation alignment in frequency-domain BSS, in *Proceedings of IEEE International Symposium on Circuits and Systems*, pp. 3247–3250.

Sawada, H., Araki, S., and Makino, S. (2011) Underdetermined convolutive blind source separation via frequency bin-wise clustering and permutation alignment. *IEEE Transactions on Audio, Speech, and Language Processing*, **19** (3), 516–527.

Sawada, H., Araki, S., Mukai, R., and Makino, S. (2007b) Grouping separated frequency components by estimating propagation model parameters in frequency-domain blind source separation. *IEEE Transactions on Audio, Speech, and Language Processing*, **15** (5), 1592–1604.

Sawada, H., Kameoka, H., Araki, S., and Ueda, N. (2013) Multichannel extensions of non-negative matrix factorization with complex-valued data. *IEEE Transactions on Audio, Speech, and Language Processing*, **21** (5), 971–982.

Sawada, H., Mukai, R., Araki, S., and Makino, S. (2003) Polar coordinate based nonlinear function for frequency domain blind source separation. *IEICE Transactions on Fundamentals of Electronics, Communications and Computer Sciences*, **E86-A** (3), 590–596.

Sawada, H., Mukai, R., Araki, S., and Makino, S. (2004) A robust and precise method for solving the permutation problem of frequency-domain blind source separation. *IEEE Transactions on Speech and Audio Processing*, **12** (5), 530–538.

Schobben, L. and Sommen, W. (2002) A frequency domain blind signal separation method based on decorrelation. *IEEE Transactions on Signal Processing*, **50** (8), 1855–1865.

Smaragdis, P. (1998) Blind separation of convolved mixtures in the frequency domain. *Neurocomputing*, **22**, 21–34.

Soon, V.C., Tong, L., Huang, Y.F., and Liu, R. (1993) A robust method for wideband signal separation, in *Proceedings of IEEE International Symposium on Circuits and Systems*, vol. 1, pp. 703–706.

Takatani, T., Nishikawa, T., Saruwatari, H., and Shikano, K. (2004) High-fidelity blind separation of acoustic signals using SIMO-model-based independent component analysis. *IEICE Transactions on Fundamentals of Electronics, Communications and Computer Sciences*, **E87-A** (8), 2063–2072.

Theis, F. and Inouye, Y. (2006) On the use of joint diagonalization in blind signal processing, in *Proceedings of IEEE International Symposium on Circuits and Systems*, pp. 3586–3589.

Tichavský, P. and Koldovský, Z. (2011) Fast and accurate methods of independent component analysis: A survey. *Kybernetika*, **47** (3), 426–438.

Tichavský, P. and Yeredor, A. (2009) Fast approximate joint diagonalization incorporating weight matrices. *IEEE Transactions on Signal Processing*, **57** (3), 878–891.

Tichavský, P., Yeredor, A., and Koldovský, Z. (2009) A fast asymptotically efficient algorithm for blind separation of a linear mixture of block-wise stationary autoregressive processes, in *Proceedings of IEEE International Conference on Audio, Speech and Signal Processing*, pp. 3133–3136.

Varanasi, M.K. and Aazhang, B. (1989) Parametric generalized gaussian density estimation. *Journal of the Acoustical Society of America*, **86** (4), 1404–1415.

Vincent, E., Gribonval, R., and Févotte, C. (2006) Performance measurement in blind audio source separation. *IEEE Transactions on Audio, Speech, and Language Processing*, **14** (4), 1462–1469.

14

Gaussian Model Based Multichannel Separation

Alexey Ozerov and Hirokazu Kameoka

The Gaussian framework for multichannel source separation consists of modeling vectors of STFT coefficients as multivariate complex Gaussian distributions. It allows specifying spatial and spectral models of the source spatial images and estimating their parameters in a joint manner. *Multichannel nonnegative matrix factorization*, illustrated in Figure 14.1, is one of the most popular such methods. It combines nonnegative matrix factorization (NMF) (see Chapter 8) and narrowband spatial modeling (see Chapter 3). Besides NMF, the Gaussian framework makes it possible to reuse many other single-channel spectral models in a multichannel scenario. It differs from the frameworks in Chapters 11, 12, and 13 in the fact that more advanced generative spectral models are typically used. Also, according to the general taxonomies introduced in Chapter 1, it covers a wide range of audio source separation scenarios, including over- or underdetermined mixtures and weakly or strongly guided separation, and a wide range of methods that are either learning-free or based on unsupervised/supervised source modeling.

In Section 14.1 we introduce the multichannel Gaussian framework. In Section 14.2 we provide a detailed list of spectral and spatial models. We explain how to estimate the parameters of these models in Section 14.3. We give a detailed presentation of a few methods in Section 14.4 and provide a summary in Section 14.5.

14.1 Gaussian Modeling

14.1.1 Joint Spectral-Spatial Local Gaussian Modeling

Let us start with the assumption that the narrowband approximation holds. Then, the $I \times 1$ spatial image $\mathbf{c}_j(n,f)$ of source j in time frame n and frequency bin f is modeled as the product of the acoustic transfer function $\mathbf{a}_j(f)$ and the short-time Fourier transform (STFT) coefficient $s_j(n,f)$ of source j:

$$\mathbf{c}_j(n,f) = \mathbf{a}_j(f)s_j(n,f). \tag{14.1}$$

Audio Source Separation and Speech Enhancement, First Edition.
Edited by Emmanuel Vincent, Tuomas Virtanen and Sharon Gannot.
© 2018 John Wiley & Sons Ltd. Published 2018 by John Wiley & Sons Ltd.
Companion Website: https://project.inria.fr/ssse/

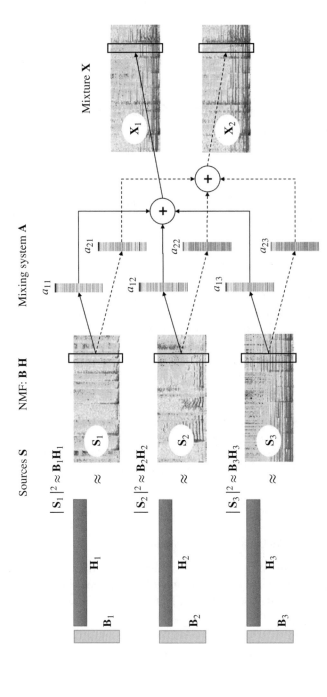

Figure 14.1 Illustration of multichannel NMF. S_j, X_i, and a_{ij} represent the complex-valued spectrograms of the sources and the mixture channels, and the complex-valued mixing coefficients, respectively. NMF factors the power spectrogram $|S_j|^2$ of each source as B_jH_j (see Chapter 8 and Section 14.2.1). The mixing system is represented by a rank-1 spatial model (see Section 14.2.2).

When $s_j(n,f)$ is assumed to follow a zero-mean complex Gaussian distribution with variance $\sigma_j^2(n,f)$

$$s_j(n,f) \sim \mathcal{N}_c(s_j(n,f) \mid 0, \sigma_j^2(n,f)), \tag{14.2}$$

$c_j(n,f)$ follows the so-called *local Gaussian model*

$$c_j(n,f) \sim \mathcal{N}_c(c_j(n,f) \mid \mathbf{0}_I, \sigma_j^2(n,f)\mathbf{R}_j(f)) \tag{14.3}$$

where $\mathbf{R}_j(f) = \mathbf{a}_j(f)\mathbf{a}_j^H(f)$ is the $I \times I$ spatial covariance matrix of source j. The narrowband approximation implies that the spatial covariance matrix has rank 1. Alternatively, $\mathbf{R}_j(f)$ can be assumed to be a full-rank matrix in (14.3). The local Gaussian model can also be defined using quadratic time-frequency representations instead of the STFT (Duong *et al.*, 2010b; Ozerov *et al.*, 2012).

Multichannel source separation problems can be formulated using this model (Pham *et al.*, 2003; Févotte and Cardoso, 2005; Vincent *et al.*, 2009; Duong *et al.*, 2010a; Sawada *et al.*, 2013; Higuchi and Kameoka, 2015). Let us study an example. The $I \times 1$ vector $\mathbf{x}(n,f)$ of STFT coefficients of the mixture signal is equal to the sum of the source spatial image vectors $c_j(n,f)$ of J sources

$$\mathbf{x}(n,f) = \sum_{j=1}^{J} c_j(n,f). \tag{14.4}$$

When the sources are assumed to be independent, $\mathbf{x}(n,f)$ follows

$$\mathbf{x}(n,f) \sim \mathcal{N}_c\left(\mathbf{x}(n,f) \,\middle|\, \mathbf{0}_I, \sum_{j=1}^{J} \sigma_j^2(n,f)\mathbf{R}_j(f)\right). \tag{14.5}$$

Hence, we obtain the log-likelihood

$$\mathcal{M}^{\mathrm{ML}}(\theta) = \sum_{nf}\left[-\log \det\left(\pi \sum_{j=1}^{J} \sigma_j^2(n,f)\mathbf{R}_j(f) \right)\right.$$
$$\left. -\mathbf{x}^H(n,f)\left(\sum_{j=1}^{J} \sigma_j^2(n,f)\mathbf{R}_j(f) \right)^{-1} \mathbf{x}(n,f)\right] \tag{14.6}$$

where $\theta = \{\{\sigma_j^2(n,f)\}_{jnf}, \{\mathbf{R}_j(f)\}_{jf}\}$ is the set of unknown model parameters. In the particular case when the narrowband approximation holds and there are as many sources as channels, i.e. $J = I$, the mixture (14.4) can be expressed as

$$\mathbf{x}(n,f) = \mathbf{A}(f)\mathbf{s}(n,f) = (\mathbf{W}^H(f))^{-1}\mathbf{s}(n,f). \tag{14.7}$$

where $\mathbf{s}(n,f) = [s_1(n,f), \dots, s_I(n,f)]^T$ is the $I \times 1$ vector of source STFT coefficients, $\mathbf{A}(f) = [\mathbf{a}_1(f), \dots, \mathbf{a}_I(f)]$ is the $I \times I$ mixing matrix, and $\mathbf{W}^H(f) = \mathbf{A}^{-1}(f)$ is the $I \times I$ separation matrix. Hence (14.5) can be rewritten as

$$\mathbf{x}(n,f) \sim \mathcal{N}_c(\mathbf{x}(n,f) \mid \mathbf{0}_I, (\mathbf{W}^H(f))^{-1}\boldsymbol{\Sigma}_s(n,f)\mathbf{W}^{-1}(f)), \tag{14.8}$$

with

$$\boldsymbol{\Sigma}_s(n,f) = \mathrm{Diag}(\sigma_1^2(n,f), \dots, \sigma_I^2(n,f)). \tag{14.9}$$

This results in the log-likelihood of frequency-domain independent component analysis (FD-ICA) based on a time-varying Gaussian source model:

$$\mathcal{M}^{\text{ML}}(\theta) = \sum_{nf} \left[-I \log \pi + 2 \log \, \det(\mathbf{W}(f)) - \sum_{j=1}^{I} \log \sigma_j^2(n,f) \right.$$

$$\left. -\mathbf{x}^H(n,f)\mathbf{W}(f)\mathbf{\Sigma}_s^{-1}(n,f)\mathbf{W}^H(f)\mathbf{x}(n,f) \right]. \tag{14.10}$$

Since all the variables are indexed by frequency f in the log-likelihood, the optimization problem can be split into frequency-wise source separation problems. The permutation problem (see Section 13.2.9) must then be solved in order to align the separated components in different frequency bins that originate from the same source. While some methods are designed to perform frequency binwise source separation followed by permutation alignment, it is preferable to solve permutation alignment and source separation in a joint manner since the clues used for permutation alignment can also be helpful for source separation.

To handle more general cases, such as when the sources outnumber the channels, or to solve the permutation and separation problems in a joint fashion, we must add further constraints to the local Gaussian model. In the following sections, we introduce assumptions and constraints that can be incorporated into the Gaussian framework in order to deal with various scenarios and to improve the source separation accuracy.

14.1.2 Source Separation: Main Steps

Multichannel source separation methods based on the local Gaussian model can be categorized according to the choices of mixing models, source spectral models, spatial models, parameter estimation schemes, and source signal estimation schemes. Here, we present the main steps to formulate these methods.

14.1.2.1 Mixing Models

Typical choices of mixing models include additive, narrowband, subband filtering, and sparse models. The first three models assume that all sources are active, while the sparse model assumes that only one source is active in each time-frequency bin. The additive model (14.4) makes no assumption about the source spatial images $\mathbf{c}_j(n,f)$, except that their sum is equal to the mixture $\mathbf{x}(n,f)$. The three latter models assume that there are J_p point sources of interest indexed by $j \in \{1, \dots, J_p\}$ and consider the other sources as background noise $\mathbf{u}(n,f) = \sum_{j=J_p+1}^{J} \mathbf{c}_j(n,f)$, which we assume to follow a zero-mean complex Gaussian distribution. The relationship between $\mathbf{x}(n,f)$, $\mathbf{s}(n,f) = [s_1(n,f), \dots, s_{J_p}(n,f)]^T$ and $\mathbf{u}(n,f)$ is defined for the narrowband mixing model by

$$\mathbf{x}(n,f) = \mathbf{A}(f)\mathbf{s}(n,f) + \mathbf{u}(n,f) \tag{14.11}$$

$$= \sum_{j=1}^{J_p} \mathbf{a}_j(f)s_j(n,f) + \mathbf{u}(n,f), \tag{14.12}$$

for the subband filtering mixing model by

$$\mathbf{x}(n,f) = \sum_{n'=0}^{N'-1} \mathbf{A}(n',f)\mathbf{s}(n-n',f) + \mathbf{u}(n,f) \tag{14.13}$$

$$= \sum_{j=1}^{J_p} \sum_{n'=0}^{N'-1} \mathbf{a}_j(n',f)s_j(n-n',f) + \mathbf{u}(n,f), \tag{14.14}$$

and for the sparse mixing model by

$$\mathbf{x}(n,f) = \mathbf{a}_{z(n,f)}(f)s_{z(n,f)}(n,f) + \mathbf{u}(n,f), \tag{14.15}$$

where $z(n,f)$ denotes the index of the predominant source, i.e. the most active source in time-frequency bin (n,f). The length N' of the subband filters is typically in the order of L/M with L the length of the time-domain mixing filters $\mathbf{a}_j(\tau)$ and M the hop size between adjacent STFT frames.

In the particular case of a determined, noiseless mixture ($I = J$ and $\mathbf{u}(n,f) = \mathbf{0}_I$), the narrowband and the subband filtering mixing models can be inverted. Hence we can alternatively consider the narrowband demixing model

$$\mathbf{s}(n,f) = \mathbf{W}^H(f)\mathbf{x}(n,f) \tag{14.16}$$

or the truncated[1] subband filtering demixing model

$$\mathbf{s}(n,f) = \sum_{n'=0}^{N'-1} \mathbf{W}^H(n',f)\mathbf{x}(n-n',f). \tag{14.17}$$

14.1.2.2   Source Spectral Models

We assume that the STFT coefficients of source j follow (14.2). With this model, it can be shown using a simple change of variables that the power and phase of $s_j(n,f)$ follow an exponential distribution with mean $\sigma_j^2(n,f)$ and a uniform distribution on the interval $[0, 2\pi)$, respectively. If there is a certain assumption, constraint or structure that we want to impose on the power spectrum of each source, we can employ a parametric model to represent $\sigma_j^2(n,f)$ instead of individually treating $\sigma_j^2(n,f)$ as a free parameter, or introduce a properly designed prior distribution over $\sigma_j^2(n,f)$. Choices include a Gaussian mixture model (GMM) (Attias, 2003), a hidden Markov model (HMM) (Higuchi and Kameoka, 2015), an autoregressive (AR) model (Dégerine and Zaïdi, 2004; Yoshioka *et al.*, 2011), a nonnegative matrix/tensor factorization (NMF) model (Ozerov and Févotte, 2010; Arberet *et al.*, 2010; Ozerov *et al.*, 2011; Sawada *et al.*, 2013; Nikunen and Virtanen, 2014; Kitamura *et al.*, 2015), an excitation-filter model (also known as the source-filter model) (Kameoka *et al.*, 2010; Ozerov *et al.*, 2012), a spectral continuity prior (Duong *et al.*, 2011), a deep neural network (DNN) model (Nugraha *et al.*, 2016), and combinations of different models (Ozerov *et al.*, 2012; Adiloğlu and Vincent, 2016), among others. These models are presented in detail in Section 14.2.1.

---

1 The inverse of a finite impulse response (FIR) subband filter is generally an infinite impulse response filter.

14.1.2.3 Spatial Models

The probability distribution of the observed signals $\mathcal{X} = \{x(n,f)\}_{nf}$, i.e. the likelihood of the unknown parameters, can be derived according to the mixing model and the source distribution. For example, we can show from (14.11) and (14.2) that a narrowband mixture follows

$$x(n,f) \sim \mathcal{N}_c\left(x(n,f) \;\middle|\; \mathbf{0}_I, \sum_{j=1}^{J_p} \sigma_j^2(n,f)\mathbf{R}_j(f) + \mathbf{\Sigma_u}(f)\right) \tag{14.18}$$

where $\mathbf{R}_j(f) = \mathbf{a}_j(f)\mathbf{a}_j^H(f)$ denotes the spatial covariance of source j and $\mathbf{\Sigma_u}(f)$ is the noise covariance matrix. We can also show that a sparse mixture follows

$$x(n,f) \mid z(n,f) \sim \mathcal{N}_c(x(n,f) \mid \mathbf{0}_I, \sigma_{z(n,f)}^2(n,f)\mathbf{R}_{z(n,f)}(f) + \mathbf{\Sigma_u}(f)). \tag{14.19}$$

As with the source power spectrum, there are several ways to model the spatial covariance $\mathbf{R}_j(f)$. These models are presented in detail in Section 14.2.2.

14.1.2.4 Parameter Estimation Schemes

Let $\boldsymbol{\theta}$ be the set of parameters of the spectral and spatial models. Once the likelihood (and the prior distribution) of $\boldsymbol{\theta}$ has been defined according to the choice of mixing, spectral, and spatial models, the next step is to derive a parameter estimation algorithm. Probabilistic parameter estimation schemes may be primarily divided into maximum likelihood (ML) or maximum a posteriori (MAP) estimation and Bayesian inference. The aim of the former is to find the estimate of $\boldsymbol{\theta}$ that maximizes the likelihood or the posterior distribution of $\boldsymbol{\theta}$ whereas the aim of the latter is to infer the posterior distribution of $\boldsymbol{\theta}$ given the observation $\mathcal{X}$. The typical choices of criteria and algorithms for parameter estimation are presented in detail in Section 14.3.

14.1.2.5 Source Signal Estimation Schemes

Once the parameters $\boldsymbol{\theta}$ have been estimated, we can estimate the source signals or their spatial images according to the assumed mixing model. In the case of the narrowband mixing model, a typical choice is the minimum mean square error (MMSE) estimator of $s(n,f)$ (see Section 10.4.2):

$$\hat{s}(n,f) = \mathbb{E}\{s(n,f) \mid x(n,f)\} = \mathbf{W}^H(n,f)x(n,f), \tag{14.20}$$

where $\mathbf{W}(n,f)$ is the well-known multichannel Wiener filter (MWF):

$$\mathbf{\Sigma_s}(n,f) = \mathrm{Diag}(\sigma_1^2(n,f), \dots, \sigma_{J_p}^2(n,f)) \tag{14.21}$$

$$\mathbf{W}(n,f) = (\mathbf{A}(f)\mathbf{\Sigma_s}(n,f)\mathbf{A}^H(f) + \mathbf{\Sigma_u}(f))^{-1}\mathbf{A}(f)\mathbf{\Sigma_s}(n,f). \tag{14.22}$$

When using a full-rank spatial covariance model, it may be convenient to use the MMSE estimator of the source spatial image $c_j(n,f)$ instead (see Section 10.4.5):

$$\hat{c}_j(n,f) = \mathbb{E}\{c_j(n,f) \mid x(n,f)\} \tag{14.23}$$

$$= \sigma_j^2(n,f)\mathbf{R}_j(f)\left(\sum_{j'=1}^{J_p} \sigma_{j'}^2(n,f)\mathbf{R}_{j'}(f) + \mathbf{\Sigma_u}(f)\right)^{-1} x(n,f). \tag{14.24}$$

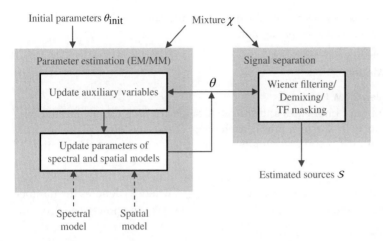

Figure 14.2 Block diagram of multichannel Gaussian model based source separation.

In the case of the narrowband and subband filtering demixing systems, we can directly use (14.16) and (14.17) (Dégerine and Zaïdi, 2004; Kameoka *et al.*, 2010; Yoshioka *et al.*, 2011; Kitamura *et al.*, 2015) once we have obtained the demixing filters $\mathbf{W}^H(f)$ or $\mathbf{W}^H(n',f)$. Algorithms for estimating the demixing filters are described in Section 14.4.3.

Finally, in the case of the sparse mixing system, one reasonable estimator is (Izumi *et al.*, 2007; Kameoka *et al.*, 2012)

$$\widehat{s}_j(n,f) = \gamma_j(n,f) \frac{\mathbf{a}_j^H(f)\mathbf{\Sigma}_{\mathbf{u}}^{-1}(n,f)\mathbf{x}(n,f)}{\mathbf{a}_j^H(f)\mathbf{\Sigma}_{\mathbf{u}}^{-1}(n,f)\mathbf{a}_j(f)} \tag{14.25}$$

which combines the source presence probability $\gamma_j(n,f) = P(z(n,f) = j \mid \mathcal{X}, \theta)$ and the minimum variance distortionless response (MVDR) beamformer. This is in fact the MMSE estimator under the modeling assumptions.

Figure 14.2 schematizes the entire process via a block diagram.

14.2 Library of Spectral and Spatial Models

As mentioned in Section 14.1, the Gaussian framework reduces to FD-ICA in the case of determined noiseless mixtures if no specific constraint or structure is assumed on the power spectra and the spatial covariances of the sources. The advantages of incorporating source spectral models and spatial models into the Gaussian framework are twofold. First, they can help solving frequency-wise separation and permutation alignment in a joint fashion since the spectral and spatial properties represented by the models are useful for permutation alignment.[2] Second, they allow us to deal with a larger range of mixtures, such as reverberant and/or underdetermined mixtures, by exploiting additional

---

2 Indeed, permutation alignment methods typically exploit the fact that the frequency components originating from the same source emanate from the same direction and that their magnitudes are correlated.

reasonable assumptions besides the independence of the sources. In this section, we present examples of spectral and spatial models.

14.2.1 Spectral Models

14.2.1.1 GMM, Scaled GMM, HMM

In speech, the number of phonemes and the pitch range are both usually limited during an entire utterance. In Western music, each piece of music is often played by only a handful of musical instruments or sung by one or a few singers and the number of musical notes is usually limited. It is thus reasonable to assume that the spectra of a real-world sound source can be described using a limited number of templates. By writing the spectral templates of source j as $b_{j,1}(f), \ldots, b_{j,K_j}(f)$, where K_j denotes the number of spectral templates assigned to source j, one way to express the power spectrogram $\sigma_j^2(n,f)$ would be

$$\sigma_j^2(n,f) = b_{j,k_j(n)}(f), \tag{14.26}$$

where $k_j(n)$ denotes the index of the spectral template selected at frame n. If we assume $k_j(n)$ to be a latent variable generated according to a categorical distribution with probabilities $\pi_{j,1}, \ldots, \pi_{j,K_j}$ such that $\sum_k \pi_{jk} = 1$, the generative process of the spatial image $\mathbf{c}_j(n,f)$ of source j is described as a GMM (Attias, 2003):[3]

$$\mathbf{c}_j(n,f) \mid k_j(n) \sim \mathcal{N}_c(\mathbf{c}_j(n,f) \mid \mathbf{0}_I, b_{j,k_j(n)}(f)\mathbf{R}_j(f)), \tag{14.27}$$

$$k_j(n) \sim \pi_{j,k_j(n)}. \tag{14.28}$$

Note that the spectral templates can be either trained on isolated signals of that source type in an unsupervised or a supervised manner or estimated from the mixture signal in a learning-free manner.

While the above model uses each template to represent a different power spectrum, it would be more reasonable to let each template represent all the power spectra that are equal up to a scale factor and treat the the scale factor as an additional parameter. Here, we use $b_{jk}(f)$ as the kth "normalized" spectral template and describe $\sigma_j^2(n,f)$ as

$$\sigma_j^2(n,f) = b_{j,k_j(n)}(f)h_j(n), \tag{14.29}$$

where $h_j(n)$ denotes the time-varying amplitude. Note that this *scaled GMM* model has been employed by Benaroya *et al.* (2006) for single-channel source separation.

Furthermore, since the probability of a particular template being selected may depend on the templates selected at the previous frames, it is natural to extend the generative process of $k_j(n)$ using a Markov chain. These two extensions lead to an HMM (Vincent and Rodet, 2004; Ozerov *et al.*, 2009, 2012; Higuchi and Kameoka, 2015), namely

$$\mathbf{c}_j(n,f) \mid k_j(n) \sim \mathcal{N}_c(\mathbf{c}_j(n,f) \mid \mathbf{0}_I, b_{j,k_j(n)}(f)h_j(n)\mathbf{R}_j(f)), \tag{14.30}$$

$$k_j(n) \mid k_j(n-1) \sim \pi_{j,k_j(n-1),k_j(n)}, \tag{14.31}$$

---

3 Note that the use of GMM as a spectral model in multichannel Gaussian model based separation differs from its typical use in single-channel separation: in Section 7.2.1, the GMM is nonzero-mean and it represents the distribution of the log-power spectrum, while here the GMM is zero-mean and it represents the distribution of the complex-valued STFT coefficients.

where (14.30) can be seen as the state emission probability, $k_j(n)$ as the hidden state, and $\pi_{jkk'}$ as the state transition probability from state k to state k' (see Figure 14.3, top). By properly designing the state transition network, we can flexibly assign probabilities to state durations (the durations of the self-transitions). In addition, by incorporating states associated with speech absence or silence into the state transition network, assuming a state-dependent generative process of the scale factor $h_j(n)$ as

$$h_j(n) \mid k_j(n) \sim \mathcal{G}(h_j(n) \mid \alpha_{k_j(n)}, \beta_{k_j(n)}), \tag{14.32}$$

where $\mathcal{G}(\cdot \mid \alpha, \beta)$ denotes the gamma distribution with shape parameter $\alpha > 0$ and scale parameter $\beta > 0$

$$\mathcal{G}(h \mid \alpha, \beta) = \frac{h^{\alpha-1}e^{-h/\beta}}{\Gamma(\alpha)\beta^\alpha}, \tag{14.33}$$

and setting the hyperparameters α_k and β_k so that $h_j(n)$ tends to be near zero for the states associated with speech absence, this model makes it possible to estimate voice activity segments along with solving the separation problem (Higuchi and Kameoka, 2015).

14.2.1.2 NMF, NTF

While the above models assume that only one of the spectral templates is activated at a time, another way to model the power spectrogram $\sigma_j^2(n,f)$ is to express it via NMF as the sum of the spectral templates $b_{j,1}(f), \ldots, b_{j,K_j}(f)$ scaled by time-varying amplitudes $h_{j,1}(n), \ldots, h_{j,K_j}(n)$ (see Figure 14.3, bottom):

$$\sigma_j^2(n,f) = \sum_{k=1}^{K_j} b_{jk}(f)h_{jk}(n). \tag{14.34}$$

Equation (14.34) can be interpreted by expressing the matrix $\hat{\mathbf{V}}_j = [\sigma_j^2(n,f)]_{fn}$ as a product of two matrices $\mathbf{B}_j = [b_{jk}(f)]_{fk}$ and $\mathbf{H}_j = [h_{jk}(n)]_{kn}$. This leads to generative models of the STFT coefficients $s_j(n,f)$ (Févotte *et al.*, 2009) and the spatial images $\mathbf{c}_j(n,f)$ (Ozerov and Févotte, 2010)

$$s_j(n,f) \sim \mathcal{N}_c\left(s_j(n,f) \mid 0, \sum_k b_{jk}(f)h_{jk}(n) \right), \tag{14.35}$$

$$\mathbf{c}_j(n,f) \sim \mathcal{N}_c\left(\mathbf{c}_j(n,f) \mid \mathbf{0}_I, \sum_k b_{jk}(f)h_{jk}(n)\mathbf{R}_j(f) \right). \tag{14.36}$$

Multichannel source separation methods using this model or its variants are called multichannel NMF (Ozerov and Févotte, 2010; Kameoka *et al.*, 2010; Sawada *et al.*, 2013; Nikunen and Virtanen, 2014; Kitamura *et al.*, 2015). They generalize the single-channel Itakura-Saito (IS) NMF methods reviewed in Chapters 8 and 9 to the multichannel case.

With this model, the entire set of spectral templates is partitioned into subsets associated with individual sources. It is also possible to allow all the spectral templates to be shared by every source and let the contribution of the kth spectral template to source j be determined in a learning-free manner (Ozerov *et al.*, 2011; Sawada *et al.*, 2013; Nikunen and Virtanen, 2014; Kitamura *et al.*, 2015). To do so, we drop the index j from $b_{jk}(f)$ and $h_{jk}(n)$, and instead introduce a continuous indicator variable $\phi_{jk} \geq 0$ such

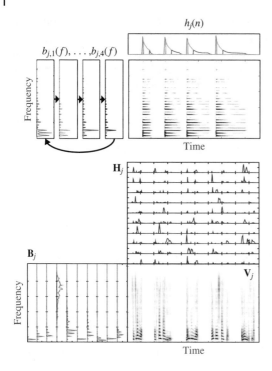

Figure 14.3 Illustration of the HMM and multichannel NMF spectral models.

that $\sum_j \phi_{jk} = 1$. ϕ_{jk} can be interpreted as the expectation of a binary indicator variable, describing to which of the J sources the kth template is assigned. The power spectrogram $\sigma_j^2(n,f)$ of source j can thus alternatively be modeled as

$$\sigma_j^2(n,f) = \sum_{k=1}^{K} \phi_{jk} b_k(f) h_k(n). \tag{14.37}$$

This spectral model is a form of nonnegative tensor factorization (NTF), which results in multichannel NTF.

14.2.1.3 AR and Variants

Another reasonable assumption we can make about source power spectrograms is spectral continuity. This amounts to the assumption that the magnitudes of the STFT coefficients in all frequency bands originating from the same source tend to vary coherently over time. The most naive way would be to assume a flat spectrum with a time-varying scale

$$\sigma_j^2(n,f) = h_j(n). \tag{14.38}$$

This is actually a particular case of the NMF model (14.34) where $K_j = 1$ and $b_{j,1}(f) = 1$, which means each source has only one flat-shaped template. Under this constraint, assuming that $s_j(n,0), \dots, s_j(n, F-1)$ independently follow (14.35) can be interpreted as assuming that the ℓ_2 norm $\|[s_j(n,0), \dots, s_j(n,F-1)]^T\|_2 = \sqrt{\sum_f |s_j(n,f)|^2}$ follows a Gaussian distribution with time-varing variance $h_j(n)$. This is analogous to the assumption employed by independent vector analysis (IVA) (see Section 13.3) where the ℓ_2 norm is assumed to follow a supergaussian distribution, which is shown to be effective in eliminating the inherent permutation indeterminacy of FD-ICA.

Other representations ensuring spectral continuity include the AR model (also known as the all-pole model) (Dégerine and Zaïdi, 2004; Yoshioka *et al.*, 2011)

$$\sigma_j^2(n,f) = \frac{\sigma_j^2(n)}{|1 - \alpha_1(n)e^{-2j\pi f/F} - \cdots - \alpha_{N'}(n)e^{-2j\pi N'f/F}|^2},\tag{14.39}$$

where $\alpha_1(n), \dots, \alpha_{N'}(n)$ denote the AR parameters at time n and N' is the number of poles. This expression is justified by the fact that the power spectrum of speech can be approximated fairly well by an excitation-filter representation using an all-pole model as the vocal tract filter. A combination of the AR model and the NMF model has also been proposed (Kameoka and Kashino, 2009; Kameoka *et al.*, 2010). With this model, the power spectrum of a source is expressed as the sum of all possible pairs of excitation and filter templates scaled by time-varying amplitudes

$$\sigma_j^2(n,f) = \sum_k \sum_{l=1}^{L} \frac{b_{jk}(f)h_{jkl}(n)}{|1 - \alpha_{jl,1}e^{-2j\pi f/F} - \cdots - \alpha_{jl,N'}e^{-2j\pi N'f/F}|^2},\tag{14.40}$$

where $b_{jk}(f)$ denotes the kth excitation spectral template, the denominator is the lth all-pole vocal tract spectral template, and $h_{jkl}(n)$ denotes the time-varying amplitude of the (k, l)th excitation-filter pair of source j. We can easily confirm that when $L = 1$ and $N' = 0$, this model reduces to the NMF model (14.34). Note that these spectral templates can be either pretrained using training samples or estimated from the mixture signal in a learning-free manner.

Another way to impose a certain structure on $\sigma_j^2(n,f)$ is to place a prior distribution over $\sigma_j^2(n,f)$. For example, a prior distribution for ensuring spectral continuity can be designed using an inverse-gamma chain (Duong *et al.*, 2011)

$$\sigma_j^2(n,f) \mid \sigma_j^2(n,f-1) \sim \mathcal{IG}(\sigma_j^2(n,f) \mid \alpha, (\alpha-1)\sigma_j^2(n,f-1)),\tag{14.41}$$

where $\mathcal{IG}(\cdot \mid \alpha, \beta)$ denotes the inverse gamma distribution with shape parameter $\alpha > 0$ and scale parameter $\beta > 0$

$$\mathcal{IG}(v \mid \alpha, \beta) = \frac{\beta^\alpha}{\Gamma(\alpha)} v^{-\alpha-1} e^{-\beta/v},\tag{14.42}$$

whose mean is $\beta/(\alpha - 1)$. It is also possible to ensure temporal continuity by assuming

$$\sigma_j^2(n,f) \mid \sigma_j^2(n-1,f) \sim \mathcal{IG}(\sigma_j^2(n,f) \mid \alpha, (\alpha-1)\sigma_j^2(n-1,f)).\tag{14.43}$$

These priors can be combined with NMF (see Section 9.3).

14.2.1.4 Composite Models and DNN

A general flexible framework with various combinations of these spectral models is presented by Ozerov *et al.* (2012) and Adiloğlu and Vincent (2016). It should also be noted that a DNN-based approach has been proposed recently (Nugraha *et al.*, 2016), where DNNs are used to model the source power spectrograms and combined with the Gaussian framework to exploit the spatial information (refer to Section 12.4 for details).

14.2.2   Spatial Models

Spatial modeling consists of constraining the spatial covariances $\mathbf{R}_j(f)$ in (14.3). Constraints are usually introduced by reparameterizing $\mathbf{R}_j(f)$, by imposing some prior distribution on it, or both.

Assuming that the narrowband approximation (14.11) holds, the spatial covariance may simply be constrained as

$$\mathbf{R}_j(f) = \mathbf{a}_j(f)\mathbf{a}_j^H(f), \tag{14.44}$$

which restricts the rank of $\mathbf{R}_j(f)$ to 1. This model is called the rank-1 model. It was used by Févotte and Cardoso (2005), Ozerov and Févotte (2010) and many other authors. Alternatively, it was later proposed by Duong *et al.* (2010a) and Sawada *et al.* (2013) to consider an unconstrained full-rank model for $\mathbf{R}_j(f)$. This model partly overcomes the limitations of the narrowband approximation, and it better handles mixtures with long reverberation times (RT60s).

A popular way of constraining spatial models is to introduce constraints related to the source direction of arrival (DOA). Izumi *et al.* (2007) simply constrained the rank-1 model to $\mathbf{R}_j(f) = \tilde{\mathbf{d}}(\alpha_j, f)\tilde{\mathbf{d}}^H(\alpha_j, f)$, with $\tilde{\mathbf{d}}(\alpha_j, f)$ the relative steering vector (3.15) corresponding to the jth source DOA α_j. The unknown source DOAs α_j are then inferred from the mixture. Duong *et al.* (2010a) extended this expression to the full-rank model by adding the covariance matrix of the diffuse reverberation field (see Section 3.4.3.2). Duong *et al.* (2013) allowed some deviation from this constraint by setting an inverse-Wishart prior on $\mathbf{R}_j(f)$ whose mean is parameterized by the DOA.

Alternatively, Nikunen and Virtanen (2014) considered a DOA grid $\{\alpha_k\}_{k=1}^K$ and constrained the spatial covariance matrix of each source as

$$\mathbf{R}_j(f) = \sum_{k=1}^{K} q_{jk}\tilde{\mathbf{d}}(\alpha_k, f)\tilde{\mathbf{d}}^H(\alpha_k, f), \tag{14.45}$$

with nonnegative weights q_{jk}. Such a combination of DOA-based rank-1 models (also called DOA kernels) makes it possible to model not only the direct path, but also reflections. Kameoka *et al.* (2012) and Higuchi and Kameoka (2015) proposed similar *DOA mixture models* where the acoustic transfer function $\mathbf{a}_j(f)$ within the rank-1 model (Kameoka *et al.*, 2012) or the full-rank model $\mathbf{R}_j(f)$ (Higuchi and Kameoka, 2015) are distributed as K-component mixture models and the distribution of each component constrains the corresponding spatial model to be close to a predefined DOA α_k.

14.3   Parameter Estimation Criteria and Algorithms

14.3.1   Parameter Estimation Criteria

Once a spectral model and a spatial model have been specified for each source, a criterion must be chosen for model parameter estimation. Let θ denote the full set of parameters of the chosen models. For example, in the case of NMF spectral models and full-rank spatial models (Arberet *et al.*, 2010) θ consists of the NMF parameters and the full-rank spatial covariance matrices of all sources. Specifying a parameter estimation criterion resides in defining a cost or an objective function to be optimized over θ given a multichannel mixture $\mathcal{X}$.

ML is one of the most popular criteria (Ozerov and Févotte, 2010; Duong *et al.*, 2010a; Sawada *et al.*, 2013). It consists of maximizing the log-likelihood

$$\mathcal{M}^{\text{ML}}(\theta) = \log p(\mathcal{X} \mid \theta). \tag{14.46}$$

In case of the local Gaussian model (14.3), this is equivalent to minimizing the cost

$$C^{\text{IS}}(\theta) = \sum_{nf} \text{tr}(\hat{\boldsymbol{\Sigma}}_{\mathbf{x}}(n,f)\boldsymbol{\Sigma}_{\mathbf{x}}^{-1}(n,f)) - \log \, \det(\hat{\boldsymbol{\Sigma}}_{\mathbf{x}}(n,f)\boldsymbol{\Sigma}_{\mathbf{x}}^{-1}(n,f)) - I, \tag{14.47}$$

where $\boldsymbol{\Sigma}_{\mathbf{x}}(n,f) = \sum_j \sigma_j^2(n,f)\mathbf{R}_j(f)$ is the model covariance and $\hat{\boldsymbol{\Sigma}}_{\mathbf{x}}(n,f) = \mathbb{E}\{\mathbf{x}(n,f)\mathbf{x}^H(n,f)\}$ is an estimate of the data covariance (Ozerov *et al.*, 2012).[4] This cost is a multichannel extension of the IS divergence (see Section 8.2.2.4).

We see that optimizing criterion (14.47) consists in minimizing a measure of fit between the model covariance $\boldsymbol{\Sigma}_{\mathbf{x}}(n,f)$ and the data covariance $\hat{\boldsymbol{\Sigma}}_{\mathbf{x}}(n,f)$. Sawada *et al.* (2013) and Nikunen and Virtanen (2014) proposed to replace this measure of fit by the Frobenius norm. This leads to the cost

$$C^{\text{EUC}}(\theta) = \sum_{nf} \|\hat{\boldsymbol{\Sigma}}_{\mathbf{x}}(n,f) - \boldsymbol{\Sigma}_{\mathbf{x}}(n,f)\|_F^2 \tag{14.48}$$

that is a multichannel generalization of the squared Euclidean (EUC) distance (see Section 8.2.2.1). The computation of the model and data covariances is then modified such that they scale with the magnitude of the data, since the EUC distance is usually applied to magnitude spectra rather than power spectra in the single-channel case.

MAP estimation is an alternative to ML (14.46) that maximizes the log-posterior

$$\mathcal{M}^{\text{MAP}}(\theta) = \log p(\mathcal{X}, \theta) = \log p(\mathcal{X} \mid \theta) + \log p(\theta) \tag{14.49}$$

with a suitable prior distribution $p(\theta)$ on the model parameters. Using MAP instead of ML results in the additional term $-\log p(\theta)$ in the corresponding cost functions (14.47) or (14.48).

As opposed to ML (14.46) and MAP (14.49), where a point estimate of the model parameters θ is sought, *variational Bayesian* (VB) inference (Kameoka *et al.*, 2012; Adiloğlu and Vincent, 2016; Kounades-Bastian *et al.*, 2016) aims to estimate the posterior distribution of the source spatial images $C = \{\mathbf{c}_j(n,f)\}_{jnf}$ while marginalizing over all possible model parameters:

$$p(C \mid \mathcal{X}) = \frac{p(C, \mathcal{X})}{p(\mathcal{X})} = \frac{\int p(C, \theta, \mathcal{X})d\theta}{\int\int p(C, \theta, \mathcal{X})dCd\theta}. \tag{14.50}$$

Since the integrals in (14.50) are computationally intractable, a factored approximation of the joint posterior $p(C, \theta, \mathcal{X})$ is assumed. The criterion to be minimized is then the Kullback–Leibler (KL) divergence between the true posterior and the factored approximation (see Section 14.3.2.3 for details).

---

4 Contrary to (14.46), (14.47) takes finite values only when $\hat{\boldsymbol{\Sigma}}_{\mathbf{x}}(n,f)$ is full-rank. To cope with this, the term $-\log \, \det \hat{\boldsymbol{\Sigma}}_{\mathbf{x}}(n,f)$ may be removed from the cost, since it is independent from θ.

14.3.2 Parameter Estimation Algorithms

14.3.2.1 EM Algorithm

We here formulate the *expectation-maximization* (EM) algorithm (Dempster *et al.*, 1977) as applied to optimize the MAP criterion (14.49), since it is more general than the ML criterion (14.46) and reduces to ML in the case of a noninformative prior $p(\theta) \propto 1$. In most cases $\mathcal{M}^{\mathrm{MAP}}(\theta)$ has several local and global maxima, and there is no closed-form solution for a global maximum.

To find a local maximum the EM algorithm consists of first defining so-called *latent data* (also called *hidden data*) $\mathcal{Z}$ and then iterating the following two steps:

- *E-step*: Compute the posterior distribution of the latent data $p(\mathcal{Z} \mid \mathcal{X}, \theta^{(m)})$ and derive the auxiliary function[5]

$$Q(\theta, \theta^{(m)}) = \mathbb{E}_{\mathcal{Z}|\mathcal{X},\theta^{(m)}} \left\{ \log \frac{p(\mathcal{X}, \mathcal{Z}, \theta)}{p(\mathcal{Z} \mid \mathcal{X}, \theta^{(m)})} \right\}. \tag{14.51}$$

- *M-step*: Update the model parameter estimates to maximize the auxiliary function:

$$\theta^{(m+1)} = \underset{\theta}{\mathrm{argmax}} \, Q(\theta, \theta^{(m)}), \tag{14.52}$$

where $\theta^{(m)}$ are the model parameter estimates obtained in the mth iteration.

It can be shown that $Q(\theta, \theta^{(m)}) \leq \mathcal{M}^{\mathrm{MAP}}(\theta)$ and $Q(\theta^{(m)}, \theta^{(m)}) = \mathcal{M}^{\mathrm{MAP}}(\theta^{(m)})$, i.e. the auxiliary function is a lower bound of the log-posterior that is tight at the current solution $\theta^{(m)}$. With this property, it can be proved that each iteration of the above EM algorithm does not decrease the value of $\mathcal{M}^{\mathrm{MAP}}(\theta)$ (Dempster *et al.*, 1977). This can be intuitively understood from the graphical illustration in Figure 14.4. A relaxed variant of the EM algorithm called generalized EM consists of replacing the M-step's closed-form maximization of the auxiliary function $Q(\theta, \theta^{(m)})$ by any update that makes it nondecreasing, i.e. $Q(\theta^{(m+1)}, \theta^{(m)}) \geq Q(\theta^{(m)}, \theta^{(m)})$.

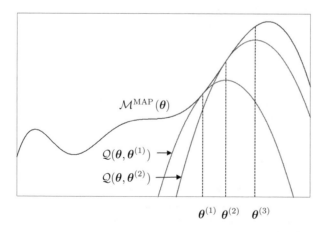

$\theta^{(1)} \quad \theta^{(2)} \quad \theta^{(3)}$

Figure 14.4 Graphical illustration of the EM algorithm for MAP estimation.

---

5 The term $-\mathbb{E}_{\mathcal{Z}}\{\log p(\mathcal{Z} \mid \mathcal{X}, \theta^{(m)})\}$ does not depend on θ hence it is often omitted in the expression of the Q function.

It is worth noting that even for the same ML or MAP criterion there may be various ways of implementing the EM algorithm. Indeed, each implementation, and as a consequence the final result, depends on the choice of the latent data $\mathcal{Z}$, the M-step parameter update in the case of the generalized EM algorithm, and the initial parameter values $\theta^{(0)}$.

In the case of multichannel NMF, Ozerov and Févotte (2010) and Arberet *et al.* (2010) defined the NMF components $\mathbf{y}_{jk}(n,f)$ such that $\mathbf{c}_j(n,f) = \sum_{k=1}^{K_j} \mathbf{y}_{jk}(n,f)$ and each component $\mathbf{y}_{jk}(n,f)$ has a zero-mean Gaussian distribution with covariance $b_{jk}(f)h_{jk}(n)\mathbf{R}_j(f)$[6] and they consider these components as latent data. Alternatively, Duong *et al.* (2010a) considered directly the source images C as latent data. Ozerov *et al.* (2012) showed that a source image with a rank-r spatial model can be represented as the sum of r *subsources*, each modeled by a rank-1 spatial model. Considering those subsources as latent data makes it possible to specify a unified EM algorithm suitable for spatial models of any rank (Ozerov *et al.*, 2012). In the case of the sparse mixing model (14.15), the indices $z(n,f)$ of the active sources are typically considered then as latent data instead, which allows considerable computational savings in the resulting EM algorithm (Thiemann and Vincent, 2013).

Several approaches (Ozerov *et al.*, 2011, 2012) considered the source spatial images or subsources as latent data and employed the multiplicative update rules of single-channel IS-NMF (see Section 8.2.3.1) within the M-step, which results in variants of the generalized EM algorithm. These approaches, which are usually referred to as (generalized) EM with multiplicative updates, often allow speeding up the algorithm's convergence (Ozerov *et al.*, 2011).

14.3.2.2 MM Algorithm

The *majorization-minimization* (MM) algorithm (also known as *auxiliary function-based* optimization) (Leeuw and Heiser, 1977; Hunter and Lange, 2004) is a generalization of the EM algorithm. When constructing an MM algorithm for a given minimization problem, the main issue is to design an auxiliary function called a majorizer that is guaranteed to never go below the cost function. If such a majorizer is properly designed, an algorithm that iteratively minimizes the majorizer is guaranteed to converge to a stationary point of the cost function. The MM algorithm was used for single-channel NMF by several authors (Lee and Seung, 2000; Nakano *et al.*, 2010; Févotte and Idier, 2011). In general, if we can build a tight majorizer that is easy to optimize, we can expect to obtain a fast converging algorithm.

Suppose $C(\theta)$ is a cost function that we want to minimize with respect to θ. A majorizer $Q(\theta, \alpha)$ is defined as a function satisfying

$$C(\theta) = \min_{\alpha} Q(\theta, \alpha) \tag{14.53}$$

where α is an auxiliary variable. $C(\theta)$ can then be shown to be nonincreasing under the updates

$$\theta \leftarrow \underset{\theta}{\arg\min}\, Q(\theta, \alpha) \tag{14.54}$$

$$\alpha \leftarrow \underset{\alpha}{\arg\min}\, Q(\theta, \alpha). \tag{14.55}$$

---

6 This model is equivalent to (14.36).

This can be proved as follows. Let us denote the iteration number by m, set θ to an arbitrary value $\theta^{(m)}$ and define $\alpha^{(m+1)} = \text{argmin}_\alpha Q(\theta^{(m)}, \alpha)$ and $\theta^{(m+1)} = \text{argmin}_\theta Q(\theta, \alpha^{(m+1)})$. First, it is obvious that $C(\theta^{(m)}) = Q(\theta^{(m)}, \alpha^{(m+1)})$. Next, we can confirm that $Q(\theta^{(m)}, \alpha^{(m+1)}) \geq Q(\theta^{(m+1)}, \alpha^{(m+1)})$ since $\theta^{(m+1)}$ is the minimizer of $Q(\theta, \alpha^{(m+1)})$ with respect to θ. By definition, it is obvious that $Q(\theta^{(m+1)}, \alpha^{(m+1)}) \geq C(\theta^{(m+1)})$ and so we can finally show that $C(\theta^{(m)}) \geq C(\theta^{(m+1)})$.

Here, we briefly show that the EM algorithm is a special case of the MM algorithm. Let $\mathcal{X}$ be the observed data, $C^{\text{MAP}}(\theta) = -\log p(\mathcal{X}, \theta)$ the cost function that we want to minimize with respect to the parameters θ, and $\mathcal{Z}$ the latent data. The latent data can be either discrete or continuous. While we consider the continuous case here, the following also applies to the discrete case by simply replacing the integral over $\mathcal{Z}$ with a summation. First, we can show that

$$C^{\text{MAP}}(\theta) = -\log \int p(\mathcal{X}, \mathcal{Z}, \theta) d\mathcal{Z} \tag{14.56}$$

$$= -\log \int \lambda(\mathcal{Z}) \frac{p(\mathcal{X}, \mathcal{Z}, \theta)}{\lambda(\mathcal{Z})} d\mathcal{Z} \tag{14.57}$$

$$\leq -\int \lambda(\mathcal{Z}) \log \frac{p(\mathcal{X}, \mathcal{Z}, \theta)}{\lambda(\mathcal{Z})} d\mathcal{Z}, \tag{14.58}$$

where $\lambda(\mathcal{Z})$ is an arbitrary nonnegative weight function that is subject to the normalization constraint

$$\int \lambda(\mathcal{Z}) d\mathcal{Z} = 1. \tag{14.59}$$

Expression (14.58) follows from Jensen's inequality by using the fact that the negative logarithm is a convex function. We can use the right-hand side of this inequality as the majorizer of $C^{\text{MAP}}(\theta)$. Thus, we can show that $C^{\text{MAP}}(\theta)$ is nonincreasing under the updates

$$\lambda(\mathcal{Z}) \leftarrow \text{argmin}_{\lambda(\mathcal{Z})} -\int \lambda(\mathcal{Z}) \log \frac{p(\mathcal{X}, \mathcal{Z}, \theta)}{\lambda(\mathcal{Z})} d\mathcal{Z} = p(\mathcal{Z} \mid \mathcal{X}, \theta) \tag{14.60}$$

$$\theta \leftarrow \text{argmin}_\theta -\int \lambda(\mathcal{Z}) \log \frac{p(\mathcal{X}, \mathcal{Z}, \theta)}{\lambda(\mathcal{Z})} d\mathcal{Z}. \tag{14.61}$$

Expression (14.60) stems from the fact that the inequality in (14.58) becomes an equality when

$$\frac{p(\mathcal{X}, \mathcal{Z}, \theta)}{\lambda(\mathcal{Z})} = \xi(\mathcal{X}, \theta), \tag{14.62}$$

is independent of $\mathcal{Z}$, which yields

$$\lambda(\mathcal{Z}) = \frac{p(\mathcal{X}, \mathcal{Z}, \theta)}{\xi(\mathcal{X}, \theta)} \tag{14.63}$$

$$\Rightarrow \int \lambda(\mathcal{Z}) d\mathcal{Z} = \frac{1}{\xi(\mathcal{X}, \theta)} \int p(\mathcal{X}, \mathcal{Z}, \theta) d\mathcal{Z} = 1 \tag{14.64}$$

$$\Rightarrow \xi(\mathcal{X}, \theta) = \int p(\mathcal{X}, \mathcal{Z}, \theta) d\mathcal{Z} = p(\mathcal{X}, \theta) \tag{14.65}$$

$$\Rightarrow \lambda(\mathcal{Z}) = \frac{p(\mathcal{X}, \mathcal{Z}, \theta)}{p(\mathcal{X}, \theta)} = p(\mathcal{Z} \mid \mathcal{X}, \theta). \tag{14.66}$$

We can confirm that (14.60) and (14.61) correspond to the expectation and maximization steps, respectively.

14.3.2.3 VB Algorithm

The VB approach is another extension of EM which aims to estimate the posterior distribution of all the random variables involved in the generative model. Let θ be the entire set of variables of interest (including, e.g., the source STFT coefficients, the model parameters, and the latent data) and $\mathcal{X}$ be the observed data. Our goal is to compute the posterior

$$p(\theta \mid \mathcal{X}) = \frac{p(\theta, \mathcal{X})}{p(\mathcal{X})}. \tag{14.67}$$

The joint distribution $p(\theta, \mathcal{X})$ can usually be written explicitly according to the assumed generative model. However, to obtain the exact posterior $p(\theta \mid \mathcal{X})$, we must compute $p(\mathcal{X})$, which involves an intractable integral. Instead of obtaining the exact posterior, the VB approach considers approximating this posterior variationally by minimizing

$$C^{\mathrm{VB}}(q(\theta)) = C^{\mathrm{KL}}(q(\theta) \mid p(\theta \mid \mathcal{X})), \tag{14.68}$$

with respect to $q(\theta)$ with

$$\int q(\theta)\mathrm{d}\theta = 1, \tag{14.69}$$

where $C^{\mathrm{KL}}(\cdot \mid \cdot)$ denotes the KL divergence

$$C^{\mathrm{KL}}(q(\theta) \mid p(\theta \mid \mathcal{X})) = \int q(\theta) \log \frac{q(\theta)}{p(\theta \mid \mathcal{X})} \mathrm{d}\theta. \tag{14.70}$$

By partitioning the set of variables as $\theta = \{\theta_k\}_k$ and restricting the class of approximate distributions to those that factorize into

$$q(\theta) = \prod_k q(\theta_k) \quad \text{with} \quad \int q(\theta_k)\mathrm{d}\theta_k = 1, \tag{14.71}$$

we can use a simple block coordinate descent algorithm to find a local minimum of (14.68) for each factor in turn. It can be shown using the calculus of variations that the optimal distribution for each factor is

$$q(\theta_k) \propto \exp[\mathbb{E}_{q(\theta \setminus \theta_k)}\{\log p(\theta, \mathcal{X})\}], \tag{14.72}$$

where $\mathbb{E}_{q(\theta \setminus \theta_k)}\{\log p(\theta, \mathcal{X})\}$ is the expectation of the joint probability of the data and the variables, taken over all variables except θ_k.

14.3.3 Categorization of Existing Methods

Table 14.1 categorizes the various approaches discussed above according to the underlying mixing model, spectral model, spatial model, estimation criterion, and algorithm.

14.4 Detailed Presentation of Some Methods

We now give detailed descriptions of two popular parameter estimation algorithms. For both algorithms, we consider the narrowband mixing model (14.11), the full-rank unconstrained spatial model (see Section 14.2.2), and the NTF spectral model (14.37).

Table 14.1 Categorization of existing approaches according to the underlying mixing model, spectral model, spatial model, estimation criterion, and algorithm.

Method	Mixing	Spatial model	Spectral model	Criterion	Algorithm
Attias (2003)	Subband filter mix	Rank-1	GMM	VB	VB
Izumi *et al.* (2007)	Sparse	Rank-1	Unconstrained	EUC	EM
Duong *et al.* (2010b)	Additive	Full-rank	Unconstrained	IS	Generalized EM
Kameoka *et al.* (2010)	Subband filter demix	Rank-1	NMF+AR	IS	EM
Ozerov and Févotte (2010)	Narrowband mix	Rank-1	NMF	IS	Generalized EM
Yoshioka *et al.* (2011)	Subband filter demix	Rank-1	AR	IS	Block coord. descent
Kameoka *et al.* (2012)	Sparse	Rank-1 DOA mixture	Unconstrained	VB	VB
Ozerov *et al.* (2012)	Additive	Any rank	NMF/GMM/ excit.-filter/...	IS	Generalized EM
Duong *et al.* (2013)	Additive	Full-rank DOA prior	Unconstrained	IS (MAP)	Generalized EM
Sawada *et al.* (2013)	Additive	Full-rank	NMF	IS/ EUC	MM
Nikunen and Virtanen (2014)	Narrowband mix	Rank-1 DOA kernels	NMF	EUC	MM
Higuchi and Kameoka (2015)	Subband filter mix	Full-rank DOA mixture	HMM	IS	MM
Kitamura *et al.* (2015)	Subband filter demix	Rank-1	NMF	IS	MM
Adiloğlu and Vincent (2016)	Narrowband mix	Rank-1	NMF/ excit.-filter	VB	VB
Nugraha *et al.* (2016)	Additive	Full-rank	DNN	IS	EM

14.4.1 IS Multichannel NTF EM Algorithm

The EM algorithm presented below is a combination of those presented by Ozerov and Févotte (2010), Arberet *et al.* (2010), and Ozerov *et al.* (2011). More specifically the spatial full-rank model is that of Arberet *et al.* (2010), the spectral NTF model is that of Ozerov *et al.* (2011), and the choice of the NTF components as latent data follows Ozerov and Févotte (2010).

Let us introduce the NTF components $\mathbf{y}_{jk}(n,f)$ such that $\mathbf{c}_j(n,f) = \sum_{k=1}^{K_j} \mathbf{y}_{jk}(n,f)$ and each component $\mathbf{y}_{jk}(n,f)$ is distributed as

$$\mathbf{y}_{jk}(n,f) \sim \mathcal{N}_c(\mathbf{y}_{jk}(n,f) \mid \mathbf{0}_I, \boldsymbol{\Sigma}_{\mathbf{y}_{jk}}(n,f)) \tag{14.73}$$

with

$$\boldsymbol{\Sigma}_{\mathbf{y}_{jk}}(n,f) = \phi_{jk}b_k(f)h_k(n)\mathbf{R}_j(f). \tag{14.74}$$

This formulation is strictly equivalent to the original model. We denote the full set of model parameters as $\theta = \{\{\phi_{jk}\}_{jk}, \{b_k(f)\}_{kf}, \{h_k(n)\}_{kn}, \{\mathbf{R}_j(f)\}_{jf}\}$.

Following Ozerov and Févotte (2010), we consider the set of NTF components $\mathcal{Y} = \{\mathbf{y}_{jk}(n,f)\}_{jknf}$ as latent data. Assuming a noninformative prior $p(\theta) \propto 1$, the auxiliary function (14.51) for the ML criterion can be written as

$$Q(\theta, \theta^{(m)}) = \mathbb{E}_{\mathcal{Y}|\mathcal{X},\theta^{(m)}}\{\log p(\mathcal{Y} \mid \theta)\} + \mathrm{cst}(\theta^{(m)})$$

$$= \sum_{jknf} -\log\det(\pi\boldsymbol{\Sigma}_{\mathbf{y}_{jk}}(n,f))$$

$$- \mathrm{tr}(\widehat{\boldsymbol{\Sigma}}_{\mathbf{y}_{jk}}(n,f)\boldsymbol{\Sigma}_{\mathbf{y}_{jk}}^{-1}(n,f)) + \mathrm{cst}(\theta^{(m)}) \tag{14.75}$$

where the term $\mathrm{cst}(\theta^{(m)})$ depends only on $\theta^{(m)}$ and is independent of θ, thus has no influence on the optimization in (14.52), and

$$\widehat{\boldsymbol{\Sigma}}_{\mathbf{y}_{jk}}(n,f) = \mathbb{E}_{\mathcal{Y}|\mathcal{X},\theta^{(m)}}\{\mathbf{y}_{jk}(n,f)\mathbf{y}_{jk}^H(n,f)\}$$

$$= \widehat{\mathbf{y}}_{jk}(n,f)\widehat{\mathbf{y}}_{jk}^H(n,f) + (\mathbf{I}_I - \mathbf{W}_{jk}^H(n,f))\boldsymbol{\Sigma}_{\mathbf{y}_{jk}}^{(m)}(n,f) \tag{14.76}$$

with $\mathbf{I}_I$ the $I \times I$ identity matrix and

$$\mathbf{W}_{jk}(n,f) = \left(\sum_{j'k'} \boldsymbol{\Sigma}_{\mathbf{y}_{j'k'}}^{(m)}(n,f)\right)^{-1}\boldsymbol{\Sigma}_{\mathbf{y}_{jk}}^{(m)}(n,f) \tag{14.77}$$

$$\widehat{\mathbf{y}}_{jk}(n,f) = \mathbf{W}_{jk}^H(n,f)\mathbf{x}(n,f). \tag{14.78}$$

Note that the maximum of $Q(\theta, \theta^{(m)})$ over θ has no closed-form solution. However, it is possible to compute a closed-form maximum for each of the four parameter subsets given the other three subsets. Alternately maximizing each subset guarantees that the auxiliary function is nondecreasing. Thus, we obtain a generalized EM algorithm that can be summarized as follows:

- *E-step*: Compute the statistics $\widehat{\boldsymbol{\Sigma}}_{\mathbf{y}_{jk}}(n,f)$ as in (14.76).
- *M-step*: Update the model parameters θ as

$$\mathbf{R}_j(f) = \frac{1}{N}\sum_n \frac{1}{\sum_{jk}\phi_{jk}b_k(f)h_k(n)}\widehat{\boldsymbol{\Sigma}}_{\mathbf{y}_{jk}}(n,f) \tag{14.79}$$

$$\widehat{v}_{jk}(n,f) = \frac{1}{I}\mathrm{tr}(\mathbf{R}_j^{-1}(f)\widehat{\boldsymbol{\Sigma}}_{\mathbf{y}_{jk}}(n,f)) \tag{14.80}$$

$$\phi_{jk} = \frac{1}{NF}\sum_{nf} \frac{\widehat{v}_{jk}(n,f)}{b_k(f)h_k(n)} \tag{14.81}$$

$$b_k(f) = \frac{1}{JN} \sum_{jn} \frac{\hat{v}_{jk}(n,f)}{\phi_{jk} h_k(n)} \qquad (14.82)$$

$$h_k(n) = \frac{1}{JF} \sum_{jf} \frac{\hat{v}_{jk}(n,f)}{\phi_{jk} b_k(f)}. \qquad (14.83)$$

14.4.2 IS Multichannel NMF MM Algorithm

We now give a detailed description of the MM algorithm for multichannel NMF of Sawada *et al.* (2013). This algorithm is an extension of the MM algorithm originally developed by Kameoka *et al.* (2006) for solving general model fitting problems using the IS divergence. We show first how to derive the MM algorithm for single-channel NMF with the IS divergence and then how to extend it to the multichannel case.

The cost function for single-channel NMF with the IS divergence can be written as

$$C^{IS}(\theta) = \sum_{nf} \left(\frac{|x(n,f)|^2}{\sigma^2(n,f)} - \log \frac{|x(n,f)|^2}{\sigma^2(n,f)} - 1 \right), \qquad (14.84)$$

where $x(n,f)$ are the observed STFT coefficients, $\sigma^2(n,f) = \sum_k b_k(f) h_k(n)$ and θ is a set consisting of $\mathbf{B} = [b_k(f)]_{kf}$ and $\mathbf{H} = [h_k(n)]_{kn}$ (Févotte *et al.*, 2009). Although it is difficult to obtain a closed-form expression of the global minimum, a majorizer of $C^{IS}(\theta)$ can be obtained as follows (Kameoka *et al.*, 2006). First, by using the fact that the function $f(x) = 1/x$ is convex for $x > 0$, we can use Jensen's inequality to obtain

$$\frac{|x(n,f)|^2}{\sigma^2(n,f)} \leq \sum_k \rho_k(n,f) \frac{|x(n,f)|^2}{b_k(f)h_k(n)/\rho_k(n,f)} = \sum_k \rho_k^2(n,f) \frac{|x(n,f)|^2}{b_k(f)h_k(n)}, \qquad (14.85)$$

where $\rho_k(n,f) \geq 0$ is an arbitrary weight that must satisfy $\sum_k \rho_k(n,f) = 1$. It can be shown that the equality holds when

$$\rho_k(n,f) = \frac{b_k(f)h_k(n)}{\sum_{k'} b_{k'}(f)h_{k'}(n)}. \qquad (14.86)$$

Next, since the function $f(x) = \log x$ is concave for $x > 0$, the tangent to $f(x)$ is guaranteed to never lie below $f(x)$. Thus, we have

$$\log \sigma^2(n,f) \leq \frac{\sigma^2(n,f) - \kappa(n,f)}{\kappa(n,f)} + \log \kappa(n,f) \qquad (14.87)$$

for any $\kappa(n,f) > 0$. The equality holds when

$$\kappa(n,f) = \sigma^2(n,f). \qquad (14.88)$$

By combining these inequalities, we have

$$C^{IS}(\theta) \leq \sum_{nf} \left(\sum_k \rho_k^2(n,f) \frac{|x(n,f)|^2}{b_k(f)h_k(n)} + \frac{\sigma^2(n,f) - \kappa(n,f)}{\kappa(n,f)} - \log \frac{|x(n,f)|^2}{\kappa(n,f)} - 1 \right). \qquad (14.89)$$

Hence, we can use the right-hand side of this inequality as a majorizer for $C^{IS}(\theta)$ where $\{\rho_k(n,f)\}_{knf}$ and $\{\kappa(n,f)\}_{nf}$ are auxiliary variables. Here, (14.86) and (14.88) correspond to the update rules for the auxiliary variables. What is particularly notable about this majorizer is that while $C^{IS}(\theta)$ involves nonlinear interaction of $b_1(f)h_1(n), \ldots, b_K(f)h_K(n)$, it is given in a separable form expressed as a sum of the $1/b_k(f)h_k(n)$ and $b_k(f)h_k(n)$ terms, which are relatively easy to optimize with respect to $b_k(f)$ and $h_k(n)$. By differentiating this majorizer with respect to $b_k(f)$ and $h_k(n)$, and setting the results to zero, we obtain the following update rules for $b_k(f)$ and $h_k(n)$:

$$b_k(f) = \sqrt{\frac{\sum_n \rho_k^2(n,f)|x(n,f)|^2/h_k(n)}{\sum_n h_k(n)/\kappa(n,f)}} \tag{14.90}$$

$$h_k(n) = \sqrt{\frac{\sum_f \rho_k^2(n,f)|x(n,f)|^2/b_k(f)}{\sum_f b_k(f)/\kappa(n,f)}}. \tag{14.91}$$

with $\rho_k(n,f)$ and $\kappa(n,f)$ computed as in (14.86) and (14.88).

Now, let us turn to the cost function (14.47) for multichannel NMF where

$$\Sigma_x(n,f) = \sum_{jk} \phi_{jk} b_k(f) h_k(n) \mathbf{R}_j(f). \tag{14.92}$$

We can confirm that when the number of channels and sources is $I = 1$ and $J = 1$, respectively, and $\phi_{jk} = 1$, this cost function reduces to the cost (14.84). We can obtain a majorizer given in a separable form in the same way as the single-channel case. By analogy with (14.85), we have

$$\text{tr}(\hat{\Sigma}_x(n,f)\Sigma_x^{-1}(n,f)) \leq \sum_{jk} \frac{\text{tr}(\hat{\Sigma}_x(n,f)\mathbf{P}_{jk}(n,f)\mathbf{R}_j^{-1}(f)\mathbf{P}_{jk}(n,f))}{\phi_{jk}b_k(f)h_k(n)} \tag{14.93}$$

for the first term with an arbitrary $I \times I$ complex-valued matrix $\mathbf{P}_{jk}(n,f)$ such that $\sum_{jk} \mathbf{P}_{jk}(n,f) = \mathbf{I}_I$, and

$$\log \det(\Sigma_x(n,f)) \leq \text{tr}(\mathbf{K}^{-1}(n,f)\Sigma_x(n,f)) + \log \det \mathbf{K}(n,f) - I \tag{14.94}$$

for the second term with a positive definite matrix $\mathbf{K}(n,f)$ (Sawada *et al.*, 2013). We can show that the equalities in (14.93) and (14.94) hold when

$$\mathbf{P}_{jk}(n,f) = \phi_{jk}b_k(f)h_k(n)\mathbf{R}_j(f)\Sigma_x^{-1}(n,f) \tag{14.95}$$

$$\mathbf{K}(n,f) = \Sigma_x(n,f). \tag{14.96}$$

By combining these inequalities, we have

$$C^{IS}(\theta) \leq \sum_{nf} \left[\sum_{jk} \frac{\text{tr}(\hat{\Sigma}_x(n,f)\mathbf{P}_{jk}(n,f)\mathbf{R}_j^{-1}(f)\mathbf{P}_{jk}(n,f))}{\phi_{jk}b_k(f)h_k(n)} \right.$$

$$\left. + \text{tr}(\mathbf{K}^{-1}(n,f)\Sigma_x(n,f)) - \log \det(\mathbf{K}^{-1}(n,f)\hat{\Sigma}_x(n,f)) - I \right]. \tag{14.97}$$

Hence, we can use the right-hand side of this inequality as a majorizer for $C^{\mathrm{IS}}(\theta)$ where $\mathcal{P} = [\mathbf{P}_{jk}(n,f)]_{jknf}$ and $\mathcal{K} = [\mathbf{K}(n,f)]_{nf}$ are auxiliary variables. Here, (14.95) and (14.96) correspond to the update rules for the auxiliary variables. Similarly to the single-channel case, this majorizer is given in a separable form, which is relatively easy to optimize with respect to $\mathbf{\Phi} = [\phi_{jk}]_{jk}$, $\mathbf{B} = [b_k(f)]_{kf}$, $\mathbf{H} = [h_k(n)]_{kn}$, and $\mathcal{R} = [\mathbf{R}_j(f)]_{jf}$. By differentiating this majorizer with respect to $b_k(f)$ and $h_k(n)$ and setting the results to zero, we obtain the following update rules for $b_k(f)$ and $h_k(n)$:

$$b_k(f) = \sqrt{\frac{\sum_{jn} \frac{1}{\phi_{jk} h_k(n)} \mathrm{tr}(\widehat{\mathbf{\Sigma}}_{\mathbf{x}}(n,f) \mathbf{P}_{jk}(n,f) \mathbf{R}_j^{-1}(f) \mathbf{P}_{jk}(n,f))}{\sum_{jn} \phi_{jk} h_k(n) \mathrm{tr}(\mathbf{K}^{-1}(n,f) \mathbf{\Sigma}_{\mathbf{x}}(n,f))}} \qquad (14.98)$$

$$h_k(n) = \sqrt{\frac{\sum_{jf} \frac{1}{\phi_{jk} b_k(f)} \mathrm{tr}(\widehat{\mathbf{\Sigma}}_{\mathbf{x}}(n,f) \mathbf{P}_{jk}(n,f) \mathbf{R}_j^{-1}(f) \mathbf{P}_{jk}(n,f))}{\sum_{jf} \phi_{jk} b_k(f) \mathrm{tr}(\mathbf{K}^{-1}(n,f) \mathbf{\Sigma}_{\mathbf{x}}(n,f))}}. \qquad (14.99)$$

with $\mathbf{P}_{jk}(n,f)$ and $\mathbf{K}(n,f)$ computed as in (14.95) and (14.96).

As regards ϕ_{jk}, although it is necessary to take the unit sum constraint into account, here we describe a convenient approach that consists of updating ϕ_{jk} as

$$\phi_{jk} = \sqrt{\frac{\sum_{nf} \frac{1}{b_k(f) h_k(n)} \mathrm{tr}(\widehat{\mathbf{\Sigma}}_{\mathbf{x}}(n,f) \mathbf{P}_{jk}(n,f) \mathbf{R}_j^{-1}(f) \mathbf{P}_{jk}(n,f))}{\sum_{nf} b_k(f) h_k(n) \mathrm{tr}(\mathbf{K}^{-1}(n,f) \mathbf{\Sigma}_{\mathbf{x}}(n,f))}}, \qquad (14.100)$$

which minimizes the majorizer, and projecting it onto the constraint space as $\phi_{jk} \leftarrow \phi_{jk} / \sum_{j'} \phi_{j'k}$, followed by rescaling of $b_k(f)$ and $h_k(n)$. As regards $\mathbf{R}_j(f)$, the optimal update is given as the solution of the algebraic Riccati equation

$$\mathbf{R}_j(f) \mathbf{\Psi}_j(f) \mathbf{R}_j(f) = \mathbf{\Omega}_j(f), \qquad (14.101)$$

where the coefficient matrices are given by

$$\mathbf{\Psi}_j(f) = \sum_{kn} \phi_{jk} b_k(f) h_k(n) \mathbf{K}^{-1}(n,f) \qquad (14.102)$$

$$\mathbf{\Omega}_j(f) = \sum_{kn} \frac{\mathbf{P}_{jk}(n,f) \widehat{\mathbf{\Sigma}}_{\mathbf{x}}(n,f) \mathbf{P}_{jk}(n,f)}{\phi_{jk} b_k(f) h_k(n)}. \qquad (14.103)$$

Since there is a scale indeterminacy between $\mathbf{R}_j(f)$ and $\phi_{jk} b_k(f) h_k(n)$, a convenient way to eliminate the indeterminacy is to update $\mathbf{R}_j(f)$ using the above equation and then perform unit trace normalization: $\mathbf{R}_j(f) \leftarrow \mathbf{R}_j(f) / \mathrm{tr}(\mathbf{R}_j(f))$.

Sawada *et al.* (2013) compared the convergence of the EM algorithm and the MM algorithm for IS multichannel NMF.

14.4.3 Other Algorithms for Demixing Filter Estimation

For the narrowband (14.16) and subband filtering (14.17) demixing models, one popular way for estimating the demixing filters $\mathbf{W}^H(f)$ involves the natural gradient method (Amari *et al.*, 1996). Here, we show other useful methods using block coordinate descent (Ono, 2011; Kameoka *et al.*, 2010; Yoshioka *et al.*, 2011).

First, let us consider the narrowband case (14.16). Recall that the log-likelihood of $\mathbf{W}(f) = [\mathbf{w}_1(f), \dots, \mathbf{w}_J(f)]$ is given by (14.10). When the log-likelihood is given in this form, it can be maximized analytically with respect to one of the column vectors of $\mathbf{W}(f)$. Thus, we use a block coordinate descent algorithm to estimate $\mathbf{W}(f)$ by iteratively minimizing the negative log-likelihood with respect to each column vector while keeping the other column vectors fixed (Ono, 2011; Kitamura *et al.*, 2015). By keeping only the terms that depend on $\mathbf{W}(f)$ in the negative log-likelihood, the cost function for $\mathbf{W}(f)$ can be written as

$$C^{\mathrm{ML}}(\mathbf{W}(f)) = N \sum_j \mathbf{w}_j^H(f)\boldsymbol{\Sigma}_{\mathbf{x}/\sigma_j}(f)\mathbf{w}_j(f) - 2N \log \det(\mathbf{W}(f)) + \mathrm{cst}, \tag{14.104}$$

where $\boldsymbol{\Sigma}_{\mathbf{x}/\sigma_j}(f) = \frac{1}{N}\sum_n \frac{\mathbf{x}(n,f)\mathbf{x}^H(n,f)}{\sigma_j^2(n,f)}$. By computing the complex derivative of $C^{\mathrm{ML}}(\mathbf{W}(f))$ with respect to the conjugate of one column vector $\mathbf{w}_j^*(f)$,[7] and setting the result to zero, we have

$$\boldsymbol{\Sigma}_{\mathbf{x}/\sigma_j}(f)\mathbf{w}_j(f) - 2\frac{\partial}{\partial \mathbf{w}_j^*(f)} \log \det(\mathbf{W}(f)) = \mathbf{0}_J. \tag{14.105}$$

By using the matrix formula $(\partial/\partial \mathbf{W}^*) \det(\mathbf{W}) = (\mathbf{W}^{-1})^H \det(\mathbf{W})$, (14.105) can be rearranged in the following simultaneous vector equations

$$\mathbf{w}_j^H(f)\boldsymbol{\Sigma}_{\mathbf{x}/\sigma_j}(f)\mathbf{w}_j(f) = 1 \tag{14.106}$$

$$\mathbf{w}_{j'}^H(f)\boldsymbol{\Sigma}_{\mathbf{x}/\sigma_j}(f)\mathbf{w}_j(f) = 0 \text{ for } j' \neq j. \tag{14.107}$$

A solution to (14.106) and (14.107) can be found by the following updates:

$$\mathbf{w}_j(f) \leftarrow (\mathbf{W}^H(f)\boldsymbol{\Sigma}_{\mathbf{x}/\sigma_j}(f))^{-1}\mathbf{e}_j \tag{14.108}$$

$$\mathbf{w}_j(f) \leftarrow \frac{\mathbf{w}_j(f)}{\sqrt{\mathbf{w}_j^H(f)\boldsymbol{\Sigma}_{\mathbf{x}/\sigma_j}(f)\mathbf{w}_j(f)}}, \tag{14.109}$$

where $\mathbf{e}_j$ denotes the jth column of the $J \times J$ identity matrix $\mathbf{I}_J$.

Next, let us turn to the subband filtering case (14.17). When $\mathbf{W}^H(0,f)$ is invertible, (14.17) can be written equivalently as the following process

$$\mathbf{y}(n,f) = \mathbf{x}(n,f) - \sum_{n'=1}^{N'-1} \tilde{\mathbf{W}}^H(n',f)\mathbf{x}(n-n',f), \tag{14.110}$$

$$\mathbf{s}(n,f) = \mathbf{W}^H(0,f)\mathbf{y}(n,f), \tag{14.111}$$

---

7 For complex-valued differentiation and matrix formulas, see Petersen and Pedersen (2005).

where $\tilde{\mathbf{W}}^H(n',f) = -(\mathbf{W}^H(0,f))^{-1}\mathbf{W}^H(n',f)$ (Yoshioka *et al.*, 2011; Kameoka *et al.*, 2010). (14.110) can be seen as a dereverberation process of the observed mixture signal $\mathbf{x}(n,f)$ described as a multichannel AR system with regression matrices $\tilde{\mathcal{W}} = \{\tilde{\mathbf{W}}^H(n',f)\}_{n'f}$ whereas (14.111) can be seen as a narrowband demixing process of the dereverberated mixture signal $\mathbf{y}(n,f)$. When $\mathbf{W}^H(0,f)$ is fixed, it can be shown that the log-likelihood of $\tilde{\mathcal{W}}$ becomes equal up to a sign to the objective function of a vector version of the linear prediction problem (also called multichannel linear prediction), which can be maximized with respect to $\tilde{\mathcal{W}}$ by solving a Yule–Walker equation. When $\tilde{\mathcal{W}}$ is fixed, on the other hand, the log-likelihood of $\mathbf{W}^H(0,f)$ becomes equal up to a sign and constant terms to (14.104) with $\mathbf{x}(n,f)$ replaced with $\mathbf{y}(n,f)$, namely $\mathbf{\Sigma}_{\mathbf{y}/\sigma_j}(f) = \frac{1}{N}\sum_n \mathbf{y}(n,f)\mathbf{y}^H(n,f)/\sigma_j^2(n,f)$, which can be locally maximized with respect to $\mathbf{W}^H(0,f)$ using the natural gradient method or the method described above. Thus, we can find estimates of $\mathbf{W}^H(0,f)$ and $\tilde{\mathcal{W}}$ by optimizing each of them in turn (Yoshioka *et al.*, 2011; Kameoka *et al.*, 2010).

14.5 Summary

The Gaussian framework for multichannel source separation is particularly noteworthy in that it provides a flexible way to incorporate source spectral models and spatial covariance models into a generative model of multichannel signals so that it can combine various clues to handle reverberation, underdetermined mixtures, and permutation alignment problems. It is also remarkable in that it allows us to develop powerful and efficient algorithms for parameter inference and estimation, taking advantage of the properties of Gaussian random variables. In this chapter, we presented the main steps to formulate Gaussian model-based methods, examples of source spectral models and spatial models along with the motivations behind them, and detailed derivations of several popular algorithms for multichannel NMF. For extensions to moving sources or microphones, refer to Chapter 19.

Acknowledgment

We thank E. Vincent for help with writing this chapter.

Bibliography

Adiloğlu, K. and Vincent, E. (2016) Variational Bayesian inference for source separation and robust feature extraction. *IEEE/ACM Transactions on Audio, Speech, and Language Processing*, **24** (10), 1746–1758.

Amari, S., Cichocki, A., and Yang, H.H. (1996) A new learning algorithm for blind signal separation, in *Proceedings of Neural Information Processing Systems*, pp. 757–763.

Arberet, S., Ozerov, A., Duong, N., Vincent, E., Gribonval, R., Bimbot, F., and Vandergheynst, P. (2010) Nonnegative matrix factorization and spatial covariance model for under-determined reverberant audio source separation, in *Proceedings of*

International Conference on Information Sciences, Signal Processing and their Applications, pp. 1–4.

Attias, H. (2003) New EM algorithms for source separation and deconvolution with a microphone array, in *Proceedings of IEEE International Conference on Audio, Speech and Signal Processing*, vol. V, pp. 297–300.

Benaroya, L., Bimbot, F., and Gribonval, R. (2006) Audio source separation with a single sensor. *IEEE Transactions on Audio, Speech, and Language Processing*, **14** (1), 191–199.

Dégerine, S. and Zaïdi, A. (2004) Separation of an instantaneous mixture of Gaussian autoregressive sources by the exact maximum likelihood approach. *IEEE Transactions on Signal Processing*, **52** (6), 1499–1512.

Dempster, A.P., Laird, N.M., and Rubin., D.B. (1977) Maximum likelihood from incomplete data via the EM algorithm. *Journal of the Royal Statistical Society: Series B*, **39** (1), 1–38.

Duong, N.Q.K., Tachibana, H., Vincent, E., Ono, N., Gribonval, R., and Sagayama, S. (2011) Multichannel harmonic and percussive component separation by joint modeling of spatial and spectral continuity, in *Proceedings of IEEE International Conference on Audio, Speech and Signal Processing*, pp. 205–208.

Duong, N.Q.K., Vincent, E., and Gribonval, R. (2010a) Under-determined reverberant audio source separation using a full-rank spatial covariance model. *IEEE Transactions on Audio, Speech, and Language Processing*, **18** (7), 1830–1840.

Duong, N.Q.K., Vincent, E., and Gribonval, R. (2010b) Under-determined reverberant audio source separation using local observed covariance and auditory-motivated time-frequency representation, in *Proceedings of International Conference on Latent Variable Analysis and Signal Separation*, pp. 73–80.

Duong, N.Q.K., Vincent, E., and Gribonval, R. (2013) Spatial location priors for Gaussian model based reverberant audio source separation. *EURASIP Journal on Advances in Signal Processing*, **2013**, 149.

Févotte, C., Bertin, N., and Durrieu, J.L. (2009) Nonnegative matrix factorization with the Itakura-Saito divergence. With application to music analysis. *Neural Computation*, **21** (3), 793–830.

Févotte, C. and Cardoso, J.F. (2005) Maximum likelihood approach for blind audio source separation using time-frequency Gaussian models, in *Proceedings of IEEE Workshop on Applications of Signal Processing to Audio and Acoustics*, pp. 78–81.

Févotte, C. and Idier, J. (2011) Algorithms for nonnegative matrix factorization with the β-divergence. *Neural Computation*, **23** (9), 2421–2456.

Higuchi, T. and Kameoka, H. (2015) Unified approach for audio source separation with multichannel factorial HMM and DOA mixture model, in *Proceedings of European Signal Processing Conference*, pp. 2043–2047.

Hunter, D.R. and Lange, K. (2004) A tutorial on MM algorithms. *The American Statistician*, **58** (1), 30–37.

Izumi, Y., Ono, N., and Sagayama, S. (2007) Sparseness-based 2ch BSS using the EM algorithm in reverberant environment, in *Proceedings of IEEE Workshop on Applications of Signal Processing to Audio and Acoustics*, pp. 147–150.

Kameoka, H., Goto, M., and Sagayama, S. (2006) Selective amplifier of periodic and non-periodic components in concurrent audio signals with spectral control envelopes, in *IPSJ SIG Technical Reports*, vol. 2006-MUS-66-13, pp. 77–84. In Japanese.

Kameoka, H. and Kashino, K. (2009) Composite autoregressive system for sparse source-filter representation of speech, in *Proceedings of IEEE International Symposium on Circuits and Systems*, pp. 2477–2480.

Kameoka, H., Sato, M., Ono, T., Ono, N., and Sagayama, S. (2012) Blind separation of infinitely many sparse sources, in *Proceedings of International Workshop on Acoustic Echo and Noise Control*.

Kameoka, H., Yoshioka, T., Hamamura, M., Le Roux, J., and Kashino, K. (2010) Statistical model of speech signals based on composite autoregressive system with application to blind source separation, in *Proceedings of International Conference on Latent Variable Analysis and Signal Separation*, pp. 245–253.

Kitamura, D., Ono, N., Sawada, H., Kameoka, H., and Saruwatari, H. (2015) Efficient multichannel nonnegative matrix factorization exploiting rank-1 spatial model, in *Proceedings of IEEE International Conference on Audio, Speech and Signal Processing*, pp. 276–280.

Kounades-Bastian, D., Girin, L., Alameda-Pineda, X., Gannot, S., and Horaud, R. (2016) A variational EM algorithm for the separation of time-varying convolutive audio mixtures. *IEEE/ACM Transactions on Audio, Speech, and Language Processing*, **24** (8), 1408–1423.

Lee, D.D. and Seung, H.S. (2000) Algorithms for non-negative matrix factorization, in *Proceedings of Neural Information Processing Systems*, vol. 13, pp. 556 –562.

Leeuw, J.D. and Heiser, W.J. (1977) Convergence of correction matrix algorithms for multidimensional scaling, in *Geometric Representations of Relational Data*, Mathesis Press.

Nakano, M., Kameoka, H., Le Roux, J., Kitano, Y., Ono, N., and Sagayama, S. (2010) Convergence-guaranteed multiplicative algorithms for non-negative matrix factorization with beta-divergence, in *Proceedings of IEEE International Workshop on Machine Learning for Signal Processing*, pp. 283–288.

Nikunen, J. and Virtanen, T. (2014) Direction of arrival based spatial covariance model for blind sound source separation. *IEEE/ACM Transactions on Audio, Speech, and Language Processing*, **22** (3), 727–739.

Nugraha, A.A., Liutkus, A., and Vincent, E. (2016) Multichannel audio source separation with deep neural networks. *IEEE/ACM Transactions on Audio, Speech, and Language Processing*, **24** (9), 1652 –1664.

Ono, N. (2011) Stable and fast update rules for independent vector analysis based on auxiliary function technique, in *Proceedings of IEEE Workshop on Applications of Signal Processing to Audio and Acoustics*, pp. 189–192.

Ozerov, A. and Févotte, C. (2010) Multichannel nonnegative matrix factorization in convolutive mixtures for audio source separation. *IEEE Transactions on Audio, Speech, and Language Processing*, **18** (3), 550–563.

Ozerov, A., Févotte, C., Blouet, R., and Durrieu, J.L. (2011) Multichannel nonnegative tensor factorization with structured constraints for user-guided audio source separation, in *Proceedings of IEEE International Conference on Audio, Speech and Signal Processing*, pp. 257–260.

Ozerov, A., Févotte, C., and Charbit, M. (2009) Factorial scaled hidden Markov model for polyphonic audio representation and source separation, in *Proceedings of IEEE Workshop on Applications of Signal Processing to Audio and Acoustics*, pp. 121–124.

Ozerov, A., Vincent, E., and Bimbot, F. (2012) A general flexible framework for the handling of prior information in audio source separation. *IEEE Transactions on Audio, Speech, and Language Processing*, **20** (4), 1118–1133.

Petersen, K.B. and Pedersen, M.S. (2005) The matrix cookbook. Version 3.

Pham, D.T., Servière, C., and Boumaraf, H. (2003) Blind separation of speech mixtures based on nonstationarity, in *Proceedings of International Conference on Information Sciences, Signal Processing and their Applications*, pp. II–73–II–76.

Sawada, H., Kameoka, H., Araki, S., and Ueda, N. (2013) Multichannel extensions of non-negative matrix factorization with complex-valued data. *IEEE Transactions on Audio, Speech, and Language Processing*, **21** (5), 971–982.

Thiemann, J. and Vincent, E. (2013) A fast EM algorithm for Gaussian model-based source separation, in *Proceedings of European Signal Processing Conference*.

Vincent, E., Arberet, S., and Gribonval, R. (2009) Underdetermined instantaneous audio source separation via local Gaussian modeling, in *Proceedings of International Conference on Independent Component Analysis and Signal Separation*, pp. 775 –782.

Vincent, E. and Rodet, X. (2004) Underdetermined source separation with structured source priors, in *Proceedings of International Conference on Independent Component Analysis and Signal Separation*, pp. 327–332.

Yoshioka, T., Nakatani, T., Miyoshi, M., and Okuno, H.G. (2011) Blind separation and dereverberation of speech mixtures by joint optimization. *IEEE Transactions on Audio, Speech, and Language Processing*, **19** (1), 69–84.

15

Dereverberation
Emanuël A.P. Habets and Patrick A. Naylor

The degrading effects of reverberation on speech are a major concern in applications including speech telecommunications and automatic speech recognition (ASR). In this chapter, we review some of the main *dereverberation* approaches that aim to cancel or suppress reverberation. We also discuss other approaches that aim to enable *dry* speech signals, free from degrading reverberation, to be obtained from one or more reverberant microphone signals.

After introducing the topic in Section 15.1, the chapter then discusses cancellation-based approaches in Section 15.2 and suppression approaches in Section 15.3. Subsequently, alternative approaches that are specifically designed for the dereverberation of speech are considered in Section 15.4 and evaluation metrics in Section 15.5. Section 15.6 concludes the chapter.

15.1 Introduction to Dereverberation

The effects of reverberation on sound are widely experienced by many people in daily life. Sounds heard in situations containing widely distributed acoustically reflecting surfaces are often described as sounding "spacious" and containing "echoes". Examples of reverberant situations include classrooms, auditoriums, cathedrals, and subway tunnels, to name but a few. These acoustic effects were well known from the times of early cave dwellings and were also exploited, for example, in the design of theaters in ancient times to improve audibility in large open-air spaces.

A historical text of key importance is the *Collected Papers on Acoustics* by Wallace Clement Sabine (Sabine, 1922), published three years after his death. Sabine's research on room acoustics was apparently motivated in part at least by the need to improve the acoustic properties of lecture rooms associated with Harvard University in Cambridge, Massachusetts, USA. This work provided a mathematical formulation describing the physics of reverberation, as well as substantial contributions related to architectural acoustics. Other relevant work towards improved understanding of reverberation on speech include that of Bolt and MacDonald (1949) and Haas (1972), which addressed the detrimental effects on speech intelligibility of reverberation and single echoes, respectively.

Audio Source Separation and Speech Enhancement, First Edition.
Edited by Emmanuel Vincent, Tuomas Virtanen and Sharon Gannot.
© 2018 John Wiley & Sons Ltd. Published 2018 by John Wiley & Sons Ltd.
Companion Website: https://project.inria.fr/ssse/

Although reverberation occurs as a physical principle of the natural world, artificial reverberation is also of interest. Artificial reverberation can be generated by a number of alternative means (Valimaki *et al.*, 2012), including the application of delay networks, convolution-based algorithms normally involving convolution of the source signal with an acoustic impulse response of finite length, and methods reliant on physical room models. This type of reverberation has important application in simulating virtual acoustic environments for gaming and training applications. Very importantly, artificial reverberation is also widely exploited in the music and movie production sectors to create the perceived impression of "naturalness" and "spaciousness" in music and soundtracks.

Dereverberation refers to the process of reducing the effect of reverberation on sound, usually by signal processing means. Study of dereverberation became of interest in sound recording and speech telecommunications for cases in which speech is sensed using microphones placed some significant distance from the talker. Pioneering work was undertaken at Bell Laboratories, Murray Hill, New Jersey, USA for single and multiple microphones (Allen, 1973; Allen *et al.*, 1977). The approaches followed in this early work included the exploitation of concepts such as linear prediction (Makhoul, 1975) and short-time Fourier analysis, both of which are still of paramount importance to the techniques being researched today.

In the period since about 2005, interest in dereverberation has been steadily increasing (Naylor and Gaubitch, 2005). Recent research (Naylor and Gaubitch, 2010) has targeted applications including hands-free telecommunications and desktop and video conferencing systems. In both these application areas, the speech or other signal should be enhanced in terms of the perceptual assessment of a human listener, so that improvements in both speech quality and speech intelligibility in noise are important.

An alternative application of dereverberation is in ASR systems. It has been widely observed that reverberation has a highly detrimental effect on ASR performance, particularly when the reverberation time (RT60) is sufficiently long compared to the ASR feature analysis frame duration as to cause speech energy from one frame to be smeared into the next, or later, frames. This application is discussed comprehensively by, for example, Yoshioka *et al.* (2012). It is shown that the effect of reverberation can increase the word error rate substantially towards 100%. It is also shown that dereverberation can be beneficially applied to reduce the word error rate in reverberant ASR, in an example presented there from over 80% to close to 20%, though in general the improvement may be less depending on the situation tested.

The study of the acoustic properties of rooms and other enclosed spaces is helpful to inform both the design of artificial reverberation and also methods for dereverberation through concepts of physical modeling. For a more detailed explanation of the underlying concepts and prerequisites, see Part I of this book. The distortion caused by reverberation can be modeled as a convolutive distortion or an additive distortion. Depending on the distortion model, one can develop reverberation *cancellation* or *suppression* approaches.

Measurement of reverberation is a multidimensional and complex task. Key metrics relevant to reverberation are presented in detail in Chapter 1. Signal-based measures can be obtained pertinent to the application, such as the word error rate for ASR. For perceived quality, it is not uncommon to employ the so-called PESQ (ITU-T, 2001a), POLQA (ITU-T, 2001b), or STOI (Taal *et al.*, 2011) metrics even though they have not been formally validated for dereverberation. System-based measures are usually derived from the acoustic impulse response and include most commonly the direct-to-reverberant ratio (DRR) and the clarity index in combination with RT60 (del Vallado *et al.*, 2013). Useful methods exist to obtain estimates of these system-based measures directly from the signals (Eaton *et al.*, 2016), and also for the normalized signal-to-reverberation ratio (SRR) (Naylor *et al.*, 2010).

In this chapter we describe dereverberation in three broad classes of approaches. In Section 15.2 we consider approaches that aim to equalize the reverberant acoustic channel. Then in Section 15.3 we focus on approaches that aim to reduce the level of reverberation in a signal using an estimate of the reverberant component. Lastly in Section 15.4 we discuss alternative approaches to estimate the nonreverberant speech signal from the reverberant observed signal(s) while treating the acoustic system as unknown. We highlight throughout whether the approaches being presented are single-channel or multichannel approaches.

15.2 Reverberation Cancellation Approaches

15.2.1 Signal Models

Reverberation cancellation approaches are based on a convolutive distortion model. In the following, two signal models are described in the time domain and in the short-time Fourier transform (STFT) domain.

The signal observed at microphone $i \in \{1, \dots, I\}$ due to a sound source $s(t)$ is the convolution of the source signal with an acoustic channel, plus typically some additive noise. In the time domain, the signal at microphone i can be expressed as

$$x_i(t) = \sum_{\tau=0}^{L-1} a_i(\tau) s(t - \tau) + u_i(t), \tag{15.1}$$

where $a_i(\tau)$ is the acoustic impulse response from the source to microphone i, and $u_i(t)$ denotes the noise present in microphone i.

An example acoustic impulse response is illustrated in Figure 15.1. After a bulk delay due to the propagation time from the source to the microphone, the coefficients associated with direct path propagation can be seen to be strong in this case, indicating that the source is relatively close to the microphone compared to other reflecting surfaces in the room. The early reflections that are due to, for example, the first reflections from the walls and other reflective surfaces in the room, are shown to follow the direct path and have somewhat lower amplitude. At times later than the mixing time (Kuttruff, 2000), the effect of the reflections loses discernible structure so that the late reflections can be modeled as a random process under a decaying envelope (Pelorson *et al.*, 1992).

In the STFT domain the signal model depends on the analysis frame length. When the analysis frame is of the same order as the length of the acoustic impulse response then the signal at microphone i can be expressed as

$$x_i(n, f) = a_i(f) \, s(n, f) + u_i(n, f). \tag{15.2}$$

However, when the analysis frame is shorter, the subband filtering model (Avargel and Cohen, 2007) can be used as described in Chapter 2, and the signal at microphone i can

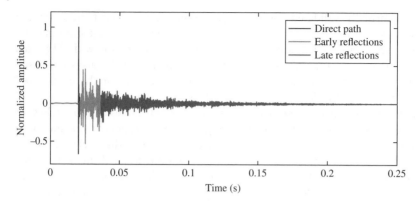

Figure 15.1 Example room acoustic impulse response.

be expressed as

$$x_i(n,f) = \sum_{n'=0}^{N'_f-1} a_i(n',f)s(n-n',f) + u_i(n,f), \tag{15.3}$$

where N'_f denotes the length of the filter in the STFT domain for frequency index f. This model was, for example, used by Schwartz *et al.* (2015a).

Alternatively, the received time-domain signal at the first microphone can be described in the absence of noise as (Gesbert and Duhamel, 1997; Nakatani *et al.*, 2010)

$$x_1(t) = \tilde{s}_1(t) + \sum_{i=1}^{I} \sum_{\tau=L_{\min}}^{L_{\max}} \alpha_i(\tau)\, x_i(t-\tau), \tag{15.4}$$

where $\alpha_i(\tau)$ denotes the τth prediction coefficient at microphone i, $L_{\min}$ the prediction delay, $L_{\max}$ the prediction order, and $\tilde{s}_1(t)$ the signal that cannot be predicted from the delayed observations. It should be noted that under certain conditions the multichannel prediction coefficients $\alpha_i(\tau)$ are directly related to the coefficients of the acoustic impulse response (see, for example, Triki and Slock (2005)).

In the STFT domain the microphone signal can be written in the absence of noise as (Yoshioka *et al.*, 2008)

$$x_1(n,f) = \tilde{s}_1(n,f) + \sum_{i=1}^{I} \sum_{n'=N'_{\min,f}}^{N'_{\max,f}} \alpha_i^*(n',f)\, x_i(n-n',f), \tag{15.5}$$

where $(\cdot)^*$ denotes the conjugate operation, $\alpha_i(n',f)$ denotes the n'th prediction coefficient at microphone i, and $N'_{\min,f}$ and $N'_{\max,f}$ are the prediction delay and prediction order for the frequency index f, respectively. The signal $\tilde{s}_1(n,f)$ is the prediction residual, i.e. the signal in the STFT domain that cannot be predicted from the previously observed STFT coefficients. With the appropriate choice of $N'_{\min,f}$, $\tilde{s}_1(n,f)$ contains the direct speech component plus early reflections. One major advantage of the model in (15.5) is that it can also be used in the presence of multiple sources (Yoshioka and Nakatani, 2012). One disadvantage is that the models in (15.4) and (15.5) do not include additive noise, and that adding such noise complicates the signal model.

15.2.2 Identification and Equalization Approaches

The approach of channel equalization is widely known in diverse areas of communications (Tong and Perreau, 1998) as a way to improve the quality of a signal after propagation via a convolutive channel. The approach requires, first, an estimate of the channel and, second, an effective way to design an equalizer, usually based on the inverse of the channel. In the case of dereverberation, the channel describes the effect of sound propagation from the source to the microphone. Several microphones can be employed so that the acoustic system has a single-input multiple-output structure and the dereverberation system has a multiple-input single-output structure.

We now focus on a two-step procedure in which blind system identification is first used to identify the single-input multiple-output acoustic system and then a multiple-input single-output equalizer is designed and applied to the received microphone signals to suppress the reverberation effects due to the acoustic propagation in an enclosure. In this context identification approaches for two or more microphones are discussed. This approach has the advantage of offering the potential for very high quality dereverberation but in practice this can only be realized if the blind system identification and the equalizer design are sufficiently precise.

15.2.2.1 Cross-Relation Based Blind System Identification

Minimization of the *cross-relation error* (Liu *et al.*, 1993; Tong *et al.*, 1994; Gürelli and Nikias, 1995; Xu *et al.*, 1995) is one of the best-known approaches to blind system identification. The cross-relation error is defined for acoustic channels by first considering the convolutions of the source signal $s(t)$ with acoustic impulse responses $a_i(\tau)$ and $a_{i'}(\tau)$ in which the subscripts i, $i' \in \{1, \ldots, I\}$ and $i \neq i'$, are the channel indices. For microphone signals $x_i(t)$ and $x_{i'}(t)$ free from noise, denoting the convolution operator as $\star$ and temporarily dropping the time index t, we can write

$$s \star a_i \star a_{i'} = x_i \star a_{i'} = x_{i'} \star a_i. \tag{15.6}$$

Using this relation, an error term can then be defined as

$$e_{ii'}(t) = \sum_{\tau=0}^{\hat{L}-1} a_{i'}(\tau) x_i(t - \tau) - \sum_{\tau=0}^{\hat{L}-1} a_i(\tau) x_{i'}(t - \tau), \tag{15.7}$$

where $\hat{L}$ is the estimated length of the acoustic impulse response.

In the case where the length of the true acoustic channels is known (i.e., $\hat{L} = L$) and the acoustic channels are coprime (i.e., they do not share any common zeros), an estimate of the true impulse response can be obtained up to an unknown scaling factor by minimizing the total squared error across all unique microphone pairs. Mathematically, this can be formulated as

$$\hat{\mathbf{a}} = \underset{\mathbf{a}}{\arg\min} \sum_{i=1}^{I} \sum_{i'=1}^{i-1} \sum_{t} e_{ii'}^2(t) \quad \text{s.t.} \quad \|\mathbf{a}\|_2 = 1 \tag{15.8}$$

where $\mathbf{a} = [\mathbf{a}_1^T, \ldots, \mathbf{a}_I^T]^T$, with $\mathbf{a}_i = [a_i(0), \ldots, a_i(\hat{L} - 1)]^T$, $\hat{\mathbf{a}}$ is an estimate of $\mathbf{a}$. The unit-norm constraint in (15.8) serves to avoid the trivial solution $\hat{\mathbf{a}} = \mathbf{0}_{I\hat{L}}$. In the case where the length of the acoustic channel is overestimated (i.e., $\hat{L} > L$), the estimated impulse responses are filtered versions of the true impulse responses.

There are several methods to obtain least squares solutions for the acoustic impulse responses **a**, subject to some identifiability conditions as described by Abed-Meraim *et al.* (1997b). Besides closed-form solutions, various adaptive solutions have been also proposed. These include multichannel least mean squares (Huang and Benesty, 2002) and multichannel Newton (Huang and Benesty, 2002) time-domain adaptive algorithms, normalized multichannel frequency-domain least mean squares (Huang and Benesty, 2003) and noise robust multichannel frequency-domain least mean squares (Haque and Hasan, 2008), as well as multichannel quasi-Newton (Habets and Naylor, 2010) and state-space frequency-domain adaptive filters (Malik *et al.*, 2012).

15.2.2.2 Noise Subspace Based Blind System Identification

Rather than minimizing the cross-relation error (Xu *et al.*, 1995), Moulines *et al.* (1995) directly identified the channels that are embedded in the null space of the correlation matrix of the data using an eigenvalue decomposition. As the null subspace vectors are shown to be filtered versions of the actual channels, extraneous roots should be eliminated. Gürelli and Nikias (1995) presented an algorithm to recursively eliminate the extraneous zeros, and hence obtain the correct filters. Gannot and Moonen (2003) extracted the null space using the generalized singular or eigenvalue decomposition of, respectively, the data or correlation matrix in the presence of colored noise. An acoustic impulse response estimation procedure is obtained by exploiting the special Silvester structure of the corresponding filtering matrix by using total least squares fitting. A computationally more efficient method, although slightly less accurate, is proposed based on the same structure and on the QR decomposition.

Given the signals at two microphones $x_1(t)$ and $x_2(t)$, the correlation matrix can be written as

$$\widehat{\boldsymbol{\Sigma}}_{\bar{X}} = \frac{\bar{X}\bar{X}^T}{T+1}, \quad \text{where} \quad \bar{X} = \begin{bmatrix} \bar{X}_2 \\ -\bar{X}_1 \end{bmatrix} \tag{15.9}$$

is the data matrix formed from the samples of $x_i(t)$, $i \in \{1, 2\}$, and T is the number of observed samples. The matrix $\bar{X}_i$ of size $\widehat{L} \times (T + \widehat{L})$, where $\widehat{L}$ is the length of the estimated acoustic impulse response, is a Sylvester structure matrix constructed using $x_i(t)$. Gürelli and Nikias (1995) showed that the rank of the null space of the correlation matrix $\widehat{\boldsymbol{\Sigma}}_{\bar{X}}$ is $\widehat{L} - L + 1$, where L is the length of the true impulse response.

In the case where the length is known so that $\widehat{L} = L$, the eigenvector of $\widehat{\boldsymbol{\Sigma}}_{\bar{X}}$ corresponding to the zero eigenvalue (or singular value) can be partitioned into $[\widehat{\mathbf{a}}_1, \ \widehat{\mathbf{a}}_2]^T$. In the case where the length is not known but $\widehat{L} \geq L$, the null space has a dimension greater than one but the filters $\widehat{a}_i(\tau) = a_i \star e(\tau)$ can still be found, up to a channel-independent filter $e(\tau)$. When the channel length and hence the null space dimension are known, $a_1(\tau)$ and $a_2(\tau)$ can still be recovered using the method proposed by Gannot and Moonen (2003). The method is also extended to handle more than two microphones as well as additive noise in the microphone signals. It should be pointed out that any roots in the z plane that are common to a_1 and a_2 cannot be identified because they will be treated as extraneous roots and will therefore be included in e.

15.2.2.3 Multichannel Equalization for Dereverberation

The second step in this two-step dereverberation procedure is to design a *multichannel equalizer* $\mathbf{g} = [\mathbf{g}_1^T, \ldots, \mathbf{g}_I^T]^T$ for which $\mathbf{g}_i = [g_i(0), \ldots, g_i(\tilde{L} - 1)]^T$ is of length $\tilde{L}$, utilizing the previously estimated acoustic channels $\mathbf{a}_i$. This multichannel equalizer can then be applied to the microphone signals to obtain the dereverberated speech signal using

$$\hat{s}(t) = \sum_{i=1}^{I} \sum_{\tau=0}^{\tilde{L}-1} g_i(\tau) \, x_i(t - \tau). \tag{15.10}$$

Several relevant characteristics of acoustic impulse responses need to be considered when formulating an equalizer design procedure. In general, acoustic impulse responses are nonminimum phase (Naylor and Gaubitch, 2010), they comprise thousands of coefficients in realistic room acoustic systems and the true order of the impulse responses is not known. Direct inversion of each impulse response is therefore normally infeasible. Additionally, although it is common to assume stationarity at least over the impulse response duration, acoustic impulse responses are general slowly time-varying. As a result of these characteristics and the presence of noise in the microphone signals, the estimation of $\hat{\mathbf{a}}_i$ is often degraded by high levels of estimation error and it is therefore not appropriate to invert the multichannel impulse responses directly. It can be said that, considering the equalizer as an inverse filtering operation, the inverse of the estimate $\hat{\mathbf{a}}_i$ is not a good estimate of the inverse of $\mathbf{a}_i$ (Hasan and Naylor, 2006).

To evaluate the performance of multichannel equalizers we define the *equalized impulse response*, denoted by $\mathbf{r}$, of length $L + \tilde{L} - 1$, which is obtained by

$$\mathbf{r} = \sum_{i=1}^{I} \mathbf{A}_i \mathbf{g}_i \tag{15.11}$$

where $\mathbf{A}_i$ is a Sylvester matrix of size $(L + \tilde{L} - 1) \times \tilde{L}$ formed from the true channel coefficients $\mathbf{a}_i$ as

$$\mathbf{A}_i^T = \begin{bmatrix} a_i(0) & a_i(1) & \ldots & a_i(L-1) & 0 & \ldots & 0 \\ 0 & a_i(0) & \ldots & a_i(L-2) & a_i(L-1) & \ldots & 0 \\ \vdots & \ddots & & \vdots & \vdots & \ddots & \vdots \\ 0 & \ldots & 0 & a_i(0) & a_i(1) & \ldots & a_i(L-1) \end{bmatrix} \tag{15.12}$$

The equalized impulse response is also instrumental in the design of multichannel equalizers, where it is computed using $\hat{\mathbf{A}}_i$ of size $\tilde{L} \times (\hat{L} + \tilde{L} - 1)$ rather than $\mathbf{A}_i$.

Exact equalization aims to shorten the equalized impulse response to an impulse function, possibly with an arbitrary delay, thereby setting all except one coefficient of $\mathbf{r}$ to zero. The so-called *MINT* solution (Miyoshi and Kaneda, 1988) is obtained by minimizing the least squares cost function

$$J_{\mathrm{MINT}}(\mathbf{g}) = \|\hat{\mathbf{A}}\mathbf{g} - \mathbf{d}\|_2^2 \tag{15.13}$$

where $\hat{\mathbf{A}} = [\hat{\mathbf{A}}_1, \ldots, \hat{\mathbf{A}}_I]$ and

$$\mathbf{d} = [0 \; \ldots \; 0 \, 1 \, 0 \; \ldots \; 0]^T. \tag{15.14}$$

$$\underbrace{}_{\tau}$$

Miyoshi and Kaneda (1988) showed that when impulse responses do not share common zeros in the z plane and $\tilde{L} \geq \left\lceil \frac{L-1}{l-1} \right\rceil$, where $\lceil \cdot \rceil$ denotes the ceiling function, the solution is given by

$$\mathbf{g}_{\text{MINT}} = \hat{\mathbf{A}}^+\mathbf{d}, \tag{15.15}$$

where $\hat{\mathbf{A}}^+$ denotes the pseudo-inverse of $\hat{\mathbf{A}}$.

It is well known that $\mathbf{g}_{\text{MINT}}$ is not robust to estimation errors (Lim *et al.*, 2014). A more advantageous approach is to target only an approximate inversion of the multichannel system. Such a solution is, for example, obtained by *channel shortening*, which aims to set only the late coefficients of the equalized impulse response to zero. The number of the nonzero coefficients in channel shortening is a parameter to be chosen.

A large number of methods have been proposed to find a suitable multichannel equalizer $\mathbf{g}$. Key considerations in comparing these methods include (i) robustness to system identification errors in the impulse responses, (ii) robustness to common zeros or near-common zeros in the channels, (iii) the perceptual quality of the speech after applying the multichannel equalizer, and finally (iv) computational complexity and numerical issues.

Channel shortening solutions for $\mathbf{g}$ can be formulated as the maximization of the generalized Rayleigh quotient, i.e.

$$\mathbf{g}_{\text{CS}} = \max_{\mathbf{g}} \frac{\mathbf{g}^T \hat{\mathbf{A}}_\text{d} \mathbf{g}}{\mathbf{g}^T \hat{\mathbf{A}}_\text{u} \mathbf{g}} \tag{15.16}$$

where $\hat{\mathbf{A}}_\text{d}$ and $\hat{\mathbf{A}}_\text{u}$ are formed from the desired and undesired coefficients of the estimated impulse responses respectively (Zhang *et al.*, 2010). In this context, undesired coefficients are those in the late temporal region of the impulse response that are intended to be suppressed under channel shortening.

The key objective of Zhang *et al.* (2010) and Lim *et al.* (2014) was robustness to system identification errors and the approach was to relax the optimization over the impulse response coefficients corresponding to the early reflections. The rationale for this relaxation is to aim to reduce the effects of the late reverberant tail in the impulse response at the expense of unconstrained early reflections, given that the perceptual effects of early reflections are often considered less damaging to speech intelligibility and ASR than the late reverberation. A weight vector $\mathbf{h}$ is defined as

$$\mathbf{h} = [\underbrace{1\,1\,\ldots\,1}_{\tau}\,\underbrace{1\,0\,\ldots\,0}_{L_h}\,1\,\ldots\,1]^T \tag{15.17}$$

with τ representing an arbitrary delay and L_h determining the length in samples of the relaxation window. The relaxation window can be set to correspond to the region of the early reflections where the coefficients in the equalized impulse response are unconstrained. Next, defining $\mathbf{d}$ as the target equalized impulse response the relaxed multichannel least squares solution is obtained (Zhang *et al.*, 2010; Lim *et al.*, 2014) by minimizing

$$J_{\text{RMCLS}}(\mathbf{g}) = \|\text{Diag}(\mathbf{h})(\hat{\mathbf{A}}\mathbf{g} - \mathbf{d})\|_2^2 \tag{15.18}$$

with respect to $\mathbf{g}$. A further development of this approach (Lim and Naylor, 2012) explored the use of additional constraints on the initial taps to achieve partial relaxation of the channel shortening problem to tradeoff the increased robustness to channel estimation errors against any undesirable perceptual coloration effects in the equalized signal. Besides these closed-form solutions, Zhang *et al.* (2008) described an adaptive inverse filtering approach in the context of channel shortening.

Following the same aim of approximate, or partial, channel inversion from estimated impulse responses, the *Partial MINT* (P-MINT) algorithm (Kodrasi and Doclo, 2012) uses a further modification of the target equalized impulse response. In the case of Partial MINT, the early coefficients of the estimated impulse response for a chosen channel $i \in \{1, \dots, I\}$ are employed, so that the equalizer $\mathbf{g}_{\text{P-MINT}}$ is found by minimizing

$$J_{\text{P-MINT}}(\mathbf{g}) = \|\hat{\mathbf{A}}\mathbf{g} - \mathbf{d}_{\text{P-MINT}}\|_2^2 \tag{15.19}$$

with

$$\mathbf{d}_{\text{P-MINT}} = [\underbrace{0 \ \dots \ 0}_{\tau} \ \underbrace{\hat{a}_i(0) \ \dots \ \hat{a}_i(L_d - 1)}_{L_d} \ 0 \ \dots \ 0]^T. \tag{15.20}$$

The length L_d needs to be chosen large enough to span the direct path and early reflections, which is typically considered to be between 50 and 80 ms. Inverse filters that partially equalize the multichannel system can then be computed using

$$\mathbf{g}_{\text{P-MINT}} = \hat{\mathbf{A}}^+ \mathbf{d}_{\text{P-MINT}}. \tag{15.21}$$

The method of Partial MINT and was further developed to include an advantageous regularization scheme (Kodrasi *et al.*, 2012).

The concept of "relaxed" criteria for equalizer design is also discussed by Mertins *et al.* (2010), who considered the mathematically dual problems of listening room compensation and postfiltering recorded microphone signals for channel equalization. A key aim is to shorten the effective room reverberation in a way well able to exploit the masking effects present in the human perceptual system (Fielder, 2003). When designing equalizing filters, it is typical to employ a least squares criterion. However, Mertins *et al.* (2010) investigated the use of the alternative criteria of ℓ_p and ℓ_∞ norms. These have the stated benefit over least squares approaches that the error distribution with time in the designed equalized impulse response can be controlled, and this can be used to avoid, for example, residual late reverberation, which is perceptually harmful.

It is interesting to review comparative results for MINT, Partial MINT and relaxed multichannel least squares approaches such as those by Lim *et al.* (2014). A key question is to consider the tradeoff in these methods, and their tuning, of channel shortening performance versus perceptual quality of the output equalized signal.

Finally, it should be noted that the aforementioned equalizers do not take into account the presence of additive noise. As a consequence, the equalizer might amplify the noise. Kodrasi and Doclo (2016) proposed two spatial filters for joint dereverberation and noise reduction. The first spatial filter is obtained by incorporating the noise statistics into the cost function of the regularized Partial MINT. An additional parameter was introduced to enable a tradeoff between dereverberation and noise reduction. The second spatial filter is a multichannel Wiener filter (MWF) that takes into account the statistics of the reverberant speech signal and the noise. In this case the output of the regularized Partial MINT filter was used as a reference signal for the MWF. An additional parameter was introduced to enable a tradeoff between speech distortion and noise reduction.

15.2.3 Identification and Estimation Approaches

Dereverberation can alternatively be achieved by jointly identifying the acoustic system and estimating the clean speech signal.

Schmid (2014) and Schmid *et al.* (2014) developed various algorithms for joint dereverberation and noise reduction using a multichannel state-space model for the acoustic channels. In particular, a first-order Markov model was used to describe the time-varying nature of acoustic impulse responses. The variational Bayesian (VB) inference approach of Schmid *et al.* (2014) combines frame-based observation equations in the frequency domain with a first-order Markov model. By modeling the channels and the source signal as latent random variables, they formulated a lower bound on the log-likelihood of the model parameters given the observed microphone signals and iteratively maximized it using an online expectation-maximization (EM) approach. The obtained update equations then jointly estimated the channel and source posterior distributions and the remaining model parameters. It was shown that the resulting algorithm for blind equalization and channel identification includes previously proposed methods as special cases.

Schwartz *et al.* (2015a) modeled the reverberation as a moving average process in the STFT domain and developed an algorithm to jointly estimate the clean speech signal and the acoustic system in the presence of spatially white noise. In particular, a recursive EM scheme was employed to obtain both the clean speech signal and the parameters of the acoustic system in an online manner. In the expectation step, the Kalman filter is applied to extract a new sample of the clean signal and, in the maximization step, the parameters of the acoustic system are updated according to the output of the Kalman filter. Schwartz *et al.* (2015a) also showed that this approach can be used effectively when the acoustic scenario is slowly time-varying.

15.2.4 Multichannel Linear Prediction Approaches

Multichannel linear prediction approaches were originally developed to blind identify wireless communication channels (Slock, 1994; Abed-Meraim *et al.*, 1997a), where the source signal is assumed to be independent and identically distributed. Compared to the identification approaches in Section 15.2.2, the multichannel linear prediction approach can robustly estimate the channels in the presence of a channel order mismatch and noise. A multistep prediction approach was proposed in the context of blind channel identification and equalization by Gesbert and Duhamel (1997). More recently, multichannel linear prediction has been adopted in the context of dereverberation, which we review here.

Delcroix *et al.* (2004) and Delcroix *et al.* (2006) developed a time-domain multichannel linear prediction approach, known as linear-predictive multiple-input equalization, for speech dereverberation. With reference to Figure 15.2, the first step is to determine linear prediction filters that minimize the prediction error for the reverberant microphone signals with respect to, for example, the first channel corresponding to the microphone closest to the source signal. The solution for the prediction filters was formulated by Delcroix *et al.* (2004, 2006) as a matrix pseudoinverse problem, and the prediction error signal can subsequently be obtained. In the second step, the input speech signal is modeled as an autoregressive (AR) process. Since the first step has the undesired effect

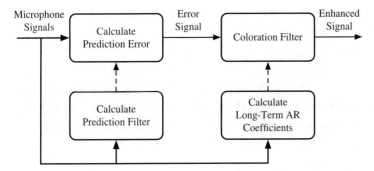

Figure 15.2 Schematic view of the linear-predictive multiple-input equalization method, after Habets (2016).

of whitening the signal, the second step filters the prediction error from the first step with the inverse of the estimated AR source process. In this way, it aims to recover the enhanced signal without the undesired whitening effect.

An alternative approach proposed by Triki and Slock (2005) uses a three-step procedure: a source whitening step, a multichannel linear prediction step, and an equalization step. The source whitening step relies on multichannel spatial diversity such that the source spectrum can be estimated up to a multiplicative constant from the microphone signals. The multichannel linear prediction is then applied to the prewhitened microphone signals to obtain the equalization filter. Lastly, the dereverberation step applies the equalizer to the original microphone signals. An advantage of this approach is that no coloration filter, which can introduce undesired artifacts, is required.

Nakatani *et al.* (2010) developed a statistical model-based dereverberation approach referred to as the variance-normalized delayed linear prediction approach. A time-domain as well as an STFT-domain method were developed to estimate the prediction coefficients and the statistics of the dereverberated signal in the maximum likelihood (ML) sense. In the time domain, the observed signal needs to be prewhitened before estimating the regression coefficients. In the STFT the early speech component for a given frequency can be assumed to be uncorrelated across time such that the method, commonly referred to as the weighted prediction error method, can be applied without prewhitening. In the following, we adopt the STFT signal model given by (15.5).

Given an estimate of the prediction coefficients, denoted by $\hat{a}_i(n',f)$ and using (15.5), an estimate of the dereverberated signal at the first microphone[1] is given by

$$\hat{s}_1(n,f) = x_1(n,f) - \sum_{i=1}^{I} \sum_{n'=N'_{\min,f}}^{N'_{\max,f}} \hat{a}_i^*(n',f)\, x_i(n-n',f). \tag{15.22}$$

The STFT coefficients of $\tilde{s}_1(n,f)$ can be modeled using a zero-mean complex Gaussian distribution with time-varying variance $\sigma_{\tilde{s}_1}^2(n,f) = \mathbb{E}\{|\tilde{s}_1(n,f)|^2\}$, such that its probability distribution is given by

$$p(\tilde{s}_1(n,f)) = \mathcal{N}_c(\tilde{s}_1(n,f) \mid 0, \sigma_{\tilde{s}_1}^2(n,f)) \tag{15.23}$$

$$= \frac{1}{\pi \sigma_{\tilde{s}_1}^2(n,f)} e^{-\frac{|\tilde{s}_1(n,f)|^2}{\sigma_{\tilde{s}_1}^2(n,f)}}. \tag{15.24}$$

---

1 With no loss of generality, the first microphone is used in the following as a reference microphone.

An ML estimate of the prediction coefficients and the spectral variances can be found for each frequency by maximizing the log-likelihood

$$\mathcal{M}^{\mathrm{ML}}(\alpha(f), \sigma^2_{\tilde{s}_1}(f)) = \sum_n \log \mathcal{N}_c(\tilde{s}_1(n,f) \mid 0, \ \sigma^2_{\tilde{s}_1}(n,f)), \tag{15.25}$$

where $\sigma^2_{\tilde{s}_1}(f) = [\sigma^2_{\tilde{s}_1}(0,f), \dots, \sigma^2_{\tilde{s}_1}(N-1,f)]^T$ is a vector containing the N spectral variances for frequency index f, and $\alpha(f) = [\alpha_1(N'_{\min,f},f), \dots, \alpha_1(N'_{\max,f},f), \dots, \alpha_I(N'_{\min,f},f),$ $\dots \alpha_I(N'_{\max,f},f)]^T$ is a vector containing all prediction coefficients for frequency index f. This leads to the following optimization problem (Nakatani *et al.*, 2010)

$$\min_{\sigma^2_{\tilde{s}_1}(f)>0, \ \alpha(f)} \sum_n \left(\frac{|\tilde{s}_1(n,f)|^2}{\sigma^2_{\tilde{s}_1}(n,f)} + \log(\pi \ \sigma^2_{\tilde{s}_1}(n,f)) \right). \tag{15.26}$$

This problem can be solved, for example, using an alternating optimization procedure in which the parameters are estimated iteratively (Nakatani *et al.*, 2010). Even though no additive noise has been taken into account, Nakatani *et al.* (2010) showed that the procedure of maximizing the log-likelihood (15.25) is robust to moderate levels of additive noise.

Jukić *et al.* (2015) modeled the early speech signal using a sparse prior, with a special emphasis on circular priors from the complex generalized Gaussian family. The proposed model can be interpreted as a generalization of the time-varying Gaussian model, with an additional hyperprior on the spectral variance $\sigma^2_{\tilde{s}_1}(n,f)$. For the circular sparse prior, the following optimization problem was obtained

$$\min_{\sigma^2_{\tilde{s}_1}(f)>0, \ \alpha(f)} \sum_n \left(\frac{|\tilde{s}_1(n,f)|^2}{\sigma^2_{\tilde{s}_1}(n,f)} + \log(\pi \ \sigma^2_{\tilde{s}_1}(n,f)) - \log \Psi(\sigma^2_{\tilde{s}_1}(n,f)) \right), \tag{15.27}$$

where $\Psi(\cdot)$ denotes a scaling function that can be interpreted as a hyperprior on the spectral variance $\sigma^2_{\tilde{s}_1}(n,f)$. This problem can be solved also using an alternating optimization procedure. It has been shown also that the underlying prior in the conventional weighted prediction error method strongly promotes sparsity of the early speech component, and can be obtained as a special case of the proposed weighted prediction error method. Furthermore, the proposed weighted prediction error method has been reformulated as an optimization problem with the cost function using the ℓ_p norm on the early speech component. It was shown that using this signal model the performance can be increased and the number of iterations required to converge can be decreased.

The spectral variances of the early speech component within a single time frame can also be modeled by exploiting the low-rank structure of the speech spectrogram, which can be modeled using nonnegative matrix factorization (NMF) as shown for example by Mohammadiha *et al.* (2013). Jukić *et al.* (2016a) incorporated the NMF in the multichannel linear prediction-based dereverberation approach. The authors showed that is possible to reduce the speech distortion, using either supervised or unsupervised NMF dictionaries.

The statistical model-based dereverberation approach can be used to directly compute I output signals, which can be postprocessed to reduce the effect of early reflections or used to perform direction of arrival (DOA) estimation. In addition, Yoshioka and Nakatani (2012) and Togami *et al.* (2013) showed that the multichannel linear prediction approach can be extended to include multiple sources. Yoshioka and

Nakatani (2012) derived a generalized weighted prediction error method to estimate the model parameters by minimizing the Hadamard–Fischer mutual correlation. The proposed method can take into account the spatial correlation of the early speech component. Togami *et al.* (2013) assumed time-varying acoustic transfer functions, introduced hyperparameter models for the early speech component and the late reverberant component, and then used an EM scheme to find the model parameters.

An adaptive version, which can be used for online processing, was developed by Yoshioka and Nakatani (2013) by incorporating an exponential weighting function into the cost function of the iterative optimization algorithm of Yoshioka and Nakatani (2012). By assuming that the early speech components $\tilde{s}(n,f)$ are spatially uncorrelated and their spectral variances are channel-independent, a computational efficient solution was obtained. Jukić *et al.* (2016b) showed that insufficient temporal smoothing leads to overestimation of the late reverberation and hence distortion of the dereverberated signal. To mitigate this problem, the authors added an inequality constraint to the method of Yoshioka and Nakatani (2013) to ensure that the instantaneous powers of the estimated reverberation do not exceed an estimate of the late reverberant spectral variances that were obtained using I single-channel spectral variance estimators, as described in Section 15.3.3. Braun and Habets (2016) modeled the prediction coefficients of the prediction filter as a first-order Markov process to account for possible changes in the acoustic environment. The coefficients were then estimated using a Kalman filter, and an approach for estimating the covariance matrix of the early speech component was proposed.

When the approach of multichannel linear prediction is used in a multiple-input multiple-output structure, it is advantageous that the phase of the multichannel signals is not perturbed. This is particularly important when it is desirable to include both spatial filtering and dereverberation in a processing pipeline so that even after dereverberation processing the multichannel signals are still suitable as inputs to a beamformer.

15.3 Reverberation Suppression Approaches

15.3.1 Signal Models

Reverberation suppression approaches are based on an additive distortion model and are commonly formulated in the STFT domain. In this case the microphone signal vector $\mathbf{x}(n,f) = [x_1(n,f), \ldots, x_I(n,f)]^T$ at time index n and frequency index f can be written as

$$\mathbf{x}(n,f) = \mathbf{x}_e(n,f) + \mathbf{x}_r(n,f) + \mathbf{u}(n,f) \tag{15.28}$$

where $\mathbf{x}_e(n,f)$ denotes the early sound component, $\mathbf{x}_r(n,f)$ denotes the reverberant sound component, and $\mathbf{u}(n,f)$ denotes the ambient noise component, which are defined similarly to the vector $\mathbf{x}(n,f)$ and are assumed to be mutually uncorrelated. In the following, we assume that the covariance of the noise component, denoted by $\Sigma_u(n,f) = \mathbb{E}\{\mathbf{u}(n,f)\mathbf{u}^H(n,f)\}$, is known.

The early speech component consists of the direct sound and possibly early reflections, and can be modeled in the time domain as a filtered version of the anechoic speech signal

$s(t)$. In the STFT domain, it can be written as

$$\mathbf{x}_e(n,f) = \tilde{\mathbf{a}}(f)x_{e,i}(n,f), \tag{15.29}$$

where $\tilde{\mathbf{a}}(f)$ denotes the relative early transfer function vector with respect to microphone i such that by definition $\tilde{a}_i(f) = 1$, and $x_{e,i}(n,f)$ denotes the early speech component as received by the ith microphone. It should be noted that in many reverberation suppression approaches the early reflections are not taken into account and it is assumed that the microphone signals consist of a direct sound component plus a reverberant sound component. In the latter case, $\tilde{\mathbf{a}}(f)$ denotes the relative direct transfer function vector. Assuming that the received direct path can be modeled as a plane wave, the relative direct transfer function depends on the DOA of the plane wave, the inter-microphone distances, the speed of sound, and the frequency of interest. The early source component $x_{e,i}(n,f)$ is commonly modeled as a zero-mean complex Gaussian random variable with variance $\sigma_{e,i}^2(n,f) = \mathbb{E}\{|x_{e,i}(n,f)|^2\}$. The covariance matrix of the early sound component is then given by

$$\mathbf{\Sigma}_{\mathbf{x}_e}(n,f) = \mathbb{E}\{\mathbf{x}_e(n,f)\mathbf{x}_e^H(n,f)\} = \sigma_{e,i}^2(n,f)\tilde{\mathbf{a}}(f)\tilde{\mathbf{a}}^H(f) \tag{15.30}$$

and has rank 1.

The reverberant sound field can be modeled as an ideal cylindrical or spherical isotropic and homogeneous sound field with a time-varying level. The signal vector $\mathbf{x}_r(n,f)$ is commonly modeled as a zero-mean multivariate Gaussian random variable with probability distribution

$$\mathbf{x}_r(n,f) \sim \mathcal{N}_c(\mathbf{x}_r(n,f) \mid \mathbf{0}_I, \sigma_r^2(n,f)\,\mathbf{\Omega}(f)), \tag{15.31}$$

where $\sigma_r^2(n,f)$ denotes the time-varying level of the reverberant sound field, and $\mathbf{\Omega}(f)$ denotes the time-invariant spatial coherence of the reverberant sound field. In an ideal spherically isotropic and homogeneous sound field the elements of the spatial coherence matrix $\mathbf{\Omega}(f)$ are given by (3.6), as described in Chapter 3.

Under the additive distortion model, the dereverberation problem now reduces to estimating the early speech component at one of the microphones, i.e. $x_{e,i}(n,f)$. Alternatively, a spatially filtered version of the early speech components can be defined as the desired signal (Habets, 2007; Schwartz *et al.*, 2015b). In Section 15.3.2 we review some approaches to estimate $x_{e,i}(n,f)$ assuming that the variances $\sigma_{e,i}^2(n,f)$ and $\sigma_r^2(n,f)$, as well as the transfer function vector $\tilde{\mathbf{a}}(f)$, the spatial coherence matrix $\mathbf{\Omega}(f)$, and the covariance of the ambient noise $\mathbf{\Sigma}_u(n,f)$ are known. In Sections 15.3.3 and 15.3.4 we then review methods to estimate the variance $\sigma_r^2(n,f)$.

15.3.2 Early Signal Component Estimators

Based on the additive distortion model, various reverberation suppression approaches can be derived. In the following we discuss three different approaches, namely data-independent spatial filtering, data-dependent spatial filtering, and spectral enhancement. Hereafter, we omit the time and frequency indices when possible for brevity.

Data-independent spatial filtering: A distortionless estimate of the early speech component $x_{e,i}(n,f)$ can be obtained using the well-known minimum variance distortionless response (MVDR) filter, i.e.

$$\hat{x}_{e,i}(n,f) = \mathbf{w}_{MVDR}^H(f)\,\mathbf{x}(n,f) \tag{15.32}$$

where

$$\mathbf{w}_{\text{MVDR}}(f) = \underset{\mathbf{w}(f)}{\arg\min}\ \mathbf{w}^H(f)\boldsymbol{\Omega}(f)\mathbf{w}(f) \quad \text{s.t.} \quad \mathbf{w}^H(f)\tilde{\mathbf{a}}(f) = 1 \tag{15.33}$$

$$= \frac{\boldsymbol{\Omega}^{-1}(f)\tilde{\mathbf{a}}(f)}{\tilde{\mathbf{a}}^H(f)\boldsymbol{\Omega}^{-1}(f)\tilde{\mathbf{a}}(f)}. \tag{15.34}$$

An equivalent spatial filter is obtained when maximizing the directivity factor. This specific MVDR filter, also known as the maximum directivity filter, is data-independent as it does not depend on the level of desired signal and reverberant sound component. The dereverberation performance of the MVDR filter strongly depends on the number of microphones, array geometry, and transfer functions $\tilde{\mathbf{a}}$, which in the far-field case and in the absence of early reflections only depends on the DOA of the direct sound.

Data-dependent spatial filtering: For the signal model described in Section 15.3.1, the minimum mean square error (MMSE) estimator for the early speech component as received by microphone i is given by

$$\hat{x}_{\text{e},i}(n,f) = \mathbb{E}\{x_{\text{e},i}(n,f) \mid \mathbf{x}(n,f), \tilde{\mathbf{a}}(f), \sigma^2_{\text{e},i}(n,f), \sigma^2_{\text{r}}(n,f), \boldsymbol{\Omega}(f), \boldsymbol{\Sigma}_{\text{u}}(n,f)\}. \tag{15.35}$$

Since all signals are modeled as complex Gaussian random variables, $\hat{x}_{\text{e},i}(n,f)$ can be obtained using

$$\hat{x}_{\text{e},i}(n,f) = \mathbf{w}^H_{\text{MWF}}(n,f)\,\mathbf{x}(n,f) \tag{15.36}$$

where $\mathbf{w}_{\text{MWF}}(n,f)$ denotes the MWF that is given by (dropping time-frequency indexes for legibility)

$$\mathbf{w}_{\text{MWF}} = (\sigma^2_{\text{e},i}\tilde{\mathbf{a}}\tilde{\mathbf{a}}^H + \sigma^2_{\text{r}}\boldsymbol{\Omega} + \boldsymbol{\Sigma}_{\text{u}})^{-1}\tilde{\mathbf{a}}\,\sigma^2_{\text{e},i}. \tag{15.37}$$

Using the Woodbury identify, the MWF can be written as

$$\mathbf{w}_{\text{MWF}} = \underbrace{\frac{\boldsymbol{\Omega}^{-1}\tilde{\mathbf{a}}}{\tilde{\mathbf{a}}^H\boldsymbol{\Omega}^{-1}\tilde{\mathbf{a}}}}_{\mathbf{w}_{\text{MVDR}}(f)} \cdot \underbrace{\frac{\sigma^2_{\text{e},i}}{\sigma^2_{\text{e},i} + [\tilde{\mathbf{a}}^H(\sigma^2_{\text{r}}\boldsymbol{\Omega} + \boldsymbol{\Sigma}_{\text{u}})^{-1}\tilde{\mathbf{a}}]^{-1}}}_{w_{\text{SWF}}(n,f)}, \tag{15.38}$$

where $\mathbf{w}_{\text{MVDR}}(f)$ and $w_{\text{SWF}}(n,f)$ denote the MVDR filter and the single-channel Wiener filter, respectively. The data-dependent spatial filtering approach is illustrated in Figure 15.3.

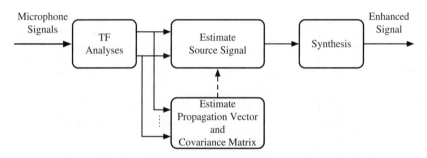

Figure 15.3 Data-dependent spatial filtering approach to perform reverberation suppression.

The MWF has been used in the context of noise reduction (Lefkimmiatis and Maragos, 2007) as well as dereverberation (Braun *et al.*, 2013; Thiergart *et al.*, 2014b; Schwartz *et al.*, 2015b). The single-channel Wiener filter is known to provide a tradeoff between interference suppression and speech distortion. To control this tradeoff, the single-channel parametric Wiener filter (Benesty *et al.*, 2011) can be used instead of the single-channel Wiener filter. Here the tradeoff is controlled by under- or overestimation of the interference level.

Spectral enhancement: In the single-channel case (i.e., $I = 1$), the MWF reduces to $w_{\text{SWF}}(n,f)$, which can be written as

$$w_{\text{SWF}}(n,f) = \frac{\sigma_{e,1}^2(n,f)}{\sigma_{e,1}^2(n,f) + \sigma_{r,1}^2(n,f) + \sigma_{u_1}^2(n,f)} \tag{15.39}$$

$$= \frac{1}{1 + \text{SRR}^{-1}(n,f) + \text{SNR}^{-1}(n,f)}, \tag{15.40}$$

where $\sigma_{e,1}^2(n,f) = \mathbb{E}\{|x_{e,1}(n,f)|^2\}$ denotes the variance of the early speech component, $\sigma_{r,1}^2(n,f) = \mathbb{E}\{|x_{r,1}(n,f)|^2\}$ denotes the variance of the reverberant sound component, $\sigma_{u_1}^2(n,f) = \mathbb{E}\{|u_1(n,f)|^2\}$ denotes the variance of the noise, $\text{SRR}(n,f) = \sigma_{e,1}^2(n,f)/\sigma_{r,1}^2(n,f)$ denotes the SRR, and $\text{SNR}(n,f) = \sigma_{e,1}^2(n,f)/\sigma_{u_1}^2(n,f)$ denotes the signal-to-noise ratio (SNR). For the single-channel case there exist estimators for $\sigma_{e,1}^2(n,f)$, $\sigma_{r,1}^2(n,f)$ and $\sigma_{u_1}^2(n,f)$. The single-channel spectral enhancement approach is illustrated in Figure 15.4.

In some works, it was proposed to estimate the SRR using multiple microphones and subsequently apply a single-channel Wiener filter to one of the microphones. Particularly when the number of microphones is small, and the MVDR filter cannot provide a significant amount of reverberation reduction, this approach is very attractive in terms of computational complexity and robustness.

We have seen how an estimate of the early sound component or direct sound component can be obtained using different approaches. The main remaining challenge is how to estimate the variance of the reverberant sound component $\sigma_r^2(n,f)$. In the following we focus on the estimation of $\sigma_r^2(n,f)$, and assume that estimates of the noise covariance $\Sigma_u(n,f)$ and the relative early or relative direct transfer functions $\tilde{\mathbf{a}}(f)$ are given. It should be noted that under these assumptions and given an estimate of $\sigma_r^2(n,f)$, one can obtain an estimate of $\sigma_{e,i}^2(n,f)$ using the decision-directed approach (Ephraim and Malah, 1985; Benesty *et al.*, 2009).

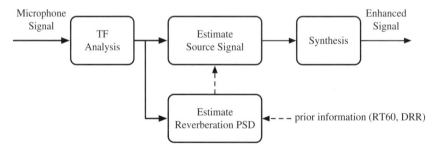

Figure 15.4 Single-channel spectral enhancement approach to perform reverberation suppression.

15.3.3 Single-Channel Spectral Variance Estimators

Existing single-channel estimators for the variance $\sigma_r^2(n,f)$ exploit the spectro-temporal structure of the acoustic impulse response. One of the best-known single-channel estimators is based on a simple exponentially decaying model, which was proposed by Polack (1993), and is fully characterized by the RT60. It should be noted that this model assumes that the source-to-microphone distance is larger than the critical distance. Using this model, Lebart *et al.* (2001) showed that in the absence of noise an estimate of $\sigma_r^2(n,f)$ could be obtained using

$$\hat{\sigma}_r^2(n,f) = e^{-2\alpha MN'} \, \hat{\sigma}_x^2(n - N', f), \tag{15.41}$$

where N' corresponds to the number of frames between the direct sound and start time of the late reverberation, $\alpha = 3\log(10)/(\text{RT60}f_s)$ is the reverberation decay constant, M is the number of samples between subsequent STFT frames, and f_s is the sampling frequency. Since the RT60 is normally frequency-dependent, Habets (2004) proposed to make α frequency-dependent. Interestingly, it can be shown that this estimator is also valid in the presence of multiple sound sources as long as all source-to-microphone distances are larger than the critical distance.

The exponentially decaying model of Lebart *et al.* (2001) is valid only when the DRR is approximately zero, i.e. when the source-to-microphone distance is beyond the critical distance. A more general model that takes into account the DRR was proposed by Habets *et al.* (2009). Based on this model, an estimator was derived that depends on the frequency-dependent RT60, denoted by RT60(k), and the inverse of the frequency-dependent DRR, denoted by DRR(k). In the absence of noise, an estimate of $\sigma_r^2(n,f)$ is then given by

$$\hat{\sigma}_r^2(n,f) = (1 - \text{DRR}(k)) \, e^{-2\alpha(k)MN'} \, \hat{\sigma}_r^2(n - N', f)$$
$$+ \, \text{DRR}(k) \, e^{-2\alpha(k)MN'} \, \hat{\sigma}_x^2(n - N', f). \tag{15.42}$$

By replacing $\hat{\sigma}_x^2(n,f)$ by $\hat{\sigma}_x^2(n,f) - \hat{\sigma}_u^2(n,f)$, the estimator can be used also in the presence of noise. Habets *et al.* (2009) showed that the estimator in (15.42) yields a smaller estimation error compared to the estimator in (15.41) in the case when the source-to-microphone distance is smaller than the critical distance.

The great interest in these estimators has led to new challenges, such as blindly estimating the RT60 and DRR. In the 2015 ACE Challenge, various blind estimators for the RT60 and DRR were evaluated. For more information about the challenge, corpus, estimators, and results the reader is referred to Eaton *et al.* (2016).

Alternative single-channel estimators for $\sigma_r^2(n,f)$ that do not require an estimate of such acoustic parameters can be found in (Kinoshita *et al.*, 2009; Erkelens and Heusdens, 2010; Braun *et al.*, 2016).

15.3.4 Multichannel Spectral Variance Estimators

In the last decade, a variety of multichannel estimators for the variance $\sigma_r^2(n,f)$ have been developed assuming that $\mathbf{x}_e(n,f)$ results from a plane wave. These estimators can be divided into several classes as follows.

Estimators in the first class utilize the DOA information of the direct sound source to block the direct sound. This simplifies the estimation procedure as the resulting signal after the blocking process only contains diffuse sound and noise. Habets and Gannot (2007) proposed a two-channel estimator using an adaptive algorithm. Another two-channel estimator that employs a blind source separation (BSS) algorithm to automatically steer a null constraint toward the desired source was proposed by Schwarz *et al.* (2012). Braun *et al.* (2013) minimized the error matrix of the blocked signals, whereas Kuklasinski *et al.* (2016) and Schwartz *et al.* (2015b) derived an ML estimator given the blocked signals. Kuklasinski *et al.* (2016) obtained the solution by a root-finding procedure, whereas Schwartz *et al.* (2015b) used the iterative Newton method. Thiergart *et al.* (2014c) derived a specific spatial filter to estimate the spectral variance of the diffuse sound by blocking the direct sound and by maximizing the diffuse-to-noise ratio.

Estimators in the second class do not block the direct sound, and estimate the variance of direct sound and diffuse sound jointly. The method presented by Schwartz *et al.* (2016c) obtains the ML estimate of the direct and diffuse variances using the iterative Newton method. Schwartz *et al.* (2016a) derived a batch EM algorithm to estimate the direct and diffuse sound variances as well as the spatial coherence matrix $\mathbf{\Omega}(f)$ in the ML sense. Schwartz *et al.* (2016b) jointly estimated the direct and diffuse sound variances in the least squares sense by minimizing the Frobenius norm of an error matrix. This results in a relatively simple estimator for $\sigma_r^2(n,f)$ which is given by (dropping time-frequency indexes for legibility)

$$\hat{\sigma}_r^2 = \frac{(\tilde{\mathbf{a}}^H\tilde{\mathbf{a}})^2\,\mathfrak{R}(\mathrm{tr}((\hat{\mathbf{\Sigma}}_x - \hat{\mathbf{\Sigma}}_u)\mathbf{\Omega}^H)) - \tilde{\mathbf{a}}^H\mathbf{\Omega}\tilde{\mathbf{a}}\,\mathfrak{R}(\tilde{\mathbf{a}}^H(\hat{\mathbf{\Sigma}}_x - \hat{\mathbf{\Sigma}}_u)\tilde{\mathbf{a}})}{(\tilde{\mathbf{a}}^H\tilde{\mathbf{a}})^2\,\mathrm{tr}(\mathbf{\Omega}^H\mathbf{\Omega}) - (\tilde{\mathbf{a}}^H\mathbf{\Omega}\tilde{\mathbf{a}})^2}. \tag{15.43}$$

Estimators in the third class estimate the SRR that can be used directly to compute the single-channel Wiener filter, or to estimate $\sigma_r^2(n,f)$ using

$$\hat{\sigma}_r^2(n,f) = \frac{\hat{\sigma}_{x_i}^2(n,f) - \hat{\sigma}_{u_i}^2(n,f)}{1 + \mathrm{SRR}(n,f)} \quad \text{for} \quad i \in \{1,\dots,I\}, \tag{15.44}$$

where $\hat{\sigma}_{x_i}^2(n,f)$ and $\hat{\sigma}_{u_i}^2(n,f)$ are estimates of $\sigma_{x_i}^2(n,f) = \mathbb{E}\{|x_i(n,f)|^2\}$ and $\sigma_{u_i}^2(n,f)$, respectively. Ahonen and Pulkki (2009) derived an SRR estimator for B-format microphones. Jeub *et al.* (2011) proposed an SRR estimator for omnidirectional microphones and a source positioned at broadside. More general SRR estimators, based on the complex spatial coherence of the microphone signals, were proposed by Thiergart *et al.* (2011, 2012). An SRR estimator that is strictly based on the power of directional microphones was proposed by Thiergart *et al.* (2014a). More recently, an unbiased SRR estimator was proposed by Schwarz and Kellermann (2015). Aforementioned coherence-based SRR estimators are derived for two channels. In the case where more channels are available, the different estimates can be combined to reduce the variance of the estimate.

Estimators in the fourth class use estimates obtained per channel using one of the estimators outlined in Section 15.3.3. Assuming that the reverberant sound field is spatially homogeneous, the individual estimates per microphone can, for example, be averaged to obtain a multichannel estimate, i.e.

$$\hat{\sigma}_r^2(n,f) = \frac{1}{I}\sum_{i=1}^{I}\hat{\sigma}_{r,i}^2(n,f). \tag{15.45}$$

Estimators in the fifth class aim to estimate the late reverberant spectral variance based on an estimate of one or more late reverberant signals. These signals can be obtained, for example, using a linearly constrained minimum variance (LCMV) filter (Thiergart and Habets, 2014), a long-term multistep linear prediction technique (Kinoshita *et al.*, 2009), or a partial acoustic system equalizer (Cauchi *et al.*, 2015).

15.4 Direct Estimation

In much of the preceding discussion in this chapter the methods have followed an overall scheme in which first an estimate is obtained of the parameters that characterize the reverberation (e.g., linear prediction coefficients or statistics of the additive reverberation), and then the dereverberated signal is obtained by exploiting these parameters. In contrast, there are several techniques to estimate the dereverberated signal directly while treating the acoustic system as unknown. These methods rely explicitly or implicitly on the characteristics of the source signal.

As well as the methods described below, other significant methods include the regional weighting function method of Yegnanarayana (1998) and the method using wavelet extrema clustering (Griebel and Brandstein, 1999; Gillespie, 2002) that exploits kurtosis maximization.

15.4.1 Synthesizing a Clean Residual Signal

One of the first methods for direct estimation of dereverberated speech was presented by Allen (1974). This method is based on synthesizing speech using the well-known source-filter model of speech production (Rabiner and Schafer, 2011). The operation of the method begins with classical frame-based linear predictive coding analysis of the reverberant speech to obtain the vocal tract parameters. The prediction residual is obtained by inverse filtering the reverberant speech with the vocal tract filter. From the prediction residual, the method obtains a voicing decision parameter, the voiced period, and a gain parameter. Dereverberated speech is then synthesized from the model using a gain-matched excitation signal formed from either noise in the unvoiced case or periodic pulses in the voiced case. The approach relies on two requirements for successful operation: first, that the source-filter model is sufficiently accurate to be able to synthesize natural sounding speech signals and, second, that the parameters of the model estimated from reverberant speech are not influenced significantly by the reverberation.

15.4.2 Linear Prediction Residual Processing

The estimation of speech model parameters from reverberant microphone signals was further studied by Gaubitch *et al.* (2003, 2004). It was shown that the linear predictive coding residual signal for reverberant voiced speech includes not only the expected quasi-periodic pulses associated with glottal closure events but also additional pulse-like features due to multipath early reflections. The approach taken is to perform dereverberation processing on the residual signal and then resynthesize the dereverberated speech by filtering the processed residual with the vocal tract filter.

Regarding the estimation of the vocal tract linear predictive coding parameters, Gaubitch *et al.* (2006) showed using statistical room acoustics that the linear predictive coding parameters of the clean speech can be estimated from the reverberant speech by taking the spatial expectation. The spatial expected values of the parameters can be found from sufficiently many microphone signals distributed over the reverberant sound field.

After inverse filtering with the vocal tract filter to obtain the prediction residual, the method then exploits the observation that the residual signal varies slowly across larynx cycles. It is therefore possible to enhance the residual by performing intercycle averaging to reduce the level of signal components due to multipath early reflections. In this way, each enhanced larynx cycle of the prediction residual is obtained from the weighted average of a chosen number of neighboring cycles in time. This process has the advantage of maintaining the overall structure of the prediction residual and so maintains the naturalness of the processed speech. Averaging over five cycles, for example, centered on the cycle of interest, has been found to be effective. Further details on the weighting scheme employed in the residual averaging and a discussion of processing for unvoiced speech are given by Gaubitch *et al.* (2004).

15.4.3 Deep Neural Networks

The use of machine learning techniques is particularly attractive for the solution of multidimensional estimation problems in the case where substantial quantities of training data are available. In the context of dereverberation, one approach is to learn the mapping from reverberant speech to dry speech (or vice versa) in some suitable domain.

In a recent example, Han *et al.* (2015) used such a mapping employed in the log spectral magnitude domain, computed using overlapping STFT time frames of 20 ms duration. The mapping is learned using a feedfoward deep neural network (DNN) with input feature vector at time frame n defined as

$$\tilde{\mathbf{o}}(n) = [\mathbf{o}(n - N_{\text{past}}), \ldots, \mathbf{o}(n), \ldots, \mathbf{o}(n + N_{\text{future}})]^T \tag{15.46}$$

where $\mathbf{o}(n)$ is the log spectral magnitude of the reverberant speech and N_{past} and N_{future} determine the number of context frames which are essential to take into account the temporal dynamics of speech. The authors found that $N_{\text{past}} = N_{\text{future}} = 5$ provides adequate support for training the mapping. The desired output feature vector at time frame n is the log spectral magnitude of the clean speech. The method is also extended to reduce the level of both reverberation and noise and shows benefits in terms of PESQ and STOI. Instead of using a feedforward DNN and context frames, Weninger *et al.* (2014) used a long short-term memory (LSTM) network, which is a special variant of recurrent neural network (RNN), to model the time dependencies implicitly.

15.5 Evaluation of Dereverberation

The task of measuring the effectiveness of dereverberation processing currently lacks standardized metrics and methods. It is common to address the task in two subcategories: for ASR and for speech enhancement. In the ASR task, dereverberation is assessed as a preprocessing stage to improve the robustness of ASR to large speaker-microphone distances. Measurement of the effect of dereverberation can be quantified in terms of word error rate reduction but consideration has to be given to what degree of retraining or adaptation is employed in the recognizer. In the speech enhancement task, dereverberation is assessed in the role of improving the quality and intelligibility of speech as perceived by the human auditory system. This can, in principle, be measured using a listening test (Cauchi *et al.*, 2016). However, such tests are difficult to construct for absolute measurement and with broad scope.

Results of benchmarking initiatives such as the REVERB Challenge aim to give insight into the current state-of-the-art performance levels on realistic dereverberation scenarios (Kinoshita *et al.*, 2013). One reproducible approach for testing algorithms is to apply them on simulated reverberant speech formed from the convolution of measured impulse responses with clean speech, together with representative levels of noise added. An alternative approach, which is nevertheless closer to real applications, is to apply dereverberation algorithms on speech recorded in real reverberant rooms. Typically, an individual headset microphone is additionally required to obtain an indication of the dry speech for evaluation purposes. It is not generally true that dereverberation performance on simulated reverberant speech is an accurate indicator of the performance that could be expected on recorded reverberant speech. Results in the REVERB Challenge for simulated and recorded reverberant speech show significant differences.

15.6 Summary

The topic of dereverberation is closely related to, and also draws many theoretical concepts from, other topics of acoustic signal processing, including acoustic modeling, acoustic source separation, interference suppression, etc. In this chapter we have presented techniques for dereverberation within a structure of three broad classes of reverberation cancellation, reverberation suppression, and direct estimation approaches. We have reviewed a representative set of currently available algorithms and, where possible, we have aimed to give relevant comparative explanations. Research is ongoing on this highly active topic, with exciting new approaches being discussed in the community as they emerge.

Bibliography

Abed-Meraim, K., Moulines, E., and Loubaton, P. (1997a) Prediction error method for second-order blind identification. *IEEE Transactions on Signal Processing*, **45** (3), 694–705.

Abed-Meraim, K., Qiu, W., and Hua, Y. (1997b) Blind system identification. *Proceedings of the IEEE*, **85** (8), 1310–1322.

Ahonen, J. and Pulkki, V. (2009) Diffuseness estimation using temporal variation of intensity vectors, in *Proceedings of IEEE Workshop on Applications of Signal Processing to Audio and Acoustics*, pp. 285–288.

Allen, J.B. (1973) Speech dereverberation. *Journal of the Acoustical Society of America*, **53** (1), 322.

Allen, J.B. (1974) Synthesis of pure speech from a reverberant signal, U.S. Patent No. 3786188.

Allen, J.B., Berkley, D.A., and Blauert, J. (1977) Multimicrophone signal-processing technique to remove room reverberation from speech signals. *Journal of the Acoustical Society of America*, **62** (4), 912–915.

Avargel, Y. and Cohen, I. (2007) System identification in the short-time Fourier transform domain with crossband filtering. *IEEE Transactions on Audio, Speech, and Language Processing*, **15** (4), 1305–1319.

Benesty, J., Chen, J., and Habets, E.A.P. (2011) *Speech Enhancement in the STFT Domain*, Springer.

Benesty, J., Chen, J., Huang, Y., and Cohen, I. (2009) *Noise Reduction in Speech Processing*, Springer.

Bolt, R.H. and MacDonald, A.D. (1949) Theory of speech masking by reverberation. *Journal of the Acoustical Society of America*, **21** (6), 577–580.

Braun, S. and Habets, E.A.P. (2016) Online dereverberation for dynamic scenarios using a Kalman filter with an autoregressive model. *IEEE Signal Processing Letters*, **23** (12), 1741–1745.

Braun, S., Jarrett, D., Fischer, J., and Habets, E.A.P. (2013) An informed spatial filter for dereverberation in the spherical harmonic domain, in *Proceedings of IEEE International Conference on Audio, Speech and Signal Processing*, pp. 669–673.

Braun, S., Schwartz, B., Gannot, S., and Habets, E.A.P. (2016) Late reverberation PSD estimation for single-channel dereverberation using relative convolutive transfer functions, in *Proceedings of International Workshop on Acoustic Signal Enhancement*.

Cauchi, B., Javed, H., Gerkmann, T., Doclo, S., Goetze, S., and Naylor, P.A. (2016) Perceptual and instrumental evaluation of the perceived level of reverberation, in *Proceedings of IEEE International Conference on Audio, Speech and Signal Processing*, pp. 629–633.

Cauchi, B., Kodrasi, I., Rehr, R., Gerlach, S., Jukić, A., Gerkmann, T., Doclo, S., and Goetze, S. (2015) Combination of MVDR beamforming and single-channel spectral processing for enhancing noisy and reverberant speech. *EURASIP Journal on Advances in Signal Processing*, **2015** (1), 1–12.

del Vallado, J.M.F., de Lima, A.A., Prego, T.d.M., and Netto, S.L. (2013) Feature analysis for the reverberation perception in speech signals, in *Proceedings of IEEE International Conference on Audio, Speech and Signal Processing*, pp. 8169–8173.

Delcroix, M., Hikichi, T., and Miyoshi, M. (2004) Dereverberation of speech signals based on linear prediction, in *Proceedings of the International Conference on Spoken Language Processing*, vol. 2, pp. 877–881.

Delcroix, M., Hikichi, T., and Miyoshi, M. (2006) On the use of lime dereverberation algorithm in an acoustic environment with a noise source, in *Proceedings of IEEE International Conference on Audio, Speech and Signal Processing*, vol. 1, pp. 825–828.

Eaton, J., Gaubitch, N.D., Moore, A.H., and Naylor, P.A. (2016) Estimation of room acoustic parameters: The ACE challenge. *IEEE/ACM Transactions on Audio, Speech, and Language Processing*, **24** (10), 1681–1693.

Ephraim, Y. and Malah, D. (1985) Speech enhancement using a minimum mean-square error log-spectral amplitude estimator. *IEEE Transactions on Acoustics, Speech, and Signal Processing*, **33** (2), 443–445.

Erkelens, J. and Heusdens, R. (2010) Correlation-based and model-based blind single-channel late-reverberation suppression in noisy time-varying acoustical environments. *IEEE Transactions on Audio, Speech, and Language Processing*, **18** (7), 1746–1765.

Fielder, L.D. (2003) Analysis of traditional and reverberation-reducing methods of room equalization. *Journal of the Acoustical Society of America*, **51** (1/2), 3–26.

Gannot, S. and Moonen, M. (2003) Subspace methods for multimicrophone speech dereverberation. *EURASIP Journal on Applied Signal Processing*, **2003** (11), 1074–1090.

Gaubitch, N.D., Naylor, P.A., and Ward, D.B. (2003) On the use of linear prediction for dereverberation of speech, in *Proceedings of International Workshop on Acoustic Echo and Noise Control*, pp. 99–102.

Gaubitch, N.D., Naylor, P.A., and Ward, D.B. (2004) Multi-microphone speech dereverberation using spatio-temporal averaging, in *Proceedings of European Signal Processing Conference*, pp. 809–812.

Gaubitch, N.D., Ward, D.B., and Naylor, P.A. (2006) Statistical analysis of the autoregressive modeling of reverberant speech. *Journal of the Acoustical Society of America*, **120** (6), 4031–4039.

Gesbert, D. and Duhamel, P. (1997) Robust blind channel identification and equalization based on multi-step predictors, in *Proceedings of IEEE International Conference on Audio, Speech and Signal Processing*, pp. 3621–3624.

Gillespie, B.W. (2002) Acoustic diversity for improved speech recognition in reverberant environments, in *Proceedings of IEEE International Conference on Audio, Speech and Signal Processing*, pp. 557–560.

Griebel, S.M. and Brandstein, M.S. (1999) Wavelet transform extrema clustering for multi-channel speech dereverberation, in *Proceedings of International Workshop on Acoustic Echo and Noise Control*, pp. 52–55.

Gürelli, M.I. and Nikias, C.L. (1995) EVAM: An eigenvector-based algorithm for multichannel blind deconvolution of input colored signals. *IEEE Transactions on Signal Processing*, **43** (1), 134–149.

Haas, H. (1972) The influence of a single echo on the audibility of speech. *Journal of the Audio Engineering Society*, **20**, 145–159.

Habets, E.A.P. (2004) Single-channel speech dereverberation based on spectral subtraction, in *Proceedings of Workshop on Circuits, Systems, and Signal Processing*, pp. 250–254.

Habets, E.A.P. (2007) *Single- and Multi-Microphone Speech Dereverberation using Spectral Enhancement*, Ph.D. thesis, Technische Universiteit Eindhoven.

Habets, E.A.P. (2016) Fifty years of reverberation reduction: From analog signal processing to machine learning, in *Proceedings of the Audio Engineering Society International Conference*.

Habets, E.A.P. and Gannot, S. (2007) Dual-microphone speech dereverberation using a reference signal, in *Proceedings of IEEE International Conference on Audio, Speech and Signal Processing*, vol. 4, pp. 901–904.

Habets, E.A.P., Gannot, S., and Cohen, I. (2009) Late reverberant spectral variance estimation based on a statistical model. *IEEE Signal Processing Letters*, **16** (9), 770–773.

Habets, E.A.P. and Naylor, P. (2010) An online quasi-Newton algorithm for blind SIMO identification, in *Proceedings of IEEE International Conference on Audio, Speech and Signal Processing*.

Han, K., Wang, Y., Wang, D., Woods, W.S., Merks, I., and Zhang, T. (2015) Learning spectral mapping for speech dereverberation and denoising. *IEEE/ACM Transactions on Audio, Speech, and Language Processing*, **23** (6), 982–992.

Haque, M. and Hasan, M. (2008) Noise robust multichannel frequency-domain LMS algorithms for blind channel identification. *IEEE Signal Processing Letters*, **15**, 305–308.

Hasan, M.K. and Naylor, P.A. (2006) Effect of noise on blind adaptive multichannel identification algorithms: Robustness issue, in *Proceedings of European Signal Processing Conference.*

Huang, Y. and Benesty, J. (2002) Adaptive multi-channel least mean square and Newton algorithms for blind channel identification. *Signal Processing*, **82**, 1127–1138.

Huang, Y. and Benesty, J. (2003) A class of frequency-domain adaptive approaches to blind multichannel identification. *IEEE Transactions on Signal Processing*, **51** (1), 11–24.

ITU-T (2001a) Recommendation P.862. perceptual evaluation of speech quality (PESQ): An objective method for end-to-end speech quality assessment of narrow-band telephone networks and speech codecs.

ITU-T (2001b) Recommendation P.863. perceptual objective listening quality assessment.

Jeub, M., Nelke, C., Beaugeant, C., and Vary, P. (2011) Blind estimation of the coherent-to-diffuse energy ratio from noisy speech signals, in *Proceedings of European Signal Processing Conference.*

Jukić, A., van Waterschoot, T., Gerkmann, T., and Doclo, S. (2015) Multi-channel linear prediction-based speech dereverberation with sparse priors. *IEEE/ACM Transactions on Audio, Speech, and Language Processing*, **23** (9), 1509–1520.

Jukić, A., van Waterschoot, T., Gerkmann, T., and Doclo, S. (2016a) A general framework for incorporating time-frequency domain sparsity in multi-channel speech dereverberation. *Journal of the Audio Engineering Society*, **65** (1/2), 17–30.

Jukić, A., Wang, Z., van Waterschoot, T., Gerkmann, T., and Doclo, S. (2016b) Constrained multi-channel linear prediction for adaptive speech dereverberation. *Proceedings of International Workshop on Acoustic Echo and Noise Control*, pp. 1–5.

Kinoshita, K., Delcroix, M., Nakatani, T., and Miyoshi, M. (2009) Suppression of late reverberation effect on speech signal using long-term multiple-step linear prediction. *IEEE Transactions on Audio, Speech, and Language Processing*, **17** (4), 534–545.

Kinoshita, K., Delcroix, M., Yoshioka, T., Nakatani, T., Sehr, A., Kellermann, W., and Maas, R. (2013) The Reverb Challenge: A common evaluation framework for dereverberation and recognition of reverberant speech, in *Proceedings of IEEE Workshop on Applications of Signal Processing to Audio and Acoustics*, pp. 1–4.

Kodrasi, I. and Doclo, S. (2012) Robust partial multichannel equalization techniques for speech dereverberation, in *Proceedings of IEEE International Conference on Audio, Speech and Signal Processing.*

Kodrasi, I. and Doclo, S. (2016) Joint dereverberation and noise reduction based on acoustic multi-channel equalization. *IEEE/ACM Transactions on Audio, Speech, and Language Processing*, **24** (4), 680–693.

Kodrasi, I., Goetze, S., and Doclo, S. (2012) Increasing the robustness of acoustic multichannel equalization by means of regularization, in *Proceedings of International Workshop on Acoustic Signal Enhancement*, pp. 161–164.

Kuklasinski, A., Doclo, S., Jensen, S., and Jensen, J. (2016) Maximum likelihood PSD estimation for speech enhancement in reverberation and noise. *IEEE/ACM Transactions on Audio, Speech, and Language Processing*, **24** (9), 1599–1612.

Kuttruff, H. (2000) *Room Acoustics*, Taylor & Francis, 4th edn.

Lebart, K., Boucher, J.M., and Denbigh, P.N. (2001) A new method based on spectral subtraction for speech de-reverberation. *Acta Acustica*, **87** (3), 359–366.

Lefkimmiatis, S. and Maragos, P. (2007) A generalized estimation approach for linear and nonlinear microphone array post-filters. *Speech Communication*, **49** (7-8), 657–666.

Lim, F. and Naylor, P.A. (2012) Relaxed multichannel least squares with constrained initial taps for multichannel dereverberation, in *Proceedings of International Workshop on Acoustic Signal Enhancement*.

Lim, F., Zhang, W., Habets, E.A.P., and Naylor, P.A. (2014) Robust multichannel dereverberation using relaxed multichannel least squares. *IEEE/ACM Transactions on Audio, Speech, and Language Processing*, **22** (9), 1379–1390.

Liu, H., Xu, G., and Tong, L. (1993) A deterministic approach to blind equalization, in *Proceedings of Asilomar Conference on Signals, Systems, and Computers*, vol. 1, pp. 751–755.

Makhoul, J. (1975) Linear prediction: A tutorial review. *Proceedings of the IEEE*, **63** (4), 561–580.

Malik, S., Schmid, D., and Enzner, G. (2012) A state-space cross-relation approach to adaptive blind SIMO system identification. *IEEE Signal Processing Letters*, **19** (8), 511–514.

Mertins, A., Mei, T., and Kallinger, M. (2010) Room impulse response shortening/reshaping with infinity- and p-norm optimization. *IEEE Transactions on Audio, Speech, and Language Processing*, **18** (2), 249–259.

Miyoshi, M. and Kaneda, Y. (1988) Inverse filtering of room acoustics. *IEEE Transactions on Acoustics, Speech, and Signal Processing*, **36** (2), 145–152.

Mohammadiha, N., Smaragdis, P., and Leijon, A. (2013) Supervised and unsupervised speech enhancement using nonnegative matrix factorization. *IEEE/ACM Transactions on Audio, Speech, and Language Processing*, **21** (10), 2140–2151.

Moulines, E., Duhamel, P., Cardoso, J.F., and Mayrargue, S. (1995) Subspace methods for the blind identification of multichannel FIR filters. *IEEE Transactions on Signal Processing*, **43** (2), 516–525.

Nakatani, T., Yoshioka, T., Kinoshita, K., Miyoshi, M., and Juang, B.H. (2010) Speech dereverberation based on variance-normalized delayed linear prediction. *IEEE/ACM Transactions on Audio, Speech, and Language Processing*, **18** (7), 1717–1731.

Naylor, P.A. and Gaubitch, N.D. (2005) Speech dereverberation, in *Proceedings of International Workshop on Acoustic Echo and Noise Control*.

Naylor, P.A. and Gaubitch, N.D. (eds) (2010) *Speech Dereverberation*, Springer.

Naylor, P.A., Gaubitch, N.D., and Habets, E.A.P. (2010) Signal-based performance evaluation of dereverberation algorithms. *Journal of Electrical and Computer Engineering*, **2010**, 1–5.

Pelorson, X., Vian, J.P., and Polack, J.D. (1992) On the variability of room acoustical parameters: Reproducibility and statistical validity. *Applied Acoustics*, **37**, 175–198.

Polack, J.D. (1993) Playing billiards in the concert hall: the mathematical foundations of geometrical room acoustics. *Acta Acustica*, **38** (2), 235–244.

Rabiner, L.R. and Schafer, R.W. (2011) *Theory and Applications of Digital Speech Processing*, Pearson.

Sabine, W.C. (1922) *Collected Papers on Acoustics*, Harvard University Press.

Schmid, D. (2014) *Multichannel Dereverberation and Noise Reduction for Hands-Free Speech Communication Systems*, Ph.D. thesis, Fakultät für Elektrotechnik und Informationstechnik, Ruhr-University Bochum, Germany.

Schmid, D., Enzner, G., Malik, S., Kolossa, D., and Martin, R. (2014) Variational Bayesian inference for multichannel dereverberation and noise reduction. *IEEE/ACM Transactions on Audio, Speech, and Language Processing*, **22** (8), 1320–1335.

Schwartz, B., Gannot, S., and Habets, E.A.P. (2015a) Online speech dereverberation using Kalman filter and EM algorithm. *IEEE/ACM Transactions on Audio, Speech, and Language Processing*, **23** (2), 394–406.

Schwartz, O., Gannot, S., and Habets, E.A.P. (2015b) Multi-microphone speech dereverberation and noise reduction using relative early transfer functions. *IEEE/ACM Transactions on Audio, Speech, and Language Processing*, **23** (2), 240–251.

Schwartz, O., Gannot, S., and Habets, E.A.P. (2016a) An expectation-maximization algorithm for multi-microphone speech dereverberation and noise reduction with coherence matrix estimation. *IEEE/ACM Transactions on Audio, Speech, and Language Processing*, **24** (9), 1495–1510.

Schwartz, O., Gannot, S., and Habets, E.A.P. (2016b) Joint estimation of late reverberant and speech power spectral densities in noisy environments using Frobenius norm, in *Proceedings of European Signal Processing Conference*.

Schwartz, O., Gannot, S., and Habets, E.A.P. (2016c) Joint maximum likelihood estimation of late reverberant and speech power spectral density in noisy environments, in *Proceedings of IEEE International Conference on Audio, Speech and Signal Processing*.

Schwarz, A. and Kellermann, W. (2015) Coherent-to-diffuse power ratio estimation for dereverberation. *IEEE/ACM Transactions on Audio, Speech, and Language Processing*, **23** (6), 1006–1018.

Schwarz, A., Reindl, K., and Kellermann, W. (2012) A two-channel reverberation suppression scheme based on blind signal separation and Wiener filtering, in *Proceedings of IEEE International Conference on Audio, Speech and Signal Processing*, IEEE.

Slock, D.T.M. (1994) Blind fractionally-spaced equalization, perfectre- construction filter-banks and multichannel linear prediction, in *Proceedings of IEEE International Conference on Audio, Speech and Signal Processing*, pp. 585–588.

Taal, C.H., Hendriks, R.C., Heusdens, R., and Jensen, J. (2011) An algorithm for intelligibility prediction of time-frequency weighted noisy speech. *IEEE Transactions on Audio, Speech, and Language Processing*, **19** (7), 2125–2136.

Thiergart, O., Ascherl, T., and Habets, E.A.P. (2014a) Power-based signal-to-diffuse ratio estimation using noisy directional microphones, in *Proceedings of IEEE International Conference on Audio, Speech and Signal Processing*, pp. 7440–7444.

Thiergart, O., Galdo, G.D., and Habets, E.A.P. (2011) Diffuseness estimation with high temporal resolution via spatial coherence between virtual first-order microphones, in *Proceedings of IEEE Workshop on Applications of Signal Processing to Audio and Acoustics*, pp. 217–220.

Thiergart, O., Galdo, G.D., and Habets, E.A.P. (2012) On the spatial coherence in mixed sound fields and its application to signal-to-diffuse ratio estimation. *Journal of the Acoustical Society of America*, **132** (4), 2337–2346.

Thiergart, O. and Habets, E.A.P. (2014) Extracting reverberant sound using a linearly constrained minimum variance spatial filter. *IEEE Signal Processing Letters*, **21** (5), 630–634.

Thiergart, O., Taseska, M., and Habets, E.A.P. (2014b) An informed parametric spatial filter based on instantaneous direction-of-arrival estimates. *IEEE/ACM Transactions on Audio, Speech, and Language Processing*, **22** (12), 2182–2196.

Thiergart, O., Taseska, M., and Habets, E.A.P. (2014c) An informed parametric spatial filter based on instantaneous direction-of-arrival estimates. *IEEE/ACM Transactions on Audio, Speech, and Language Processing*, **22** (12), 2182–2196.

Togami, M., Kawaguchi, Y., Takeda, R., Obuchi, Y., and Nukaga, N. (2013) Optimized speech dereverberation from probabilistic perspective for time varying acoustic transfer function. *IEEE Transactions on Audio, Speech, and Language Processing*, **21** (7), 1369–1380.

Tong, L. and Perreau, S. (1998) Multichannel blind identification: from subspace to maximum likelihood methods. *Proceedings of the IEEE*, **86** (10), 1951–1968.

Tong, L., Xu, G., and Kailath, T. (1994) Blind identification and equalization based on second-order statistics: A time domain approach. *IEEE Transactions on Information Theory*, **40** (2), 340–349.

Triki, M. and Slock, D.T.M. (2005) Blind dereverberation of quasi-periodic sources based on multichannel linear prediction, in *Proceedings of International Workshop on Acoustic Echo and Noise Control*.

Valimaki, V., Parker, J.D., Savioja, L., Smith, J.O., and Abel, J.S. (2012) Fifty years of artificial reverberation. *IEEE Transactions on Audio, Speech, and Language Processing*, **20** (5), 1421–1448.

Weninger, F., Geiger, J., Wöllmer, M., Schuller, B., and Rigoll, G. (2014) Feature enhancement by deep LSTM networks for ASR in reverberant multisource environments. *Computer Speech and Language*, **28** (4), 888–902.

Xu, G., Liu, H., Tong, L., and Kailath, T. (1995) A least-squares approach to blind channel identification. *IEEE Transactions on Signal Processing*, **43** (12), 2982–2993.

Yegnanarayana, B. (1998) Enhancement of reverberant speech using LP residual, in *Proceedings of IEEE International Conference on Audio, Speech and Signal Processing*, vol. 1, pp. 405–408.

Yoshioka, T. and Nakatani, T. (2012) Generalization of multi-channel linear prediction methods for blind MIMO impulse response shortening. *IEEE Transactions on Audio, Speech, and Language Processing*, **20** (10), 2707–2720.

Yoshioka, T. and Nakatani, T. (2013) Dereverberation for reverberation-robust microphone arrays, in *Proceedings of European Signal Processing Conference*, pp. 1–5.

Yoshioka, T., Nakatani, T., Hikichi, T., and Miyoshi, M. (2008) Maximum likelihood approach to speech enhancement for noisy reverberant signals, in *Proceedings of IEEE International Conference on Audio, Speech and Signal Processing*, pp. 4585–4588.

Yoshioka, T., Sehr, A., Delcroix, M., Kinoshita, K., Maas, R., Nakatani, T., and Kellermann, W. (2012) Making machines understand us in reverberant rooms: Robustness against reverberation for automatic speech recognition. *IEEE Signal Processing Magazine*, **29** (6), 114–126.

Zhang, W., Habets, E.A.P., and Naylor, P.A. (2010) On the use of channel shortening in multichannel acoustic system equalization, in *Proceedings of International Workshop on Acoustic Echo and Noise Control*.

Zhang, W., Khong, A.W.H., and Naylor, P.A. (2008) Adaptive inverse filtering of room acoustics, in *Proceedings of Asilomar Conference on Signals, Systems, and Computers*.

Part IV

Application Scenarios and Perspectives

16

Applying Source Separation to Music

Bryan Pardo, Antoine Liutkus, Zhiyao Duan, and Gaël Richard

Separation of existing audio into remixable elements is useful in many contexts, especially in the realm of music and video remixing. Much musical audio content, including audio tracks for video, is available only in mono (e.g., 1940s movies and records) or stereo (YouTube videos, commercially released music where the source tracks are not available). Separated sources from such tracks would be useful to repurpose this audio content. Applications include upmixing video soundtracks to surround sound (e.g., home theater 5.1 systems), facilitating music transcription by separating into individual instrumental tracks, allowing better mashups and remixes for disk jockeys, and rebalancing sound levels after multiple instruments or voices were recorded simultaneously to a single track (e.g., turning up only the dialog in the movie, not the music). Effective separation would also let producers edit out individual musician's note errors in a live recording without the need for an individual microphone on each musician, or apply audio effects (equalization, reverberation) to individual instruments recorded on the same track. Given the large number of potential applications and their impact, it is no surprise that many researchers have focused on the application areas of music recordings and movie soundtracks. In this chapter we provide an overview of the algorithms and approaches designed specifically for music. Where applicable, we will also introduce commonalities and links to source separation for video soundtracks, since many musical scenarios involve video soundtracks (e.g., YouTube recordings of live concerts, movie sound tracks).

We discuss the challenges and opportunities related to music in Section 16.1. We show how to constrain nonnegative matrix factorization (NMF) models for music in Section 16.2 and how to use musical instrument timbre models or musical scores in Section 16.3. We then explain how to take advantage of redundancies within a given musical recording in Section 16.4 or between multiple recordings in Section 16.5. We discuss how to involve the user in interactive source separation in Section 16.6 and in crowd-based evaluation in Section 16.7. We provide some examples of applications in Section 16.8 and conclude in Section 16.9.

Audio Source Separation and Speech Enhancement, First Edition.
Edited by Emmanuel Vincent, Tuomas Virtanen and Sharon Gannot.
© 2018 John Wiley & Sons Ltd. Published 2018 by John Wiley & Sons Ltd.
Companion Website: https://project.inria.fr/ssse/

16.1   Challenges and Opportunities

Music, in particular, provides a unique set of challenges and opportunities that have led algorithm designers to create methods particular to the task of separating out elements of a musical scene.

16.1.1   Challenges

In many nonmusical scenarios (e.g., recordings in a crowded street, multiple conversations in a cocktail party) sound sources are uncorrelated in their behavior and have relatively little overlap in time and frequency. In music, sources are often strongly correlated in onset and offset times, such as a choir singing together. This kind of correlation makes approaches that depend on independent component analysis (ICA) unreliable. Sources also often have strong frequency overlap. For example, voices singing in unison, octaves, or fifths produce many harmonics that overlap in frequency. Therefore, time-frequency overlap is a significant issue in music.

Unrealistic mixing scenarios are also common in both music and commercial videos. Both music and video sound tracks are often recorded in individual tracks that have individual equalization, panning, and reverberation applied to each track prior to the mixdown. Therefore, systems that depend on the assumption that sources share a reverberant space and have self-consistent timing and phase cues resulting from placement in a real environment will encounter problems.

These problems can be exacerbated in pop music, where it may be difficult to say what constitutes a source, as the sound may never have been a real physical source, such as the output of a music synthesizer. A single sonic element may be composed of one or more recordings of other sources that have been manipulated and layered together (e.g., drum loops with effects applied to them, or a voice with reverb and octave doubling applied to it).

Finally, evaluation criteria for music are different than for problems such as speech separation. Often, intelligibility is the standard for speech. For music, it is often an artistic matter. For some applications the standard may be that a separated source must sound perfect, while for others perhaps the separation need not be perfect, if the goal is to simply modify the relative loudness of a source within the mixture. Therefore, the choice of an evaluation measure is more task dependent than for some other applications.

16.1.2   Opportunities

Music also provides opportunities that can be exploited by source separation algorithms. Music that has a fully notated score provides knowledge of the relative timing and fundamental frequency of events. This can be exploited to guide source separation (e.g., seeding the activation matrix and spectral templates in NMF (Ewert and Muller, 2012)). Acoustic music typically has a finite set of likely sound producing sources (e.g., a string quartet almost always has two violins, one viola and one cello). This provides an opportunity to seed source models with timbre information, such as from libraries of sampled instrument sounds (Rodriguez-Serrano *et al.*, 2015a). Even when a score or list of instruments is not available, knowledge of musical rules can be used to construct a language model to constrain likely note transitions. Common sound engineering (Ballou, 2013) and panning techniques can be used to perform vocal isolation on many music recordings. For example, vocals are frequently center panned. One can retrieve the center panned elements of a two-channel recording by phase inverting the left channel and subtracting it from the right channel.

Often, the musical structure itself can be used to guide separation, as composers frequently group elements working together and present them in ways to teach the listener what the important groupings are. This can be used to guide source separation (Seetharaman and Pardo, 2016). If the goal is to separate the background music from the speech in a recording, having multiple examples of the recording with different speakers is very helpful. This is often possible with music concert recordings and with commercially released video. It is common for multiple people to record the same musical performance (e.g., YouTube concert videos). This can often provide multiple channels that allow a high-quality audio recording to be constructed from multiple low-quality ones (Kim and Smaragdis, 2013). Commercially released video content that has been dubbed into multiple languages provides a similar opportunity.

Having laid out the challenges and opportunities inherent in musical source separation, we now move on to discuss the ways that source separation techniques have been adapted to or designed for separation of music. We begin with NMF.

16.2 Nonnegative Matrix Factorization in the Case of Music

The previous chapters have already included much discussion on NMF and its use in audio source separation, particularly in the case of speech. NMF has been widely used for music source separation since it was introduced to the musical context in 2003 (Smaragdis and Brown, 2003). The characteristics of music signals have called for dedicated models and numerous developments and extension of the basic "blind" NMF decomposition. In this section we discuss some of the extensions of the models described in Chapters 8 and 9 for the particular case of music signals.

16.2.1 Shift-Invariant NMF

While the relative strengths of harmonics in vocals are constantly changing, many musical instruments (e.g., flutes, saxophones, pianos) produce a harmonic overtone series that keeps relationships relatively constant between the amplitudes of the harmonics even as the fundamental frequency changes. This shift invariance is made clear when the audio is represented using the log of the frequency, rather than representing frequency linearly, as the widely used short-time Fourier transform (STFT) does. Therefore, many instruments may well be approximated as frequency shift-invariant in a log-frequency representation, such as the constant-Q transform (Brown, 1991). The constant-Q transform allows one to represent the signal with a single frequency pattern, shifted in frequency for different pitches. This is illustrated in Figure 16.1 on a musical signal composed of three successive musical notes of different pitch. With the STFT, these three notes would need at least three frequency patterns to represent them.

Based on such a log-frequency transformation, several shift-invariant approaches have been proposed for audio signal analysis, especially in the framework of probabilistic

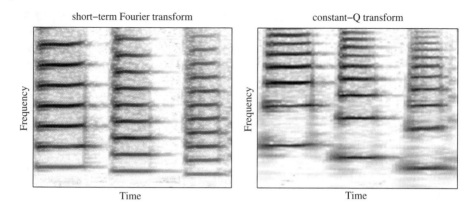

short–term Fourier transform constant–Q transform

Figure 16.1 STFT and constant-Q representation of a trumpet signal composed of three musical notes of different pitch.

latent component analysis (PLCA), the probabilistic counterpart of NMF (Smaragdis *et al.*, 2008). An extension of the initial framework, called blind harmonic adaptive decomposition, was presented by Fuentes *et al.* (2012) to better model real music signals. In this model, each musical note may present fundamental frequency and spectral envelope variations across repetitions. These characteristics make the model particularly efficient for real musical signals.

More precisely, the absolute value of the normalized constant-Q representation is modeled as the sum of a harmonic component and a noise component. The noise component is modeled as the convolution of a fixed smooth narrowband frequency window and a noise time-frequency distribution. The polyphonic harmonic component is modeled as a weighted sum of different harmonic spectra (to account for harmonics from multiple sources), each one having its own spectral envelope and pitch. The spacing between partials in a harmonic sound is a function of the fundamental frequency. As the frequency increases, the spacing between partials increases. When one uses a constant-Q representation, the spacing between channels is also a function of the frequency. Therefore, a pitch modulation in a constant-Q representation can be seen as a simple shifting of the partials with no worry that the spacing between partials needs to be adjusted as they would in a linear frequency representation.

The original approach is entirely unsupervised, unlike most other approaches in this framework, which rely on prior generic information or signal models to obtain a semantically meaningful decomposition. It was shown in some controlled cases that improved performance can be obtained by integrating a learning stage or by adapting the pretrained models using a multistage transcription strategy (see Benetos *et al.* (2014), for example) or by involving a user during the transcription process (Kirchhoff *et al.*, 2013; Bryan *et al.*, 2013). It was in particular shown by de Andrade Scatolini *et al.* (2015) that a partial annotation brings further improvement by providing a better model initiation with adapted or learned spectral envelope models.

16.2.2 Constrained and Structured NMF

The success of NMF for music source separation is largely due to its flexibility to take advantage of important characteristics of music signals in the decomposition. Indeed, those characteristics can be used to adequately constrain the NMF decomposition. Numerous extensions have been proposed to adapt the basic "blind" NMF model to incorporate constraints or data source structures (see Chapters 8 and 9, as well as Wang and Zhang (2013)). In this section, we will only focus on some extensions that are specific to the case of music signals, namely the incorporation of constraints deduced from music instrument models or from general music signal properties.

16.2.2.1 Exploiting Music Instrument Models

In acoustic music, the individual sources are typically produced by acoustic music instruments whose physical properties can be learned or modeled. We illustrate below three instrument models that have been successfully applied to NMF for music source separation.

- **Harmonic and inharmonic models in NMF:** A large set of acoustic music instruments produce well-defined pitches composed of partials or tones in relatively pure harmonic relations. For some other instruments, e.g. piano, due to the string stiffness, the frequencies of the partials slightly deviate from a purely harmonic relationship. Depending on the type of music signals at hand, specific parametric models can be built and used in NMF decompositions to model the spectra of the dictionary atoms. Harmonic models have been exploited in multiple works (Hennequin *et al.*, 2010; Bertin *et al.*, 2010). As an extension of this idea, Rigaud *et al.* (2013) proposed an additive model for which three different constraints on the partial frequencies were introduced to obtain a blend between a strict harmonic and a strict inharmonic relation given by specific physical models.
- **Temporal evolution models in NMF:** The standard NMF is shown to be efficient when the elementary components (notes) of the analyzed music signal are nearly stationary. However, in several situations elementary components can be strongly nonstationary and the decomposition will need multiple basis templates to represent a single note. Incorporating temporal evolution models in NMF is then particularly attractive to represent each elementary component with a minimal number of basis spectra. Several models were proposed, including the autoregressive (AR) moving average time-varying model introduced by Hennequin *et al.* (2011) to constrain the activation coefficients, which allowed an efficient single-atom decomposition for a single audio event with strong spectral variations to be obtained. Another strategy to better model the temporal structure of sounds is to rely on hidden Markov models (HMMs) that aim at describing the structure of changes between subsequent templates or dictionaries used in the decomposition (Mysore *et al.*, 2010).
- **Source-filter models in NMF:** Some instruments, e.g. the singing voice, are well represented by a source-filter production model with negligible interaction between the source (e.g., the vocal cords) and the filter (e.g., the vocal tract). Using such a production model has several advantages. First, it adds meaningful constraints to the decomposition which help to converge to an efficient separation. Second, it allows the choice of appropriate initializations, e.g. by predefining acceptable shapes for source and filter spectra. Third, it avoids the usual permutation problem of frequency-domain source separation methods since the component corresponding to the source-filter model is well identified. This motivated the use of source filter models in music source separation for singing voice extraction (Durrieu *et al.*, 2010, 2011) but also for other music instruments (Heittola *et al.*, 2009). Durrieu *et al.* (2011) modeled the singing

voice by a specific NMF source-filter model and the background by a regular and unconstrained NMF model as expressed below:

$$|\mathbf{X}|^2 \approx \underbrace{\mathbf{W}^F}_{\text{filter}} \circ \underbrace{\mathbf{W}^{f_0}}_{\text{source}} + \underbrace{\mathbf{B}^M \mathbf{H}^M}_{\text{background}} \tag{16.1}$$

where $|\mathbf{X}|^2$ denotes the elementwise exponentiation, $\mathbf{B}^M$ is the dictionary, $\mathbf{H}^M$ is the activation matrices of the NMF decomposition[1] of the background component, and $\mathbf{W}^F$ (respectively $\mathbf{W}^{f_0}$) represents the filter part (resp. the source part) of the singing voice component. Both the filter and source parts are further parameterized to allow good model expressivity. In particular, the filter is defined as a weighted combination of basic filter shapes, themselves built as a linear weighted combination of atomic elements (e.g., a single resonator symbolizing a formant filter). The filter part is then given by:

$$\mathbf{W}^F = \mathbf{B}^F \mathbf{H}^F \mathbf{H}^{\Phi} \tag{16.2}$$

where $\mathbf{B}^F$ is the filter atomic elements, $\mathbf{H}^F$ is the weighting coefficients of the filter atomic elements to build basic filter shapes, and $\mathbf{H}^{\Phi}$ is the weighting coefficients of the basic filter shapes (see Figure 16.2).

Concurrently, the source part is modeled as a positive linear combination of a number (in the ideal case reduced to one) of frequency patterns which represent basic source power spectra obtained by a source production model. The source part is then expressed as:

$$\mathbf{W}^{f_0} = \mathbf{B}^{f_0} \mathbf{H}^{f_0} \tag{16.3}$$

where $\mathbf{B}^{f_0}$ is the basic source power spectra for a predefined range of fundamental frequencies f_0 and $\mathbf{H}^{f_0}$ is the weighting (or activation) coefficients.

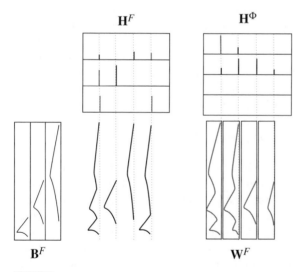

Figure 16.2 Schematic illustration of the filter part $\mathbf{W}^F$. Durrieu *et al.* (2011) defined the dictionary $\mathbf{B}^F$ of filter atomic elements as a set of 30 Hann functions, with 75% overlap.

---

1 See Chapter 8 for more details on NMF.

16.2.2.2   Exploiting Music Signal Models

Another strategy for adapting the raw NMF decompositions to music signals is to rely on dedicated signal models. Contrary to the music instrument models described above, these models are more generic and generally apply to a large class of music signals. Two examples of such signal models are briefly described below.

- **Harmonic/percussive models in NMF:** In general, harmonic instruments tend to produce few tones simultaneously which are slowly varying in times. Enforcing the temporal smoothness and sparsity of the decomposition is therefore an efficient strategy for separating harmonic instruments (Virtanen, 2007). Recently, Canadas-Quesada *et al.* (2014) used four constraints to achieve a specific harmonic/percussive decomposition with NMF. An alternative strategy is to rely on specific decomposition models that will automatically highlight the underlying harmonic/percussive musical concepts. Orthogonal NMF and *projective NMF* (Choi, 2008) are typical examples of such decompositions. For example, Laroche *et al.* (2015) used projective NMF to obtain an initial nearly orthogonal decomposition well adapted to represent harmonic instruments. This decomposition is further extended by a nonorthogonal component that is particularly relevant to represent percussive or transient signals. The so-called structured projective NMF model is then given by:

$$|\mathbf{X}|^2 \approx \underbrace{\mathbf{B}^h \mathbf{H}^h}_{\text{Harmonic}} + \underbrace{\mathbf{B}^p \mathbf{H}^p}_{\text{Percussive}} \tag{16.4}$$

where $\mathbf{B}^h$ (resp. $\mathbf{B}^p$) gathers the harmonic (resp. percussive) atoms and $\mathbf{H}^h$ (resp. $\mathbf{H}^p$) the activation coefficients of the harmonic (resp. percussive) component. Note that in this model, the harmonic part is obtained by projective NMF and the percussive part by a regular NMF.

- **Musical constraints in NMF models:** To adapt the decomposition to music signals, it is also possible to integrate constraints deduced from high-level musical concepts such as temporal evolution of sounds, rhythm structure or timbre similarity. For example, Nakano *et al.* (2010, 2011) constrain the model by using a Markov chain that governs the order in which the basis spectra appear for the representation of a musical note. This concept is extended by Kameoka *et al.* (2012), where a beat structure constraint is included in the NMF model. This constraint let them better represent note onsets. It is also possible to rely on timbre similarity for grouping similar spectra that describe the same instrument (e.g., the same source). An appropriate grouping of basis spectra allows one to improve the decomposition of the audio mixture into meaningful sources, since multiple basis spectra can be used for a single source, improving reproduction audio quality.

When additional information is available (such as the score), it is possible to use better musical constraints, as further discussed in Section 16.3.3. It is also worth mentioning that a number of applications may benefit from a two-component model such as the one briefly sketch for singing voice separation or harmonic/percussive decomposition (see, for example, Section 16.5.1 in the context of movie content remastering using multiple recordings).

16.3 Taking Advantage of the Harmonic Structure of Music

Harmonic sound sources (e.g., strings, woodwind, brass and vocals) are widely present in music signals and movie sound tracks. The spectrum of a harmonic sound source shows a harmonic structure: prominent spectral components are located at integer multiples of the fundamental frequency of the signal and hence are called harmonics; relative amplitudes of the harmonics are related to the spectral envelope and affect the timbre of the source. Modeling the harmonic structure helps to organize the frequency components of a harmonic source and separate it from the mixture.

16.3.1 Pitch-Based Harmonic Source Separation

The most intuitive idea for taking advantage of the harmonic structure is to organize and separate spectral components of a harmonic source according to its fundamental frequency (F0), which is also referred to as the pitch. Estimating the concurrent fundamental frequencies in each time frame, i.e. *multipitch estimation*, is the first important step. Multipitch estimation is a challenging research problem on its own, and many different approaches have been proposed. Time-domain approaches try to estimate the period of harmonic sources using autocorrelation functions (Tolonen and Karjalainen, 2000) or probabilistic sinusoidal modeling (Davy *et al.*, 2006). Frequency-domain approaches attempt to model the frequency regularity of harmonics (Klapuri, 2003; Duan *et al.*, 2010). Spectrogram decomposition methods use fixed (Ari *et al.*, 2012) or adaptive (Bertin *et al.*, 2010) harmonic templates to recognize the harmonic structure of harmonic sources. There are also approaches that fuse time-domain and frequency-domain information towards multipitch estimation (Su and Yang, 2015).

With the fundamental frequency of a harmonic source estimated in a time frame, a harmonic mask can be constructed to separate the source from the mixture. This mask can be binary so that all the spectral energy located at the harmonic frequencies is extracted for the source (Li and Wang, 2008). However, when harmonics of different sources overlap, a soft mask is needed to allocate the mixture signal's spectral energy to these overlapping sources appropriately. This overlapping harmonic issue is very common in music signals. This is due to the fact that tonal harmony composition rules prefer small integer ratios among the F0s of concurrent harmonic sources. For example, the frequency ratio of the F0s of a C major chord is C:E:G = 4:5:6. This causes 46.7% of the harmonics of the C note, 33.3% of E and 60% of G, being overlapped with the other two notes.

A harmonic index is the integer multiple that one must apply to the fundamental frequency to give the frequency of a harmonic. One simple effective method to allocate the spectral energy to overlapping harmonics is to consider the harmonic indexes. Higher harmonics tend to be softer than lower harmonics for most harmonic sounds. Therefore, the spectral energy tends to decrease when the harmonic index increases. Thus, it is reasonable to allocate more energy to the source whose low-indexed harmonic is overlapped by a higher-indexed harmonic of another source. Duan and Pardo (2011b) proposed to build a soft mask on the magnitude spectrum inverse proportional to the square of the harmonic index:

$$w_j = \frac{1/m_j^2}{\sum_{j'=1}^{J} 1/m_{j'}^2} \tag{16.5}$$

where w_j is the mask value for source j at the overlapping harmonic frequency bins and m_j is the harmonic index of the jth source. This is based on the assumptions that (i) overlapping sources have roughly the same energy and (ii) the spectral energy decays at a rate of 12 dB per octave regardless of the pitch and instrument. Although this simple method achieves a decent result, the assumptions are obviously oversimplified. For example, it does not model the timbre of the sources, which we will discuss in the next section.

16.3.2 Modeling Timbre

Multipitch estimation and harmonic masking allow us to separate harmonic sources in each individual frame. However, how can we organize the sources over time? In other words, which pitch (and its harmonics) belongs to which source? In auditory scene analysis (Bregman, 1994) this is called sequential grouping or streaming. Commonly used grouping cues include time and frequency proximity (i.e., sounds that are close in both time and frequency are likely to belong to the same source) and timbre and location consistency (i.e., sounds from the same source tend to have similar timbre and location while sounds from different sources often have distinct timbre and location). Time and frequency proximity cues, however, only help to group pitches within the same note because there is often a gap in time and/or frequency between two successive notes from the same source. In addition, the location consistency cue can only be exploited in stereo or multichannel recordings, where the interchannel level difference (ILD) and interchannel time difference (ITD) can be calculated to localize sound sources. The timbre consistency cue, on the other hand, is more universal.

One widely adopted approach to disentangle the pitch and timbre of a harmonic sound is the *source-filter model*, which is also called the excitation-resonance model. As shown in Figure 16.3, a harmonic sound, such as one produced by a clarinet, can be modeled as the convolution of an excitation signal (vibration of the clarinet reed) with a filter's impulse response (determined by the clarinet body). In the frequency domain, the magnitude spectrum of the harmonic sound is then the multiplication of the magnitude spectrum of the excitation signal and that of the filter's frequency response. The excitation spectrum is often modeled as a harmonic comb with flat or decaying amplitudes; it determines the F0 and harmonic frequencies. The filter's frequency response, on the other hand, characterizes the slowly varying envelope of the signal's spectrum; it is considered to affect the timbre of the signal.

Various ways for representing the spectral envelope have been proposed. Mel-frequency cepstral coefficients (MFCC) (Davis and Mermelstein, 1980) are a commonly used representation. However, MFCCs must be calculated from the full spectrum of a signal. Therefore, this representation cannot be used to represent the

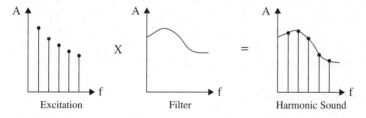

Figure 16.3 The source-filter model in the magnitude frequency domain.

spectral envelope of a single source, if the source has not been separated from the mixture. It is reasonable, though, to assume that some subset of the desired source's spectrum is not obscured by other sources in the mixture. This subset can be sparse in frequency and the spectral values can be noisy. For example, with the F0 of a source being estimated, the spectral points at the harmonic positions are likely to belong to the source spectrum, although the spectral values can be contaminated by the overlapping harmonics. Based on this observation, timbre representations that can be calculated from isolated spectral points have been proposed, such as the discrete cepstrum (Galas and Rodet, 1990), the regularized discrete cepstrum (Cappé *et al.*, 1995), and the uniform discrete cepstrum (Duan *et al.*, 2014b). The uniform discrete cepstrum and its Mel-frequency variant have been shown to achieve good results in instrument recognition in polyphonic music mixtures as well as multipitch streaming in speech mixtures (Duan *et al.*, 2014a).

Another timbre representation that does not directly characterize the spectral envelope but can be calculated from harmonics is the harmonic structure feature. It is defined as the relative logarithmic amplitude of harmonics (Duan *et al.*, 2008). Within a narrow pitch range (e.g., two octaves), the harmonic structure feature of musical instruments has been shown to be quite invariant to pitch and dynamic and also quite discriminative among different instruments. Based on this observation, Duan *et al.* (2008) proposed an unsupervised music source separation method that clusters the harmonic structure features of all pitches detected in a piece of music. Each cluster corresponds to one instrumental source and the average of the harmonic structures within the cluster is calculated and defined as the average harmonic structure model. Sound sources are then separated using the average harmonic structure models.

Duan *et al.* (2014a) further pursued the clustering idea by incorporating pitch locality constraints into the clustering process. A must-link constraint is imposed between two pitches that are close in both time and frequency to encourage them to be assigned to the same cluster. A cannot-link constraint is imposed between simultaneous pitches to encourage them to be assigned to different clusters, under the assumption of monophonic sound sources. These constraints implement the time and frequency proximity cue in auditory scene analysis. With these constraints, the clustering problem becomes a constrained clustering problem. A greedy algorithm is proposed to solve this problem and is shown to achieve good results in multipitch streaming for both music and speech sound mixtures. Although this method was designed to address multipitch streaming instead of source separation, sound sources can be separated by harmonic masking on the pitch streams, as discussed in Section 16.3.1.

This constrained clustering idea has also been pursued by others. Arora and Behera (2015) designed a hidden Markov random field framework for multipitch streaming and source separation of music signals. The likelihood accounts for timbre similarity between pitches and the priors for pitch locality constraints. MFCCs of the separated spectrum of each pitch are used as the timbre representation. Hu and Wang (2013) proposed an approach to cluster time-frequency units according to their gammatone frequency cepstral coefficients features for speech separation of two simultaneous talkers.

16.3.3 Training and Adapting Timbre Models

When isolated training recordings of the sound sources are available, one can train timbre models beforehand and then apply these models to separate sources from the audio mixture. Various NMF-based source separation approaches rely on this assumption

(Smaragdis *et al.*, 2007) (see Section 16.2 for various types of NMF models). To take advantage of the harmonic structure of musical signals, the dictionary templates (basis spectra) can be designed as harmonic combs to correspond to the quantized musical pitches (Bay *et al.*, 2012) whose amplitudes are learned from training materials. To account for minor pitch variations such as vibrato, shift-invariant NMF has been proposed to shift the basis spectra along the frequency axis (see Section 16.2.1). When the shift invariance is used at its maximum strength, different pitches of the same instrument are assumed to share the same basis spectrum (Kim and Choi, 2006). This is similar to the harmonic structure idea in Section 16.3.2. This significantly reduces the number of parameters in the timbre models, but the shift invariance assumption is only valid within a narrow pitch range.

Another way to reduce the number of parameters in the pitch-dependent NMF dictionaries is to adopt a source-filter model (Section 16.2.2.1 describes source filter models in the context of NMF). The simplest approach is to model each basis function as the product of a pitch-dependent excitation spectrum and an instrument-dependent filter (Virtanen and Klapuri, 2006). This model can be further simplified to make the excitation spectrum always be a flat harmonic comb (Klapuri *et al.*, 2010). This simple model is able to represent some instruments with a smooth envelope of their spectral peaks. However, the spectral envelopes of other instruments, such as the clarinet, are not smooth and they cannot be well represented with a flat excitation function. For example, the second harmonic of a clarinet note is often very soft, no matter what pitch the note has. This makes it impossible to represent the spectral envelopes of different clarinet notes with a single filter.

To deal with this issue, Carabias-Orti *et al.* (2011) proposed a *multi-excitation per instrument* model. This model defines the excitation spectrum of each pitch as a linear combination of a few pitch-independent excitation basis vectors with pitch-dependent weights. The excitation basis vectors are instrument dependent but are not pitch dependent. The weights in the linear combination, however, are both instrument dependent and pitch dependent. This multi-excitation model is a good comprise between the regular source filter model and the flat harmonic comb model. Compared to the regular source-filter model, the multi-excitation model significantly reduces the number of parameters. Compared to the flat harmonic comb model, it preserves the flexibility of modeling sources whose excitation spectra are not flat, such as the clarinet.

To adapt the pretrained timbre models to the sources in the music mixture, the source dictionaries can be first initialized with the pretrained dictionaries and then kept updated during the separation process (Ewert and Muller, 2012). This approach, however, only works well when the initialization is very good or strong constraints of the dictionary and/or the activation coefficients are imposed (e.g., *score-informed* constraints in Section 16.3.4). Another way is to set the pretrained dictionary as a prior (e.g., Dirichlet prior) of the source dictionary (Rodriguez-Serrano *et al.*, 2015a). In this way, the updating of the source dictionaries can be guided by the pretrained models throughout the separation process, hence it is more robust when strong constraints are not available.

16.3.4 Score-Informed Source Separation

When available, the musical score can significantly help music source separation (Ewert *et al.*, 2014), First, it helps pitch estimation, as it indicates likely pitches, helps resolve octave errors, and indicates the likely number of sources and even the likely timbres of the sources (if instrumentation is notated in the score). Second, it helps note activity detection, which is especially important for NMF-based approaches. Third, it helps to stream pitches of the same source across time, which is a key step for pitch-based source separation.

To utilize the score information, *audio-to-score alignment* is needed to synchronize the audio with the score. Various approaches for polyphonic audio-to-score alignment have been proposed. There are two key components of audio-to-score alignment: the feature representation and the alignment method. Commonly used feature representations include the chromagram (Fujishima, 1999), multipitch representations (Duan and Pardo, 2011a), and auditory filterbank responses (Montecchio and Orio, 2009). Commonly used alignment methods include dynamic time warping (Orio and Schwarz, 2001), HMMs (Duan and Pardo, 2011a), and conditional random fields (Joder and Schuller, 2013). Audio-to-score alignment can be performed offline or online. Offline methods require the access of the entire audio recording beforehand, while online methods do not need to access future frames when aligning the current frame. Therefore, offline methods are often more robust while online methods are suitable for real-time applications including real-time score-informed source separation (Duan and Pardo, 2011b).

Once the audio and score are synchronized, the score provides information about what notes are supposed to be played in each audio frame by each source. This information is very helpful for pitch estimation. Duan and Pardo (2011b) estimated the actually performed audio pitches within one semitone of the score-indicated pitches. This significantly improves the multipitch estimation accuracy, which is essential for pitch-based source separation. Finally, harmonic masking is employed to separate the signal of each source. Rodriguez-Serrano *et al.* (2015b) further improved this approach by replacing harmonic masking with a multi-excitation source-filter NMF model to adapt pretrained timbre models for the sources in the mixture. Ewert and Muller (2012) proposed to employ the score information through constraints for an NMF-based source separation model to separate sounds played by the two hands in a piano recording. The basis spectra are constrained to be harmonic combs where values outside a frequency range of each nominal musical pitch are set to zero. The activation coefficients of the basis spectra are set according to the note activities in the score. Values outside a time range of the note duration are set to zero. As the multiplicative update rule is used in the NMF algorithm, zero values in the basis spectra or the activation coefficients will not be updated. Therefore, this initialization imposes strong constraints that are guided by the score on the NMF update process.

16.4 Nonparametric Local Models: Taking Advantage of Redundancies in Music

As we have seen above, music signals come with a particular structure that may be exploited to constraint separation algorithms for better performance. In the previous section, we discussed an approach where separation models are described explicitly with two main ingredients. The first one is a musicologically meaningful information, the score, that indicates which sounds are to be expected at each time instant. The second one is a parametric signal model that describes each sound independently of when it is activated in the track. For this purpose, we considered harmonic and NMF models.

Apart from their good performance when correctly estimated, the obvious advantage of such parametric models for music separation is their interpretability. They make it possible to help the algorithm with specific high-level information such as the score or user input, as we will see shortly.

However, explicit parameterization of the musical piece using a multilevel approach is sometimes not the most natural nor the most efficient solution. Its most demanding constraint, which is critical for good performance, is that the superimposed signals obey their parametric models. While this may be verified in some cases, it may also fail in others, especially for sources found in real full-length popular songs. A first option to address this issue may be to use more realistic acoustic models such as deep neural networks (DNNs) instead of NMF, but this comes with the need to gather a whole development database to learn them.

An alternative option we present in this section is to avoid explicit parameterization of the power spectrum of each source but rather to focus on a nonparametric approach that exploits only its local regularities. This research area has led to numerous efficient algorithms, such as HPSS (Fitzgerald, 2010), REPET (Rafii and Pardo, 2011, 2013; Liutkus *et al.*, 2012), and KAM (Liutkus *et al.*, 2014), that we discuss now.

16.4.1   HPSS: Harmonic-Percussive Source Separation

As an introductory example for nonparametric audio modeling, consider again the scenario where we want to separate the drums section of the music piece that was mentioned in Section 16.2.2.2 above. This can happen, for instance, when a musician wants to reuse some drum loops to mix them with another instrumental accompaniment. We already discussed parametric approaches where both the drum signal and the harmonic accompaniment would be described explicitly through particular spectral templates or the smoothness of their activation parameters, or thanks to their score. However, a very different approach considered by Fitzgerald (2010) directly concentrates on the raw spectrograms of these different sources and uses a simple fact: drum sounds tend to be located in time (percussive), while the accompaniment is composed of very narrowband sinusoidal partials located in frequency (harmonic). As an example, consider the two spectrograms in Figure 16.4. We can see that both percussive and harmonic sounds are characterized by vertical or horizontal lines in their spectrograms. Given a music

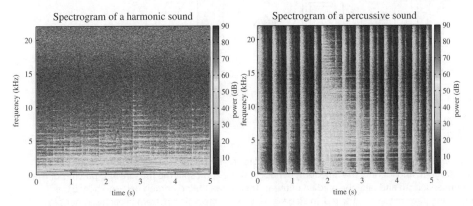

Figure 16.4 Local regularities in the spectrograms of percussive (vertical) and harmonic (horizontal) sounds.

signal, we may hence safely assume that the vertical lines in its spectrogram are mostly due to drums sound, while the horizontal ones pertain to accompaniment.

This simple idea leads to the celebrated *HPSS* algorithm (Fitzgerald, 2010): given a mixture spectrogram $|\mathbf{X}|$, we apply a median filter on it along the time (resp. frequency) dimension to keep only the harmonic (resp. percussive) contribution. This straightforwardly provides an estimate of the power spectra $v_j(n, f)$ of the two sources to use for source separation, as in Chapter 14. The originality of the approach is that no explicit parametric model was picked as in NMF: each power spectrum was only defined as locally constant on some vertical or horizontal neighborhood depicted in Figure 16.6 as kernels (a) and (b).

Whatever the harmonic complexity of the musical piece, this method achieves excellent performance provided the harmonic sounds remain constant for a few frames, typically 200 ms. The model is indeed not on the actual position of the partials, but rather on their duration and steadiness. This proves extremely robust in practice when only separating drum sounds is required.

16.4.2 REPET: Separating Repeating Background

HPSS exploits local regularity in the most straightforward way: the power spectrum of a source is assumed constant in neighboring time-frequency bins, either horizontally or vertically. However, a natural extension would be to consider longer term dependencies. In this respect, Rafii and Pardo (2011) observed that the musical accompaniment of a song often has a spectrogram that is periodic. In other words, whatever the particular drum loop or guitar riff to be separated from vocals, it often comes as repeated over and over again in the song, while the vocals are usually not very repetitive.

Taking the repeating nature of musical accompaniment into account in a nonparametric way leads to the *REPET* algorithm presented originally by Rafii and Pardo (2011, 2013) and summarized in Figure 16.5. Its first step is to identify the period at which the

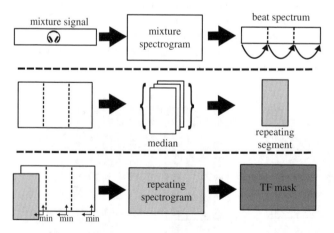

Figure 16.5 REPET: building the repeating background model. In stage 1, we analyze the mixture spectrogram and identify the repeating period. In stage 2, we split the mixture into patches of the identified length and take the median of them. This allows their common part and hence the repeating pattern to be extracted. In stage 3, we use this repeating pattern in each segment to construct a mask for separation.

power spectrum of the accompaniment is repeating. In practice, this is done by picking the most predominant peaks of a tempo detection feature such as the beat spectrogram. Then, the power spectrum of the accompaniment is estimated as a robust averaging of its different repetitions. Time-domain signals are produced by a classical spectral subtraction method, as presented in Chapter 5.

The most interesting feature of REPET is that it allows a wide variety of musical signals to be captured using only a single parameter, which is the period of the accompaniment. Most other separation models (e.g., NMF) require a much larger number of parameters to be estimated and have many more meta parameters to adjust (e.g., number of basis functions, loss function, method of seeding the activation matrix, etc.). In its original form, REPET assumes a strictly repetitive accompaniment, which is not realistic for full-length tracks except for some electronic songs. It was extended to slowly varying accompaniment patterns by Liutkus *et al.* (2012), yielding the adaptive aREPET algorithm.

16.4.3   REPET-Sim: Exploiting Self-Similarity

REPET and aREPET are approaches to modeling the musical accompaniment with the only assumption that its power spectrum will be locally periodic. From a more general perspective, this may be seen as assuming that each part of the accompaniment can be found elsewhere in the song, only superimposed with incoherent parts of the vocals. The specificity of REPET in this context is to provide a way to identify these similar parts as juxtaposed one after the other. In some cases, like a rapidly varying tempo or complex rhythmic structures, this simple strategy may be inappropriate. Instead, when attempting to estimate the accompaniment at each time frame it may be necessary to adaptively look for the similar parts of the song, no longer assumed as located fixed periods away.

The idea behind the *REPET-sim* method (Fitzgerald, 2012) is to exploit this *self-similarity* of the accompaniment. It works by first constructing a $N \times N$ similarity matrix that indicates which frames are close to one another under some spectral similarity criterion. Then, the power spectrum of the accompaniment is estimated for each frame as a robust average of all frames in the song that were identified as similar. REPET and aREPET appear as special cases of this strategy when the neighbors are constrained to be regularly located over time.

In practice, REPET-sim leads to good separation performance as long as the similarity matrix for the accompaniment is correctly estimated. The way similarity between two frames is computed hence appears as a critical choice for it to work. While more sophisticated methods may be proposed in the future, simple correlations between each frame of the mixture spectrogram $|\mathbf{X}|$ have already been shown as giving good results in practice. In any case, REPET-sim is a method of bridging music information retrieval with audio source separation.

16.4.4   KAM: Nonparametric Modeling for Spectrograms

The REPET algorithm and its variants focus on a model for only one of the signals to separate: the accompaniment. They can be understood as ways to estimate the power spectrum of this one source given the mixture spectrogram $|\mathbf{X}|$. Then, separation is performed by spectral subtraction or some variant, producing only two separated signals. Similarly, HPSS may only be used to separate harmonic and percussive components.

To improve the performance of REPET, Rafii *et al.* (2013, 2014) proposed to combine it with a parametric spectrogram model based on NMF for the vocals. In practice, a first NMF decomposition of the mixture is performed that produces a model of the power spectra of both vocals and accompaniment. Then, this preliminary accompaniment model is further processed using REPET to enforce local repetition. A clear advantage of this approach is that it adds the constraints of a vocal model, which often leads to increased performance over REPET alone. In the same vein, the methodology proposed by Tachibana *et al.* (2014) and Driedger and Müller (2015) sequentially applies HPSS with varying length scales in a cascade fashion. This allows the separation not only of harmonic and percussive parts using HPSS, but also vocals and residual.

A general framework to gather both HPSS and REPET under the same umbrella and to allow for arbitrary combinations between them and also other models was introduced by Liutkus *et al.* (2014) and named *KAM*. It was then instantiated for effective music separation by Liutkus *et al.* (2015). In essence, the basic building blocks of KAM for music separation are the same as the Gaussian probabilistic model presented in Chapter 14. The only fundamental difference lies in the way the power spectra $v_j(n,f)$ of the sources are modeled and estimated.

The specificity of KAM for spectral modeling is to avoid picking one single and global parametric model to describe $v_j(n,f)$ as in the various approaches such as NMF described in Chapter 14. Instead, the power spectrum $v_j(n,f)$ of each source j at time-frequency bin (n,f) is simply assumed constant on some neighborhood $\mathcal{I}_j(n,f)$:

$$\forall (n',f') \in \mathcal{I}_j(n,f), v_j(n',f') \approx v_j(n,f). \tag{16.6}$$

For each time-frequency bin (n,f), the neighborhood $\mathcal{I}_j(n,f)$ is thus the set of all the time-frequency bins for which the power spectrum should have a value close to that found at (n,f). This kind of model is typical of a *nonparametric kernel method*: it does not impose a global fixed form to the power spectrum, but does constrain it locally. Several examples of such kernels are given in Figure 16.6 that correspond to the different methods discussed in the previous sections. REPETsim can easily be framed in this context by introducing the similarity matrix and thresholding it to construct $\mathcal{I}_j$.

As can be seen, KAM generalizes the REPET and HPSS methods by enabling their combination in a principled framework. Each source is modeled using its own kernel, or alternatively using a more classical parametric model such as a NMF. Then, the iterative estimation procedure is identical as in the expectation-maximization (EM) method presented in Chapter 14, except for the maximization step, which needs to be adapted

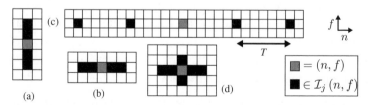

Figure 16.6 Examples of kernels to use in KAM for modeling (a) percussive, (b) harmonic, (c) repetitive, and (d) spectrally smooth sounds.

for sources modeled with a kernel. Indeed, lacking a unique global parametric model, the concept of maximum likelihood (ML) estimation of the parameters does not make sense anymore. Hence, an approach is to replace it by a model cost function that accounts for the discrepancies between the estimated power spectrum and its value in neighboring bins. Liutkus *et al.* (2015) picked the absolute error:

$$v_j(n,f) \leftarrow \underset{v}{\mathrm{argmin}} \sum_{(n',f') \in \mathcal{I}_j(n,f)} |v - \hat{v}_j(n',f')|, \tag{16.7}$$

where $\hat{v}_j(n,f)$ is defined similarly to (14.80) as the unconstrained estimate of the power spectrum of source j obtained during the preceding E-step of the algorithm. It is straightforward that this choice amounts to simply process $\hat{v}_j$ with a *median filter* to estimate v_j: the previous equation is equivalent to:

$$v_j(n,f) \leftarrow \underset{(n',f') \in \mathcal{I}_j(n,f)}{\mathrm{median}} \hat{v}_j(n',f'). \tag{16.8}$$

In the case where the neighborhoods $\mathcal{I}_j$ are shift-invariant, operation 16.8 can be implemented efficiently as a running median filter with linear complexity, yielding a computationally cheap parameter estimation method.

The KAM framework provides a common umbrella for methods exploiting local regularities in music source separation, as well as for their combination with parametric models. It furthermore allows their straightforward extension to the multichannel case, benefiting from the Gaussian framework presented in depth in Chapter 14.

16.5 Taking Advantage of Multiple Instances

The previous section focused on exploiting redundancies within the same song to perform source separation. Doing so, the rhythmic structure inherent to music signals can be leveraged to yield efficient algorithms for fast source separation.

In this section, we show how some application scenarios in music signal processing come with further redundancies that may be exploited to improve separation. In some cases indeed, apart from the mixture to separate, other signals are available that feature some of the sources to separate, possibly distorted and superimposed with other interfering signals. In general, this scenario can be referred to as source separation using *deformed references* (Souviraa-Labastie *et al.*, 2015a).

16.5.1 Common Signal Separation

Back-catalog exploitation and remastering of legacy movie content may come with specific difficulties concerning the audio soundtrack. For very old movies, adverse circumstances such as fires in archive buildings may have led the original music and dialogue separated soundtracks to be lost. In this case, all that is available are the stereophonic or monophonic downmixes. However, making a restored version of the movie at professional quality comes with the requirement of not only remastering its audio content, but also upmixing it (taking one or two tracks and turning them into many more tracks) to modern standards such as 5.1 surround sound. In such a situation, it is desirable to recover estimates of the music and dialogue soundtracks using source separation techniques.

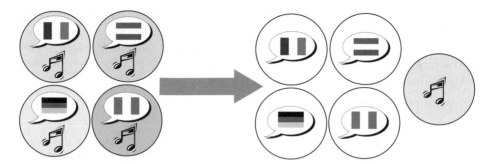

Figure 16.7 Using the audio tracks of multiple related videos to perform source separation. Each circle on the left represents a mixture containing music and vocals in the language associated with the flag. The music is the same in all mixtures. Only the language varies. Given multiple copies of a mixture where one element is fixed lets one separate out this stable element (the music) from the varied elements (the speech in various languages).

Even if the original tracks have been lost, a particular feature of movie soundtracks is that they often come in several languages. These many versions are likely to feature the same musical content, superimposed with a language-specific dialogue track. This situation is depicted in Figure 16.7. The objective of common signal separation in this context becomes to recover the separated dialogues and common music tracks based on the international versions of the movie. An additional difficulty of this scenario is that the music soundtrack is not identical in all international versions. On the contrary, experience shows that mixing and mastering was usually performed independently for all of them. We generally arbitrarily pick one of the international versions as the one we wish to separate, taking others as deformed references.

Leveau *et al.* (2011) presented a method for common signal separation. It is a straightforward application of the coupled NMF methodology presented in Chapter 8. Let us consider for now the case of I very old monophonic international versions of the same movie, denoted x_i, with STFT $x_i(n,f)$. We assume all versions have the same length. The model of Leveau *et al.* (2011) decomposes each version x_i as the sum of a common component c_{i1} and a specific component c_{i2}. Then, a local Gaussian model is chosen for each c_{ij}, as in Chapter 14:

$$c_{ij}(n,f) \sim \mathcal{N}_c(c_{ij}(n,f) \mid 0, v_{ij}(n,f)). \tag{16.9}$$

The method then assigns a standard NMF model for the version-specific power spectrum $\mathbf{V}_{i2} = \mathbf{B}_i\mathbf{H}_i$, and constrains the common signals $c_{i1}(n,f)$ to share the NMF model, up to a version-specific filter: $\mathbf{V}_{i1} = \mathrm{Diag}(\mathbf{g_i})\mathbf{B}_0\mathbf{H}_0$, where $\mathbf{g}_i$ is a nonnegative $F \times 1$ vector modeling the deformation applied on reference i. As usual, all $\mathbf{B}$ and $\mathbf{H}$ matrices gather NMF parameters and are of a user-specified dimension.

Putting this all together, the ith international version $x_i(n,f)$ is modeled as Gaussian with variance $\mathbf{V}_i$ taken as:

$$\mathbf{V}_i = \underbrace{\mathbf{B}_i\mathbf{H}_i}_{\text{dialogues for } i} + \underbrace{\mathrm{Diag}(\mathbf{g_i})\mathbf{B}_0\mathbf{H}_0}_{\text{filtered common part}}. \tag{16.10}$$

Given the model 16.10, inference of the parameters using the standard methodology presented in Chapter 8 is straightforward. Once the parameters have been estimated, the components can be separated through Wiener filtering.

16.5.2 Multireference Bleeding Separation

Studio recordings often consist of different musicians all playing together, with a set of microphones capturing the scene. In a typical setup, each musician or orchestral instrumental group will be recorded by at least one dedicated microphone. Although a good sound engineer does his best to acoustically shield each microphone from sources other than its target, interferences are inevitable and often denote *bleeding* in the sound engineering parlance. The situation is depicted in Figure 16.8, whose left part shows an exemplary recording setup, while the right part illustrates the fact that each recording will feature interferences from all sources in varying proportions.

Dealing with multitrack recordings featuring some amount of bleeding has always been one of the daily duties of professional sound engineers. In the recent years, some research was conducted to design engineering tools aimed at reducing these interferences (Kokkinis *et al.*, 2012; Prätzlich *et al.*, 2015). In the remainder of this section, we briefly present the so-called *MIRA* model of Prätzlich *et al.* (2015).

As usual, let J be the number of sources – the different musicians – and let x_i be the single channel signal captured by microphone i. The MIRA model comes with two main assumptions. First, all the recordings x_i are assumed to be independent. This amounts to totally discarding any phase dependency that could be present in the recordings and proved a safe choice in case of very complex situations such as the real-world orchestral recordings considered by Prätzlich *et al.* (2015). Second, each recording is modeled using the Gaussian model presented in Chapter 14:

$$x_i(n,f) \sim \mathcal{N}_c(x_i(n,f) \mid 0, v_i(n,f)), \tag{16.11}$$

 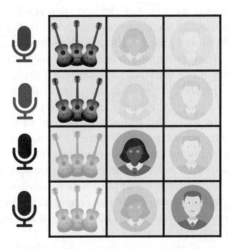

Figure 16.8 Interferences of different sources in real-world multitrack recordings. Left: microphone setup. Right: interference pattern. (Courtesy of R. Bittner and T. Prätzlich.)

where v_i is the power spectrum of recording i. This power spectrum is decomposed as a sum of contributions originating from all the J sources in the following way:

$$v_i(n,f) = \sum_{j=1}^{J} \lambda_{ij} v_j(n,f), \tag{16.12}$$

where v_j stands for the power spectrum of the latent source j and the entry $\lambda_{ij} \geq 0$ of the interference matrix indicates the amount of bleeding of source j in recording i. If desired, the source power spectra v_j can be further constrained using either an NMF or a kernel model as presented above.

Although it is very simple and discards phase dependencies, this model has the nice feature of providing parameters that are readily interpretable by sound engineers. Given a diagram of the acoustic setup, it is indeed straightforward for a user to initialize the interference matrix, then the MIRA algorithm iterates estimation of v_j and λ_{ij}. A critical point for this algorithm to work is that we know that each source is predominant in only a few recordings. Only those are therefore used to estimate v_j. Then, given the source power spectra v_j, the interference matrix can readily be estimated using classical multiplicative updates, as presented in Chapter 8.

Once the parameters have been estimated, separation can be achieved by estimating the contribution c_{ij} of source j in any desired recording x_i using the classical Wiener filter:

$$\hat{c}_{ij}(n,f) = \frac{\lambda_{ij} v_j(n,f)}{\sum_{j'} \lambda_{ij'} v_{j'}(n,f)} x_i(n,f). \tag{16.13}$$

MIRA was shown to be very effective for interference reduction in real orchestral recordings featuring more than 20 microphones and sources (Prätzlich *et al.*, 2015).

16.5.3 A General Framework: Reference-Based Separation

As can be seen, the methods described in the preceding sections may all be gathered under the same general methodology that brings together the recent advances made in NMF and probabilistic modeling for audio (see Chapters 8 and 14, respectively).

In this common framework for separation using multiple deformed references presented by Souviraa-Labastie *et al.* (2015a), the signals analyzed are not only those to be separated, but also auxiliary observations that share some information with the sources to recover. The main idea of the approach is parameter sharing among observations.

In short, this framework models all observations using a NMF model where some parameters may be assumed common to several signals, but deformed from one to another by the introduction of time and frequency deformation matrices. The whole approach is furthermore readily extended to multichannel observations.

Apart from the applications briefly mentioned above, it is noticeable that the same kind of framework may be applied in other application settings, such as text-informed music separation (Magoarou *et al.*, 2014) or separation of songs guided by covers (Souviraa-Labastie *et al.*, 2015b).

In all these cases, a very noticeable feature of the approach is that it fully exploits music-specific knowledge for assessing which parameters of the target mixture may benefit from the observation of the reference signals. For instance, while the pitch information of a spoken reference may be useless to model a sung target, its acoustic envelope can prove valuable for this purpose. In the case of covers, this same pitch information may be useful, although the particular timbre of the sources may change completely from one version to another.

16.6   Interactive Source Separation

Commonly used algorithms for source separation, such as NMF, REPET, and sinusoidal modeling, are not designed to let the user specify which source from the mixture is of interest. As a result, even a successful separation may be a failure from the user's perspective if the algorithm separates the wrong thing from the mixture (you gave me the voice, but I wanted the percussion). Researchers working on audio source separation in a musical context have been at the forefront of making *interactive* systems that let the user guide separation in an interactive way.

Smaragdis and Mysore (2009) presented a novel approach to guiding source separation. The user is asked to provide a sound that mimics the desired source. For example, to separate a saxophone solo from a jazz recording, the user plays the recording over headphones and sings or hums along with the saxophone. This information is used to guide PLCA (Smaragdis *et al.*, 2006), a probabilistic framing of NMF.

The user records a sound that mimics the target sound (in frequency and temporal behavior) to be extracted from the mixture. The system estimates a K-component PLCA model from the user imitation. Then, the system learns a new $K' + K$ element PLCA model for the mixture to be separated. Here, K' components are learned from scratch, while the K components start with the values learned from the user recording. Once the system converges, the target of interest can be separated by using only the K elements seeded by the user vocalizations to reconstruct the signal.

Since not all sources are easily imitated by vocalization, Bryan *et al.* (2014) followed up this work by building a visual editor for NMF/PLCA source separation called the Interactive Source Separation Editor. This editor displays the audio as a time-frequency visualization of the sound (e.g., a magnitude spectrogram). The user is given drawing and selection tools so they can roughly paint on time-frequency visualizations of sound, to select the portions of the audio for which dictionary elements will be learned and marked as the source of interest. They conducted users studies on both inexperienced and expert users and found that both groups can achieve good quality separation with this tool.

Rafii *et al.* (2015) followed this work with an interactive source separation editor for the REPET algorithm. In this work, an audio recording is displayed as a time-frequency visualization (a log-frequency spectrogram). The user then selects a rectangular region containing the element to be removed. The selected region is then cross-correlated with the remainder of the spectrogram to find other instances of the same pattern. The identified regions are averaged together to generate a canonical repeating pattern. This canonical pattern is used as a mask to remove the repeating elements.

16.7   Crowd-Based Evaluation

The performance of today's source separation algorithms are typically measured in terms of intelligibility of the output (in the case of speech separation) or with the commonly used measures in the BSS-EVAL toolkit (Févotte *et al.*, 2005), namely the signal-to-distortion ratio (SDR), the signal-to-interference ratio (SIR), and the signal-to-artifacts ratio (SAR). Several international campaigns regularly assess the performance of different algorithms, and the interested reader is referred to SiSEC (Ono *et al.*, 2015) and MIREX[2] for more pointers.

More recently, automated approaches to estimating the perceptual quality of signals have been applied, the most prominent being PEASS (Vincent, 2012). As discussed at the start of this chapter, commonly used measures for source separation quality may not always be appropriate in a musical context since the goal of the separation may vary, depending on the application. An obvious alternative to using automated evaluation metrics with fixed definitions of "good" is to use humans as the evaluators. This lets the researcher define "good" in a task-dependent way.

The gold standard evaluation measure for audio is a lab-based listening test, such as the *MUSHRA* protocol (ITU, 2014). Subjective human ratings collected in the lab are expensive, slow, and require significant effort to recruit subjects and run evaluations. Moving listening tests from the lab to the micro-task labor market of Amazon Mechanical Turk can greatly speed the process and reduce effort on the part of the researcher.

Cartwright *et al.* (2016) compared MUSHRA performed by expert listeners in a laboratory to a MUSHRA test performed over the web on a population of 530 participants drawn from Amazon Mechanical Turk. The resulting perceptual evaluation scores were highly correlated to those estimated in the controlled lab environment.

16.8 Some Examples of Applications

16.8.1 The Good Vibrations Problem

The majority of the Beach Boys material recorded and released between 1965 and 1967 was long available in mono only. Some of the band's most popular songs were released in that period and are included in albums as famous as *Pet Sounds* or *Wild Honey*. As explained by Fitzgerald (2013), the choice of the monophonic format first comes out of a preference of the band's main songwriter and producer Brian Wilson, and then is also due to the way some tracks were recorded and produced. As extraordinary as it may seem today, overdubs were in some cases directly recorded live during mixdown. Of course, this gives a flavor of uniqueness and spontaneity to the whole recording process, but also means that simply no version of the isolated stems would ever be available, at least not until 2012.

In 2012, the Beach Boys' label was willing to release stereo versions of those tracks for the first time. Some of the tracks for which the multitrack session was available had already been released in stereo during the 1990s. However, for some tracks as famous as *Good Vibrations*, only the monophonic downmix was available. Hence, Capitol records decided to contact researchers in audio signal processing to see if something could be done to get back the stems from the downmix, permitting stereo remastering of the songs. Music source separation appeared as the obvious way to go.

As mentioned by Fitzgerald (2013), who reports the event, the various techniques used for upmixing those legacy songs included various combinations of the methods we have

---

2 http://www.music-ir.org.

discussed in this chapter. For instance, both the accompaniment alone and the downmix were available in some cases, but not isolated stems such as vocals that were over-dubbed during mixdown. In that case, common signal separation, as discussed above in Section 16.5.1, could be applied to recover the isolated lost stems. In other cases, only the downmix was available and blind separation was the only option to recover separated signals. Depending on the track, various techniques were considered. To separate drums signals, HPSS as described in Section 16.4.1 was used. For vocals, Fitzgerald (2013) reports good performance of the REPET-sim method presented in Section 16.4.3. Then, instrument-specific models such as the variants of NMF presented in Section 16.2.2.1 were involved to recover further decomposition of the accompaniment track.

Of course, the separated tracks that were obtained using all those techniques are in no way as clean as original isolated stems would have been. However, they were good enough for the sound engineers involved in the project to produce very good stereophonic upmixed versions of tracks that had never been available in a format other than mono. These remastered versions were released in 2012.

16.8.2 Reducing Drum Leakage: Drumatom

As mentioned in Section 16.5.2, multitrack recording sessions often feature leakage of the different sounds into all microphones, even in professional quality recordings. This problem is particularly noticeable for drum signals, which routinely end up as perfectly audible in most of the simultaneous recordings of the session. This is due to the very large bandwidth of those signals. Such leakage is an important issue that sound engineers have to face because it raises many difficulties when mixing the tracks or when slightly modifying the time alignment of the stems.

The MIRA algorithm briefly summarized in Section 16.5.2 builds on previous research on interference reduction (Kokkinis *et al.*, 2012). The fact is that this previous research actually developed out of academia into a very successful commercial audio engineering tool called Drumatom, produced by the Accusonus company.[3] Drumatom notably includes many enhancements of the algorithm specialized for percussive signals, as well as real-time implementation and good-looking professional graphical user interface (Kokkinis *et al.*, 2013).

As highlighted by Kokkinis *et al.* (2013) but also by Vaneph *et al.* (2016), the objective of devising professional audio engineering software based on source separation is not to create completely automated tools. Indeed, experience shows that source separation methods in the case of music systematically feature some kind of tradeoff between separation and distortion. In essence, it is possible to process the recordings so that interference is not audible, but this may come at the price of too large a distortion, leading to perceptually poor results. For this reason and depending on the application, the tuning of parameters must be performed by an end user. In the case of leakage reduction, these facts are briefly discussed in the perceptual evaluation presented by Prätzlich *et al.* (2015).

It is hence clear that apart from research conducted in music signal processing, an important research and development effort is also required in the field of human–computer interaction to create tools that end up as actually being useful to practitioners.

---

3 See the product webpage at www.drumatom.com.

16.8.3 Impossible Duets Made Real

Since music is a powerful means for artistic creativity, managing the demixing of musical content through source separation necessarily has deep artistic consequences. Indeed, the impact of source separation is not limited to upmixing and remastering songs to reach unprecedented levels of quality. It also makes it possible to use separated signals from existing tracks as raw material for artistic creativity, yielding a whole new range of possibilities.

In the electronic music scene, including existing recordings in new tracks has long been known and acknowledged under the name of sampling and this practice is now widely spread across all musical genres. Still, sampling has always been limited either to excerpts of downmixes or to already available isolated sounds. Source separation makes it possible for an artist to reuse drum loops or vocals from within another song to produce original creations.

This scenario is presented by Vaneph *et al.* (2016) as a groundbreaking opportunity for artists to create exciting new music. In particular, "impossible duets" between two artists that never met in real life are made possible by separating vocals and embedding them into new recordings.

Of course, going from the laboratory to actual artistic creation and to methods that may be used in practice by musicians is far from trivial. Apart from the already mentioned important effort required in the design of graphical user interfaces, the separated results often greatly benefit from postprocessing using dedicated audio engineering techniques. This is, for instance, what motivates Audionamix[4] to have sound engineers working in close collaboration with researchers in signal processing for the design of their separation software. The interested reader is referred to Vaneph *et al.* (2016) for more details.

16.9 Summary

In this chapter we have covered a broad range of source separation algorithms applied to music and, often, to audio tracks of videos. While we have attempted to also provide broad coverage in our referencing of work applied to music, space prohibits describing every method in detail. Where appropriate, we reference other chapters to provide algorithmic detail of approaches that are used across many types of audio. We have also attempted to include sufficient detail on representative music-specific algorithms and approaches not covered in other chapters. The intent is to give the reader a high-level understanding of the workings of key exemplars of the source separation approaches applied in this domain. We strongly encourage the reader to explore the works cited in the bibliography for further details.

Bibliography

Ari, I., Simsekli, U., Cemgil, A., and Akarun, L. (2012) Large scale polyphonic music transcription using randomized matrix decompositions, in *Proceedings of European Signal Processing Conference*, pp. 2020–2024.

---

4 See www.audionamix.com.

Arora, V. and Behera, L. (2015) Multiple f0 estimation and source clustering of polyphonic music audio using PLCA and HMRFs. *IEEE/ACM Transactions on Audio, Speech, and Language Processing*, **23** (2), 278–287.

Ballou, G. (2013) *Handbook for Sound Engineers*, Taylor & Francis.

Bay, M., Ehmann, A.F., Beauchamp, J.W., Smaragdis, P., and Downie, J.S. (2012) Second fiddle is important too: pitch tracking individual voices in polyphonic music, in *Proceedings of International Society for Music Information Retrieval Conference*, pp. 319–324.

Benetos, E., Badeau, R., Weyde, T., and Richard, G. (2014) Template adaptation for improving automatic music transcription, in *Proceedings of International Society for Music Information Retrieval Conference*.

Bertin, N., Badeau, R., and Vincent, E. (2010) Enforcing harmonicity and smoothness in bayesian non-negative matrix factorization applied to polyphonic music transcription. *IEEE Transactions on Audio, Speech, and Language Processing*, **18** (3), 538–549.

Bregman, A.S. (1994) *Auditory Scene Analysis: The Perceptual Organization of Sound*, MIT Press.

Brown, J.C. (1991) Calculation of a constant q spectral transform. *Journal of the Acoustical Society of America*, **89** (1), 425–434.

Bryan, N.J., Mysore, G.J., and Wang, G. (2013) Source separation of polyphonic music with interactive user-feedback on a piano-roll display, in *Proceedings of International Society for Music Information Retrieval Conference*.

Bryan, N.J., Mysore, G.J., and Wang, G. (2014) ISSE: an interactive source separation editor, in *Proceedings of SIGCHI Conference on Human Factors in Computing Systems*, pp. 257–266.

Canadas-Quesada, F., Vera-Candeas, P., Ruiz-Reyes, N., Carabias-Orti, J., and Cabanas-Molero, P. (2014) Percussive/harmonic sound separation by non-negative matrix factorization with smoothness/sparseness constraints. *EURASIP Journal on Audio, Speech, and Music Processing*, **2014** (1), 1–17.

Cappé, O., Laroche, J., and Moulines, E. (1995) Regularized estimation of cepstrum envelope from discrete frequency points, in *Proceedings of IEEE Workshop on Applications of Signal Processing to Audio and Acoustics*.

Carabias-Orti, J., Virtanen, T., Vera-Candeas, P., Ruiz-Reyes, N., and Canadas-Quesada, F. (2011) Musical instrument sound multi-excitation model for non-negative spectrogram factorization. *IEEE Journal of Selected Topics in Signal Processing*, **5** (6), 1144–1158.

Cartwright, M., Pardo, B., Mysore, G.J., and Hoffman, M. (2016) Fast and easy crowdsourced perceptual audio evaluation, in *Proceedings of IEEE International Conference on Audio, Speech and Signal Processing*, pp. 619–623.

Choi, S. (2008) Algorithms for orthogonal nonnegative matrix factorization, in *Proceedings of International Joint Conference on Neural Networks*, pp. 1828–1832.

Davis, S. and Mermelstein, P. (1980) Comparison of parametric representations for monosyllabic word recognition in continuously spoken sentences. *IEEE Transactions on Acoustics, Speech, and Signal Processing*, **28** (4), 357–366.

Davy, M., Godsill, S.J., and Idier, J. (2006) Bayesian analysis of polyphonic Western tonal music. *Journal of the Acoustical Society of America*, **119**, 2498–2517.

de Andrade Scatolini, C., Richard, G., and Fuentes, B. (2015) Multipitch estimation using a PLCA-based model: Impact of partial user annotation, in *Proceedings of IEEE International Conference on Audio, Speech and Signal Processing*, pp. 186–190.

Driedger, J. and Müller, M. (2015) Extracting singing voice from music recordings by cascading audio decomposition techniques, in *Proceedings of IEEE International Conference on Audio, Speech and Signal Processing*, pp. 126–130.

Duan, Z., Han, J., and Pardo, B. (2014a) Multi pitch streaming of harmonic sound mixtures. *IEEE Transactions on Audio, Speech, and Language Processing*, **22** (1), 138–150.

Duan, Z. and Pardo, B. (2011a) Aligning semi-improvised music audio with its lead sheet, in *Proceedings of International Society for Music Information Retrieval Conference*.

Duan, Z. and Pardo, B. (2011b) Soundprism: An online system for score-informed source separation of music audio,. *IEEE Journal of Selected Topics in Signal Processing*, **5** (6), 1205–1215.

Duan, Z., Pardo, B., and Daudet, L. (2014b) A novel cepstral representation for timbre modeling of sound sources in polyphonic mixtures, in *Proceedings of IEEE International Conference on Audio, Speech and Signal Processing*.

Duan, Z., Pardo, B., and Zhang, C. (2010) Multiple fundamental frequency estimation by modeling spectral peaks and non-peak areas. *IEEE Transactions on Audio, Speech and Language Processing*, **18** (8), 2121–2133.

Duan, Z., Zhang, Y., Zhang, C., and Shi, Z. (2008) Unsupervised single-channel music source separation by average harmonic structure modeling. *IEEE Transactions on Audio, Speech, and Language Processing*, **16** (4), 766–778.

Durrieu, J.L., David, B., and Richard, G. (2011) A musically motivated mid-level representation for pitch estimation and musical audio source separation. *IEEE Journal of Selected Topics in Signal Processing*, **5** (6), 1180–1191.

Durrieu, J.L., Richard, G., David, B., and Févotte, C. (2010) Source/filter model for unsupervised main melody extraction from polyphonic audio signals. *IEEE Transactions on Audio, Speech, and Language Processing*, **18** (3), 564–575.

Ewert, S. and Muller, M. (2012) Using score-informed constraints for NMF-based source separation, in *Proceedings of IEEE International Conference on Audio, Speech and Signal Processing*.

Ewert, S., Pardo, B., Muller, M., and Plumbley, M. (2014) Score-informed source separation for musical audio recordings: An overview. *IEEE Signal Processing Magazine*, **31** (3), 116–124.

Févotte, C., Gribonval, R., and Vincent, E. (2005) BSS_EVAL toolbox user guide — Revision 2.0.

Fitzgerald, D. (2010) Harmonic/percussive separation using median filtering, in *Proceedings of International Conference on Digital Audio Effects*.

Fitzgerald, D. (2012) Vocal separation using nearest neighbours and median filtering, in *Proceedings of IET Irish Signals and Systems Conference*.

Fitzgerald, D. (2013) The good vibrations problem, in *Proceedings of the Audio Engineering Society Convention*.

Fuentes, B., Badeau, R., and Richard, G. (2012) Blind harmonic adaptive decomposition applied to supervised source separation, in *Proceedings of European Signal Processing Conference*, pp. 2654–2658.

Fujishima, T. (1999) Realtime chord recognition of musical sound: A system using common lisp music, in *Proceedings of International Computer Music Conference*.

Galas, T. and Rodet, X. (1990) An improved cepstral method for deconvolution of source-filter systems with discrete spectra: Application to musical sounds, in *Proceedings of International Computer Music Conference*, pp. 82–84.

Heittola, T., Klapuri, A., and Virtanen, T. (2009) Musical instrument recognition in polyphonic audio using source-filter model for sound separation, in *Proceedings of International Society for Music Information Retrieval Conference*, pp. 327–332.

Hennequin, R., Badeau, R., and David, B. (2010) Time-dependent parametric and harmonic templates in non-negative matrix factorization, in *Proceedings of International Conference on Digital Audio Effects*, pp. 246–253.

Hennequin, R., Badeau, R., and David, B. (2011) NMF with time-frequency activations to model nonstationary audio events. *IEEE Transactions on Audio, Speech, and Language Processing*, **19** (4), 744–753.

Hu, K. and Wang, D. (2013) An unsupervised approach to cochannel speech separation. *IEEE Transactions on Audio, Speech, and Language Processing*, **21** (1), 122–131.

ITU (2014) Recommendation ITU-R BS.1534-2: Method for the subjective assessment of intermediate quality level of audio systems.

Joder, C. and Schuller, B. (2013) Off-line refinement of audio-to-score alignment by observation template adaptation, in *Proceedings of IEEE International Conference on Audio, Speech and Signal Processing*.

Kameoka, H., Nakano, M., Ochiai, K., Imoto, Y., Kashino, K., and Sagayama, S. (2012) Constrained and regularized variants of non-negative matrix factorization incorporating music-specific constraints, in *Proceedings of IEEE International Conference on Audio, Speech and Signal Processing*, pp. 5365–5368.

Kim, M. and Choi, S. (2006) Monaural music source separation: Nonnegativity, sparseness, and shift-invariance, in *Proceedings of International Conference on Independent Component Analysis and Signal Separation*, pp. 617–624.

Kim, M. and Smaragdis, P. (2013) Collaborative audio enhancement using probabilistic latent component sharing, in *Proceedings of IEEE International Conference on Audio, Speech and Signal Processing*, pp. 896–900.

Kirchhoff, H., Dixon, S., and Klapuri, A. (2013) Missing template estimation for user-assisted music transcription, in *Proceedings of IEEE International Conference on Audio, Speech and Signal Processing*, pp. 26–30.

Klapuri, A. (2003) Multiple fundamental frequency estimation based on harmonicity and spectral smoothness. *IEEE Transactions on Speech and Audio Processing*, **11** (6), 804–815.

Klapuri, A., Virtanen, T., and Heittola, T. (2010) Sound source separation in monaural music signals using excitation-filter model and EM algorithm, in *Proceedings of IEEE International Conference on Audio, Speech and Signal Processing*, pp. 5510–5513.

Kokkinis, E., Reiss, J., and Mourjopoulos, J. (2012) A Wiener filter approach to microphone leakage reduction in close-microphone applications. *IEEE Transactions on Audio, Speech, and Language Processing*, **20** (3), 767–779.

Kokkinis, E., Tsilfidis, A., Kostis, T., and Karamitas, K. (2013) A new DSP tool for drum leakage suppression, in *Proceedings of the Audio Engineering Society Convention*.

Laroche, C., Kowalski, M., Papadopoulos, H., and Richard, G. (2015) A structured nonnegative matrix factorization for source separation, in *Proceedings of European Signal Processing Conference*.

Leveau, P., Maller, S., Burred, J.J., and Jaureguiberry, X. (2011) Convolutive common audio signal extraction, in *Proceedings of IEEE Workshop on Applications of Signal Processing to Audio and Acoustics*, pp. 165–168.

Li, Y. and Wang, D. (2008) Musical sound separation using pitch-based labeling and binary time-frequency masking, in *Proceedings of IEEE International Conference on Audio, Speech and Signal Processing*, pp. 173–176.

Liutkus, A., Fitzgerald, D., and Rafii, Z. (2015) Scalable audio separation with light kernel additive modelling, in *Proceedings of IEEE International Conference on Audio, Speech and Signal Processing*, pp. 76–80.

Liutkus, A., Fitzgerald, D., Rafii, Z., Pardo, B., and Daudet, L. (2014) Kernel additive models for source separation. *IEEE Transactions on Signal Processing*, **62** (16), 4298–4310.

Liutkus, A., Rafii, Z., Badeau, R., Pardo, B., and Richard, G. (2012) Adaptive filtering for music/voice separation exploiting the repeating musical structure, in *Proceedings of IEEE International Conference on Audio, Speech and Signal Processing*.

Magoarou, L.L., Ozerov, A., and Duong, N. (2014) Text-informed audio source separation. Example-based approach using non-negative matrix partial co-factorization. *Journal of Signal Processing Systems*, p. 13.

Montecchio, N. and Orio, N. (2009) A discrete filter bank approach to audio to score matching for polyphonic music., in *Proceedings of International Society for Music Information Retrieval Conference*.

Mysore, G.J., Smaragdis, P., and Raj, B. (2010) Non-negative hidden Markov modeling of audio with application to source separation, in *Proceedings of International Conference on Latent Variable Analysis and Signal Separation*, pp. 140–148.

Nakano, M., Le Roux, J., Kameoka, H., Nakamura, T., Ono, N., and Sagayama, S. (2011) Bayesian nonparametric spectrogram modeling based on infinite factorial infinite hidden Markov model, in *Proceedings of IEEE Workshop on Applications of Signal Processing to Audio and Acoustics*.

Nakano, M., Le Roux, J., Kameoka, H., Ono, N., and Sagayama, S. (2010) Nonnegative matrix factorization with Markov-chained bases for modeling time-varying patterns in music spectrograms, in *Proceedings of International Conference on Latent Variable Analysis and Signal Separation*.

Ono, N., Rafii, Z., Kitamura, D., Ito, N., and Liutkus, A. (2015) The 2015 signal separation evaluation campaign, in *Proceedings of International Conference on Latent Variable Analysis and Signal Separation*, pp. 387–395.

Orio, N. and Schwarz, D. (2001) Alignment of monophonic and polyphonic music to a score, in *Proceedings of International Computer Music Conference*.

Prätzlich, T., Bittner, R., Liutkus, A., and Müller, M. (2015) Kernel additive modeling for interference reduction in multi-channel music recordings, in *Proceedings of IEEE International Conference on Audio, Speech and Signal Processing*, pp. 584–588.

Rafii, Z., Duan, Z., and Pardo, B. (2014) Combining rhythm-based and pitch-based methods for background and melody separation. *IEEE/ACM Transactions on Audio, Speech, and Language Processing*, **22** (12), 1884–1893.

Rafii, Z., Germain, F.G., Sun, D.L., and Mysore, G.J. (2013) Combining modeling of singing voice and background music for automatic separation of musical mixtures, in *Proceedings of International Society for Music Information Retrieval Conference*.

Rafii, Z., Liutkus, A., and Pardo, B. (2015) A simple user interface system for recovering patterns repeating in time and frequency in mixtures of sounds, in *Proceedings of IEEE International Conference on Audio, Speech and Signal Processing*, pp. 271–275.

Rafii, Z. and Pardo, B. (2011) A simple music/voice separation method based on the extraction of the repeating musical structure, in *Proceedings of IEEE International Conference on Audio, Speech and Signal Processing*.

Rafii, Z. and Pardo, B. (2013) Repeating pattern extraction technique (REPET): A simple method for music/voice separation. *IEEE Transactions on Audio, Speech, and Language Processing*, **21** (13), 71–82.

Rigaud, F., Falaize, A., David, B., and Daudet, L. (2013) Does inharmonicity improve an NMF-based piano transcription model?, in *Proceedings of IEEE International Conference on Audio, Speech and Signal Processing*, pp. 11–15.

Rodriguez-Serrano, F.J., Duan, Z., Vera-Candeas, P., Pardo, B., and Carabias-Orti, J.J. (2015a) Online score-informed source separation with adaptive instrument models. *Journal of New Music Research*, **44** (2), 83–96.

Rodriguez-Serrano, F.J., Duan, Z., Vera-Candeas, P., Pardo, B., and Carabias-Orti, J.J. (2015b) Online score-informed source separation with adaptive instrument models. *Journal of New Music Research*, **44** (2), 83–96.

Seetharaman, P. and Pardo, B. (2016) Simultaneous separation and segmentation in layered music, in *Proceedings of International Society for Music Information Retrieval Conference*.

Smaragdis, P. and Brown, J.C. (2003) Non-negative matrix factorization for polyphonic music transcription, in *Proceedings of IEEE Workshop on Applications of Signal Processing to Audio and Acoustics*, pp. 177–180.

Smaragdis, P. and Mysore, G.J. (2009) Separation by humming: User guided sound extraction from monophonic mixtures, in *Proceedings of IEEE Workshop on Applications of Signal Processing to Audio and Acoustics*, pp. 69–72.

Smaragdis, P., Raj, B., and Shashanka, M. (2006) A probabilistic latent variable model for acoustic modeling, in *Proceedings of Neural Information Processing Systems Workshop on Advances in Models for Acoustic Processing*, pp. 1–8.

Smaragdis, P., Raj, B., and Shashanka, M. (2007) Supervised and semi-supervised separation of sounds from single-channel mixtures, in *Proceedings of International Conference on Independent Component Analysis and Signal Separation*.

Smaragdis, P., Raj, B., and Shashanka, M.V.S. (2008) Sparse and shift-invariant feature extraction from non-negative data, in *Proceedings of IEEE International Conference on Audio, Speech and Signal Processing*, pp. 2069–2072.

Souviraa-Labastie, N., Olivero, A., Vincent, E., and Bimbot, F. (2015a) Multi-channel audio source separation using multiple deformed references. *IEEE/ACM Transactions on Audio, Speech, and Language Processing*, **23** (11), 1775–1787.

Souviraa-Labastie, N., Vincent, E., and Bimbot, F. (2015b) Music separation guided by cover tracks: designing the joint nmf model, in *Proceedings of IEEE International Conference on Audio, Speech and Signal Processing*, pp. 484–488.

Su, L. and Yang, Y.H. (2015) Combining spectral and temporal representations for multipitch estimation of polyphonic music. *IEEE/ACM Transactions on Audio, Speech, and Language Processing*, **23** (10), 1600–1612.

Tachibana, H., N. Ono, N., and Sagayama, S. (2014) Singing voice enhancement in monaural music signals based on two-stage harmonic/percussive sound separation on multiple resolution spectrograms. *IEEE/ACM Transactions on Audio, Speech, and Language Processing*, **22** (1), 228–237.

Tolonen, T. and Karjalainen, M. (2000) A computationally efficient multipitch analysis model. *IEEE Transactions on Speech and Audio Processing*, **8** (6), 708–716.

Vaneph, A., McNeil, E., Rigaud, F., and Silva, R. (2016) An automated source separation technology and its practical applications, in *Proceedings of the Audio Engineering Society Convention*.

Vincent, E. (2012) Improved perceptual metrics for the evaluation of audio source separation, in *Proceedings of International Conference on Latent Variable Analysis and Signal Separation*, pp. 430–437.

Virtanen, T. (2007) Monaural sound source separation by nonnegative matrix factorization with temporal continuity and sparseness criteria. *IEEE Transactions on Audio, Speech, and Language Processing*, **15** (3), 1066–1074.

Virtanen, T. and Klapuri, A. (2006) Analysis of polyphonic audio using source-filter model and non-negative matrix factorization, in *Proceedings of Neural Information Processing Systems Workshop on Advances in Models for Acoustic Processing*.

Wang, Y.X. and Zhang, Y.J. (2013) Nonnegative matrix factorization: A comprehensive review. *IEEE Transactions on Knowledge and Data Engineering*, **25** (6), 1336–1353.

17

Application of Source Separation to Robust Speech Analysis and Recognition

Shinji Watanabe, Tuomas Virtanen, and Dorothea Kolossa

This chapter describes applications of source separation techniques to robust speech analysis and recognition, including *automatic speech recognition* (ASR), speaker/language identification, emotion and paralinguistic analysis, and audiovisual analysis. These are the most successful applications in audio and speech processing, with various commercial products including Google Voice Search, Apple Siri, Amazon Echo, and Microsoft Cortana. Robustness against noise or nontarget speech still remains a challenging issue, and source separation and speech enhancement techniques are gathering much attention in the speech community.

This chapter systematically describes how source separation and speech enhancement techniques are applied to improve the robustness of these applications. It first describes the challenges and opportunities in Section 17.1, and defines the considered speech analysis and recognition applications with basic formulations in Section 17.2. Section 17.3 describes the current state-of-the-art system using source separation as a front-end method for speech analysis and recognition. Section 17.4 introduces a way of tightly integrating these methods by preserving the uncertainties between them. Section 17.5 provides another possible solution to the robustness issues with the help of cross-modality information. Section 17.6 concludes the chapter.

17.1 Challenges and Opportunities

17.1.1 Challenges

To use source separation and enhancement techniques in real-world speech analysis and recognition we have to tackle the following challenges:

1) Speech is highly nonstationary and consists of very different kinds of acoustic elements (e.g., harmonic, transient, and noise-like components). This breaks stationarity assumptions used in basic source separation and enhancement algorithms, and the straightforward application of source separation and enhancement might not improve the speech analysis and recognition performance in real applications.

Audio Source Separation and Speech Enhancement, First Edition.
Edited by Emmanuel Vincent, Tuomas Virtanen and Sharon Gannot.
© 2018 John Wiley & Sons Ltd. Published 2018 by John Wiley & Sons Ltd.
Companion Website: https://project.inria.fr/ssse/

2) Most speech analysis and recognition applications, such as ASR, are based on data-driven methods, which must handle from 10 to 1000 or even more hours of speech training data depending on the task (amounting to millions or billions of frames with 10 ms frame shift). Therefore, in many cases when we use them in training, the algorithms have to process such orders of magnitude of data, and must be scalable and parallelized.

3) It is very hard to integrate front-end source separation techniques with back-end speech analysis and recognition since distortions caused by source separation techniques drastically degrade the performance. This degradation comes from the different objectives targeted when designing the front-end (mainly SNR or speech quality/intelligibility) and the back-end (mainly classification performance).

4) Speech is affected by environments depending on noise types, room characteristics, microphone properties, and positions of sources, microphones, and noises. Thus, speech involves huge (combinatorial) variations, and since back-end speech analysis and recognition systems are trained on specific conditions of speech data that we obtain in advance, the systems must cope with mismatches between training and test data.

Table 17.1 summarizes several conditions of recent speech recognition tasks focusing on source separation and enhancement issues. These tasks are designed for a few major scenarios in real speech applications, including home, public space, and meeting environments, and we can find many acoustic/linguistic variations, even when we select a few examples.

17.1.2 Opportunities

We have various opportunities to overcome some of the challenges above and to further improve the separation and enhancement performance:

- In speech analysis and recognition applications, large amounts of speech data are available with corresponding annotations, including transcriptions, speaker/language information, and other paralinguistic labels. Thus, we can build a robust system by training statistical models with large amounts of data.[1] In addition, with the large-scale supervised data above, we can leverage various machine learning techniques, including emergent deep-learning techniques.

- We can use application-specific knowledge and models. For example, language models can provide quite a strong regularization/constraint for a speech recognizer, and possibly for separation and enhancement. In a multimodal scenario (typically audio and visual modalities), we can further utilize the location of the sources and user states obtained from visual information.

---

1 However, the data annotation of minor languages or ambiguous paralinguistic categories, including human emotions, is still fairly costly, and the application of data-driven techniques for such cases is rather limited.

Table 17.1 Recent noise robust speech recognition tasks: ASpIRE (Harper, 2015), AMI (Hain *et al.*, 2007), CHiME-1 (Barker *et al.*, 2013), CHiME-2 (Vincent *et al.*, 2013), CHiME-3 (Barker *et al.*, 2015), CHiME-4 (Vincent *et al.*, 2017), and REVERB (Kinoshita *et al.*, 2016). See Le Roux and Vincent (2014)[a] for a more detailed list of robust speech processing datasets.

Task	Vocabulary	Amount of training data	Real/simulated	Type of distortions	# microphones	Microphone–speaker distance	Ground truth
ASpIRE	100k	~2000 h	Real	Reverberation	8/1	N/A	N/A
AMI	11k	~107k utt. (~75 h)	Real	Multispeaker conversations Reverberation and noise	8	N/A	Headset
CHiME-1	50	~17k utt.	Simulated	Nonstationary noise recorded in a living room (SNR -6 dB to 9 dB) Reverberation from recorded impulse responses	2	2 m	Clean
CHiME-2 (Track 2)	5k	7,138 utt. (~15 h)	Simulated	Same as CHiME-1	2	2 m	Clean
CHiME-3	5k	8,738 utt. (~18 h)	Simulated + real	Nonstationary noise in four environments	6	0.5 m	Clean/close talk microphone
CHiME-4	5k	8,738 utt. (~18 h)	Simulated + real	Nonstationary noise in four environments	6/2/1	0.5 m	Clean/close talk microphone
REVERB	5k	7,861 utt. (~15 h)	Simulated + real	Reverberation in different living rooms (RT60 from 0.25 to 0.7 s) + stationary noise (SNR ~20 dB)	8/2/1	0.5 m to 2 m	Clean /headset

a) https://wiki.inria.fr/rosp/Datasets.

- There exist clear objective measures: the accuracy, word error rate, and F-measure. These measures directly relate to the performance of real speech applications, and can provide appropriate assessments of separation and enhancement techniques. In addition, we can also develop discriminative methods for separation and enhancement techniques with these measures.
- We can use various statistical inference and deep-learning techniques, which help to integrate the front-end and back-end systems coherently through normalization, transformation, and adaptation, and enable joint training of both systems with a consistent discriminative measure.

Consequently, these opportunities make source separation and enhancement techniques powerful and attractive for a wide range of speech applications.

17.2 Applications

In this section we introduce the most common speech analysis tasks: ASR, speaker identification, paralinguistic speech analysis, and audiovisual speech analysis.

17.2.1 Automatic Speech Recognition

One of the most successful applications of speech analysis and recognition is ASR, which converts an audio signal containing speech into a sequence of words. In ASR, a length-N sequence $\mathbf{O} = [\mathbf{o}(0), \dots, \mathbf{o}(N-1)]$ of $D \times 1$ speech feature vectors $\mathbf{o}(n)$ is extracted from a length-T single channel speech signal $[x(t)]_t$ as

$$\mathbf{O} = f([x(t)]_t). \tag{17.1}$$

$\mathbf{O}$ is mainly used as an input. The transformation $f(\cdot)$ is a usually deterministic function. *Mel-frequency cepstral coefficients* (MFCCs), perceptual linear prediction coefficients, and the log Mel-filterbank spectrum are often used as speech features. The details of feature extraction are described below. The output is a length-M word sequence $\mathbf{v} = [v_1, \dots, v_M]$, $v_m \in \mathcal{V}$ with vocabulary $\mathcal{V}$. Therefore, ASR is considered as a structured classification problem, where both input and output are represented by sequences $\mathbf{O}$ and $\mathbf{v}$ of different lengths.

ASR is mathematically formulated in a probabilistic framework. Given a speech feature sequence $\mathbf{O}$, the goal is to estimate the word sequence $\mathbf{v}$ according to Bayes' decision theory:

$$\hat{\mathbf{v}} = \underset{\mathbf{v} \in \mathcal{V}*}{\operatorname{argmax}} P(\mathbf{v} \mid \mathbf{O}). \tag{17.2}$$

This equation means that ASR attempts to find the most probable word sequence among all possible word sequences in $\mathcal{V}*$. In the training and test phases, an ASR dataset usually provides paired word and speech sequences for each utterance, i.e. $\{\mathbf{v}_u^{\text{ref}}, \mathbf{O}_u\}_u$ where u is the utterance index. The reference word sequences are obtained by human annotation and used to compute the performance by using some metrics (e.g., edit distance) between reference $\mathbf{v}_u^{\text{ref}}$ and the hypothesis $\hat{\mathbf{v}}_u$ or used for supervised training. Thus, the primary goal of ASR is to build an accurate inference system finding the closest hypothesis $\hat{\mathbf{v}}_u$ to the reference $\mathbf{v}_u^{\text{ref}}$. Note that the utterance index u in the chapter is often omitted for simplicity.

(17.2) itself is simple, but directly obtaining the posterior distribution $P(\mathbf{v} \mid \mathbf{O})$ is quite difficult, as all possible word sequences have to be considered.[2] Instead, $P(\mathbf{v} \mid \mathbf{O})$ is factorized by using the phoneme representation and introducing *hidden Markov models* (HMMs) as follows (Rabiner and Juang, 1993; Huang *et al.*, 2001):

$$
\begin{aligned}
\hat{\mathbf{v}} &= \underset{\mathbf{v} \in \mathcal{V}_*}{\arg\max}\; p(\mathbf{O} \mid \mathbf{v})P(\mathbf{v}), \\
&= \underset{\mathbf{v} \in \mathcal{V}_*}{\arg\max} \sum_{\mathbf{q}, \mathbf{v}'} p(\mathbf{O}, \mathbf{q} \mid \mathbf{v}')P(\mathbf{v}' \mid \mathbf{v})P(\mathbf{v}), \\
&\approx \underset{\mathbf{v} \in \mathcal{V}_*}{\arg\max} \max_{\mathbf{q}, \mathbf{v}'} \underbrace{p(\mathbf{O}, \mathbf{q} \mid \mathbf{v}')}_{\text{acoustic model}} \underbrace{P(\mathbf{v}' \mid \mathbf{v})}_{\text{lexicon}} \underbrace{P(\mathbf{v})}_{\text{language model}},
\end{aligned}
\tag{17.3}
$$

where $\mathbf{q} = [q(0), \ldots, q(N-1)]$, $q(n) \in \{1, \ldots, R\}$, is a length-$N$ HMM state sequence and $\mathbf{v}' = [v'(0), \ldots, v'(N-1)]$ is a subword (phoneme) sequence given $\mathbf{v}$. R is the number of distinct HMM states. The final approximation by replacing the summation over $\mathbf{q}$ and $\mathbf{v}'$ to the maximization is called the Viterbi approximation, which is actually used in the ASR decoding process. A phoneme class is often extended as a context-dependent HMM state (e.g., *triphones*, where phonemes having different preceding/succeeding phonemes are regarded as different classes) and several context-dependent HMM states are clustered to share the same output distribution to make the training data balanced for each category. We use the conditional independence assumption that $p(\mathbf{O}, \mathbf{q} \mid \mathbf{v}', \mathbf{v}) \approx p(\mathbf{O}, \mathbf{q} \mid \mathbf{v}')$, i.e. speech features only depend on the phonemes. Now we have three probabilistic distributions:

- *Acoustic model* $p(\mathbf{O}, \mathbf{q} \mid \mathbf{v}')$: joint probability of the speech features and the state sequence given a phoneme sequence.
- *Lexicon* $P(\mathbf{v}' \mid \mathbf{v})$: probability of a phoneme sequence given a word sequence. This model is usually deterministic and defined using a pronunciation dictionary.
- *Language model* $P(\mathbf{v})$: prior probability of a word sequence, which is learned from large text corpora. N-gram or recurrent neural network (RNN) language models are often used.

This scheme has been widely used to develop ASR systems. Figure 17.1 depicts the scheme, which is composed of the three models above, feature extraction, and recognizer. The rest of the section mainly discusses feature extraction and acoustic modeling, which are also used in the applications other than ASR in this chapter.

17.2.1.1 Feature Extraction

Figure 17.2 shows a block diagram of MFCC and *log Mel-filterbank* feature extraction. The difference between them lies only in the application of a discrete cosine transform for MFCCs. We briefly describe the role of each block as follows:

- Short-time Fourier transform (STFT): for spectrum analysis.
- Power extraction: extract the power spectrum and disregard phase information.

---

2 Recently there have been end-to-end ASR attempts to model $P(\mathbf{v} \mid \mathbf{O})$ directly using an attention mechanism (Bahdanau *et al.*, 2016; Kim *et al.*, 2016) or connectionist temporal classification (Graves and Jaitly, 2014).

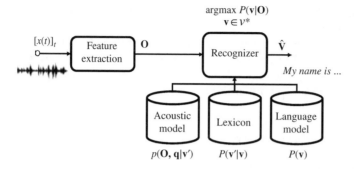

Figure 17.1 Block diagram of an ASR system.

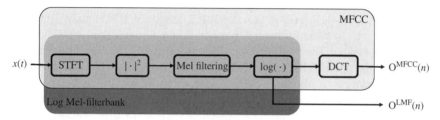

Figure 17.2 Calculation of MFCC and log Mel-filterbank features. "DCT" is the discrete cosine transform.

- Mel filtering: emphasize low-frequency power with a perceptual auditory scale (Mel scale, see Chapter 2).
- Log scaling: compress the dynamic range, which is very large in the power domain.
- Discrete cosine transform: decorrelate the log Mel-filterbank vector, which is important for *Gaussian mixture model* (GMM) based acoustic models with diagonal covariances. This block is less important (or even harmful) for deep neural network (DNN)-based acoustic models.

MFCC or log Mel-filterbank features are often augmented by additional dynamic features such as delta and double-delta features, which are computed by first- and second-order regression coefficients. Further details about feature extraction for the application of source separation to speech analysis and recognition are given in Section 17.3.3.

17.2.1.2 Acoustic Model

We provide the HMM-based acoustic model formulation with GMMs, and with DNNs as (approximations of) emission probability functions. For simplicity we omit conditioning on the phoneme sequence $\mathbf{v}'$. Dealing with the joint probability $p(\mathbf{O}, \mathbf{q})$ in (17.3) is very difficult, as we have to handle all possible state sequences. However, for HMMs we can factorize the function with the conditional independence assumption and the first-order Markovian assumption as follows:

$$p(\mathbf{O}, \mathbf{q}) = p(\mathbf{O} \mid \mathbf{q})P(\mathbf{q}) \tag{17.4}$$

$$\approx p(\mathbf{o}(0) \mid q(0))P(q(0)) \prod_{n=1}^{N-1} p(\mathbf{o}(n) \mid q(n))P(q(n) \mid q(n-1)), \tag{17.5}$$

where $P(q(0))$ and $P(q(n) \mid q(n-1))$ are the initial state probability and state transition probability, respectively. $p(\mathbf{o}(n) \mid q(n))$ is an emission probability, which can be computed using GMMs or DNNs. For HMMs, the dynamic programming principle provides an efficient way of computing the following values:

- Likelihood (conditional likelihood given a phoneme sequence $\mathbf{v}'$) $p(\mathbf{O}) = \sum_{\mathbf{q}} p(\mathbf{O}, \mathbf{q})$ based on the forward or backward algorithm.
- Most probable state sequence $\hat{\mathbf{q}} = \text{argmax}_{\mathbf{q}} \, p(\mathbf{O}, \mathbf{q})$ based on the Viterbi algorithm, which is described in Chapter 7.
- Occupation posterior probability (responsibility) $P(q(n) = r \mid \mathbf{O})$ of HMM state r at frame n based on the forward-backward algorithm.

Thus, HMM has been used as a standard acoustic model for a long time.

17.2.1.3 GMM

GMMs were used as a standard model for computing emission probabilities, since their parameters can be jointly optimized with HMM parameters in the maximum likelihood (ML) sense via the expectation-maximization (EM) algorithm. The GMM has an additional latent variable for the mixture component and when the hidden state $q(n) = r$, the conditional likelihood is represented as follows:

$$p(\mathbf{o}(n) \mid r) = \sum_{k=1}^{K} \pi_{rk} \, \mathcal{N}(\mathbf{o}(n) \mid \boldsymbol{\mu}_{rk}, \boldsymbol{\Sigma}_{rk}), \tag{17.6}$$

where k is the index of a Gaussian component, K is the number of components, $\pi_{rk} \in [0, 1]$ is a mixture weight, $\boldsymbol{\mu}_{rk}$ is a $D \times 1$ Gaussian mean vector, and $\boldsymbol{\Sigma}_{rk}$ is a $D \times D$ Gaussian covariance matrix. Diagonal or full covariance matrices are used depending on the application and the size of training data. In general, diagonal covariance matrices are not capable of faithfully modeling complex distributions due to their simple representation. On the other hand, full-covariance Gaussians potentially model such distributions, but they are difficult to train because the required data per Gaussian becomes very large, and also because of the ill-conditioning problem arising for correlated data. Finally, in GMM-HMM systems, temporal dynamics are only roughly addressed using first- and second-order derivatives of features, but due to large correlations between adjacent frames it is hard to model dynamics without sacrificing the immensely helpful first-order-Markov property.

17.2.1.4 DNN

DNNs have successfully replaced GMMs since they can deal with correlated features by learning correlation structures without severe degeneracy issues, and can hence be employed to obtain quasi-state-posteriors over multiple frames. A DNN is an artificial neural network, which is composed of several linear transformations and nonlinear activation layers. DNNs provide an alternative way of computing conditional emission probability functions $p(\mathbf{o}(n) \mid r)$ in the HMM. However, DNNs themselves cannot provide the conditional likelihood directly, unlike the GMM. Instead, the posterior probability $p(r \mid \mathbf{o}(n))$ is computed based on discriminative training. Therefore, the following *pseudo-likelihood* trick is applied, utilizing Bayes' rule:

$$p(\mathbf{o}(n) \mid r) \propto \frac{P(r \mid \mathbf{o}(n))}{P(r)}, \tag{17.7}$$

where $P(r)$ is the prior probability of the HMM state r. This way of combining HMM and DNN with the pseudo-likelihood technique above is called the hybrid approach. HMM-DNN hybrid systems are widely used in speech analysis and recognition (Hinton et al., 2012; Yu and Deng, 2015).

For the most common type of DNN that uses only feedforward connections, the following posterior probability $P(r \mid \mathbf{o}(n))$ is obtained given the observation vector $\mathbf{o}(n)$ as input

$$P(r \mid \mathbf{o}(n)) = (\mathrm{softmax}(\bar{\mathbf{y}}_H(n)))_r, \tag{17.8}$$

where $\bar{\mathbf{y}}_H(n)$ is a hidden vector after applying the matrix transformation in the last layer H, which is described in Chapter 7 in more detail. In the multiclass classification problem above, we use the *softmax* activation function, which is defined as:

$$\mathrm{softmax}(\mathbf{o}) = \frac{\exp(\mathbf{o})}{\sum_{d=1}^{D} \exp(o_d)}, \tag{17.9}$$

where $\exp(\cdot)$ is an elementwise exponential operation.

DNNs significantly improve the classification performance compared to GMMs due to their ability to learn nonlinear feature representations and to model temporal structure between adjacent frames. In addition, with their largely nonsequential property, DNN-based methods are suitable for distributed computing, e.g. with graphical processing units, and also achieve the desired data scalability. Thus, DNNs are applied to current ASR systems, including voice search technologies.

17.2.1.5 Other Network Architectures

State-of-the-art ASR systems employ various network architectures other than a simple feedforward network. The most popular extension is to include explicit time dependencies in hidden layers by using RNNs, which is also addressed in Chapter 7. The long short-term memory (LSTM) RNN (Hochreiter and Schmidhuber, 1997) has been widely used as a practical solution of the standard RNN's vanishing gradient problem. In addition, convolutional neural networks (LeCun et al., 1998) and a variant called *time-delayed neural networks* (Waibel et al., 1989) are used as alternative acoustic models.

17.2.1.6 Training Objectives

We summarize the objective/cost functions for acoustic model training:

- *ML*: An objective function for generative model including GMM-HMM:

$$\mathcal{M}^{\mathrm{ML}}(\theta) = p(\mathbf{O} \mid \theta), \tag{17.10}$$

  where θ is a set of model parameters, e.g. μ and Σ for the GMM case and affine transformation parameters for the DNN case. This chapter uses a vector representation of model parameters.
- *Bayesian criterion*: Estimate the posterior distribution of the model parameters $p(\theta \mid \mathbf{O})$ instead of the model parameters θ, which is obtained from the prior distribution $p(\theta)$ and the joint probability $p(\mathbf{O}, \theta)$. We can use maximum a posteriori (MAP) estimation, which uses the prior distribution as a regularization, or variational Bayesian (VB) inference, which obtains approximated posterior distributions based on a variational lower bound (see Watanabe and Chien (2015)).

- *Cross-entropy*: A cost function of frame-level multiclass classifiers including DNNs. This requires reference HMM state labels $q^{\text{ref}}(n)$ to be known for each time frame.

$$C^{\text{CE}}(\theta) = -\sum_n \sum_{q(n)} p(q(n) \mid q^{\text{ref}}(n)) \log p(q(n) \mid \mathbf{o}(n), \theta) \qquad (17.11)$$

$$= -\sum_n \sum_{q(n)} \delta_{q^{\text{ref}}(n)}(q(n)) \log p(q(n) \mid \mathbf{o}(n), \theta) \qquad (17.12)$$

$$= -\sum_n \log p(q^{\text{ref}}(n) \mid \mathbf{o}(n), \theta) \qquad (17.13)$$

- *Sequence discriminative training*: Alternative cost function that directly compares the word sequence hypothesis from the ASR output in (17.2) and the reference. The following is an example of sequence discriminative training criterion based on minimum Bayes risk:

$$C^{\text{MBR}}(\theta) = \sum_{\mathbf{v} \in \mathcal{V}*} C^{\text{edit}}(\mathbf{v}^{\text{ref}}, \mathbf{v}) P(\mathbf{v} \mid \mathbf{O}, \theta), \qquad (17.14)$$

where $C^{\text{edit}}(\mathbf{v}^{\text{ref}}, \mathbf{v})$ is a sequence measure (e.g., edit distance) between the hypothesis $\mathbf{v}$ and the reference $\mathbf{v}^{\text{ref}}$. When we compute the HMM state-level edit distance, this is called the *state-level minimum Bayes risk*, which is often used as a training criterion of state-of-the-art ASR systems (Veselý *et al.*, 2013).

17.2.1.7 Decoding

In the decoding stage, frame-level acoustic model scores computed by a GMM likelihood (17.6) or DNN pseudo-likelihood (17.7) are combined with word-level language model scores, and the word sequence having the highest score is selected among all possible word sequences according to (17.3). This is performed by the Viterbi algorithm with a beam search, and a weighted finite state transducer is often employed to compose a context-dependent HMM network in an acoustic model with lexicon and language model networks efficiently based on automata theory (Mohri *et al.*, 2002; Hori and Nakamura, 2013).

Thus, we have given a brief overview of the current ASR framework, mainly focusing on acoustic modeling. ASR is a mature technique in clean conditions, but faces severe degradation in adverse environments, which limits its applicability to specific conditions typically based on a close-talking scenario. Therefore, the integration of ASR techniques with source separation and enhancement techniques is highly demanded. The next section describes speaker and language recognition, which is another major application of speech analysis and recognition, and shares several common techniques with ASR, including the use of MFCC-based features and GMM-based acoustic models.

17.2.2 Speaker and Language Recognition

Speaker and language recognition refers to a set of tasks that are related to estimating the identity of a person from an audio signal automatically, and recognizing the language spoken in a speech signal. Speaker recognition has applications, e.g. in biometric authentication, where speech can be used to control the access to various devices, places, or services. It also has applications in forensics and acoustic surveillance, e.g. to recover the identity of a person from a recording or to find a specific person in a large number of telephone conversations. Automatic language recognition has applications, e.g. as a preprocessing step in multilingual speech recognition, speech-to-speech translation systems, and in the retrieval of multimedia material based on the spoken language content.

Speaker recognition consists of two problems: in speaker *identification*, the goal is to estimate the identity of a speaker, given a database of reference speakers (Reynolds, 1995). This is a closed-set problem, i.e. the set of possible speakers is known in advance and the signal to be analyzed is spoken by one of these speakers. In speaker *verification*, the goal is to determine whether a speech signal is spoken by the target speaker or not (Reynolds *et al.*, 2000). This is an open-set problem: the audio signal can originate from any speaker, not necessarily one in the database used to develop the speaker identification system.

Speaker recognition or language recognition are typically formulated as MAP classification problems. The estimated speaker (or language) index $\hat{r}$ is the one that maximizes the posterior probability:

$$\hat{r} = \operatorname*{argmax}_{r \in \{1,\dots,R\}} P(r \mid \mathbf{O}), \tag{17.15}$$

where the audio signal to be analyzed is represented as sequence $\mathbf{O}$ of feature vectors, and R is the number of speakers (languages) in the development dataset.

Speaker verification is typically formulated as a *likelihood ratio test* between hypotheses H_0 (sentence is spoken by target speaker) and H_1 (sentence is not spoken by the target speaker) given by

$$\frac{p(\mathbf{O} \mid H_0)}{p(\mathbf{O} \mid H_1)} = \begin{cases} \geq \epsilon & \text{accept } H_0 \\ < \epsilon & \text{accept } H_1, \end{cases} \tag{17.16}$$

where $p(\mathbf{O} \mid H_0)$ and $p(\mathbf{O} \mid H_1)$ are the likelihoods for hypotheses H_0 and H_1, respectively, and ϵ is the decision threshold.

In the tasks above, similar to ASR, established acoustic features used to represent audio signals are MFCCs (see Section 17.2.1.1 for the calculation of MFCCs) or other similar features that characterize the short-time spectral shape, and established acoustic models to parameterize probability distributions are GMMs (see Section 17.2.1.3). Thus, in simple speaker and language recognition systems, the feature sequence $\mathbf{O}$ would be a sequence of MFCCs, and $p(\mathbf{O} \mid r)$, $p(\mathbf{O} \mid H_0)$, and $p(\mathbf{O} \mid H_1)$ would be modeled using GMMs that are estimated from each respective training dataset. In practice, the distribution $p(\mathbf{O} \mid H_1)$ that models the set of all speakers other than the target speaker is a generic (speaker-independent) speech model (also called a *universal background model*) trained using a large set of speakers.

In speaker recognition tasks, in addition to the short-time acoustic features, it is also important to model long-term characteristics in order to be able to model speaker variability and acoustic channel distortion effects, and to be able to use more powerful classifiers such as support vector machines and artificial neural networks or probabilistic linear discriminant analysis-based speaker verification (Kenny, 2010) with a fixed-length vector representation for variable-length audio signals. A typical approach is to first represent the audio signal $[x_u(t)]_t$ at utterance u using a sequence of short-time features such as MFCCs ($\mathbf{O}_u$). Then, it models their distribution using a GMM – $p(\mathbf{O}_u) = \prod_n \sum_{k=1}^K \pi_k \mathcal{N}(\mathbf{o}_u(n) \mid \boldsymbol{\mu}_k, \boldsymbol{\Sigma}_k)$ – and represents the signal as a GMM *supervector* $\boldsymbol{\mu}_u$ at

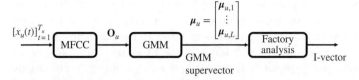

Figure 17.3 Flow chart of feature extraction in a typical speaker recognition system.

utterance u, which consists of the means of all Gaussians, i.e. $\boldsymbol{\mu}_u = [\boldsymbol{\mu}_{u,1}^T, \ldots, \boldsymbol{\mu}_{u,K}^T]^T$. Note that the dimensionality of the GMM supervector is independent of the duration of the signal and can be an input feature vector for powerful classifiers such as support vector machines.

State-of-the-art features for representing speech signals use *factor analysis* techniques in the GMM supervector space to separate the variability of the short-time features into speaker-independent, speaker-dependent, and distortion- or environment-dependent components. In the established *I-vector* approach, the latent components of the factor analysis model are not explicitly assigned to speaker or distortion properties, but are used directly as features for speaker recognition. Speaker verification can be performed simply by calculating a distance (e.g., cosine distance) between a target and a test I-vector (Dehak *et al.*, 2011).

As we can see from the explanation above, also illustrated in Figure 17.3, acoustic feature extraction used in established speaker and language recognition methods consists of a sequence of processing stages (MFCCs, GMM supervectors, I-vectors) that convert a low-level audio signal into a representation that is more robust. It should be noted, however, that factor analysis is performed in the GMM supervector space using MFCCs as features: in this domain the target speech and the other interfering acoustic sources are not additive. Therefore, in the presence of additive noise sources there is a need for explicit source separation techniques at the preprocessing stage. Most recent work on speaker and language recognition has used deep learning to obtain the feature representation (Lei *et al.*, 2014; Kenny *et al.*, 2014), but similar limitations regarding noise robustness still hold.

17.2.3 Paralinguistic Analysis

In the context of speech signal processing, *paralinguistic analysis* is used to refer to the analysis of properties of speech and other human vocalizations that are not related to linguistics or phonetics. These include speaker traits such as gender, age, and personality, and states such as emotions, sleepiness, or uncertainty (Schuller and Batliner, 2013). There is a large number of possible traits and states, and therefore a paralinguistic analysis system is developed to analyze only a small set of traits or states (e.g., emotion analysis). It is difficult to give a complete list of possible traits or states that would be subject to analysis, but we will illustrate the diversity of possible variables by listing the tasks that have been studied in the Computational Paralinguistics Challenge that has been organized annually at the Interspeech Conference between 2009 and 2016: spontaneous emotions; age, gender and affect; intoxication and sleepiness; personality, likability, and pathology; social signals, conflict, and autism; cognitive and physical load; degree of nativeness, Parkinson's condition, and eating condition; deception and sincerity.

Paralinguistic analysis has applications (Schuller and Weninger, 2012), e.g. in human–computer interaction systems, where the computer interaction can be varied depending on the analyzed human traits or states to improve the efficiency and user satisfaction of such a system. It can be used, for example, in acoustic surveillance, to monitor customer satisfaction automatically in call centers, or to identify potentially threatening persons in a public environment. In speech-to-speech translation systems, it can be used, together with a suitable speech synthesis engine, to synthesize translated speech that better matches the properties of the input speech.

Paralinguistic analysis problems can be roughly divided into two categories, categorical or continuous, depending on the target variable to be estimated. A frequently used categorical variable is, for example, the gender of a person, by simplifying gender to binary male/female categorization. Emotions can be modeled as categorical (e.g., anger, fear, sadness, happiness), but there are also several emotion models that represent emotions using a set of continuous variables such as arousal and valence.

In the case of categorical target variables, the problem is essentially a supervised classification problem, which can be formally defined similar to speaker and language recognition: from an audio signal, estimate the category index r that has the highest posterior probability $P(r \mid \mathbf{O})$ for a particular sequence of acoustic features. In the case of continuous target variables, the goal is to predict the value of the continuous target variables based on the input features, i.e. to perform regression. A paralinguistic analysis system can consider multiple target variables simultaneously, e.g. the gender and height of a person, or the arousal and valence of a speech utterance.

According to Schuller and Batliner (2013), a typical paralinguistic analysis system can include the following stages:

- Preprocessing to enhance the signal of interest, e.g. noise reduction and dereverberation.
- Low-level acoustic feature extraction. Similar to ASR and speaker/language recognition, this stage typically extracts a sequence of acoustic features such as MFCCs (see Section 17.2.1.1) in short time frames. In paralinguistics, features such as fundamental frequency, energy, and harmonic-to-noise ratio are more important than in ASR or speaker recognition, and often a large set of low-level features is used.
- Chunking, which segments the input feature sequence into shorter chunks, inside which the target variables are assumed to be fixed.
- Functional extraction, which calculates the statistics of the low-level features that characterize their distribution within a chunk.
- Feature reduction by dimensionality reduction techniques such as principal component analysis.
- Feature selection and generation, which select the features relevant for the analysis task.
- Optimization of the hyperparameters of the model used.
- Model learning, which is the typical training stage in supervised classification. In addition to the acoustic model that characterizes the dependencies between the acoustic features and the target variables, this stage can also involve a language model, similar to ASR.
- Classification or regression stage, where the learned model is used to perform the analysis for a test sample. For classification, established techniques such as decision

trees, support vector machines, or artificial neural networks are commonly used. For regression, support vector regression or neural networks provide efficient nonlinear techniques.

In addition to these processing stages above, the system can also include fusion of different classifiers and postprocessing to represent the output appropriately for specific applications.

Regarding the need for speech separation that is the scope of this chapter, many low-level acoustic features used in paralinguistic analysis such as harmonic-to-noise ratio are more easily affected by even small amounts of additive interference. Therefore, speech separation and enhancement are highly important in specific paralinguistic analysis tasks where features are affected by noise.

17.2.4 Audiovisual Analysis

In noisy scenarios, be it for paralinguistic analysis, for speaker recognition or for speech recognition, a second modality such as video – which is unaffected by acoustic noise and contains information on speaker location, speech activity, and even on phonetic content – can often help to improve robustness. However, when applying source separation to audiovisual data, a number of problems arise that are not typically encountered in audio-only signal enhancement:

- First, in standard audio-only speech enhancement, all signals that are measured are typically directly informative about the speech signal in question, i.e. the clean speech signals **s** are observed at all microphones, albeit superimposed and possibly with linear distortions such as those due to the room impulse response or, in the worst case, additional nonlinear distortions due to recording characteristics. In contrast, when dealing with multiple modalities, some observations are only informative in a rather indirect fashion.
- Secondly, an asynchrony problem arises, which is more severe than in audio-only signal processing. Not only are the audio and video signal recorded in a possibly asynchronous fashion, but the visible motion of the articulators (of the lips, the tongue, etc.) may precede the audible corresponding phonation by up to 120 ms (Luettin *et al.*, 2001).
- Finally, an entirely new subsystem, the feature extraction for the video data, needs to be considered, ideally in such a way as to make the video features maximally informative of the variables of interest (e.g., the speaker localization or the articulatory movements) while simultaneously keeping it as independent as possible from irrelevant information, e.g. from lighting and camera positioning.

Visual feature extraction is typically performed in two phases. After extracting and tracking the face and mouth region (e.g., via the Viola–Jones algorithm (Viola and Jones, 2004) and a Kalman filter), the actual feature extraction is carried out, for example, by

- the computation of mouth-opening parameters (Berthommier, 2004);
- the computation of discrete cosine transform coefficients of the gray-scale image of the mouth region, potentially followed by a linear discriminant analysis and a subsequent semi-tied covariance transform (Potamianos *et al.*, 2003);

- active shape and appearance models (Neti *et al.*, 2000), which have proven successful, albeit incurring a significant labeling effort in their training phase;
- convolutional neural networks (Noda *et al.*, 2015) as well as deep bottleneck feature extractors (Ninomiya *et al.*, 2015; Tamura *et al.*, 2015), which have become the method of choice in many recent works on audiovisual speech recognition or lipreading.

In all of these cases, the core task of feature extraction is to maximize class discrimination. This implies the need for separating irrelevant information, e.g. pertaining to the lighting and more general recording conditions, from the information of interest. This desired independence can be attained by all approaches mentioned above. For example, linear discriminant analysis and semitied covariance transform (cf. Section 17.3.3 for more details) are designed in view of class discriminability, the mouth-opening parameters abstract from the speaker and the environment, and convolutional neural networks can be trained discriminatively (Noda *et al.*, 2015).

17.3   Robust Speech Analysis and Recognition

This section provides a standard pipeline for noise-robust speech analysis and recognition. Source separation and enhancement are essential techniques to achieve robustness.

The main part of this section regards the front-end processing, including source separation and speech enhancement as a deterministic function $f'(\cdot)$, which outputs the enhanced single channel signal $[\hat{s}(t)]_t$ with the time domain representation given multichannel signals $\mathbf{X} = [x_i(t)]_{it}$. With the feature extraction, which is also a deterministic function $f(\cdot)$, we can process our speech analysis and recognition as follows:

$$[\hat{s}(t)]_t = f'(\mathbf{X}) \tag{17.17}$$

$$\Rightarrow \mathbf{O} = f([\hat{s}(t)]_t) \tag{17.18}$$

$$\Rightarrow \begin{cases} \operatorname{argmax}_{\mathbf{v} \in \mathcal{V}_*} P(\mathbf{v} \mid \mathbf{O}) & \text{for ASR,} \\ \operatorname{argmax}_{r \in \{1,\dots,R\}} P(r \mid \mathbf{O}) & \text{for speaker recognition,} \\ \vdots & \end{cases} \tag{17.19}$$

Based on this scheme, we can separate front-end and back-end modules, and simply apply the techniques developed in each module independently.

Given this modular structure, this section describes a standard pipeline established in the community through intensive studies mainly performed in the ASR field (Wölfel and McDonough, 2009; Virtanen *et al.*, 2012; Li *et al.*, 2014) and various challenge activities (Kinoshita *et al.*, 2013; Harper, 2015; Barker *et al.*, 2015). Therefore, the following explanations mainly consider ASR as an example of speech analysis and recognition with a standard pipeline as represented in Figure 17.4. We categorize the techniques used in the standard pipeline with the following four modules.

- Multichannel enhancement and separation: This module converts the multichannel signal to a single-channel enhanced signal. A typical method is beamforming: details are given in Section 17.3.2 and Chapters 10 and 11.

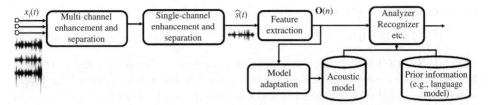

Figure 17.4 Flow chart of robust multichannel ASR with source separation, adaptation, and testing blocks.

- Single-channel enhancement and separation: This module can achieve further enhancement for the beamformed signals, as shown in Figure 17.4. The details are described in Section 17.3.1 and Chapter 7.
- Feature extraction: The feature extraction module described in Section 17.3.3 outputs speech features from enhanced (time-domain) signals, which will be used for acoustic modeling. This module also includes feature normalization and (linear) transformation techniques, and the section mainly describes the feature normalization and transformation, compared with Section 17.2.1.
- Acoustic modeling: As an acoustic modeling module, we can use either GMM-HMM or DNN-HMM as presented in Section 17.2.1. Section 17.3.4 mainly focuses on DNN-HMM acoustic modeling handling noise robustness issues.

17.3.1 Application of Single-Channel Source Separation

This section introduces single-channel source separation techniques applied to noise robust speech analysis and recognition, where source separation techniques are mainly used to separate noise and speech components in mixed signals to provide the enhanced speech signal for the back-end process. There have been various source separation techniques applied to speech analysis and recognition including computational auditory scene analysis (CASA) (Shao *et al.*, 2010) and shallow networks using support vector machines or support vector regression (Srinivasan *et al.*, 2006), but this section focuses on recent trends of using nonnegative matrix factorization (NMF) and deep learning.

17.3.1.1 Matrix Factorization

NMF and its variants have been widely applied to noise robust speech analysis and recognition (Raj *et al.*, 2010; Gemmeke *et al.*, 2011). The basic scheme of this NMF application assumes that a $F \times N$ mixture magnitude spectrum $|\mathbf{X}| = [|x(n,f)|]_{fn}$ is represented by the summation of noise and speech components with the $F \times K^s$ dictionary matrix $\mathbf{B}^s$ for target speech and the $F \times K^u$ dictionary matrix $\mathbf{B}^u$ for noise signals, as follows:

$$|\mathbf{X}| \approx \mathbf{B}^u \mathbf{H}^u + \mathbf{B}^s \mathbf{H}^s \tag{17.20}$$

$$= \underbrace{\left[\mathbf{B}^u, \mathbf{B}^s \right]}_{=\mathbf{B}} \underbrace{\begin{bmatrix} \mathbf{H}^u \\ \mathbf{H}^s \end{bmatrix}}_{=\mathbf{H}} = \mathbf{B}\mathbf{H} \tag{17.21}$$

where $\mathbf{H}^u$ and $\mathbf{H}^s$ are the $K^u \times N$ noise activation matrix and the $K^s \times N$ speech activation matrix, respectively. With the nonnegativity constraint on all $\mathbf{B}$, $\mathbf{H}$, and $|\mathbf{X}|$ matrices, the activation matrix $\mathbf{H}$ is estimated as:

$$\hat{\mathbf{H}} = \underset{\mathbf{H}}{\mathrm{argmin}}\ C(|\mathbf{X}|\ |\ \mathbf{BH}), \tag{17.22}$$

where $C(|\mathbf{X}|\ |\ \mathbf{BH})$ is the Kullback–Leibler (KL) divergence between matrices $|\mathbf{X}|$ and $\mathbf{BH}$, or another divergence such as Itakura–Saito (IS). For a description of the divergences and algorithms used to estimate the excitations, see Chapter 8.

Once we have obtained $\hat{\mathbf{H}}$, clean speech magnitude spectra $|\hat{\mathbf{S}}|$ can be obtained by masking as

$$|\hat{\mathbf{S}}| = |\mathbf{X}| \circ \frac{\mathbf{B}^s\hat{\mathbf{H}}^s}{\mathbf{B}^s\hat{\mathbf{H}}^s + \mathbf{B}^u\hat{\mathbf{H}}^u}, \tag{17.23}$$

where $\circ$ represents an elementwise multiplication.

Raj *et al.* (2010) reconstructed a time-domain signal from $|\hat{\mathbf{S}}|$ with speaker-independent and noise-independent (music in their study) dictionaries estimated in advance, which improves the ASR performance by 10% absolutely from that of the original signal (e.g., 60% to 50% in the 5 dB SNR case) in the low SNR cases. The paper also reports the effectiveness of feature-space ML linear regression feature transformation (which is discussed in Section 17.3.3) for the NMF-based enhanced speech signals above, which is extremely useful to remove a specific distortion caused by NMF enhancement.

Hurmalainen *et al.* (2015) used the activations of the NMF model directly in noise-robust speaker recognition, without converting the estimated model parameters back to the time domain. The speech dictionary used consisted of separate components for each speaker as well for the nonstationary room noise used in the evaluation. The method was shown to produce superior speaker recognition accuracy (average accuracy 96%, with SNRs of the test material ranging between -6 dB and 9 dB) in comparison to established I-vector (Dehak *et al.*, 2011) and GMM (Reynolds, 1995) methods (average accuracies 49% and 80%, respectively).

One of the important issues in NMF-based speech enhancement is how to obtain speech and noise dictionaries. In real application scenarios, we have huge varieties of speech and noise signals due to speaker and noise types, speaking styles, environments, and their combinations. In addition, in multiparty conversation scenarios, we have to remove nontarget speech as a noise. Therefore, in such scenarios it is challenging to use NMF-based speech enhancement. However, compared to the other enhancement techniques (e.g., deep-learning-based enhancement, described below), NMF-based speech enhancement does not require parallel clean and noisy data, and it potentially has wide application areas.

17.3.1.2 Deep-Learning-Based Enhancement

One of the recent trends in source separation is to deal with the problem by regression or classification where we can leverage techniques developed in machine learning, especially deep-learning techniques. For that, one must have parallel data consisting of target clean speech $\mathbf{S}$ and mixture of the target clean speech and noise $\mathbf{X}$, respectively. In normal speech recording scenarios, it is hard to obtain these parallel data except for the case of recording both close-talk and distant microphone signals or artificially generating parallel data with simulation.

Details of deep-learning-based enhancement are given in Chapter 7, where single-channel source separation has achieved great improvements with decent back-end systems, e.g. ASR with GMM-based acoustic models. However, once we use state-of-the-art back-end systems (e.g., ASR with DNN-based acoustic models trained with multicondition data), the performance improvement is marginal or sometimes the performance is slightly degraded (Hori *et al.*, 2015). This is because the DNN acoustic models learn noise-robust feature representations as if the lower layer of the networks performed single-channel source separation. The next section explains the application of multichannel source separation, which always yields significant improvements compared to single-channel cases due to the use of additional spatial information obtained by multichannel inputs. This section also combines single-channel deep-learning-based enhancement with beamforming to provide the state-of-the-art beamforming.

17.3.2 Application of Multichannel Source Separation

Multichannel source separation is considered to be very effective since it further enhances target signals from single-channel source separation with spatial information, as discussed in Chapter 10. Similar to the single-channel scenario, there have been many techniques applied to speech analysis and recognition, including frequency-domain independent component analysis (FD-ICA) (Takahashi *et al.*, 2008; Asano *et al.*, 2003; Kolossa *et al.*, 2010) and beamforming (Omologo *et al.*, 1997; Anguera *et al.*, 2007; Wölfel and McDonough, 2009). In particualr, recent distant-microphone ASR benchmarks verify the importance of beamforming techniques (Barker *et al.*, 2015; Kinoshita *et al.*, 2016).

Figure 17.5 compares the delay-and-sum (DS) and minimum variance distortionless response (MVDR) beamformers, which are described in Chapters 10 and 11, on the CHiME-3 dataset, and shows that both DS and MVDR beamformers significantly improved the performance from that obtained with original noisy speech by more than 35% error reduction rates. These results show the effectiveness of the application of beamforming to robust speech analysis and recognition.

17.3.3 Feature Extraction and Acoustic Models

This section focuses on noise robust extensions of feature extraction and acoustic models as introduced in Section 17.2.1.

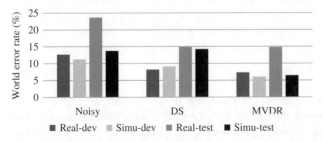

Figure 17.5 Comparison of the DS and MVDR beamformers on the CHiME-3 dataset. The results are obtained with a DNN acoustic model applied to the feature pipeline in Figure 17.6 and trained with the state-level minimum Bayes risk cost. Real-dev and Simu-dev denote the real and simulation development sets, respectively. Real-test and Simu-test denote the real and simulation evaluation sets, respectively.

17.3.3.1 Robust Feature Extraction

There are several proposals for noise robust feature extraction, including power normalized cepstral coefficients (Kim and Stern, 2012) and time-domain gammatone filter-based robust features (Schluter *et al.*, 2007; Mitra *et al.*, 2013) in addition to log Mel-filterbank and MFCCs, as presented in Section 17.2.1.1. However, many state-of-the-art systems still use standard log Mel-filterbank or MFCC features (Barker *et al.*, 2015; Kinoshita *et al.*, 2016), and this section mainly focuses on techniques based on these standard features.

17.3.3.2 Feature Normalization

In addition to the feature extraction, feature normalization based on *cepstral mean normalization* can remove channel distortion by subtracting the average cepstrum from the original cepstrum, as follows:

$$\mathbf{o}^{\mathrm{CMN}}(n) = \mathbf{o}(n) - \frac{1}{N} \sum_n \mathbf{o}(n). \tag{17.24}$$

Cepstral mean normalization can also remove the distortion caused by speech enhancement. It is performed utterance-by-utterance or speaker-by-speaker depending on the application. In addition, cepstral variance normalization is also performed, but the effect of variance normalization is limited.

17.3.3.3 Feature Transformation

Feature transformations based on linear discriminant analysis and feature-space ML linear regression are very effective for noisy speech analysis and recognition. *Linear discriminant analysis* usually transforms a high dimensional vector obtained by concatenating N_{past} past and N_{future} future context frames as follows:

$$\mathbf{o}^{\mathrm{LDA}}(n) = \mathbf{\Phi}^{\mathrm{LDA}}[\mathbf{o}^T(n - N_{\mathrm{past}}), \dots, \mathbf{o}^T(n), \dots, \mathbf{o}^T(n + N_{\mathrm{future}})]^T, \tag{17.25}$$

where $\mathbf{\Phi}^{\mathrm{LDA}}$ is a linear transformation matrix estimated with the linear discriminant analysis criterion (Haeb-Umbach and Ney, 1992). With linear discriminant analysis, we can consider long context information in the acoustic model, and handle reverberated speech and nonstationary noises, which distort speech features across multiple frames.

The *semitied covariance model*, also known as the *ML linear transform model* was originally proposed within the GMM-HMM framework as introduced in Section 17.2.1 to approximate the full covariance matrix with a diagonal covariance matrix. Let $\mathbf{\Sigma}_{rk}^{\mathrm{full}}$ be a full covariance matrix of HMM state r and mixture component k. The semitied covariance model assumes the following approximation with transformation matrix $\mathbf{\Phi}^{\mathrm{STC}}$:

$$\mathbf{\Sigma}_{rk}^{\mathrm{full}} \approx \mathbf{\Phi}^{\mathrm{STC}} \mathbf{\Sigma}_{rk}^{\mathrm{diag}} (\mathbf{\Phi}^{\mathrm{STC}})^T, \tag{17.26}$$

where $\mathbf{\Sigma}_{rk}^{\mathrm{diag}}$ is a corresponding diagonal matrix. Since $\mathbf{\Phi}^{\mathrm{STC}}$ is shared among all Gaussians, we can efficiently represent the full covariance matrix with a very small number of parameters. The equation above with the model space transformation is equivalently written with the feature space transformation as follows:

$$\mathcal{N}(\mathbf{\Phi}^{\mathrm{STC}} \mathbf{o}(n) \mid \mathbf{\Phi}^{\mathrm{STC}} \boldsymbol{\mu}_{rk}, \mathbf{\Sigma}_{rk}^{\mathrm{diag}}). \tag{17.27}$$

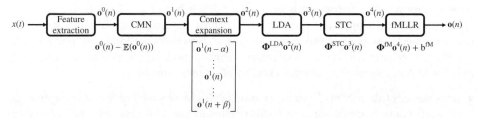

Figure 17.6 Pipeline of state-of-the-art feature extraction, normalization, and transformation procedure for noise robust speech analysis and recognition. CMN, LDA, STC, and fMLLR stand for cepstral mean normalization, linear discriminant analysis, semitied covariance transform, and feature-space ML linear regression, respectively.

Thus, we can regard the semitied covariance model as another feature transformation, which corresponds to using approximately a full covariance matrix in a diagonal covariance-based GMM-HMM, and has the effect of making the features uncorrelated. The transformation matrix $\boldsymbol{\Phi}^{\text{STC}}$ is obtained using ML estimation.

Feature-space ML linear regression considers the linear transformation and bias shift (affine transformation) of the original feature vector, i.e. $\mathbf{o}^{\text{fM}}(n) = \boldsymbol{\Phi}^{\text{fM}}\mathbf{o}(n) + \mathbf{b}^{\text{fM}}$. The transformation parameters are obtained from a GMM-HMM in the ML criterion as (Gales, 1998)

$$\widehat{\boldsymbol{\Phi}}^{\text{fM}}, \widehat{\mathbf{b}}^{\text{fM}} = \underset{\boldsymbol{\Phi}^{\text{fM}}, \mathbf{b}^{\text{fM}}}{\text{argmax}}\; Q(\boldsymbol{\Phi}^{\text{fM}}, \mathbf{b}^{\text{fM}}), \tag{17.28}$$

where

$$Q(\boldsymbol{\Phi}^{\text{fM}}, \mathbf{b}^{\text{fM}}) = \sum_{n,k,r} \gamma_{rk}(n) \left[\log(\det(\boldsymbol{\Phi}^{\text{fM}})) + \log \mathcal{N}(\boldsymbol{\Phi}^{\text{fM}}\mathbf{o}(n) + \mathbf{b}^{\text{fM}} \mid \boldsymbol{\mu}_{rk}, \boldsymbol{\Sigma}_{rk})\right].$$

$$\tag{17.29}$$

In the above, $\gamma_{rk}(n)$ is the posterior probability of HMM state r and mixture component k at frame n in the GMM-HMM framework. Feature-space ML linear regression largely mitigates the static distortion in speech features caused by stationary noise and reverberation.

Finally, we summarize the feature extraction, normalization, and transformation procedure in Figure 17.6. Note that we can cascade all feature normalizations/transformations to obtain the output features, which are used for GMM and DNN-based acoustic modeling.

17.3.4 Acoustic Model

Similar to the feature transformation with linear discriminant analysis, one of the standard methodologies of noise robust acoustic modeling is to deal with nonstationary noises and long-term reverberations. LSTM, convolutional, and time-delayed neural networks are suitable acoustic models for that purpose, and are actually used for noise robust ASR (Weng *et al.*, 2014; Chen *et al.*, 2015; Yoshioka *et al.*, 2015; Peddinti *et al.*, 2015), as introduced in Section 17.2.1. In addition, sequence discriminative training steadily improves the performance for these models (Chen *et al.*, 2015). Note that these techniques are developed not only for noise robust acoustic modeling, but are effective for generic speech analysis and recognition. However, the following remarks are specific for training DNNs for noise robust acoustic modeling, which are effective not only for ASR but speaker recognition and paralinguistic analysis.

- Importance of alignment: To train accurate DNNs we require precise time alignments of labels consisting of phonemes/HMM states in the ASR case, which is used as a ground truth for supervised training. However, precise alignment is difficult for noisy speech, which can cause the application of DNNs to fail in this scenario due to wrong supervision based on misaligned labels.
- Multicondition training: DNNs have the notable ability to make their model robust to input feature variations if we can provide them appropriately. Multicondition training fits to this DNN ability, and it is empirically known that multicondition training can obtain significant gains even if we roughly simulate to create multicondition data.

Given these remarks, if we simulate noisy speech data from relatively clean speech by applying noises and reverberations, we could obtain both multicondition training data and precise alignments from the corresponding original clean speech data by avoiding the difficulty of obtaining alignments from noisy data.[3] This method is very important for acoustic modeling in practice, when we do not have enough in-domain (matched condition) training data. In addition, since the scenario allows us to have parallel data, we can also use deep-learning-based enhancement, as discussed in Section 17.3.1. Thus, data simulation is also very important for speech analysis and recognition.

Besides the acoustic modeling above, there are several adaptation techniques developed for acoustic modeling, including DNN feature augmentation by using environment-dependent features and the re-estimation of DNN parameters with adaptation data. Feature augmentation is performed by concatenating the original feature vector $\mathbf{o}(n)$ and a vector ρ representing environmental information, including I-vector (Saon *et al.*, 2013), as discussed in Section 17.2.2, or noise statistics (Seltzer *et al.*, 2013), as follows:

$$\mathbf{o}'(n) = \begin{bmatrix} \mathbf{o}(n) \\ \rho \end{bmatrix}. \tag{17.30}$$

With this augmented information, the DNN is adapted to a specific environment characterized by ρ.

The re-estimation scheme is usually performed by undertaking the training procedure of the DNN similar to the standard one in Section 17.2.1.6. With the cross-entropy criterion, the adaptation parameter θ^{adapt} is obtained as follows:

$$\theta^{\mathrm{adapt}} = \underset{\theta}{\mathrm{argmax}} \sum_{n} \log p(\hat{q}(n) \mid \mathbf{o}^{\mathrm{adapt}}(n), \theta) \tag{17.31}$$

where $\mathbf{o}^{\mathrm{adapt}}(n)$ is a feature vector of adaptation data and $\hat{q}(n)$ is an HMM state hypothesis obtained by first-pass decoding. The most difficult issue in this re-estimation scheme is overfitting; it is avoided by regularization techniques, early stopping, or fixing a part of the parameters and only updating the rest (Liao, 2013; Yu *et al.*, 2013). Top systems

---

3 See Kinoshita *et al.* (2016) and Barker *et al.* (2015) for data simulation aimed at noise robust ASR.

in recent CHiME-3/4 challenges use this re-estimation scheme and show consistent improvements (5–10% error rate reduction) (Yoshioka *et al.*, 2015; Erdogan *et al.*, 2016).

17.4 Integration of Front-End and Back-End

17.4.1 Uncertainty Modeling and Uncertainty-Based Decoding

The above pipeline architecture suffers from an essential issue: the errors that occur in an early module in the pipeline are propagated to the following modules, and it is difficult to recover from these errors later. This is the motivation for preserving not only point estimates of quantities but also their associated uncertainties in each module. In this way, it becomes possible to mitigate the error propagation by considering the signals that are exchanged between modules in the pipeline not as fixed, but rather as random variables, which always come with a distribution or, at least, with an estimated uncertainty. This approach is briefly described in the following section.

17.4.1.1 Observation Uncertainties in the GMM-HMM Framework

In missing data approaches, e.g. Barker *et al.* (2001), each component of a feature vector is considered as either reliable or unreliable. If it is reliable, it is considered as a deterministic, known value in decoding. Otherwise, it can be disregarded completely (*marginalization*), it can be seen as a random variable, whose value is unknown but constrained to be lower than the observation (bounded marginalization), or one can attempt to recover its value based on the model (*imputation*) (Cooke *et al.*, 2001).

All these methods, however, have in common that each feature vector component is considered either completely observable or completely unobservable, which is suitable for speech enhancement based on binary masking (Srinivasan *et al.*, 2006). In contrast, observation uncertainty techniques rely on the estimates of the first- and second-order statistics of each feature vector component, i.e. on each respective estimated value and an associated estimated uncertainty. A component would then only be considered as completely unreliable if its variance tends towards infinity, and would only be completely reliable if its variance is zero.

This consideration allows for a more precise evaluation of uncertain feature information, and it is also especially amenable to the coupling of signal enhancement in one feature domain with signal decoding in another feature domain. Hence, it is of special significance for the application of source separation (typically performed in the STFT domain) to speech or speaker recognition (which can use completely different parameterizations, ranging from Mel spectra via MFCCs, cf. Section 17.2.1.1, to I-vectors).

This principle is shown in Figure 17.7. In general, three tasks need to be addressed to accomplish this:

- The uncertainties need to be estimated. For example, these could be uncertainties in the time-frequency domain, so they would be estimates of the variance of the estimated speech spectrum[4] at each point in time and frequency, $\hat{\sigma}_s^2(n,f)$.

---

4 Please note that this is different from the clean speech variance, which is also denoted by $\hat{\sigma}_s^2$ in other chapters.

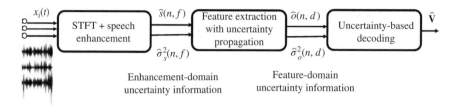

Figure 17.7 Observation uncertainty pipeline.

- If the feature domain of speech enhancement and that of speech recognition are different, the uncertain observations need to be propagated to the recognition domain, where they will be denoted by $\hat{o}(n, d)$ for the dth feature at time frame n, with its corresponding uncertainty $\hat{\sigma}_o^2(n, d)$.
- Finally, the decoder needs to consider the uncertainties in finding the optimum word sequence.

All these three issues are described, albeit briefly, below.

a) *Estimation* of the observation uncertainties can be achieved by a number of distinct considerations and approaches, often as a part of the signal estimation stage:
- A very simple model assumes that the uncertainty is proportional to the change in the signal incurred by signal enhancement (Kolossa *et al.*, 2010), so $\hat{\sigma}_s^2(n, f) = \alpha |\hat{s}(n, f) - x(n, f)|$. $\alpha > 0$ is a tunable scaling parameter.
- When speech enhancement is achieved by (possibly multichannel) Wiener filtering, a closed-form solution exists that will yield the signal estimator variance given the estimated speech and noise power (Astudillo *et al.*, 2012a; Kolossa *et al.*, 2011).
- Alternatively, uncertainties can be estimated specifically for masking-based source separation. A number of heuristic uncertainty estimates for such applications, e.g. based on the number of exemplars used to reconstruct a certain segment of speech, are introduced and compared by Gemmeke *et al.* (2010).
b) *Propagation* of the uncertainties (Kolossa and Haeb-Umbach, 2011), the second block in Figure 17.7, is only necessary if the feature domains of source separation and speech recognition differ, and if the feature uncertainty is already estimated in the source separation domain. If this is the case, uncertainty propagation can yield estimated uncertainties in the domain of speech recognition. The basic idea is to assume that the estimated signal and the associated uncertainty in the speech enhancement domain describe a Gaussian random variable, and to consider the effect of the subsequent stages of feature extraction on this random variable. If only second-order statistics are to be considered, pseudo Monte Carlo approaches like the unscented transform (Julier and Uhlmann, 2004) are applicable. Otherwise, a range of publications have considered the effect of each step of the feature extraction upon the feature distribution in a step-by-step fashion, see, for example, Astudillo *et al.* (2010).

Another general approach is to compute the uncertainty by a mapping function learned from the data. For this purpose, a range of mapping functions has been applied, e.g. linear mappings (Gemmeke *et al.*, 2010), regression trees (Srinivasan and Wang, 2007), and GMM-based mappings (Kallasjoki, 2016). These mapping functions might even be learned discriminatively using DNNs (Tran *et al.*, 2015).

Similarly, the integration of DNN-based signal estimates and uncertainty modeling has been shown to provide a clear advantage (Astudillo *et al.*, 2015) for the output of a DNN speech enhancement.

c) The final step of uncertainty-of-observation-based robust speech recognition is *uncertainty-aware decoding*. Uncertainty-aware decoding can be seen as a continuous-valued generalization of missing data techniques and, as there is a range of underlying probabilistic models that are potentially suitable, a range of methods exists.

Among these, one can distinguish state-independent uncertainty compensation techniques, such as the weighted Viterbi algorithm (Yoma and Villar, 2002), the so-called uncertainty decoding rule (Deng *et al.*, 2005), and modified imputation (Kolossa *et al.*, 2010), and state-dependent compensation techniques such as joint uncertainty decoding (Liao and Gales, 2005). Whereas the former are computationally more efficient, the latter are closer to the theoretical optimum (Liao and Gales, 2005). A range of solutions that correspond to different tradeoffs between accuracy and computational effort can be found by starting from the Bayesian decoding rule (Ion and Haeb-Umbach, 2008) and successively making simplifying assumptions, e.g. by considering only Gaussian posterior distributions or only diagonal covariance structures (Haeb-Umbach, 2011). A final concern, the training of models under observation uncertainty, is addressed by Ozerov *et al.* (2013) and leads to further improvements.

In addition to these uncertainty compensation techniques, which transform the model parameters, there is also the option to transform the features, frame-by-frame, for maximal discrimination under the current observation uncertainty, an approach that is termed noise-adaptive linear discriminant analysis (Kolossa *et al.*, 2013) due to its relationship to the originally static linear discriminant analysis that is described in Section 17.3.3.3. This adaptive variant of linear discriminant analysis has recently been applied to the task of audiovisual speech recognition in multisource environments (Zeiler *et al.*, 2016b).

An evaluation of a number of uncertainty-of-observation approaches for FD-ICA based enhancement is given by Kolossa *et al.* (2010). Table 17.2 shows some exemplary results for the application to two-speaker and three-speaker scenarios with simultaneously spoken English digit sequences. As can be seen, given oracle uncertainty information, the approaches fare very well. However, with the realistically available information in actual applications, noticeable improvements in word error rate are also achievable.

17.4.1.2 Observation Uncertainties in the DNN-HMM Framework

The idea of considering observation vectors – obtained through speech enhancement or source separation – as random variables, with time-varying reliabilities estimated feature-by-feature, has proven very helpful in GMM-HMM-based decoding of noisy speech (Kolossa and Haeb-Umbach, 2011). In DNNs, however, it is not directly clear how best to make use of uncertainties and in which way to interpret them mathematically, which, obviously, has significant consequences for the optimal algorithm design. Currently, a range of possible interpretations and algorithms is being explored.

The initial idea of using observation uncertainties in DNNs was suggested by Astudillo and da Silva Neto (2011) based on the original concept of considering

Table 17.2 ASR word accuracy achieved by a GMM-HMM acoustic model on MFCCs with delta and double-delta features. The data are enhanced by FD-ICA followed by time-frequency masking. We compare two different masking schemes, based on phase or interference estimates (Kolossa *et al.*, 2010). UD and MI stand for uncertainty decoding and modified imputation with estimated uncertainties, respectively, while UD* and MI* stand for uncertainty decoding and modified imputation with ideal uncertainties, respectively. Bold font indicates the best results achievable in practice, i.e. without the use of oracle knowledge.

No. of speakers	Phase-based masking		Interference-based masking	
	2	3	2	3
No FD-ICA	37.7	17.2	37.7	17.2
Only FD-ICA	70.0	68.6	70.0	68.6
FD-ICA + mask	03.5	03.7	38.7	35.7
FD-ICA + mask + UD	**72.9**	73.2	**76.7**	**74.7**
FD-ICA + mask + MI	72.3	**75.4**	72.0	72.2
FD-ICA + mask + UD*	89.9	82.3	92.8	88.1
FD-ICA + mask + MI*	90.3	87.4	92.1	87.5

observation vectors as random variables, described by their second-order statistics. Then, the entire processing chain of uncertainty estimation, uncertainty propagation, and (with some caveats) uncertainty-based decoding can be transferred to the neural network decoder. For the purpose of uncertainty propagation, (pseudo) Monte Carlo strategies (Astudillo *et al.*, 2012b; Abdelaziz *et al.*, 2015; Huemmer *et al.*, 2015) are then applicable, and the posteriors output by the neural network (17.8) can be approximated by their pseudo Monte Carlo expectations (Astudillo and da Silva Neto, 2011; Abdelaziz *et al.*, 2015) via

$$\widehat{P}(r \mid \mathbf{o}(n)) = (\mathbb{E}_{\mathbf{o}(n)}\{\text{softmax}(\bar{\mathbf{y}}_H(n))\})_r. \tag{17.32}$$

In approximating the expectation, it is typically assumed that the observation vector at time n, $\mathbf{o}(n)$, is distributed according to a Gaussian distribution with diagonal covariance $\widehat{\Sigma}_{\mathbf{o}}^2(n)$.

Monte Carlo sampling can also be used to train under awareness of observation uncertainties, which was introduced for HMM/DNN-systems by Tachioka and Watanabe (2015). Here, DNNs have been shown to generalize better when presented with many samples of features, with the sampling done based on the observation uncertainty. Even multiple recognition hypotheses, based on multiple samples, can be combined, once an efficient sampling strategy is available for uncertain observations.

In contrast, confidences rather than uncertainties may be employed in DNN decoding. Mallidi *et al.* (2015) proposed multistream combination for noise robust speech recognition based on the idea of training multiple models and fusing their outputs in accordance to their respective autoencoder reconstruction error. Here, the autoencoder is a particular type of DNN, which is learned to reconstruct input features with bottleneck hidden layers, and extracts their hidden vectors to reduce the dimensions of the original input features in a nonlinear manner. In a similar, but somewhat simpler, approach masking-based robust ASR has been shown by averaging the posterior output of two DNN-based ASR systems, one of which is trained on unprocessed, noisy speech, and the other on binary masked speech (Li and Sim, 2013).

Such multiple-model robust ASR seems to be a trend that is helpful in a number of successful DNN systems, and it will be interesting to see in which way source separation-based models can contribute within such frameworks – an open question at this moment.

The next section describes the trend of more closely integrating front-end source separation and enhancement and back-end speech analysis and recognition based on deep learning.

17.4.2 Joint Training Frameworks

This section describes the emergent *joint training* framework, which represents front-end and back-end systems as a single DNN and optimizes both systems simultaneously with back propagation.

In the single-channel case, deep-learning-based speech enhancement (Section 17.3.1) and acoustic modeling (Section 17.3.3) are integrated with a single DNN architecture including some deterministic computational nodes (Mel transformation, logarithm) and learnable layers (Narayanan and Wang, 2014). Here, a mask estimation-based approach, as discussed in Section 17.3.1, is used for speech enhancement. The architecture is jointly trained by using the cross-entropy cost, and Narayanan *et al.* (2015) extended the approach with sequence discriminative training and tested in large-scale datasets.

The joint training framework would be more powerful in the multichannel case, where it integrates beamforming (Section 17.3.2) and acoustic modeling (Sainath *et al.*, 2016; Xiao *et al.*, 2016; Meng *et al.*, 2017). Figure 17.8 shows the computational graph of a unified network for beamforming and acoustic modeling, as proposed by Xiao *et al.* (2016). The first DNN subnetwork estimates the time-invariant beamforming filter $\mathbf{w}(f) = [w_i(f)]_i$ from generalized cross-correlation with phase transform (GCC-PHAT) features, which is used to enhance the original multichannel STFT signal $\mathbf{x}(n,f) = [x_i(n,f)]_i$ as follows:

$$\hat{s}(n,f) = \mathbf{w}^H(f)\mathbf{x}(n,f). \tag{17.33}$$

The enhanced STFT $\hat{s}(n,f)$ is processed into a log Mel-filterbank representation by following the same steps as in Section 17.2.1 inside one unified joint network. The obtained features $\mathbf{O}$ are further processed by an acoustic modeling subnetwork to predict the state posterior $P(r \mid \mathbf{o}(n))$ for HMM state r. Since all of the modules above are differentiable and connected as one large computational graph, this unified network is jointly optimized by using back propagation with the cross-entropy criterion. Table 17.3 compares the performance achieved on a single distant microphone signals and on enhanced signals obtained by the DS beamformer and the joint beamforming network using the AMI meeting task (Hain *et al.*, 2007). The result clearly shows the improvement of the beamforming network from the conventional DS beamformer by 3.2% absolutely, and verifies the effectiveness of the joint training framework.

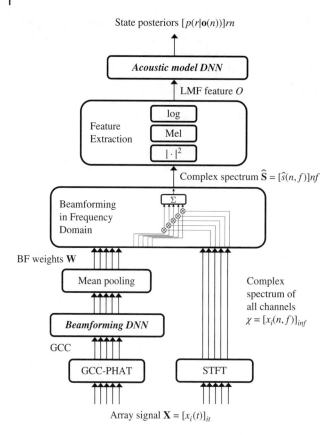

Figure 17.8 Joint training of a unified network for beamforming and acoustic modeling. Adapted from Xiao *et al.* (2016).

Table 17.3 ASR performance on the AMI meeting task using a single distant microphone and enhanced signals obtained by the DS beamformer and the joint training-based beamforming network.

	Word error rate (%)
Noisy	53.8
DS beamformer	47.9
Beamforming network	44.7

The joint training framework has the potential benefit of tightly integrating front-end and back-end systems by preserving uncertainties in a DNN manner. The method can also optimize the speech enhancement network without using parallel clean and noisy speech data, which is a requirement of deep-learning-based speech enhancement. Therefore, the joint training framework is one of the most promising directions of integrating front-end and back-end systems by leveraging deep-learning techniques.

17.5   Use of Multimodal Information with Source Separation

Although the previous sections mainly focused on single model (audio/speech) information, real-world applications can also make use of *multimodal* information. This section gives the potential benefits of multimodal audiovisual information with source separation. A natural categorization of audiovisual speech separation strategies is given by the latent variables that are considered in the estimation. The three types of algorithms are those based on localization, voice activity detection (VAD), and articulatory modeling. These are addressed, in that order, in the following three subsections.

17.5.1   Localization-Based Multimodal Source Separation

When focusing on location as the central latent variable, a natural strategy is to use audiovisual speaker tracking (Okuno *et al.*, 2001) and combine it with acoustic speech separation (Nakdadai *et al.*, 2002). In this way, by combining audition and vision for source localization, the robustness of the system can be increased with respect to visual occlusions as well as to acoustic noise. Overall, a separation gain of about 4.8 dB can thus be achieved even for moving sources, which is about 1 dB more than the audio-only system has provided for this task.

Thus, after tracking speakers audiovisually, source separation can be carried out based on the localization information. This can be performed, for example, by using the direction of arrival (DOA) information for spatially informed time-frequency masking, or by including the expected DOA of the target speaker in the localization-aware permutation correction step of FD-ICA.

17.5.2   Voice Activity Detection Based Multimodal Source Separation

Visual data supply ample information for VAD, e.g. the change in visual lip features was exploited for visual VAD (Rivet *et al.*, 2007c). The speech inactivity information can subsequently be employed for permutation correction in FD-ICA, or used with geometric source separation to recover the relevant subspace corresponding to the obtained support of the relevant speaker's time-domain signal (Rivet *et al.*, 2007b). In this way, overall, typical gains of 18–20 dB are achieved in terms of signal-to-interference ratio (SIR).

17.5.3   Joint Model-Based Multimodal Source Separation

Acoustic and visual features both depend on articulatory positions. In speech enhancement, this idea has been utilized, for example, by Abdelaziz *et al.* (2013).

The core idea is to find acoustic and visual features that are highly correlated with the underlying latent variable space of articulatory movements. Given those features, joint audiovisual speech models can be trained, where the model topology is an interesting design choice with strong implications for the algorithm design and performance.

One direct approach consists of learning a joint model of acoustic and video features in the form of a mixture model. For this purpose, kernels can be used to describe the joint probability of acoustic and visual observations (Rivet *et al.*, 2007a). Alternatively, audiovisual speech recognition can form the first stage in a two-step process, where

Table 17.4 Human recognition rates (%) of a listening test. Each score is based on averaging about 260 unique utterances. "Ephraim–Malah" refers to the log-spectral amplitude estimator of Ephraim and Malah (1985) in the implementation by Loizou (2007).

Signal	−9 dB	−6 dB	−3 dB	0 dB	Average
Noisy	48.65	60.23	70.47	77.24	64.15
Ephraim–Malah enhancement	41.53	53.88	61.41	71.08	56.97
Audiovisual speech enhancement	74.82	79.35	81.43	84.64	80.06

the inferred audiovisual state can then be utilized for speech enhancement (Zeiler *et al.*, 2016a), with notable improvements in human recognition rates, as shown in Table 17.4.

While this approach possesses the capability of considering the entire joint time series of acoustic and visual observations through the inclusion of audiovisual word and language models, the former approach is computationally more tractable for large vocabularies.

It can be expected, however, that audiovisual source separation will perform best when the sequential nature of the data is taken into account in an optimal fashion – a task that is likely of great interest in the near future based on the recent developments of deep, recurrent and convolutive networks, which have not yet been exploited for this challenging task that is carried out every time that lipreading is used to inform listening in noisy and multisource environments.

17.6 Summary

In this chapter we have discussed the application of source separation and speech enhancement to speech analysis and recognition, including ASR, speaker/language recognition, and paralinguistic analysis. With the great success of recent noise robustness studies and challenge activities, and the demand for hands-free speech products, this area has been rapidly developing and provides significant outcomes by establishing a solid pipeline of front-end and back-end systems, as described above. The chapter also pointed out the limitation of the pipeline architecture, which propagates unrecoverable errors and distortions caused by the front-end to the back-end, and discussed the importance of integrating front-end and back-end techniques by preserving uncertainties between them or by establishing joint training frameworks. Furthermore, it discussed how multimodal information provides meaningful clues based on localization and VAD, which helps to improve source separation and speech enhancement further. Although with such techniques the performance of source separation and speech enhancement techniques is becoming close to the requirements of real applications in speech analysis and recognition, these techniques could potentially be further evolved with the help of emergent techniques, including deep learning. We hope that this chapter helped to contribute to this evolution by guiding readers to tackle such challenging and attractive topics.

Bibliography

Abdelaziz, A.H., Watanabe, S., Hershey, J.R., Vincent, E., and Kolossa, D. (2015) Uncertainty propagation through deep neural networks, in *Proceedings of Interspeech*, pp. 3561–3565.

Abdelaziz, A.H., Zeiler, S., and Kolossa, D. (2013) Twin-HMM-based audiovisual speech enhancement, in *Proceedings of IEEE International Conference on Audio, Speech and Signal Processing*, pp. 3726–3730.

Anguera, X., Wooters, C., and Hernando, J. (2007) Acoustic beamforming for speaker diarization of meetings. *IEEE/ACM Transactions on Audio, Speech, and Language Processing*, **15** (7), 2011–2022.

Asano, F., Ikeda, S., Ogawa, M., Asoh, H., and Kitawaki, N. (2003) Combined approach of array processing and independent component analysis for blind separation of acoustic signals. *IEEE Transactions on Speech and Audio Processing*, **11** (3), 204–215.

Astudillo, R.F., Abad, A., and da Silva Neto, J.P. (2012a) Integration of beamforming and automatic speech recognition through propagation of the Wiener posterior, in *Proceedings of IEEE International Conference on Audio, Speech and Signal Processing*, pp. 4909–4912.

Astudillo, R.F., Abad, A., and da Silva Neto, J.P. (2012b) Uncertainty driven compensation of multi-stream-MLP acoustic models for robust ASR, in *Proceedings of Interspeech*, pp. 2606–2609.

Astudillo, R.F., Correia, J., and Trancoso, I. (2015) Integration of DNN based speech enhancement and ASR, in *Proceedings of Interspeech*, pp. 3576–3580.

Astudillo, R.F. and da Silva Neto, J.P. (2011) Propagation of uncertainty through multilayer perceptrons for robust automatic speech recognition, in *Proceedings of Interspeech*, pp. 461–464.

Astudillo, R.F., Kolossa, D., Mandelartz, P., and Orglmeister, R. (2010) An uncertainty propagation approach to robust ASR using the ETSI advanced front-end. *IEEE Journal of Selected Topics in Signal Processing*, **4**, 824–833.

Bahdanau, D., Chorowski, J., Serdyuk, D., and Bengio, Y. (2016) End-to-end attention-based large vocabulary speech recognition, in *Proceedings of IEEE International Conference on Audio, Speech and Signal Processing*, pp. 4945–4949.

Barker, J., Green, P., and Cooke, M. (2001) Linking auditory scene analysis and robust ASR by missing data techniques, in *Proceedings of Worshop on Innovation in Speech Processing*.

Barker, J., Marxer, R., Vincent, E., and Watanabe, S. (2015) The third 'CHiME' speech separation and recognition challenge: Dataset, task and baselines, in *Proceedings of IEEE Workshop on Automatic Speech Recognition and Understanding*, pp. 504–511.

Barker, J., Vincent, E., Ma, N., Christensen, H., and Green, P. (2013) The PASCAL CHiME speech separation and recognition challenge. *Computer Speech and Language*, **27** (3), 621–633.

Berthommier, F. (2004) Characterization and extraction of mouth opening parameters available for audiovisual speech enhancement, in *Proceedings of IEEE International Conference on Audio, Speech and Signal Processing*, pp. 789–792.

Chen, Z., Watanabe, S., Erdogan, H., and Hershey, J.R. (2015) Speech enhancement and recognition using multi-task learning of long short-term memory recurrent neural networks, in *Proceedings of Interspeech*, pp. 3274–3278.

Cooke, M., Green, P., Josifovski, L., and Vizinho, A. (2001) Robust automatic speech recognition with missing and unreliable acoustic data. *Speech Communication*, **34**, 267–285.

Dehak, N., Kenny, P., Dehak, R., Dumouchel, P., and Ouellet, P. (2011) Front-end factor analysis for speaker verification. *IEEE Transactions on Audio, Speech, and Language Processing*, **19** (4), 788–798.

Deng, L., Droppo, J., and Acero, A. (2005) Dynamic compensation of HMM variances using the feature enhancement uncertainty computed from a parametric model of speech distortion. *IEEE Transactions on Speech and Audio Processing*, **13** (3), 412–421.

Ephraim, Y. and Malah, D. (1985) Speech enhancement using a minimum mean-square error log-spectral amplitude estimator. *IEEE Transactions on Acoustics, Speech, and Signal Processing*, **33** (2), 443–445.

Erdogan, H., Hayashi, T., Hershey, J.R., Hori, T., Hori, C., Hsu, W.N., Kim, S., Le Roux, J., Meng, Z., and Watanabe, S. (2016) Multi-channel speech recognition: LSTMs all the way through, in *Proceedings of International Workshop on Speech Processing in Everyday Environments*.

Gales, M.J. (1998) Maximum likelihood linear transformations for HMM-based speech recognition. *Computer Speech and Language*, **12** (2), 75–98.

Gemmeke, J., Remes, U., and Palomäki, K. (2010) Observation uncertainty measures for sparse imputation, in *Proceedings of Interspeech*, pp. 2262–2265.

Gemmeke, J.F., Virtanen, T., and Hurmalainen, A. (2011) Exemplar-based sparse representations for noise robust automatic speech recognition. *IEEE Transactions on Speech and Audio Processing*, **19** (7), 2067–2080.

Graves, A. and Jaitly, N. (2014) Towards end-to-end speech recognition with recurrent neural networks., in *Proceedings of International Conference on Machine Learning*, pp. 1764–1772.

Haeb-Umbach, R. (2011) Uncertainty decoding and conditional Bayesian estimation, in *Robust Speech Recognition of Uncertain or Missing Data: Theory and Applications*, Springer.

Haeb-Umbach, R. and Ney, H. (1992) Linear discriminant analysis for improved large vocabulary continuous speech recognition, in *Proceedings of IEEE International Conference on Audio, Speech and Signal Processing*, pp. 13–16.

Hain, T., Burget, L., Dines, J., Garau, G., Wan, V., Karafiat, M., Vepa, J., and Lincoln, M. (2007) The AMI system for the transcription of speech in meetings, in *Proceedings of IEEE International Conference on Audio, Speech and Signal Processing*, pp. 357–360.

Harper, M. (2015) The automatic speech recogition in reverberant environments ASpIRE challenge, in *Proceedings of IEEE Workshop on Automatic Speech Recognition and Understanding*, pp. 547–554.

Hinton, G., Deng, L., Yu, D., Dahl, G.E., Mohamed, A.R., Jaitly, N., Senior, A., Vanhoucke, V., Nguyen, P., Sainath, T.N. *et al.* (2012) Deep neural networks for acoustic modeling in speech recognition: The shared views of four research groups. *IEEE Signal Processing Magazine*, **29** (6), 82–97.

Hochreiter, S. and Schmidhuber, J. (1997) Long short-term memory. *Neural Computation*, **9** (8), 1735–1780.

Hori, T., Chen, Z., Erdogan, H., Hershey, J.R., Le Roux, J., Mitra, V., and Watanabe, S. (2015) The MERL/SRI system for the 3rd CHiME challenge using beamforming, robust feature extraction, and advanced speech recognition, in *Proceedings of IEEE Workshop on Automatic Speech Recognition and Understanding*, pp. 475–481.

Hori, T. and Nakamura, A. (2013) *Speech Recognition Algorithms Using Weighted Finite-State Transducers*, Morgan & Claypool.

Huang, X., Acero, A., and Hon, H.W. (2001) *Spoken Language Processing: A Guide to Theory, Algorithm, and System Development*, Prentice Hall.

Huemmer, C., Maas, R., Astudillo, R., Kellermann, W., and Schwarz, A. (2015) Uncertainty decoding for DNN-HMM hybrid systems based on numerical sampling, in *Proceedings of Interspeech*, pp. 3556–3560.

Hurmalainen, A., Saeidi, R., and Virtanen, T. (2015) Noise robust speaker recognition with convolutive sparse coding, in *Proceedings of Interspeech*.

Ion, V. and Haeb-Umbach, R. (2008) A novel uncertainty decoding rule with applications to transmission error robust speech recognition. *IEEE Transactions on Audio, Speech, and Language Processing*, **16** (5), 1047–1060.

Julier, S. and Uhlmann, J. (2004) Unscented filtering and nonlinear estimation. *Proceedings of the IEEE*, **92** (3), 401–422.

Kallasjoki, H. (2016) *Feature Enhancement and Uncertainty Estimation for Recognition of Noisy and Reverberant Speech*, Ph.D. thesis, Aalto University.

Kenny, P. (2010) Bayesian speaker verification with heavy-tailed priors, in *Proceedings of Odyssey: The Speaker and Language Recognition Workshop*.

Kenny, P., Gupta, V., Stafylakis, T., Ouellet, P., and Alam, J. (2014) Deep neural networks for extracting Baum-Welch statistics for speaker recognition, in *Proceedings of Odyssey: The Speaker and Language Recognition Workshop*, pp. 293–298.

Kim, C. and Stern, R.M. (2012) Power-normalized cepstral coefficients (PNCC) for robust speech recognition, in *Proceedings of IEEE International Conference on Audio, Speech and Signal Processing*, pp. 4101–4104.

Kim, S., Hori, T., and Watanabe, S. (2016) Joint CTC-attention based end-to-end speech recognition using multi-task learning, arXiv:1609.06773.

Kinoshita, K., Delcroix, M., Gannot, S., Habets, E.A.P., Haeb-Umbach, R., Kellermann, W., Leutnant, V., Maas, R., Nakatani, T., Raj, B. *et al.* (2016) A summary of the REVERB challenge: state-of-the-art and remaining challenges in reverberant speech processing research. *EURASIP Journal on Advances in Signal Processing*, **2016** (1), 1–19.

Kinoshita, K., Delcroix, M., Yoshioka, T., Nakatani, T., Habets, E.A.P., Sehr, A., Kellermann, W., Gannot, S., Maas, R., Haeb-Umbach, R., Leutnant, V., and Raj, B. (2013) The REVERB challenge: a common evaluation framework for dereverberation and recognition of reverberant speech, in *Proceedings of IEEE Workshop on Applications of Signal Processing to Audio and Acoustics*, pp. 1–4.

Kolossa, D., Astudillo, R.F., Abad, A., Zeiler, S., Saeidi, R., Mowlaee, P., da Silva Neto, J.P., and Martin, R. (2011) CHIME challenge: Approaches to robustness using beamforming and uncertainty-of- observation techniques, in *Proceedings of International Workshop on Machine Listening in Multisource Environments*, pp. 6–11.

Kolossa, D., Astudillo, R.F., Hoffmann, E., and Orglmeister, R. (2010) Independent component analysis and time-frequency masking for speech recognition in multitalker conditions. *EURASIP Journal on Audio, Speech, and Music Processing*.

Kolossa, D. and Haeb-Umbach, R. (2011) *Robust Speech Recognition of Uncertain or Missing Data: Theory and Applications*, Springer.

Kolossa, D., Zeiler, S., Saeidi, R., and Astudillo, R. (2013) Noise-adaptive LDA: A new approach for speech recognition under observation uncertainty. *IEEE Signal Processing Letters*, **20** (11), 1018–1021.

Le Roux, J. and Vincent, E. (2014) A categorization of robust speech processing datasets, *Tech. Rep. TR2014-116*, Mitsubishi Electric Research Laboratories.

LeCun, Y., Bottou, L., Bengio, Y., and Haffner, P. (1998) Gradient-based learning applied to document recognition. *Proceedings of the IEEE*, **86** (11), 2278–2324.

Lei, Y., Scheffer, N., Ferrer, L., and McLaren, M. (2014) A novel scheme for speaker recognition using a phonetically-aware deep neural network, in *Proceedings of IEEE International Conference on Audio, Speech and Signal Processing*, pp. 1695–1699.

Li, B. and Sim, K.C. (2013) Improving robustness of deep neural networks via spectral masking for automatic speech recognition, in *Proceedings of IEEE Workshop on Automatic Speech Recognition and Understanding*, pp. 279–284.

Li, J., Deng, L., Gong, Y., and Haeb-Umbach, R. (2014) An overview of noise-robust automatic speech recognition. *IEEE/ACM Transactions on Audio, Speech, and Language Processing*, **22** (4), 745–777.

Liao, H. (2013) Speaker adaptation of context dependent deep neural networks, in *Proceedings of IEEE International Conference on Audio, Speech and Signal Processing*, pp. 7947–7951.

Liao, H. and Gales, M. (2005) Joint uncertainty decoding for noise robust speech recognition, in *Proceedings of Interspeech*, pp. 3129–3132.

Loizou, P.C. (2007) *Speech Enhancement: Theory and Practice*, CRC Press.

Luettin, J., Potamianos, G., and Neti, C. (2001) Asynchronous stream modelling for large vocabulary audio-visual speech recognition, in *Proceedings of IEEE International Conference on Audio, Speech and Signal Processing*, pp. 169–172.

Mallidi, S., Ogawa, T., Vesely, K., Nidadavolu, P., and Hermansky, H. (2015) Autoencoder based multi-stream combination for noise robust speech recognition, in *Proceedings of Interspeech*, pp. 3551–3555.

Meng, Z., Watanabe, S., Hershey, J.R., and Erdogan, H. (2017) Deep long short-term memory adaptive beamforming networks for multichannel robust speech recognition, in *Proceedings of IEEE International Conference on Audio, Speech and Signal Processing*, pp. 271–275.

Mitra, V., Franco, H., and Graciarena, M. (2013) Damped oscillator cepstral coefficients for robust speech recognition, in *Proceedings of Interspeech*, pp. 886–890.

Mohri, M., Pereira, F., and Riley, M. (2002) Weighted finite-state transducers in speech recognition. *Computer Speech and Language*, **16** (1), 69–88.

Nakdadai, K., Hidai, K., Okuno, H., and Kitano, H. (2002) Real-time speaker localization and speech separation by audio-visual integration, in *Proceedings of IEEE International Conference on Robotics and Automation*, pp. 1043–1049.

Narayanan, A., Misra, A., and Chin, K. (2015) Large-scale, sequence-discriminative, joint adaptive training for masking-based robust ASR, in *Proceedings of Interspeech*, pp. 3571–3575.

Narayanan, A. and Wang, D. (2014) Joint noise adaptive training for robust automatic speech recognition, in *Proceedings of IEEE International Conference on Audio, Speech and Signal Processing*, pp. 2504–2508.

Neti, C., Potamianos, G., Luettin, J., Matthews, I., Glotin, H., Vergyri, D., Sison, J., Mashari, A., and Zhou, J. (2000) Audio-visual speech recognition, *Tech. Rep. WS00AVSR*, Johns Hopkins University.

Ninomiya, H., Kitaoka, N., Tamura, S., Iribe, Y., and Takeda, K. (2015) Integration of deep bottleneck features for audio-visual speech recognition, in *Proceedings of Interspeech*, pp. 563–567.

Noda, K., Yamaguchi, Y., Nakadai, K., Okuno, H., and Ogata, T. (2015) Audio-visual speech recognition using deep learning. *Applied Intelligence*, **42**, 722–737.

Okuno, H.G., Nakadai, K., Hidai, K., Mizoguchi, H., and Kitano, H. (2001) Human-robot interaction through real-time auditory and visual multiple-talker tracking, in *Proceedings of IEEE/RSJ International Conference on Intelligent Robots and Systems*, pp. 1402–1409.

Omologo, M., Matassoni, M., Svaizer, P., and Giuliani, D. (1997) Microphone array based speech recognition with different talker-array positions, in *Proceedings of IEEE International Conference on Audio, Speech and Signal Processing*, pp. 227–230.

Ozerov, A., Lagrange, M., and Vincent, E. (2013) Uncertainty-based learning of acoustic models from noisy data. *Computer Speech and Language*, **27** (3), 874–894.

Peddinti, V., Povey, D., and Khudanpur, S. (2015) A time delay neural network architecture for efficient modeling of long temporal contexts, in *Proceedings of Interspeech*, pp. 2440–2444.

Potamianos, G., Neti, C., and Deligne, S. (2003) Joint audio-visual speech processing for recognition and enhancement, in *Proceedings of International Conference on Audio Visual Speech Processing*, pp. 95–104.

Rabiner, L. and Juang, B.H. (1993) *Fundamentals of Speech Recognition*, Prentice Hall.

Raj, B., Virtanen, T., Chaudhuri, S., and Singh, R. (2010) Non-negative matrix factorization based compensation of music for automatic speech recognition., in *Proceedings of Interspeech*, pp. 717–720.

Reynolds, D.A. (1995) Speaker identification and verification using Gaussian mixture speaker models. *Speech Communication*, **17** (1), 91–108.

Reynolds, D.A., Quatieri, T.F., and Dunn, R.B. (2000) Speaker verification using adapted Gaussian mixture models. *Digital Signal Processing*, **10** (1), 19–41.

Rivet, B., Girin, L., and Jutten, C. (2007a) Mixing audiovisual speech processing and blind source separation for the extraction of speech signals from convolutive mixtures. *IEEE Transactions on Audio, Speech, and Language Processing*, **15** (1), 96–108.

Rivet, B., Girin, L., and Jutten, C. (2007b) Visual voice activity detection as a help for speech source separation from convolutive mixtures. *Speech Communication*, **49**, 667–677.

Rivet, B., Girin, L., Servière, C., Pham, D.T., and Jutten, C. (2007c) Audiovisual speech source separation: a regularization method based on visual voice activity detection, in *Proceedings of International Conference on Audio Visual Speech Processing*, p. 7.

Sainath, T.N., Weiss, R.J., Wilson, K.W., Narayanan, A., and Bacchiani, M. (2016) Factored spatial and spectral multichannel raw waveform CLDNNs, in *Proceedings of IEEE International Conference on Audio, Speech and Signal Processing*, pp. 5075–5079.

Saon, G., Soltau, H., Nahamoo, D., and Picheny, M. (2013) Speaker adaptation of neural network acoustic models using i-vectors., in *Proceedings of IEEE Workshop on Automatic Speech Recognition and Understanding*, pp. 55–59.

Schluter, R., Bezrukov, I., Wagner, H., and Ney, H. (2007) Gammatone features and feature combination for large vocabulary speech recognition, in *Proceedings of IEEE International Conference on Audio, Speech and Signal Processing*, pp. 649–652.

Schuller, B. and Batliner, A. (2013) *Computational Paralinguistics: Emotion, Affect and Personality in Speech and Language Processing*, Wiley.

Schuller, B. and Weninger, F. (2012) Ten recent trends in computational paralinguistics, in *Proceedings of COST 2102 International Training School on Cognitive Behavioural Systems*, pp. 35–49.

Seltzer, M.L., Yu, D., and Wang, Y. (2013) An investigation of deep neural networks for noise robust speech recognition, in *Proceedings of IEEE International Conference on Audio, Speech and Signal Processing*, pp. 7398–7402.

Shao, Y., Srinivasan, S., Jin, Z., and Wang, D. (2010) A computational auditory scene analysis system for speech segregation and robust speech recognition. *Computer Speech and Language*, **24** (1), 77–93.

Srinivasan, S., Roman, N., and Wang, D. (2006) Binary and ratio time-frequency masks for robust speech recognition. *Speech Communication*, **48** (11), 1486–1501.

Srinivasan, S. and Wang, D. (2007) Transforming binary uncertainties for robust speech recognition. *IEEE Transactions on Audio, Speech, and Language Processing*, **15** (7), 2130–2140.

Tachioka, Y. and Watanabe, S. (2015) Uncertainty training and decoding methods of deep neural networks based on stochastic representation of enhanced features, in *Proceedings of Interspeech*, pp. 3541–3545.

Takahashi, Y., Osako, K., Saruwatari, H., and Shikano, K. (2008) Blind source extraction for hands-free speech recognition based on Wiener filtering and ICA-based noise estimation, in *Proceedings of Joint Workshop on Hands-free Speech Communication and Microphone Arrays*, pp. 164–167.

Tamura, S., Ninomiya, H., Kitaoka, N., Osuga, S., Iribe, Y., Takeda, K., and Hayamizu, S. (2015) Audio-visual speech recognition using deep bottleneck features and high-performance lipreading, in *Proceedings of APSIPA Annual Summit and Conference*, pp. 575–582.

Tran, D.T., Vincent, E., and Jouvet, D. (2015) Nonparametric uncertainty estimation and propagation for noise-robust ASR. *IEEE/ACM Transactions on Audio, Speech, and Language Processing*, **23** (11), 1835–1846.

Veselỳ, K., Ghoshal, A., Burget, L., and Povey, D. (2013) Sequence-discriminative training of deep neural networks., in *Proceedings of Interspeech*, pp. 2345–2349.

Vincent, E., Barker, J., Watanabe, S., Le Roux, J., Nesta, F., and Matassoni, M. (2013) The second 'CHiME' speech separation and recognition challenge: Datasets, tasks and baselines, in *Proceedings of IEEE International Conference on Audio, Speech and Signal Processing*, pp. 126–130.

Vincent, E., Watanabe, S., Nugraha, A., Barker, J., and Marxer, R. (2017) An analysis of environment, microphone and data simulation mismatches in robust speech recognition. *Computer Speech and Language*, **46**, 535–557.

Viola, P. and Jones, M. (2004) Robust real-time face detection. *International Journal of Computer Vision*, **57** (2), 137–154.

Virtanen, T., Singh, R., and Raj, B. (2012) *Techniques for Noise Robustness in Automatic Speech Recognition*, Wiley.

Waibel, A., Hanazawa, T., Hinton, G., Shikano, K., and Lang, K.J. (1989) Phoneme recognition using time-delay neural networks. *IEEE Transactions on Acoustics, Speech, and Signal Processing*, **37** (3), 328–339.

Watanabe, S. and Chien, J.T. (2015) *Bayesian Speech and Language Processing*, Cambridge University Press.

Weng, C., Yu, D., Watanabe, S., and Juang, B.H.F. (2014) Recurrent deep neural networks for robust speech recognition, in *Proceedings of IEEE International Conference on Audio, Speech and Signal Processing*, pp. 5532–5536.

Wölfel, M. and McDonough, J. (2009) *Distant Speech Recognition*, Wiley.

Xiao, X., Watanabe, S., Erdogan, H., Lu, L., Hershey, J., Seltzer, M.L., Chen, G., Zhang, Y., Mandel, M., and Yu, D. (2016) Deep beamforming networks for multi-channel speech recognition, in *Proceedings of IEEE International Conference on Audio, Speech and Signal Processing*, pp. 5745–5749.

Yoma, N. and Villar, M. (2002) Speaker verification in noise using a stochastic version of the weighted Viterbi algorithm. *IEEE Transactions on Speech and Audio Processing*, **10** (3), 158–166.

Yoshioka, T., Ito, N., Delcroix, M., Ogawa, A., Kinoshita, K., Fujimoto, M., Yu, C., Fabian, W.J., Espi, M., Higuchi, T., Araki, S., and Nakatani, T. (2015) The NTT CHiME-3 system: Advances in speech enhancement and recognition for mobile multi-microphone devices, in *Proceedings of IEEE Workshop on Automatic Speech Recognition and Understanding*, pp. 436–443.

Yu, D. and Deng, L. (2015) *Automatic Speech Recognition - A Deep Learning Approach*, Springer.

Yu, D., Yao, K., Su, H., Li, G., and Seide, F. (2013) Kl-divergence regularized deep neural network adaptation for improved large vocabulary speech recognition, in *Proceedings of IEEE International Conference on Audio, Speech and Signal Processing*, pp. 7893–7897.

Zeiler, S., Meutzner, H., Abdelaziz, A.H., and Kolossa, D. (2016a) Introducing the turbo-twin-HMM for audio-visual speech enhancement, in *Proceedings of Interspeech*, pp. 1750–1754.

Zeiler, S., Nickel, R., Ma, N., Brown, G., and Kolossa, D. (2016b) Robust audiovisual speech recognition using noise-adaptive linear discriminant analysis, in *Proceedings of IEEE International Conference on Audio, Speech and Signal Processing*, pp. 2797–2801.

18

Binaural Speech Processing with Application to Hearing Devices

Simon Doclo, Sharon Gannot, Daniel Marquardt, and Elior Hadad

This chapter provides an overview of multi-microphone speech enhancement algorithms for a *binaural* hearing system, consisting of a hearing device on each ear of the user. In contrast to the multi-microphone speech enhancement algorithms presented in Chapter 10, the objective of a binaural algorithm is not only to selectively enhance the desired speech source and to suppress interfering sources (e.g., competing speakers) and ambient background noise, but also to preserve the auditory impression of the acoustic scene. This can be achieved by preserving the so-called binaural cues of the desired speech source, the interfering sources and the background noise, such that the binaural hearing advantage of the auditory system can be exploited and confusions due to a mismatch between acoustic and visual information are avoided.

Section 18.1 provides a general introduction, focusing on the specific objectives and challenges encountered in binaural noise reduction. In Section 18.2 the binaural hearing properties are discussed, which play a crucial role in the design and evaluation of binaural algorithms. Section 18.3 provides an overview of the two main binaural processing paradigms: binaural spectral postfiltering and binaural spatial filtering. In the remainder of the chapter we mainly focus on binaural spatial filtering. In Section 18.4 the considered acoustic scenario is presented and a binaural version of the multichannel Wiener filter (MWF), presented in Chapter 10, is discussed. In Sections 18.5 and 18.6 several extensions of the binaural MWF are presented, aiming at combining noise reduction and binaural cue preservation both for diffuse noise as well as for coherent interfering sources. Section 18.7 concludes the chapter.

18.1 Introduction to Binaural Processing

In many everyday speech communication situations, such as social meetings or in traffic, we are immersed in unwanted noise and interfering sounds. Particularly in complex acoustic scenarios where several people are talking simultaneously, i.e. the so-called cocktail party scenario (Cherry, 1953), speech intelligibility may be substantially deteriorated. While this may be annoying for normal-hearing persons, it is particularly problematic for hearing-impaired persons, who may encounter severe difficulties in communication. Although hearing aids are able to significantly

Audio Source Separation and Speech Enhancement, First Edition.
Edited by Emmanuel Vincent, Tuomas Virtanen and Sharon Gannot.
© 2018 John Wiley & Sons Ltd. Published 2018 by John Wiley & Sons Ltd.
Companion Website: https://project.inria.fr/ssse/

increase speech intelligibility in quiet environments, in the presence of background noise, competing speakers and reverberation, speech understanding is still a challenge for most hearing aid users. Hence, in addition to standard hearing aid processing, such as frequency-dependent amplification, dynamic range compression, feedback suppression, and acoustic environment classification (Dillon, 2012; Popelka *et al.*, 2016), several single- and multi-microphone speech enhancement algorithms have been developed for hearing aids and assisted listening devices which aim at improving intelligibility in noisy environments (Hamacher *et al.*, 2005; Luts *et al.*, 2010; Doclo *et al.*, 2010, 2015). It is important to realize that not all algorithms discussed in the previous chapters are suitable to be implemented in hearing aids in practice. First, due to the restricted capacity of the batteries and hence the limited processing power of the digital signal processor, the algorithmic complexity needs to be limited. Second, in order to avoid disturbing artifacts and echo effects (especially when the hearing aid user is talking), the total input–output signal delay of the algorithms needs to be in the order of 10 ms or less (Heurig and Chalupper, 2010). Therefore, in this chapter we will focus on algorithms that satisfy both the complexity and the latency requirements that are typical for hearing aids.

In principle, using hearing devices on both ears can generate an important advantage, both from a signal processing perspective, since all microphone signals from both devices can be used, and from a perceptual perspective, since the auditory system can exploit binaural cues (cf. Chapters 3 and 12). In addition to monaural cues, binaural cues play a major role in source localization and speech intelligibility (Blauert, 1997), cf. Section 18.2. However, in a *bilateral* system where both hearing devices operate independently, this potential is not fully exploited, possibly even leading to a distortion of the binaural cues and hence localization ability (van den Bogaert *et al.*, 2006; Keidser *et al.*, 2009).

In order to achieve true binaural processing, both hearing devices need to cooperate with each other and exchange data or signals, e.g. through a wireless link (Boothroyd *et al.*, 2007). Whereas early binaural systems only had limited data rates, allowing the transmission of data in order to coordinate the parameter settings of both devices (e.g., volume, acoustic environment classification), the first commercial systems that exchange microphone signals in full-duplex mode are currently entering the market. These systems pave the way to the implementation of full-fledged binaural signal processing algorithms, where microphone signals from both devices are processed and combined in each device (Hamacher *et al.*, 2008; Doclo *et al.*, 2010; Wouters *et al.*, 2013). Although in practice the bandwidth and the latency of the wireless link play an important role, in this chapter we will only focus on the binaural algorithms themselves, i.e. assuming a near-to-perfect wireless link such that all microphone signals are assumed to be available for processing.

In contrast to the multi-microphone noise reduction algorithms presented in Chapter 10, the objective of a binaural noise reduction algorithm is twofold: (1) enhance the desired speech source and suppress interfering sources and background noise, and (2) preserve the auditory impression (spatial characteristics) of the complete acoustic scene, such that no mismatch between acoustic and visual information occurs and the binaural hearing advantage can be exploited. This last objective can be achieved by preserving the binaural cues of all sound sources, i.e. the desired speech source, the interfering sources and the background noise.

To combine noise reduction and binaural cue preservation, two different paradigms are typically adopted (cf. Section 18.3). In the first paradigm, two microphone signals (i.e., one on each hearing device) are filtered with the same real-valued spectro-temporal gain, intrinsically guaranteeing binaural cue preservation for all sound sources. In the second paradigm, all available microphone signals from both hearing devices are processed by different complex-valued spatial filters. Although the second paradigm allows for more degrees of freedom to achieve more noise reduction and less speech distortion than the first paradigm, there is typically a tradeoff between noise reduction performance and binaural cue preservation. As will be shown in Section 18.4, many well-known multi-microphone noise reduction algorithms, such as the minimum variance distortionless response (MVDR) beamformer and the MWF, can be straight-forwardly extended into a binaural version producing two output signals. Although the binaural MVDR beamformer and MWF preserve the binaural cues of the desired speech source, they typically distort the binaural cues of the interfering sources and the background noise, such that all sources are perceived as coming from the same direction.

Aiming at preserving the binaural cues of all sound sources while not degrading the noise reduction performance, different extensions of the binaural MVDR beamformer and the binaural MWF have been proposed. For the sake of conciseness, we will mainly elaborate on MWF-based techniques in this chapter, since MVDR-based techniques represent a special case. Extensions for diffuse noise are presented in Section 18.5, while extensions for coherent interfering sources are presented in Section 18.6. The performance of all algorithms is validated by simulations using measured binaural acoustic transfer functions.

18.2   Binaural Hearing

In addition to monaural cues, such as spectral pinna cues, pitch, and amplitude modulation, binaural cues play a major role in spatial awareness, i.e. for localizing sources and for determining the spatial width of auditory objects, and are very important for speech intelligibility due to so-called binaural unmasking (Bregman, 1990; Blauert, 1997). The main binaural cues used by the auditory system are:

1) the *interaural level difference* (ILD), due to the head acting as an obstacle for the sound waves traveling between the ears, i.e. the so-called head shadow effect. While for low frequencies the head barely acts as an obstacle, large ILDs up to 20–25 dB may occur at high frequencies, depending on the source position and the size of the head.
2) the *interaural phase difference* (IPD) or *interaural time difference* (ITD), due to the time difference of arrival (TDOA) of the sound waves traveling between the ears. ITD values typically lie in the range −700 to 700 μs, depending on the source position and the size of the head.
3) the *interaural coherence* (IC), defined as the normalized cross-correlation between the signals at both ears. Especially in multisource and reverberant environments, the IC is an important binaural cue for source localization and spatial perception, since it determines the reliability of the ILD and ITD cues (Faller and Merimaa, 2004; Dietz

et al., 2011). In addition, the IC is an important cue for the perception of the spatial width of sound fields (Kurozumi and Ohgushi, 1983). Based on a geometrical head model, a theoretical model for the IC has been derived for isotropic sound fields by Jeub *et al.* (2011a). Furthermore, based on experimental data, Lindevald and Benade (1986) showed that the IC of a spherically isotropic sound field can be approximated as a modified sine cardinal function.

The sensitivity to the ILD and ITD cues can be expressed by the *just noticeable differences*, which for normal-hearing listeners can be as low as 1 dB and 10 µs, respectively. For source localization it has been shown that the ITD cue plays a dominant role at lower frequencies, i.e. below 1500 Hz, while the ILD cue plays a dominant role at higher frequencies (Wightman and Kistler, 1992). Subjective listening experiments evaluating the IC discrimination ability of the auditory system in a diffuse noise field have shown that the sensitivity to changes in the IC strongly depends on the reference IC (Tohyama and Suzuki, 1989; Walther and Faller, 2013). For a reference IC close to 1 small changes can be perceived, whereas for a reference IC close to 0 the auditory system is less sensitive to changes.

Besides source localization and spatial perception, binaural cues play an important role in speech intelligibility for normal-hearing and hearing-impaired listeners due to binaural unmasking (Blauert, 1997; Hawley *et al.*, 2004; Best *et al.*, 2011). When the desired speech source is spatially separated from the interfering sources and background noise, a binaural hearing advantage compared to monaural hearing occurs. For example, in an anechoic environment with one desired speech source and one interfering source, both located in front of the listener, a *speech reception threshold* corresponding to 50% speech intelligibility of about −8 dB is obtained (Bronkhorst and Plomp, 1988). If the sources are spatially separated, i.e. if the interfering source is not located in front of the listener, the speech reception threshold decreases down to −20 dB, depending on the position of the interfering source. Although for reverberant environments this speech reception threshold difference is smaller than for anechoic environments, differences for spatially separated sources up to 6 dB have been reported (Bronkhorst, 2000). Furthermore, for a speech source located in front of the listener in a diffuse noise field, speech reception threshold improvements for binaural hearing compared to monaural hearing up to 3 dB have been reported. Consequently, in combination with the signal-to-noise ratio (SNR) the binaural ILD, ITD and IC cues are commonly used in binaural models to predict speech intelligibility in noisy and reverberant environments (Beutelmann and Brand, 2006; Lavandier and Culling, 2010; Rennies *et al.*, 2011).

18.3 Binaural Noise Reduction Paradigms

In contrast to the multi-microphone noise reduction algorithms presented in Chapter 10, which produce a single-channel output signal, binaural algorithms need to produce two different output signals in order to enable the listener to exploit binaural unmasking. Existing binaural algorithms for combined noise reduction and cue preservation can be classified into two paradigms: binaural spectral postfiltering (cf. Section 18.3.1) and binaural spatial filtering (cf. Section 18.3.2).

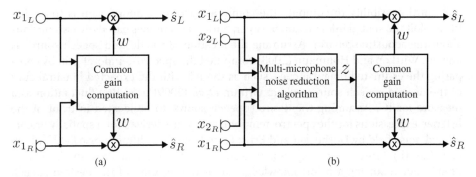

Figure 18.1 Block diagram for binaural spectral postfiltering based on a common spectro-temporal gain: (a) direct gain computation (one microphone on each hearing device) and (b) indirect gain computation (two microphones on each hearing device).

18.3.1 Paradigm 1: Binaural Spectral Postfiltering

Binaural spectral postfiltering techniques apply the same (real-valued) spectro-temporal gain w to one microphone signal on each hearing device (cf. Figure 18.1). The aim is to apply a large gain when the time-frequency bin should be retained (dominant desired speech source) and a small gain when the time-frequency bin should be suppressed (dominant interfering source or background noise). Although binaural spectral postfiltering techniques perfectly preserve the instantaneous binaural cues of all sources, in essence they can be viewed as single-channel noise reduction techniques (cf. Chapter 5), hence possibly introducing speech distortion and typical single-channel noise reduction artifacts, especially at low SNRs.

The common spectro-temporal gain can either be computed directly from the microphone signals (typically when only one microphone is available on each hearing device), cf. Figure 18.1a, or indirectly based on the single-channel output signal of a multi-microphone noise reduction algorithm (typically when multiple microphones are available on each hearing device), cf. Figure 18.1b.

- **Direct gain computation** (Figure 18.1a): A first possibility is to independently compute a spectro-temporal gain for each microphone signal, e.g. using any of the single-channel noise reduction techniques discussed in Chapter 5, and to combine both spectro-temporal gains, e.g. by taking the arithmetic or geometric mean.

  A second possibility is to compute the common spectro-temporal gain based on computational auditory scene analysis (CASA) (Wang and Brown, 2006), which typically enables the determination of which time-frequency bin the desired speech source is dominant in order to compute a spectral (soft or binary) mask. Please refer to Chapter 12 for a survey on multichannel spectral masking methods. Sound sources can be grouped based on monaural cues, such as common pitch, amplitude modulation, and onset, while for a binaural system binaural cues, such as ILD and ITD, can be exploited in addition (Nakatani and Okuno, 1999; Roman *et al.*, 2003; Raspaud *et al.*, 2010; May *et al.*, 2011; Woodruff and Wang, 2013). The binary mask generated by the CASA system is often directly applied as a spectro-temporal gain to achieve speech enhancement/segregation, typically, however, leading to speech distortion and artifacts (Madhu *et al.*, 2013).

A third possibility to compute the common spectro-temporal gain is to explicitly exploit the spatial information between both microphones based on assumptions about the acoustic scenario. Assuming the location of the desired speech source is known, Wittkop and Hohmann (2003) computed the spectro-temporal gain by comparing the estimated binaural cues, such as the IC, with the expected binaural cues of the desired speech source, where Grimm *et al.* (2009) used IPD fluctuation as a measure for IC. Assuming the desired speech source to be located in front of the listener, expressions for the spectro-temporal gain were derived for a spatially uncorrelated noise field by Dörbecker and Ernst (1996) and for a diffuse noise field by Jeub *et al.* (2011b). More general expressions were derived by Kamkar-Parsi and Bouchard (2009), not requiring a priori knowledge about the location of the desired speech source, and by Kamkar-Parsi and Bouchard (2011), allowing for an additional coherent interfering source to be present.

- **Indirect gain computation** (Figure 18.1b): The common spectro-temporal gain can also be computed indirectly based on an intermediate (single-channel) signal z, which represents either an estimate of the desired speech source or an estimate of the interfering sources and background noise. An estimate of the desired speech source can be obtained by applying any of the multi-microphone noise reduction algorithms discussed in Chapter 10 to all microphone signals on both hearing devices. Based on this concept, different algorithms have been investigated: a fixed superdirective beamformer (Lotter and Vary, 2006), an adaptive MVDR beamformer (Rohdenburg, 2008), and fixed and adaptive MVDR beamformers combined with single-channel noise reduction (Baumgärtel *et al.*, 2015). By applying this indirectly computed spectro-temporal gain to a microphone signal on each hearing device, the SNR improvement due to multi-microphone noise reduction (spatial filtering) can be approximately achieved through spectro-temporal filtering, while the instantaneous binaural cues of all sources are perfectly preserved. It should, however, be realized that applying a spectro-temporal gain may introduce undesired speech distortion and artifacts. Alternatively, an estimate of the interfering sources and background noise can be obtained by canceling the desired speech source using a blocking matrix, similarly to a generalized sidelobe canceler (GSC) structure (cf. Chapter 10, Figure 10.5). The common spectro-temporal gain can then be computed either directly using this interference and noise estimate (Reindl *et al.*, 2010a,b) or after compensating its power spectrum assuming the IC of the interference and noise field is known (Reindl *et al.*, 2013).

18.3.2 Paradigm 2: Binaural Spatial Filtering

In the second paradigm, all microphone signals from both hearing devices are processed by two different (complex-valued) spatial filters, generating two different output signals (cf. Figure 18.2). On the one hand, since more degrees of freedom are available in binaural spatial filtering than in binaural spectral postfiltering, typically more noise reduction and less speech distortion can be achieved. On the other hand, a tradeoff typically exists between the noise reduction performance and the preservation of the binaural cues for the desired speech source, the interfering sources, and the background noise.

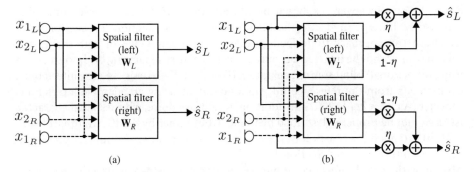

Figure 18.2 Block diagram for binaural spatial filtering: (a) incorporating constraints into spatial filter design and (b) mixing with scaled reference signals.

The first proposed binaural spatial filtering techniques were extensions of data-independent beamformers into a binaural version producing two output signals (Merks *et al.*, 1997), where Desloge *et al.* (1997) and Suzuki *et al.* (1999) incorporated additional constraints for binaural cue preservation of the desired speech source. As will be shown in Section 18.4, data-dependent multi-microphone noise reduction algorithms, such as the MVDR beamformer using relative transfer functions (RTFs) (cf. Section 10.4.4) and the MWF (cf. Section 10.4.3), can also be straightforwardly extended into a binaural version (Doclo *et al.*, 2010; Hadad *et al.*, 2015b) by estimating the speech component at a reference microphone on the left and the right hearing device. Cornelis *et al.* (2010) and van den Bogaert *et al.* (2008) showed both theoretically and using listening experiments that the binaural MWF (and hence also the binaural MVDR beamformer) preserves the binaural cues of the desired speech source, but distorts the binaural cues of the interfering sources and the background noise, such that all sources are perceived as coming from the direction of the speech source. Clearly, this is undesired since no binaural unmasking can be exploited by the auditory system and in some situations (e.g., traffic) this can even be dangerous.

To optimally benefit from binaural unmasking and to optimize the spatial awareness of the user, several extensions of the binaural MVDR beamformer and the binaural MWF have recently been proposed, which aim at also preserving the binaural cues of the interfering sources and/or the background noise. It has been proposed to either incorporate additional (soft or hard) constraints into the spatial filter design or to mix the output signals of a binaural noise reduction algorithm with a scaled version of the noisy reference microphone signals (cf. Figure 18.2 and Sections 18.5 and 18.6).

- **Incorporating constraints into spatial filter design** (Figure 18.2a): For a directional interfering source Hadad *et al.* (2016a) extended the binaural MVDR beamformer into the binaural linearly constrained minimum variance (LCMV) beamformer by incorporating an interference reduction constraint, which aims to partially suppress the interfering source while perfectly preserving the binaural cues of both the desired speech source and the interfering source. Instead of an interference reduction constraint, it is also possible to use an RTF preservation constraint for the interfering source (Hadad *et al.*, 2015b) or to use inequality constraints (Liao *et al.*, 2015). Moreover, as shown by Marquardt *et al.* (2015a), these constraints can also be incorporated into the binaural MWF cost function (cf. Section 18.6). An alternative approach to perfectly preserve the binaural cues of both the desired speech source and the interfering source consists of first separating both sources, e.g. using (directional) blind source separation (BSS), and then removing the interference components from the reference microphone signals using an adaptive interference canceler (Aichner *et al.*, 2007). In addition, for diffuse noise, whose spatial characteristics cannot be properly described by the RTF but rather by the IC, Marquardt *et al.* (2015b) extended the binaural MWF cost function with a term aiming at preserving the IC of diffuse noise (cf. Section 18.5.2).

- **Mixing with scaled reference signals** (Figure 18.2b): Instead of directly incorporating constraints into the spatial filter design, a simple – but rather generally applicable – alternative is to mix the output signals of a binaural noise reduction algorithm (preserving the binaural cues of the desired speech source) with scaled noisy reference microphone signals in the left and right hearing device. Due to mixing with the noisy microphone signals this will obviously result in a tradeoff between interference/noise reduction and binaural cue preservation of the interfering source and background noise (Cornelis *et al.*, 2010). Welker *et al.* (1997) proposed using the output signals of a GSC only for the higher frequencies (above 800 Hz), while using the noisy reference microphone signals for the lower frequencies in order to preserve the ITD cues of the desired speech source. Klasen *et al.* (2007) proposed mixing the output signals of the binaural MWF with scaled noisy reference microphone signals, which is referred to as the binaural MWF with partial noise estimation (cf. Section 18.5.1). While the original technique uses a frequency-independent mixing parameter, different approaches have recently been proposed to determine a frequency-dependent mixing parameter, based on either psychoacoustically motivated boundaries (Marquardt *et al.*, 2015c) (cf. Section 18.5.3) or the output SNR (Thiemann *et al.*, 2016).

18.4 The Binaural Noise Reduction Problem

After presenting the considered acoustic scenario in Section 18.4.1 and mathematically defining several performance measures and the binaural cues in Section 18.4.2, Section 18.4.3 discusses binaural versions of two well-known multi-microphone noise reduction algorithms, namely the MVDR beamformer and the MWF.

18.4.1 Acoustic Scenario and Signal Definitions

We consider an acoustic scenario with one desired source and one interfering source in a noisy and reverberant environment, cf. Figure 18.3. The source signals are received by two hearing devices consisting of a microphone array with I_L microphones on the left hearing device and I_R microphones on the right hearing device, such that $I = I_L + I_R$ is the total number of microphones. The indices $i_L \in \{1, \dots, I_L\}$ and $i_R \in \{1, \dots, I_R\}$ denote the microphone index of the left and the right hearing devices, respectively (compare with the definitions in Chapters 3 and 10).

Figure 18.3 Considered acoustic scenario, consisting of a desired source (s_1), an interfering source (s_2), and background noise in a reverberant room. The signals are received by the microphones on both hearing devices of the binaural hearing system.

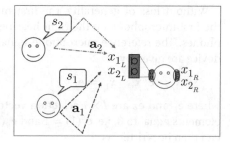

The received microphone signals in the short-time Fourier transform (STFT) domain can be represented by an $I \times 1$ vector $\mathbf{x}(n,f)$, i.e.

$$\mathbf{x}(n,f) = \begin{bmatrix} x_{1_L}(n,f) \\ \vdots \\ x_{I_L}(n,f) \\ x_{1_R}(n,f) \\ \vdots \\ x_{I_R}(n,f) \end{bmatrix}, \tag{18.1}$$

which can be decomposed as

$$\mathbf{x}(n,f) = \mathbf{c}_1(n,f) + \mathbf{c}_2(n,f) + \mathbf{u}(n,f) = \mathbf{c}_1(n,f) + \mathbf{v}(n,f), \tag{18.2}$$

where $\mathbf{c}_1(n,f)$ and $\mathbf{c}_2(n,f)$ are the spatial images of the desired source and the interfering source, respectively, $\mathbf{u}(n,f)$ is the background noise component and $\mathbf{v}(n,f) = \mathbf{c}_2(n,f) + \mathbf{u}(n,f)$ is the overall noise component. Henceforth, the frame index n and the frequency bin index f will be omitted for the sake of brevity.

The spatial images $\mathbf{c}_1$ and $\mathbf{c}_2$ can be written as

$$\mathbf{c}_1 = s_1\mathbf{a}_1, \qquad \mathbf{c}_2 = s_2\mathbf{a}_2, \tag{18.3}$$

where s_1 and s_2 denote the desired and interfering source signals, respectively, and $\mathbf{a}_1$ and $\mathbf{a}_2$ denote the vector of acoustic transfer functions between the I microphones and the desired and interfering sources, respectively.

The covariance matrices of the desired source component $\boldsymbol{\Sigma}_{\mathbf{c}_1}$, the interfering source component $\boldsymbol{\Sigma}_{\mathbf{c}_2}$, and the background noise component $\boldsymbol{\Sigma}_{\mathbf{u}}$ are given by

$$\boldsymbol{\Sigma}_{\mathbf{c}_1} = \mathbb{E}\{\mathbf{c}_1\mathbf{c}_1^H\} = \sigma_1^2\mathbf{a}_1\mathbf{a}_1^H, \quad \boldsymbol{\Sigma}_{\mathbf{c}_2} = \mathbb{E}\{\mathbf{c}_2\mathbf{c}_2^H\} = \sigma_2^2\mathbf{a}_2\mathbf{a}_2^H, \tag{18.4}$$

$$\boldsymbol{\Sigma}_{\mathbf{u}} = \mathbb{E}\{\mathbf{u}\mathbf{u}^H\}, \tag{18.5}$$

where $\sigma_1^2 = \mathbb{E}\{|s_1|^2\}$ and $\sigma_2^2 = \mathbb{E}\{|s_2|^2\}$ denote the power spectrum of the desired source and the interfering source, respectively. Assuming statistical independence between the components in (18.2), the covariance matrix of the received microphone signals $\boldsymbol{\Sigma}_{\mathbf{x}} = \mathbb{E}\{\mathbf{x}\mathbf{x}^H\}$ can be written as

$$\boldsymbol{\Sigma}_{\mathbf{x}} = \underbrace{\boldsymbol{\Sigma}_{\mathbf{c}_1} + \boldsymbol{\Sigma}_{\mathbf{c}_2} + \boldsymbol{\Sigma}_{\mathbf{u}}}_{\boldsymbol{\Sigma}_{\mathbf{v}}}, \tag{18.6}$$

with $\boldsymbol{\Sigma}_{\mathbf{v}} = \mathbb{E}\{\mathbf{v}\mathbf{v}^H\}$ the covariance matrix of the overall noise component.

Without loss of generality, the first microphone on the left hearing device 1_L and the first microphone on the right hearing device 1_R are chosen as the reference microphones. The reference microphone signals x_{1_L} and x_{1_R} of the left and the right hearing device are given by

$$x_{1_L} = \mathbf{e}_L^T \mathbf{x}, \quad x_{1_R} = \mathbf{e}_R^T \mathbf{x}, \tag{18.7}$$

where $\mathbf{e}_L$ and $\mathbf{e}_R$ are $I \times 1$ selection vectors with one element equal to 1 and all other elements equal to 0, i.e. $\mathbf{e}_L(1) = 1$ and $\mathbf{e}_R(I_L + 1) = 1$. The reference microphone signals can then be written as

$$x_{1_L} = s_1 a_{1_L,1} + s_2 a_{1_L,2} + u_{1_L}, \quad x_{1_R} = s_1 a_{1_R,1} + s_2 a_{1_R,2} + u_{1_R}, \tag{18.8}$$

where $a_{1_L,1} = \mathbf{e}_L^T \mathbf{a}_1$, $a_{1_L,2} = \mathbf{e}_L^T \mathbf{a}_2$, $u_{1_L} = \mathbf{e}_L^T \mathbf{u}$ and $a_{1_R,1} = \mathbf{e}_R^T \mathbf{a}_1$, $a_{1_R,2} = \mathbf{e}_R^T \mathbf{a}_2$, $u_{1_R} = \mathbf{e}_R^T \mathbf{u}$.

The output signals of the left and the right hearing device $\hat{s}_L$ and $\hat{s}_R$ are obtained by applying two different beamformers to the microphone signals of both hearing devices (cf. Figure 18.2a), i.e.

$$\hat{s}_L = \mathbf{w}_L^H \mathbf{x}, \quad \hat{s}_R = \mathbf{w}_R^H \mathbf{x}, \tag{18.9}$$

where $\mathbf{w}_L$ and $\mathbf{w}_R$ are $I \times 1$ complex-valued filter vectors for the left and the right hearing devices, respectively. Furthermore, we define the $2I \times 1$ stacked filter vector $\mathbf{w}$ as

$$\mathbf{w} = \begin{bmatrix} \mathbf{w}_L \\ \mathbf{w}_R \end{bmatrix}. \tag{18.10}$$

18.4.2 Performance Measures and Binaural Cues

In this section we define several performance measures and the relevant binaural cues for the coherent sources and the background noise.

The binaural speech distortion is defined as the ratio of the average input power spectrum of the desired source component in the reference microphones and the average output power spectrum of the desired source component, i.e.

$$SD = \frac{\sigma_1^2 |a_{1_L,1}|^2 + \sigma_1^2 |a_{1_R,1}|^2}{\mathbf{w}_L^H \mathbf{\Sigma}_{\mathbf{c}_1} \mathbf{w}_L + \mathbf{w}_R^H \mathbf{\Sigma}_{\mathbf{c}_1} \mathbf{w}_R}. \tag{18.11}$$

The output SNR in the left and the right hearing device is defined as the ratio of the output power spectrum of the desired source component and the background noise component, i.e.

$$SNR_L^{out} = \frac{\mathbf{w}_L^H \mathbf{\Sigma}_{\mathbf{c}_1} \mathbf{w}_L}{\mathbf{w}_L^H \mathbf{\Sigma}_{\mathbf{u}} \mathbf{w}_L}, \quad SNR_R^{out} = \frac{\mathbf{w}_R^H \mathbf{\Sigma}_{\mathbf{c}_1} \mathbf{w}_R}{\mathbf{w}_R^H \mathbf{\Sigma}_{\mathbf{u}} \mathbf{w}_R}. \tag{18.12}$$

The binaural output SNR is defined as the ratio of the average output power spectrum of the desired source component and the average output power spectrum of the background noise component in the left and the right hearing devices, i.e.

$$SNR^{out} = \frac{\mathbf{w}_L^H \mathbf{\Sigma}_{\mathbf{c}_1} \mathbf{w}_L + \mathbf{w}_R^H \mathbf{\Sigma}_{\mathbf{c}_1} \mathbf{w}_R}{\mathbf{w}_L^H \mathbf{\Sigma}_{\mathbf{u}} \mathbf{w}_L + \mathbf{w}_R^H \mathbf{\Sigma}_{\mathbf{u}} \mathbf{w}_R}. \tag{18.13}$$

Similarly, the binaural output signal-to-interference-plus-noise ratio (SINR) is defined as the ratio of the average output power spectrum of the desired source component and the average output power spectrum of the overall noise component (interfering source plus background noise) in the left and the right hearing devices, i.e.

$$\text{SINR}^{\text{out}} = \frac{\mathbf{w}_L^H \mathbf{\Sigma}_{\mathbf{c}_1} \mathbf{w}_L + \mathbf{w}_R^H \mathbf{\Sigma}_{\mathbf{c}_1} \mathbf{w}_R}{\mathbf{w}_L^H \mathbf{\Sigma}_{\mathbf{v}} \mathbf{w}_L + \mathbf{w}_R^H \mathbf{\Sigma}_{\mathbf{v}} \mathbf{w}_R}. \tag{18.14}$$

For coherent sources the binaural cues, i.e. the ILD and the ITD, can be computed from the RTF (Cornelis *et al.*, 2010), cf. (3.14) and (3.12). The input RTFs of the desired source and the interfering source between the reference microphones of the left and the right hearing devices are defined as the ratio of the acoustic transfer functions, i.e.

$$\tilde{a}_1^{\text{in}} = \frac{a_{1_L,1}}{a_{1_R,1}}, \qquad \tilde{a}_2^{\text{in}} = \frac{a_{1_L,2}}{a_{1_R,2}}. \tag{18.15}$$

The output RTFs of the desired source and the interfering source are defined as the ratio of the filtered acoustic transfer functions of the left and the right hearing devices, i.e.

$$\tilde{a}_1^{\text{out}} = \frac{\mathbf{w}_L^H \mathbf{a}_1}{\mathbf{w}_R^H \mathbf{a}_1}, \qquad \tilde{a}_2^{\text{out}} = \frac{\mathbf{w}_L^H \mathbf{a}_2}{\mathbf{w}_R^H \mathbf{a}_2}. \tag{18.16}$$

For incoherent background noise, the binaural spatial characteristics can be better described by the IC. The input IC of the background noise is defined as the normalized cross-correlation between the noise components in the reference microphone signals, i.e.

$$\text{IC}_{\mathbf{u}}^{\text{in}} = \frac{\mathbf{e}_L^T \mathbf{\Sigma}_{\mathbf{u}} \mathbf{e}_R}{\sqrt{(\mathbf{e}_L^T \mathbf{\Sigma}_{\mathbf{u}} \mathbf{e}_L)(\mathbf{e}_R^T \mathbf{\Sigma}_{\mathbf{u}} \mathbf{e}_R)}}. \tag{18.17}$$

The output IC of the background noise is defined as the normalized cross-correlation between the noise components in the output signals, i.e.

$$\text{IC}_{\mathbf{u}}^{\text{out}} = \frac{\mathbf{w}_L^H \mathbf{\Sigma}_{\mathbf{u}} \mathbf{w}_R}{\sqrt{(\mathbf{w}_L^H \mathbf{\Sigma}_{\mathbf{u}} \mathbf{w}_L)(\mathbf{w}_R^H \mathbf{\Sigma}_{\mathbf{u}} \mathbf{w}_R)}}. \tag{18.18}$$

The *magnitude squared coherence* (MSC) is defined as the squared absolute value of the IC, i.e.

$$\text{MSC} = |\text{IC}|^2. \tag{18.19}$$

18.4.3 Binaural MWF and Binaural MVDR Beamformer

Similarly to the MWF discussed in Section 10.4.3, the binaural MWF (Doclo *et al.*, 2010; Cornelis *et al.*, 2010) produces a minimum mean square error (MMSE) estimate of the desired source component in the reference microphone signals of both hearing devices. The binaural MWF cost functions to estimate the desired source components $c_{1_L,1}$ and $c_{1_R,1}$ in the left and the right hearing devices are given by

$$C^{\text{MWF},L}(\mathbf{w}_L) = \mathbb{E}\{|c_{1_L,1} - \mathbf{w}_L^H \mathbf{c}_1|^2 + \mu|\mathbf{w}_L^H \mathbf{v}|^2\}, \tag{18.20}$$

$$C^{\text{MWF},R}(\mathbf{w}_R) = \mathbb{E}\{|c_{1_R,1} - \mathbf{w}_R^H \mathbf{c}_1|^2 + \mu|\mathbf{w}_R^H \mathbf{v}|^2\}, \tag{18.21}$$

where the weighting parameter $\mu \geq 0$ allows for a tradeoff between interference/noise reduction and speech distortion. The filter vectors minimizing (18.20) and (18.21) are given by (Cornelis *et al.*, 2010)

$$\mathbf{w}_{\text{MWF},L} = \sigma_1^2 a_{1_L,1}^* \overline{\boldsymbol{\Sigma}}_{\mathbf{x}}^{-1} \mathbf{a}_1, \quad \mathbf{w}_{\text{MWF},R} = \sigma_1^2 a_{1_R,1}^* \overline{\boldsymbol{\Sigma}}_{\mathbf{x}}^{-1} \mathbf{a}_1, \tag{18.22}$$

where

$$\overline{\boldsymbol{\Sigma}}_{\mathbf{x}} = \boldsymbol{\Sigma}_{\mathbf{c}_1} + \mu \boldsymbol{\Sigma}_{\mathbf{v}} \tag{18.23}$$

is defined as the speech-distortion-weighted covariance matrix. As shown by Doclo *et al.* (2010) and similar to the derivations in Section 10.6, the binaural MWF can be decomposed into a binaural MVDR beamformer (using RTFs) and a single-channel postfilter applied to the output of the binaural MVDR beamformer, i.e.

$$\mathbf{w}_{\text{MWF},L} = \underbrace{\frac{\rho}{\mu + \rho} \frac{\boldsymbol{\Sigma}_{\mathbf{v}}^{-1} \mathbf{a}_1}{\mathbf{a}_1^H \boldsymbol{\Sigma}_{\mathbf{v}}^{-1} \mathbf{a}_1} a_{1_L,1}^*}_{\mathbf{w}_{\text{MVDR},L}}, \quad \mathbf{w}_{\text{MWF},R} = \underbrace{\frac{\rho}{\mu + \rho} \frac{\boldsymbol{\Sigma}_{\mathbf{v}}^{-1} \mathbf{a}_1}{\mathbf{a}_1^H \boldsymbol{\Sigma}_{\mathbf{v}}^{-1} \mathbf{a}_1} a_{1_R,1}^*}_{\mathbf{w}_{\text{MVDR},R}}, \tag{18.24}$$

with

$$\rho = \sigma_1^2 \mathbf{a}_1^H \boldsymbol{\Sigma}_{\mathbf{v}}^{-1} \mathbf{a}_1. \tag{18.25}$$

The filter vectors $\mathbf{w}_{\text{MVDR},L}$ and $\mathbf{w}_{\text{MVDR},R}$ are the solutions of the binaural MVDR optimization problem, minimizing the overall output noise power subject to distortionless constraints for both hearing devices, i.e.

$$\underset{\mathbf{w}_L}{\text{argmin}} \; \mathbf{w}_L^H \boldsymbol{\Sigma}_{\mathbf{v}} \mathbf{w}_L \quad \text{s.t.} \quad \mathbf{w}_L^H \mathbf{a}_1 = a_{1_L,1}, \tag{18.26}$$

$$\underset{\mathbf{w}_R}{\text{argmin}} \; \mathbf{w}_R^H \boldsymbol{\Sigma}_{\mathbf{v}} \mathbf{w}_R \quad \text{s.t.} \quad \mathbf{w}_R^H \mathbf{a}_1 = a_{1_R,1}. \tag{18.27}$$

The binaural output SINR of the binaural MWF (and the binaural MVDR beamformer) can be calculated by substituting (18.24) in (18.14) and is equal to (Cornelis *et al.*, 2010)

$$\text{SINR}_{\text{MWF}}^{\text{out}} = \rho. \tag{18.28}$$

It is important to note that the filter vectors of the left and the right hearing device in (18.24) are related as $\mathbf{w}_{\text{MWF},L} = (\tilde{a}_1^{\text{in}})^* \mathbf{w}_{\text{MWF},R}$, implying that $\mathbf{w}_{\text{MWF},L}$ and $\mathbf{w}_{\text{MWF},R}$ are parallel. Hence, the output RTFs of the desired source and the interfering source are both equal to the input RTF of the desired source, i.e.

$$\tilde{a}_1^{\text{out}} = \frac{a_{1_L,1}}{a_{1_R,1}} = \tilde{a}_1^{\text{in}}, \quad \tilde{a}_2^{\text{out}} = \frac{a_{1_L,1}}{a_{1_R,1}} = \tilde{a}_1^{\text{in}}. \tag{18.29}$$

This implies that both output components are perceived as directional sources coming from the direction of the desired source, which is obviously not desired. Furthermore, by substituting (18.24) in (18.18) and (18.19), it has been shown by Marquardt *et al.* (2015b) that the output MSC of the background noise component of the binaural MWF is equal to 1. This implies that the MSC of the background noise, which is typically frequency dependent, is not preserved and the background noise will be perceived as a directional source coming from the direction of the desired source, such that no binaural unmasking can be exploited (cf. Section 18.2).

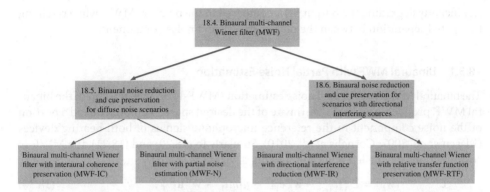

Figure 18.4 Schematic overview of this chapter.

In the remainder of this chapter we will focus on MWF-based binaural noise reduction techniques. Since the binaural MWF distorts the binaural cues of both the interfering source and the background noise component, in the next sections we will present extensions for different scenarios (see schematic overview in Figure 18.4). For diffuse noise, Section 18.5 presents two extensions of the binaural MWF, aiming to preserve the IC of the background noise based on psychoacoustically motivated boundaries. For interfering sources, Section 18.6 presents two extensions of the binaural MWF, aiming to preserve the ILD and ITD of the interfering sources, either by controlling the amount of interference reduction in both hearing devices or by imposing an RTF preservation constraint.

18.5   Extensions for Diffuse Noise

In this section we assume an acoustic scenario with one desired speech source and no directional interfering source, i.e. $\mathbf{v} = \mathbf{u}$, and we assume that the background noise is diffuse. Since the binaural MWF is not able to preserve the IC of the noise component (cf. Section 18.4.3), in this section we present two extensions, namely the binaural MWF with partial noise estimation (MWF-N) and the binaural MWF with IC preservation (MWF-IC), which both aim at preserving the IC of the noise component. While the MWF-N is a rather general approach for preserving the binaural cues of the noise component, the MWF-IC is specifically designed to preserve the IC in diffuse noise. We will show that for both algorithms a tradeoff between IC preservation and output SNR exists, depending on the selection of a tradeoff parameter. To determine this tradeoff parameter, we define frequency-dependent lower and upper boundaries for the MSC of the output noise component based on psychoacoustical data evaluating the IC discrimination ability of the human auditory system. In this way an optimal tradeoff between noise reduction performance and preservation of the spatial impression of a diffuse noise field can be obtained.

The performance of the binaural MWF and both extensions is evaluated in terms of objective performance measures. Experimental results show that incorporating psychoacoustically motivated MSC boundaries to determine the tradeoff parameters for the MWF-N and the MWF-IC yields a controllable IC/MSC preservation without significantly degrading the output SNR compared to the binaural MWF, while retaining the spatial separation between the output speech and noise component.

18.5.1 Binaural MWF with Partial Noise Estimation

The binaural MWF with partial noise estimation (MWF-N) is an extension of the binaural MWF, producing an MMSE estimate of the desired speech component and a portion of the noise component in the reference microphone signals of both hearing devices (Klasen *et al.*, 2007; Cornelis *et al.*, 2010). Similarly to (18.20) and (18.21), the MWF-N cost functions for the left and right hearing devices are given by

$$C^{\text{MWF-N},L}(\mathbf{w}_L) = \mathbb{E}\left\{|c_{1_L,1} - \mathbf{w}_L^H \mathbf{c}_1|^2 + \mu|\eta u_{1_L} - \mathbf{w}_L^H \mathbf{u}|^2\right\}, \tag{18.30}$$

$$C^{\text{MWF-N},R}(\mathbf{w}_R) = \mathbb{E}\left\{|c_{1_R,1} - \mathbf{w}_R^H \mathbf{c}_1|^2 + \mu|\eta u_{1_R} - \mathbf{w}_R^H \mathbf{u}|^2\right\}, \tag{18.31}$$

where the parameter η, with $0 \leq \eta \leq 1$, enables a tradeoff between noise reduction and preservation of the binaural cues of the noise component. If $\eta = 0$, the MWF-N cost functions reduce to the binaural MWF cost functions in (18.20) and (18.21). The filter vectors minimizing (18.30) and (18.31) are given by (Cornelis *et al.*, 2010)

$$\mathbf{w}_{\text{MWF-N},L} = (1 - \eta)\mathbf{w}_{\text{MWF},L} + \eta\mathbf{e}_L, \tag{18.32}$$

$$\mathbf{w}_{\text{MWF-N},R} = (1 - \eta)\mathbf{w}_{\text{MWF},R} + \eta\mathbf{e}_R. \tag{18.33}$$

Hence, the output signals of the MWF-N are equal to the sum of the output signals of the binaural MWF (weighted with $1 - \eta$) and the noisy reference microphone signals (weighted with η), cf. Figure 18.2b. Cornelis *et al.* (2010) showed that for $\eta > 0$ the MWF-N yields a lower speech distortion compared to the MWF, which can be intuitively explained by the fact that the mixing of the output speech components of the binaural MWF with the input speech components of the reference microphone signals partially compensates the speech distortion introduced by the postfilter in (18.24). The output SNR of the MWF-N can be calculated by substituting (18.32) and (18.33) in (18.12), yielding (Cornelis *et al.*, 2010)

$$\text{SNR}^{\text{out}}_{\text{MWF-N},L} = \rho \frac{1}{1 + \eta^2 \left(\frac{\mu+\rho}{\eta\mu+\rho}\right)^2 (\Delta\text{SNR}_{\text{MWF},L} - 1)}, \tag{18.34}$$

$$\text{SNR}^{\text{out}}_{\text{MWF-N},R} = \rho \frac{1}{1 + \eta^2 \left(\frac{\mu+\rho}{\eta\mu+\rho}\right)^2 (\Delta\text{SNR}_{\text{MWF},R} - 1)}, \tag{18.35}$$

with

$$\Delta\text{SNR}_{\text{MWF},L} = \rho \frac{\mathbf{e}_L^T \boldsymbol{\Sigma}_{\mathbf{u}} \mathbf{e}_L}{\mathbf{e}_L^T \boldsymbol{\Sigma}_{\mathbf{c}_1} \mathbf{e}_L}, \qquad \Delta\text{SNR}_{\text{MWF},R} = \rho \frac{\mathbf{e}_R^T \boldsymbol{\Sigma}_{\mathbf{u}} \mathbf{e}_R}{\mathbf{e}_R^T \boldsymbol{\Sigma}_{\mathbf{c}_1} \mathbf{e}_R}, \tag{18.36}$$

the SNR improvement of the binaural MWF in the left and the right hearing devices, respectively. Since the SNR improvement of the MWF is always larger than or equal to 1 (Doclo and Moonen, 2005), (18.34) and (18.35) imply that the output SNR of the MWF-N is always lower than or equal to the output SNR of the binaural MWF, which again can be intuitively explained by the mixing of the output signals of the binaural MWF with the reference microphone signals. Cornelis *et al.* (2010) showed that for the

MWF-N the binaural cues of the speech component are preserved for all values of the tradeoff parameter η, i.e.

$$\tilde{a}_1^{\text{out}} = \frac{(1-\eta)\frac{\rho}{\mu+\rho}a_{1_L,1} + \eta a_{1_L,1}}{(1-\eta)\frac{\rho}{\mu+\rho}a_{1_R,1} + \eta a_{1_R,1}} = \frac{a_{1_L,1}}{a_{1_R,1}} = \tilde{a}_1^{\text{in}}. \tag{18.37}$$

Substituting (18.32) and (18.33) in (18.18) and (18.19), the output MSC of the noise component of the MWF-N can be calculated as (Marquardt *et al.*, 2015c; Marquardt, 2016)

$$\text{MSC}_{\mathbf{u}}^{\text{out}} = \frac{|\psi \mathbf{e}_L^T \boldsymbol{\Sigma}_{\mathbf{c}_1} \mathbf{e}_R + \eta^2 \mathbf{e}_L^T \boldsymbol{\Sigma}_{\mathbf{u}} \mathbf{e}_R|^2}{(\psi \mathbf{e}_L^T \boldsymbol{\Sigma}_{\mathbf{c}_1} \mathbf{e}_L + \eta^2 \mathbf{e}_L^T \boldsymbol{\Sigma}_{\mathbf{u}} \mathbf{e}_L)(\psi \mathbf{e}_R^T \boldsymbol{\Sigma}_{\mathbf{c}_1} \mathbf{e}_R + \eta^2 \mathbf{e}_R^T \boldsymbol{\Sigma}_{\mathbf{u}} \mathbf{e}_R)}, \tag{18.38}$$

with

$$\psi = (1-\eta)^2 \frac{\rho}{(\mu+\rho)^2} + 2\eta(1-\eta)\frac{1}{\mu+\rho}. \tag{18.39}$$

As already shown by Marquardt *et al.* (2015b), for the binaural MWF, i.e. $\eta = 0$, the output MSC of the noise component is equal to 1. On the other hand, for $\eta = 1$, the output MSC of the noise component is equal to the input MSC of the noise component, but obviously no noise reduction is achieved, cf. (18.34) and (18.35). Hence, for the MWF-N a substantial tradeoff between preservation of the IC/MSC of the noise component and noise reduction performance exists. In order to achieve a psychoacoustically optimized tradeoff, in Section 18.5.3 a method is described to determine the (frequency-dependent) tradeoff parameter η based on the (frequency-dependent) IC discrimination ability of the human auditory system.

18.5.2 Binaural MWF with Interaural Coherence Preservation

While the MWF-N is a rather general approach for preserving the binaural cues of the noise component, in this section we present another extension of the binaural MWF, aimed at specifically preserving the IC in diffuse noise. Marquardt *et al.* (2015b) proposed to extend the binaural MWF cost function with an IC preservation term for the noise component, i.e.

$$C^{\text{IC}}(\mathbf{w}) = |\text{IC}_{\mathbf{u}}^{\text{out}} - \text{IC}_{\mathbf{u}}^{\text{tgt}}|^2 \tag{18.40}$$

$$= \left| \frac{\mathbf{w}_L^H \boldsymbol{\Sigma}_{\mathbf{u}} \mathbf{w}_R}{\sqrt{(\mathbf{w}_L^H \boldsymbol{\Sigma}_{\mathbf{u}} \mathbf{w}_L)(\mathbf{w}_R^H \boldsymbol{\Sigma}_{\mathbf{u}} \mathbf{w}_R)}} - \text{IC}_{\mathbf{u}}^{\text{tgt}} \right|^2, \tag{18.41}$$

where $\text{IC}_{\mathbf{u}}^{\text{tgt}}$ represents the desired output IC for the noise component. The desired output IC can, for example, be chosen to be equal to the input IC of the noise component or can be defined based on models of the IC in diffuse noise fields, cf. Section 18.2.

The total MWF-IC cost function is given by (Marquardt *et al.*, 2015b)

$$C^{\text{MWF-IC}}(\mathbf{w}) = C^{\text{MWF}}(\mathbf{w}) + \lambda C^{\text{IC}}(\mathbf{w}), \tag{18.42}$$

with $C^{\text{MWF}}(\mathbf{w}) = C^{\text{MWF},L}(\mathbf{w}_L) + C^{\text{MWF},R}(\mathbf{w}_R)$. Similarly to the MWF-N, a tradeoff between noise reduction and IC preservation arises for the MWF-IC, which can be controlled by the tradeoff parameter λ. Since no closed-form expression is available for the filter vector $\mathbf{w}_{\text{MWF-IC}}$ minimizing the nonlinear cost function in (18.42), an iterative numerical optimization method is required (Marquardt *et al.*, 2015b). Therefore, contrary to the MWF-N, for the MWF-IC it is unfortunately not possible to derive closed-form expressions to investigate the impact of the tradeoff parameter λ on the noise reduction, speech distortion, and MSC preservation performance. However, simulation results show that – similarly to the MWF-N – increasing the tradeoff parameter λ leads to a better preservation of the IC/MSC of the noise component at the cost of a decreased noise reduction performance (Marquardt *et al.*, 2015b). In addition, these simulation results show that the binaural cues of the output speech component of the MWF-IC are well preserved.

18.5.3 Psychoacoustically Optimized Tradeoff Parameters

Since for both the MWF-N and the MWF-IC a substantial tradeoff between IC preservation of the noise component and the output SNR exists, the selection of the tradeoff parameters η and λ is quite important. Marquardt *et al.* (2015b) proposed limiting the possible solutions of the optimization problems in (18.30), (18.31) and (18.42) by imposing a constraint on the output MSC of the noise component by means of frequency-dependent lower and upper boundaries γ_{min} and γ_{max}. These boundaries can, for example, be defined based on subjective listening experiments evaluating the IC discrimination ability of the auditory system in diffuse noise (Tohyama and Suzuki, 1989; Walther and Faller, 2013). For frequencies below 500 Hz, the boundaries γ_{min} and γ_{max} are chosen to be a function of the desired MSC, while for frequencies above 500 Hz a fixed lower and upper boundary $\gamma_{\text{min}} = 0$ and $\gamma_{\text{max}} = 0.36$ is chosen, cf. Figure 18.5.

For the MWF-IC, an exhaustive search is required to determine the optimal tradeoff parameter λ such that the output MSC is equal to the upper boundary γ_{max}, achieving an optimal tradeoff between MSC preservation and noise reduction (Marquardt *et al.*, 2015b). For the MWF-N, the optimal tradeoff parameter η can in principle be determined by solving (18.38) for $\text{MSC}_{\mathbf{u}}^{\text{out}} = \gamma_{\text{max}}$. Although no general closed-form expression for the optimal tradeoff parameter η has been found, for two important special cases

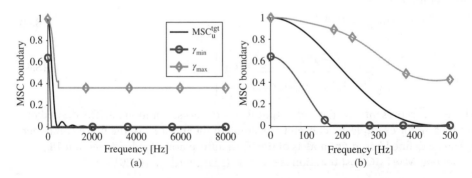

Figure 18.5 Psychoacoustically motivated lower and upper MSC boundaries: (a) frequency range 0–8000 Hz and (b) frequency range 0–500 Hz. For frequencies below 500 Hz, the boundaries depend on the desired MSC while for frequencies above 500 Hz the boundaries are independent of the desired MSC.

closed-form expressions have been derived. For $\mu = 0$, corresponding to the binaural MVDR beamformer, a closed-form expression has been derived by Marquardt (2016), while for a desired speech source in front of the listener a closed-form expression has been derived by Marquardt *et al.* (2015c).

18.5.4 Experimental Results

In this section we compare the performance of the binaural MWF, MWF-N, and MWF-IC for a speech source in a cafeteria scenario. Binaural behind-the-ear impulse responses measured on an artificial head in a cafeteria (Kayser *et al.*, 2009) were used to generate the speech component in the microphone signals. Each hearing device was equipped with two microphones, therefore in total $I = 4$ microphone signals were available. The desired speech source was located at positions of $0°$ and $-35°$ and distances of 102 cm and 117.5 cm, respectively. Recorded ambient noise from the cafeteria, including babble noise, clacking plates, and occasionally interfering speakers, was added to the speech component at an intelligibility-weighted SNR (SNR_{int}) (Greenberg *et al.*, 1993) at the left hearing device of 0 dB. The speech signals were generated using 10 sentences from the HINT dataset (Nilsson *et al.*, 1994). The signals were processed at $f_s = 16$ kHz using a weighted overlap-add framework with a block size of 512 samples and an overlap of 50% between successive blocks. The covariance matrix of the noise component Σ_u was computed during a 2 s noise-only initialization phase and the covariance matrix of the microphone signals Σ_x was computed during the duration of the 10 HINT sentences. The covariance matrix of the desired speech component was estimated as $\Sigma_{c_1} = \Sigma_x - \Sigma_u$, where a rank-1 approximation of Σ_{c_1} was used. The desired IC was calculated using the acoustic transfer functions of the anechoic behind-the-ear impulse response from the same dataset as

$$
\text{IC}_u^{\text{tgt}} = \frac{\sum_{j=1}^{J} a_{1_{L,j}}^{\text{dir}} a_{1_{R,j}}^{\text{dir}}}{\sqrt{\sum_{j=1}^{J} |a_{1_{L,j}}^{\text{dir}}|^2 \sum_{j=1}^{J} |a_{1_{R,j}}^{\text{dir}}|^2}}, \tag{18.43}
$$

where $a_{1_{L,j}}^{\text{dir}}$ and $a_{1_{R,j}}^{\text{dir}}$ denote the anechoic acoustic transfer functions for the jth angle in the dataset for the reference microphone on the left and right hearing devices, respectively, and J denotes the total number of angles ($J = 72$). The weighting parameter μ was set to 1 for all algorithms. As objective performance measures we have used the output SNR_{int} (in the left and right hearing devices), the intelligibility-weighted speech distortion (averaged over both output signals), and the global absolute MSC error of the noise component (averaged over all frequencies).

The results are depicted in Figure 18.6. It can be observed that the binaural MWF yields a very large MSC error, which is significantly smaller for the MWF-N and the MWF-IC. For both speech source positions, the MSC error is very similar for the MWF-N and the MWF-IC, which implies that for both algorithms a suitable tradeoff parameter yielding a predefined output MSC of the noise component can be determined. In terms of speech distortion, the MWF-IC shows the best performance, while the MWF-N performs slightly better than the MWF, as discussed in Section 18.5.1. The impact of preserving the MSC of the noise component on the output SNR_{int} is depicted in the lower half of Figure 18.6. As expected, the binaural MWF yields the largest

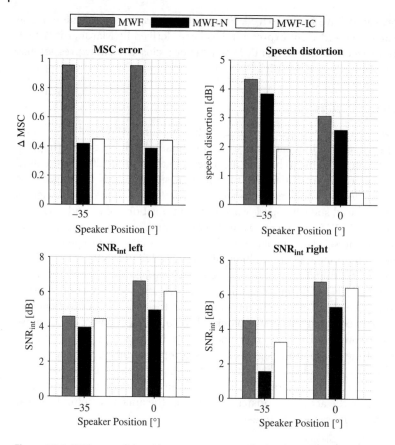

Figure 18.6 MSC error of the noise component, intelligibility-weighted speech distortion, and intelligibility-weighted output SNR for the MWF, MWF-N, and MWF-IC.

output SNR_{int} for both speech source positions, while the MWF-IC achieves a larger output SNR_{int} compared to the MWF-N. This can possibly be explained by the fact that the MWF-IC is specifically designed for IC preservation in diffuse noise, resulting in a better tradeoff between noise reduction, speech distortion, and MSC preservation compared to the more general MWF-N.

Since the explanatory power of objective performance measures on speech intelligibility and spatial quality of noise reduction algorithms is rather limited, a subjective listening test evaluating speech intelligibility and spatial quality has been conducted (Marquardt, 2016). In terms of speech intelligibility, it has been shown that the MWF-IC achieves a slightly better performance than the MWF, while the MWF-N performs slightly worse than the MWF. In terms of spatial quality, both the MWF-N and the MWF-IC outperform the MWF, while the MWF-N generally performs better than the MWF-IC. Hence, it can be concluded that for the MWF-N and the MWF-IC the loss in noise reduction performance is compensated for by the partial preservation of the binaural hearing advantage due to the preservation of the MSC of the noise component (cf. Section 18.2). This leads to a comparable speech intelligibility and a better spatial quality compared to the binaural MWF.

18.6 Extensions for Interfering Sources

In many acoustic scenarios, not only background noise but also a directional interfering source (e.g., a competing speaker) is present. Since the binaural MWF is not able to preserve the binaural cues of the directional interfering source (cf. Section 18.4.3), in this section we present two extensions of the binaural MWF (Marquardt *et al.*, 2015a; Hadad *et al.*, 2016b), namely the binaural MWF with interference RTF constraint (MWF-RTF) and the binaural MWF with interference reduction constraint (MWF-IR). In addition to minimizing the overall noise output power and limiting speech distortion, these extensions aim at preserving the binaural cues of the interfering source. For both algorithms closed-form expressions for the binaural output SINR and the binaural cues of the desired and the interfering source will be presented. This theoretical analysis is then validated using acoustic transfer functions measured on a binaural hearing device and using noisy speech signals in a reverberant environment.

18.6.1 Binaural MWF with Interference RTF Constraint

The MWF-RTF aims at preserving the binaural cues of the interfering source by adding an RTF preservation constraint to the binaural MWF cost function (Marquardt *et al.*, 2015a), i.e.

$$\min_{\mathbf{w}} C^{\mathrm{MWF}}(\mathbf{w}) \quad \text{s.t.} \quad \tilde{a}_2^{\mathrm{out}} = \frac{\mathbf{w}_L^H \mathbf{a}_2}{\mathbf{w}_R^H \mathbf{a}_2} = \frac{a_{1_L,2}}{a_{1_R,2}} = \tilde{a}_2^{\mathrm{in}}. \tag{18.44}$$

Using (18.10), the RTF preservation constraint can be written as $\mathbf{w}^H \mathbf{C} = 0$, with

$$\mathbf{C} = \begin{bmatrix} \mathbf{a}_2 \\ \alpha \mathbf{a}_2 \end{bmatrix}, \qquad \alpha = -\frac{a_{1_L,2}}{a_{1_R,2}}. \tag{18.45}$$

The filter vectors solving (18.44) are given by (Marquardt *et al.*, 2015a)

$$\mathbf{w}_{\mathrm{MWF\text{-}RTF},L} = \mathbf{w}_{\mathrm{MWF},L} - \kappa \, \overline{\boldsymbol{\Sigma}}_{\mathbf{x}}^{-1} \mathbf{a}_2 \tag{18.46}$$

$$\mathbf{w}_{\mathrm{MWF\text{-}RTF},R} = \mathbf{w}_{\mathrm{MWF},R} - \alpha \, \kappa \, \overline{\boldsymbol{\Sigma}}_{\mathbf{x}}^{-1} \mathbf{a}_2, \tag{18.47}$$

with $\overline{\boldsymbol{\Sigma}}_{\mathbf{x}}$ defined in (18.23) and

$$\kappa = \frac{\sigma_1^2 (a_{1_L,1} + \alpha a_{1_R,1})^* v_{12}^* \bar{\gamma}}{(1 + |\alpha|^2) v_2 \bar{v}}, \quad \bar{\gamma} = \frac{|\gamma_{12}|^2}{\gamma_1 \gamma_2}, \quad \bar{v} = \frac{|v_{12}|^2}{v_1 v_2}, \tag{18.48}$$

with

$$\gamma_{12} = \mathbf{a}_1^H \overline{\boldsymbol{\Sigma}}_{\mathbf{x}}^{-1} \mathbf{a}_2 \quad \gamma_1 = \mathbf{a}_1^H \overline{\boldsymbol{\Sigma}}_{\mathbf{x}}^{-1} \mathbf{a}_1, \quad \gamma_2 = \mathbf{a}_2^H \overline{\boldsymbol{\Sigma}}_{\mathbf{x}}^{-1} \mathbf{a}_2, \tag{18.49}$$

$$v_{12} = \mathbf{a}_1^H \boldsymbol{\Sigma}_{\mathbf{v}}^{-1} \mathbf{a}_2 \quad v_1 = \mathbf{a}_1^H \boldsymbol{\Sigma}_{\mathbf{v}}^{-1} \mathbf{a}_1, \quad v_2 = \mathbf{a}_2^H \boldsymbol{\Sigma}_{\mathbf{v}}^{-1} \mathbf{a}_2. \tag{18.50}$$

The binaural output SINR of the MWF-RTF can be calculated by substituting (18.46) and (18.47) in (18.14) and is given by (Marquardt *et al.*, 2015a)

$$\mathrm{SINR}_{\mathrm{MWF\text{-}RTF}}^{\mathrm{out}} = \varrho_{\mathrm{RTF}} \mathrm{SINR}_{\mathrm{MWF}}^{\mathrm{out}}, \tag{18.51}$$

with

$$\varrho_{\mathrm{RTF}} = \frac{1 + \bar{\gamma}^2 K - 2\bar{\gamma} K}{1 + \psi \bar{\gamma}^2 K - 2\bar{\gamma} K}, \quad \psi = \frac{(\mu + \rho)^2}{\mu^2 \bar{v}} - \frac{\rho^2 + 2\mu\rho}{\mu^2}. \tag{18.52}$$

Since it can be shown that ϱ_{RTF} is always smaller than or equal to 1 (Marquardt *et al.*, 2015a), the binaural output SINR of the MWF-RTF is always smaller than or equal to the binaural output SINR of the binaural MWF. This can be intuitively explained by the additional constraint in (18.44), which reduces the available degrees of freedom for noise reduction.

Due to the RTF preservation constraint in (18.44), the RTF of the interfering source at the output of the MWF-RTF is obviously equal to the input RTF. Substituting (18.46) and (18.47) into (18.16), the output RTF of the desired speech source is equal to

$$\tilde{a}_1^{\mathrm{out}} = \tilde{a}_1^{\mathrm{in}} \frac{1 - \dfrac{\bar{\gamma}}{a_{1_L,1}} \dfrac{a_{1_L,1} + \alpha a_{1_R,1}}{1 + |\alpha|^2}}{1 - \dfrac{\alpha^*\bar{\gamma}}{a_{1_R,1}} \dfrac{a_{1_L,1} + \alpha a_{1_R,1}}{1 + |\alpha|^2}}. \tag{18.53}$$

Hence, contrary to the binaural MWF the output RTF (and hence the binaural cues) of the desired speech source is not always perfectly preserved for the MWF-RTF. However, Marquardt (2016) showed that for $\mu = 0$, corresponding to the binaural MVDR beamformer with RTF preservation constraint (Hadad *et al.*, 2015b), the RTF of the desired speech source is perfectly preserved.

18.6.2 Binaural MWF with Interference Reduction Constraint

Instead of applying an RTF preservation constraint, the MWF-IR aims at both controlling the amount of interference reduction and preserving the binaural cues of the interfering source by adding an interference reduction constraint to the binaural MWF cost functions in (18.20) and (18.21). The MWF-IR cost functions for the left and right hearing devices are defined as (Hadad *et al.*, 2016b)

$$\min_{\mathbf{w}_L} C^{\mathrm{MWF}}(\mathbf{w}_L) \quad \text{s.t.} \quad \mathbf{w}_L^H \mathbf{a}_2 = \delta a_{1_L,2}, \tag{18.54}$$

$$\min_{\mathbf{w}_R} C^{\mathrm{MWF}}(\mathbf{w}_R) \quad \text{s.t.} \quad \mathbf{w}_R^H \mathbf{a}_2 = \delta a_{1_R,2}, \tag{18.55}$$

where the real-valued scaling parameter δ, with $0 \le \delta \le 1$, controls the amount of interference reduction. The filter vectors solving (18.54) and (18.55) are given by (Hadad *et al.*, 2016b)

$$\mathbf{w}_{\mathrm{MWF-IR},L} = \mathbf{w}_{\mathrm{MWF},L} - \sigma_1^2 a_{1_L,1}^* \frac{\gamma_{12}^* - \dfrac{\delta a_{2_L,1}^*}{\sigma_1^2 a_{1_L,1}^*}}{\gamma_2} \overline{\boldsymbol{\Sigma}}_{\mathbf{x}}^{-1} \mathbf{a}_2, \tag{18.56}$$

$$\mathbf{w}_{\mathrm{MWF-IR},R} = \mathbf{w}_{\mathrm{MWF},R} - \sigma_1^2 a_{1_R,1}^* \frac{\gamma_{12}^* - \dfrac{\delta a_{2_R,1}^*}{\sigma_1^2 a_{1_R,1}^*}}{\gamma_2} \overline{\boldsymbol{\Sigma}}_{\mathbf{x}}^{-1} \mathbf{a}_2, \tag{18.57}$$

which are similar to the MWF-RTF expressions in (18.46) and (18.47) in that a scaled version of $\overline{\boldsymbol{\Sigma}}_{\mathbf{x}}^{-1} \mathbf{a}_2$ is subtracted from the binaural MWF filter vectors.

Since the MWF-IR satisfies the interference reduction constraints in (18.54) and (18.55), the RTF of the interfering source at the output of the MWF-IR is equal to the input RTF, i.e.

$$\tilde{a}_2^{\mathrm{out}} = \frac{\mathbf{w}_L^H \mathbf{a}_2}{\mathbf{w}_R^H \mathbf{a}_2} = \frac{a_{1_L,2}}{a_{1_R,2}} = \tilde{a}_2^{\mathrm{in}}, \tag{18.58}$$

such that the MWF-IR perfectly preserves the binaural cues of the interfering source. Substituting (18.57) into (18.16), the output RTF of the desired speech source is equal to

$$\tilde{a}_1^{\text{out}} = \frac{\mathbf{w}_L^H \mathbf{a}_1}{\mathbf{w}_R^H \mathbf{a}_1} = \frac{a_{1_L,1}(1-\bar{\gamma}) + \delta a_{2_L,1} \frac{\bar{\gamma}}{\sigma_1^2 \gamma_{12}}}{a_{1_R,1}(1-\bar{\gamma}) + \delta a_{2_R,1} \frac{\bar{\gamma}}{\sigma_1^2 \gamma_{12}}}. \tag{18.59}$$

Hence, similarly to the MWF-RTF, the output RTF of the desired speech source is not always perfectly preserved for the MWF-IR, although for small values of δ the output RTF of the desired speech source is very close to the input RTF (cf. simulations in Section 18.6.4). Note that in contrast to the MWF-RTF, for the MWF-IR it is possible to control the amount of interference reduction and to set δ in accordance with the amount of RTF estimation errors. A similar approach for preserving the RTF of the interfering source by controlling the amount of interference reduction based on the binaural MVDR beamformer has been proposed by Hadad *et al.* (2015b, 2016a). In the next section we will investigate the MWF-IR for the special case $\delta = 0$.

18.6.3 Special Case: Binaural MWF-IR for $\delta = 0$

For $\delta = 0$, the MWF-IR filter vectors in (18.56) and (18.57) reduce to

$$\mathbf{w}_{\text{MWF-IR-0},L} = \sigma_1^2 a_{1_L,1}^* \left[\overline{\Sigma}_{\mathbf{x}}^{-1} \mathbf{a}_1 - \frac{\gamma_{12}^*}{\gamma_2} \overline{\Sigma}_{\mathbf{x}}^{-1} \mathbf{a}_2 \right], \tag{18.60}$$

$$\mathbf{w}_{\text{MWF-IR-0},R} = \sigma_1^2 a_{1_R,1}^* \left[\overline{\Sigma}_{\mathbf{x}}^{-1} \mathbf{a}_1 - \frac{\gamma_{12}^*}{\gamma_2} \overline{\Sigma}_{\mathbf{x}}^{-1} \mathbf{a}_2 \right]. \tag{18.61}$$

Hadad *et al.* (2015a) showed that the MWF-IR-0 can be decomposed into a binaural LCMV beamformer and a single-channel postfilter applied to the output of the binaural LCMV beamformer, i.e.

$$\mathbf{w}_{\text{MWF-IR-0},L} = \frac{\rho_{\text{LCMV}}}{\mu + \rho_{\text{LCMV}}} \underbrace{\frac{a_{1_L,1}^*}{v_1(1-\bar{v})} \left[\Sigma_{\mathbf{v}}^{-1} \mathbf{a}_1 - \bar{v} \frac{v_1}{v_{12}} \Sigma_{\mathbf{v}}^{-1} \mathbf{a}_2 \right]}_{\mathbf{w}_{\text{LCMV},L}}, \tag{18.62}$$

$$\mathbf{w}_{\text{MWF-IR-0},R} = \frac{\rho_{\text{LCMV}}}{\mu + \rho_{\text{LCMV}}} \underbrace{\frac{a_{1_R,1}^*}{v_1(1-\bar{v})} \left[\Sigma_{\mathbf{v}}^{-1} \mathbf{a}_1 - \bar{v} \frac{v_1}{v_{12}} \Sigma_{\mathbf{v}}^{-1} \mathbf{a}_2 \right]}_{\mathbf{w}_{\text{LCMV},R}}, \tag{18.63}$$

where $\mathbf{w}_{\text{LCMV},L}$ and $\mathbf{w}_{\text{LCMV},R}$ are the filter vectors solving the binaural LCMV optimization problem for the special case of $\delta = 0$ (Hadad *et al.*, 2016a), i.e.

$$\underset{\mathbf{w}_L}{\text{argmin}}\ \mathbf{w}_L^H \Sigma_{\mathbf{v}} \mathbf{w}_L \quad \text{s.t.} \quad \mathbf{w}_L^H \mathbf{a}_1 = a_{1_L,1}, \mathbf{w}_L^H \mathbf{a}_2 = 0, \tag{18.64}$$

$$\underset{\mathbf{w}_R}{\text{argmin}}\ \mathbf{w}_R^H \Sigma_{\mathbf{v}} \mathbf{w}_R \quad \text{s.t.} \quad \mathbf{w}_R^H \mathbf{a}_1 = a_{1_R,1}, \mathbf{w}_R^H \mathbf{a}_2 = 0, \tag{18.65}$$

and $\rho_{\text{LCMV}} = \sigma_1^2 v_1 (1 - \bar{v})$ is the output SINR of the binaural LCMV beamformer. As shown by Marquardt *et al.* (2015a), for the MWF-IR-0 the RTF of the desired speech source is perfectly preserved, which can be directly seen from (18.59) by setting $\delta = 0$, while the output RTF of the interfering source cannot be calculated since the interfering source is completely suppressed. In addition, Marquardt *et al.* (2015a) showed that the output SINR for the MWF-IR-0 can be calculated by setting $K = 1$ in the expression for the output SINR of the MWF-RTF in (18.51) and (18.52), i.e.

$$\text{SINR}^{\text{out}}_{\text{MWF-IR}} = \varrho_{\text{IR}}\text{SINR}^{\text{out}}_{\text{MWF}} \tag{18.66}$$

with

$$\varrho_{\text{IR}} = \frac{1 + \bar{\gamma}^2 - 2\bar{\gamma}}{1 + \psi\bar{\gamma}^2 - 2\bar{\gamma}}, \tag{18.67}$$

and ψ defined in (18.52). Since it can be shown that ϱ_{IR} is always smaller than or equal to ϱ_{RTF} (Marquardt *et al.*, 2015a), the output SINR of the MWF-IR-0 is always smaller than or equal to the output SINR of the binaural MWF-RTF. Hence, the output SINRs of the discussed algorithms in this section are related as

$$\text{SINR}^{\text{out}}_{\text{MWF}} \geq \text{SINR}^{\text{out}}_{\text{MWF-RTF}} \geq \text{SINR}^{\text{out}}_{\text{MWF-IR-0}}. \tag{18.68}$$

This can be intuitively explained by the fact that compared to the MWF, the MWF-RTF needs to satisfy one additional constraint, whereas the MWF-IR-0 needs to satisfy two additional constraints.

18.6.4 Simulations with Measured Acoustic Transfer Functions

In this section we validate the closed-form expressions derived in the previous sections. We present experimental results comparing the performance of the binaural MWF, MWF-RTF, MWF-IR with $\delta = 0.2$ (MWF-IR-0.2) and MWF-IR-0 using behind-the-ear impulse responses measured on an artificial head in an office environment (Kayser *et al.*, 2009) with a reverberation time (RT60) of approximately 300 ms. In order to analyze the full potential of the presented algorithms, we assume that perfect estimates of the covariance matrices and the acoustic transfer functions of the desired and interfering sources are available. All experiments were carried out at $f_s = 16$ kHz using $I = 4$ microphones, i.e. two microphones on the left and right hearing devices. The acoustic scenario comprised one desired source at an angle of $-5°$ and 1 m from the artificial head, one interfering source at different angles[1] and 1 m from the artificial head, and diffuse background noise. The angle $0°$ corresponds to the frontal direction and the angle $90°$ corresponds to the right side of the head. The power spectra of the desired and interfering sources σ_1^2 and σ_2^2 were calculated from two different speech signals using an STFT size of 512 and a Hann window. For the background noise a cylindrically isotropic noise field was assumed, i.e. the (i, i')th element $\sigma_{u_i u_{i'}}$ of the noise covariance matrix $\boldsymbol{\Sigma}_{\mathbf{u}}$ was calculated using the anechoic acoustic transfer functions from the same dataset as

$$\sigma_{u_i u_{i'}} = \sigma_u^2 \frac{\sum_{j=1}^{J} a_{ij}^{\text{dir}} a_{i'j}^{\text{dir}}}{\sqrt{\sum_{j=1}^{J} |a_{ij}^{\text{dir}}|^2 \sum_{j=1}^{J} |a_{i'j}^{\text{dir}}|^2}}, \tag{18.69}$$

where a_{ij}^{dir} denotes the *ith* element of the anechoic acoustic transfer function for the *j*th angle in the dataset, J denotes the total number of angles ($J = 72$), and the power spectrum of the background noise σ_u^2 is equal to the power spectrum of speech-shaped

---

1 Note that the interfering source angle of $-5°$ has not been evaluated.

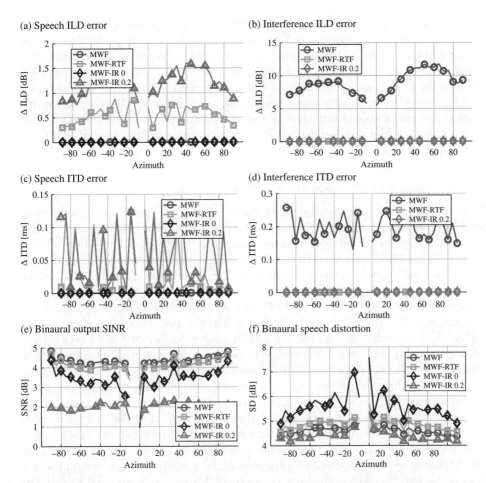

Figure 18.7 Performance measures for the binaural MWF, MWF-RTF, MWF-IR-0, and MWF-IR-0.2 for a desired speech source at −5° and different interfering source positions. The global input SINR was equal to −3 dB.

noise. The weighting parameter μ was set to 1 for all algorithms and the input SINR was equal to −3 dB. As objective performance measures, we have used the global binaural output SINR in (18.14), the global binaural speech distortion in (18.11), and the global absolute ILD and ITD errors for the desired and the interfering source[2] (averaged over all frequencies).

The results are depicted in Figure 18.7. On the one hand, for the desired speech source the MWF-RTF and the MWF-IR-0.2 introduce a small ILD error (up to 1.6 dB) and a small ITD error (up to 0.13 ms), depending on the position of the interfering source, while the binaural MWF and the MWF-IR-0 perfectly preserve the binaural cues of the desired speech source. On the other hand, for the interfering source the binaural MWF introduces a large ILD error (up to 12 dB) and the ITD error varies around

---

2 Note that the ILD and ITD errors for the interfering source for the MWF-IR-0 are not depicted since the interfering source is entirely suppressed in this theoretical validation.

0.2 ms, while the MWF-RTF and the MWF-IR-0.2 perfectly preserve the binaural cues of the interfering source. In terms of binaural output SINR, while the performance of the binaural MWF and the MWF-RTF are very similar, the performance of the MWF-IR is significantly lower (for both values of δ), in particular for interfering source positions close to the desired source position. It can also be observed that the SINR relations in (18.68) hold. In terms of binaural speech distortion, the MWF-IR-0 introduces the highest amount of speech distortion, especially for interfering source positions close to the speech source position, while the speech distortion of the binaural MWF, the MWF-RTF and the MWF-IR-0.2 are very similar.

18.6.5 Simulations with Noisy Speech Signals

In this section we compare the performance of the considered algorithms using simulated signals in a noisy and reverberant environment. Similarly to Section 18.5.4, binaural behind-the-ear impulse responses measured on an artificial head in a cafeteria (Kayser *et al.*, 2009) were used to generate the signal components, and each hearing device was equipped with two microphones. The desired speech source was located at −35° and a distance of 117.5 cm, while different positions for the interfering speech source were considered: 0°, −90°, 135°, and 90°, and distances of 102 cm, 52 cm, 129 cm, and 162 cm, respectively. Recorded ambient noise from the cafeteria was added to the speech components. The signals were processed at $f_s = 16$ kHz using a weighted overlap-add framework with a block size of 512 samples and an overlap of 50% between successive blocks. For the estimation procedure, three training sections were used. The first training section consisted of a segment of 2 s in which none of the speech sources was active. This segment was used to estimate the covariance matrix of the noise component $\mathbf{\Sigma_u}$. The second training section consisted of a segment of 2.5 s in which the desired source was active but the interfering source was inactive. This segment was used to estimate the noisy desired source covariance matrix $\mathbf{\Sigma_x}$. The third training section consisted of a segment of 2.5 s in which the interfering source was active but the desired source was inactive. This segment was used to estimate the overall noise covariance matrix $\mathbf{\Sigma_v}$. The covariance matrix of the desired speech

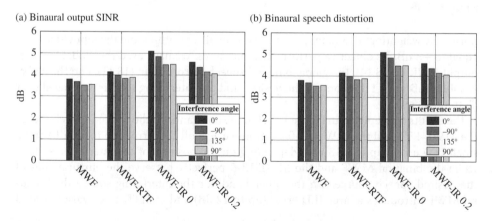

Figure 18.8 Performance measures for the binaural MWF, MWF-RTF, MWF-IR-0.2, and MWF-IR-0 for a desired speech source at −35° and different interfering source positions. The global input SINR was equal to −3 dB.

component was estimated as $\Sigma_{c_1} = \Sigma_x - \Sigma_u$, where a rank-1 approximation of Σ_{c_1} was used. The RTF vector of the interfering source was estimated using the generalized eigenvalue decomposition of Σ_v and Σ_u (cf. Section 11.3).

As objective performance measures, we have used the global binaural output SINR and the global binaural speech distortion (averaged over all frequencies). The results are depicted in Figure 18.8. It can be observed that the SINR relations in (18.68) hold, i.e. for all interfering source positions the SINR of the MWF is larger than the SINR of the MWF-RTF and the MWF-IR (for both values of δ). Although the MWF also introduces the lowest amount of speech distortion compared to the MWF-RTF and the MWF-IR (for both values of δ), it should be realized that the MWF-RTF and the MWF-IR preserve the binaural cues of the interfering source, which is not the case for the MWF.

18.7 Summary

In this chapter we have explored multi-microphone noise reduction algorithms for binaural hearing devices. The objective of a binaural algorithm is not only to enhance the desired speech source, but also to preserve the auditory impression of the complete acoustic scene. To incorporate binaural cue preservation into noise reduction algorithms, two different paradigms are typically adopted. In the first paradigm, the same real-valued spectro-temporal gain is applied to two reference microphone signals, one on each hearing device. In the second paradigm, all available microphone signals from both hearing devices are processed by different complex-valued spatial filters. This chapter focused on the second paradigm and explored four MWF-based methods for preserving the binaural cues of the background noise and/or an interfering source, either by directly incorporating constraints into the spatial filter design or by mixing with scaled reference microphone signals.

The development of comprehensive solutions for combined binaural noise reduction, dereverberation (not reviewed in this chapter), and acoustic scene preservation, under strict requirements on the computational complexity, algorithmic latency, and bandwidth of the wireless link, is still an active topic of research.

Bibliography

Aichner, R., Buchner, H., Zourub, M., and Kellermann, W. (2007) Multi-channel source separation preserving spatial information, in *Proceedings of IEEE International Conference on Audio, Speech and Signal Processing*, pp. 5–8.

Baumgärtel, R., Krawczyk-Becker, M., Marquardt, D., Völker, C., Hu, H., Herzke, T., Coleman, G., Adiloğlu, K., Ernst, S., Gerkmann, T., Doclo, S., Kollmeier, B., Hohmann, V., and Dietz, M. (2015) Comparing binaural pre-processing strategies I: Instrumental evaluation. *Trends in Hearing*, **19**, 1–16.

Best, V., Carlile, S., Kopco, N., and van Schaik, A. (2011) Localization in speech mixtures by listeners with hearing loss. *JASA Express Letters*, **129** (5), 210–215.

Beutelmann, R. and Brand, T. (2006) Prediction of speech intelligibility in spatial noise and reverberation for normal-hearing and hearing-impaired listeners. *Journal of the Acoustical Society of America*, **120** (1), 331–342.

Blauert, J. (1997) *Spatial Hearing: The Psychophysics of Human Sound Localisation*, MIT Press.

Boothroyd, A., Fitz, K., Kindred, J., Kochkin, S., Levitt, H., Moore, B., and Yantz, J. (2007) Hearing aids and wireless technology. *Hearing Review*, **14** (6), 44–48.

Bregman, A.S. (1990) *Auditory Scene Analysis*, MIT Press.

Bronkhorst, A. (2000) The cocktail party phenomenon: A review of research on speech intelligibility in multiple-talker conditions. *Acta Acustica*, **86** (1), 117–128.

Bronkhorst, A. and Plomp, R. (1988) The effect of head-induced interaural time and level differences on speech intelligibility in noise. *Journal of the Acoustical Society of America*, **83** (4), 1508–1516.

Cherry, E.C. (1953) Some experiments on the recognition of speech, with one and with two ears. *Journal of the Acoustical Society of America*, **25** (5), 975–979.

Cornelis, B., Doclo, S., van den Bogaert, T., Wouters, J., and Moonen, M. (2010) Theoretical analysis of binaural multi-microphone noise reduction techniques. *IEEE Transactions on Audio, Speech, and Language Processing*, **18** (2), 342–355.

Desloge, J., Rabinowitz, W., and Zurek, P. (1997) Microphone-array hearing aids with binaural output — part I: Fixed-processing systems. *IEEE Transactions on Speech and Audio Processing*, **5** (6), 529–542.

Dietz, M., Ewert, S.D., and Hohmann, V. (2011) Auditory model based direction estimation of concurrent speakers from binaural signals. *Speech Communication*, **53**, 592–605.

Dillon, H. (2012) *Hearing Aids*, Thieme.

Doclo, S., Gannot, S., Moonen, M., and Spriet, A. (2010) Acoustic beamforming for hearing aid applications, in *Handbook on Array Processing and Sensor Networks*, Wiley, pp. 269–302.

Doclo, S., Kellermann, W., Makino, S., and Nordholm, S. (2015) Multichannel signal enhancement algorithms for assisted listening devices. *IEEE Signal Processing Magazine*, **32** (2), 18–30.

Doclo, S. and Moonen, M. (2005) On the output SNR of the speech-distortion weighted multichannel Wiener filter. *IEEE Signal Processing Letters*, **12** (12), 809–811.

Dörbecker, M. and Ernst, S. (1996) Combination of two-channel spectral subtraction and adaptive Wiener post-filtering for noise reduction and dereverberation, in *Proceedings of European Signal Processing Conference*, pp. 995–998.

Faller, C. and Merimaa, J. (2004) Source localization in complex listening situations: Selection of binaural cues based on interaural coherence. *Journal of the Acoustical Society of America*, **116** (5), 3075–3089.

Greenberg, J.E., Peterson, P.M., and Zurek, P.M. (1993) Intelligibility-weighted measures of speech-to-interference ratio and speech system performance. *Journal of the Acoustical Society of America*, **94** (5), 3009–3010.

Grimm, G., Hohmann, V., and Kollmeier, B. (2009) Increase and subjective evaluation of feedback stability in hearing aids by a binaural coherence-based noise reduction scheme. *IEEE Transactions on Audio, Speech, and Language Processing*, **17**, 1408–1419.

Hadad, E., Doclo, S., and Gannot, S. (2016a) The binaural LCMV beamformer and its performance analysis. *IEEE/ACM Transactions on Audio, Speech, and Language Processing*, **24** (3), 543–558.

Hadad, E., Marquardt, D., Doclo, S., and Gannot, S. (2015a) Binaural multichannel Wiener filter with directional interference rejection, in *Proceedings of IEEE International Conference on Audio, Speech and Signal Processing*, pp. 644–648.

Hadad, E., Marquardt, D., Doclo, S., and Gannot, S. (2015b) Theoretical analysis of binaural transfer function MVDR beamformers with interference cue preservation constraints. *IEEE/ACM Transactions on Audio, Speech, and Language Processing*, **23** (12), 2449–2464.

Hadad, E., Marquardt, D., Doclo, S., and Gannot, S. (2016b) Extensions of the binaural MWF with interference reduction preserving the binaural cues of the interfering source, in *Proceedings of IEEE International Conference on Audio, Speech and Signal Processing*, pp. 241–245.

Hamacher, V., Chalupper, J., Eggers, J., Fischer, E., Kornagel, U., Puder, H., and Rass, U. (2005) Signal processing in high-end hearing aids: State of the art, challenges, and future trends. *EURASIP Journal on Applied Signal Processing*, **2005** (18), 2915–2929.

Hamacher, V., Kornagel, U., Lotter, T., and Puder, H. (2008) Binaural signal processing in hearing aids: Technologies and algorithms, in *Advances in Digital Speech Transmission*, Wiley, pp. 401–429.

Hawley, M.L., Litovsky, R.Y., and Culling, J.F. (2004) The benefit of binaural hearing in a cocktail party: Effect of location and type of interferer. *Journal of the Acoustical Society of America*, **115** (2), 833–843.

Heurig, R. and Chalupper, J. (2010) Acceptable processing delay in digital hearing aids. *Hearing Review*, **17** (1), 28–31.

Jeub, M., Dörbecker, M., and Vary, P. (2011a) A semi-analytical model for the binaural coherence of noise fields. *IEEE Signal Processing Letters*, **18** (3), 197–200.

Jeub, M., Nelke, C., Krüger, H., Beaugeant, C., and Vary, P. (2011b) Robust dual-channel noise power spectral density estimation, in *Proceedings of European Signal Processing Conference*, pp. 2304–2308.

Kamkar-Parsi, A. and Bouchard, M. (2009) Improved noise power spectrum density estimation for binaural hearing aids operating in a diffuse noise field environment. *IEEE Transactions on Audio, Speech, and Language Processing*, **17** (4), 521–533.

Kamkar-Parsi, A.H. and Bouchard, M. (2011) Instantaneous binaural target PSD estimation for hearing aid noise reduction in complex acoustic environments. *IEEE Transactions on Instrumentation and Measurement*, **60** (4), 1141–1154.

Kayser, H., Ewert, S., Annemüller, J., Rohdenburg, T., Hohmann, V., and Kollmeier, B. (2009) Database of multichannel in-ear and behind-the-ear head-related and binaural room impulse responses. *EURASIP Journal on Advances in Signal Processing*, **2009**, 10.

Keidser, G., O'Brien, A., Hain, J.U., McLelland, M., and Yeend, I. (2009) The effect of frequency-dependent microphone directionality on horizontal localization performance in hearing-aid users. *International Journal of Audiology*, **48** (11), 789–803.

Klasen, T., van den Bogaert, T., Moonen, M., and Wouters, J. (2007) Binaural noise reduction algorithms for hearing aids that preserve interaural time delay cues. *IEEE Transactions on Signal Processing*, **55** (4), 1579–1585.

Kurozumi, K. and Ohgushi, K. (1983) The relationship between the cross-correlation coefficient of two-channel acoustic signals and sound image quality. *Journal of the Acoustical Society of America*, **74** (6), 1726–1733.

Lavandier, M. and Culling, J.F. (2010) Prediction of binaural speech intelligibility against noise in rooms. *Journal of the Acoustical Society of America*, **127**, 387–399.

Liao, W.C., Luo, Z.Q., Merks, I., and Zhang, T. (2015) An effective low complexity binaural beamforming algorithm for hearing aids, in *Proceedings of IEEE Workshop on Applications of Signal Processing to Audio and Acoustics*, pp. 1–5.

Lindevald, I. and Benade, A. (1986) Two-ear correlation in the statistical sound fields of rooms. *Journal of the Acoustical Society of America*, **80** (2), 661–664.

Lotter, T. and Vary, P. (2006) Dual-channel speech enhancement by superdirective beamforming. *EURASIP Journal on Applied Signal Processing*, **2006** (1), 175–175.

Luts, H., Eneman, K., Wouters, J., Schulte, M., Vormann, M., Buechler, M., Dillier, N., Houben, R., Dreschler, W.A., Froehlich, M., Puder, H., Grimm, G., Hohmann, V., Leijon, A., Lombard, A., Mauler, D., and Spriet, A. (2010) Multicenter evaluation of signal enhancement algorithms for hearing aids. *Journal of the Acoustical Society of America*, **127** (3), 2054–2063.

Madhu, N., Spriet, A., Jansen, S., Koning, R., and Wouters, J. (2013) The potential for speech intelligibility improvement using the ideal binary mask and the ideal Wiener filter in single channel noise reduction systems: Application to auditory prostheses. *IEEE Transactions on Audio, Speech, and Language Processing*, **21** (1), 63–72.

Marquardt, D. (2016) *Development and evaluation of psychoacoustically motivated binaural noise reduction and cue preservation techniques*, Ph.D. thesis, University of Oldenburg.

Marquardt, D., Hadad, E., Gannot, S., and Doclo, S. (2015a) Theoretical analysis of linearly constrained multi-channel Wiener filtering algorithms for combined noise reduction and binaural cue preservation in binaural hearing aids. *IEEE/ACM Transactions on Audio, Speech, and Language Processing*, **23** (12), 2384–2397.

Marquardt, D., Hohmann, V., and Doclo, S. (2015b) Interaural coherence preservation in multi-channel Wiener filtering based noise reduction for binaural hearing aids. *IEEE/ACM Transactions on Audio, Speech, and Language Processing*, **23** (12), 2162–2176.

Marquardt, D., Hohmann, V., and Doclo, S. (2015c) Interaural coherence preservation in MWF-based binaural noise reduction algorithms using partial noise estimation, in *Proceedings of IEEE International Conference on Audio, Speech and Signal Processing*, pp. 654–658.

May, T., van de Par, S., and Kohlrausch, A. (2011) A probabilistic model for robust localization based on a binaural auditory front-end. *IEEE Transactions on Audio, Speech, and Language Processing*, **19** (1), 1–13.

Merks, I., Boone, M., and Berkhout, A. (1997) Design of a broadside array for a binaural hearing aid, in *Proceedings of IEEE Workshop on Applications of Signal Processing to Audio and Acoustics*, pp. 1–4.

Nakatani, T. and Okuno, H.G. (1999) Harmonic sound stream segregation using localisation and its application to speech stream segregation. *Speech Communication*, **27** (3-4), 209–222.

Nilsson, M., Soli, S.D., and Sullivan, J.A. (1994) Development of the hearing in noise test for the measurement of speech reception thresholds in quiet and in noise. *Journal of the Acoustical Society of America*, **95** (2), 1085–1099.

Popelka, G.R., Moore, B.C.J., Fay, R.R., and Popper, A.N. (eds) (2016) *Hearing Aids*, Springer.

Raspaud, M., Viste, H., and Evangelista, G. (2010) Binaural source localization by joint estimation of ILD and ITD. *IEEE Transactions on Audio, Speech, and Language Processing*, **18**, 68–77.

Reindl, K., Zheng, Y., and Kellermann, W. (2010a) Analysis of two generic Wiener filtering concepts for binaural speech enhancement in hearing aids, in *Proceedings of European Signal Processing Conference*, pp. 989–993.

Reindl, K., Zheng, Y., and Kellermann, W. (2010b) Speech enhancement for binaural hearing aids based on blind source separation, in *Proceedings of International Symposium on Communications, Control and Signal Processing*, pp. 1–6.

Reindl, K., Zheng, Y., Schwarz, A., Meier, S., Maas, R., Sehr, A., and Kellermann, W. (2013) A stereophonic acoustic signal extraction scheme for noisy and reverberant environments. *Computer Speech and Language*, **27** (3), 726–745.

Rennies, J., Brand, T., and Kollmeier, B. (2011) Prediction of the influence of reverberation on binaural speech intelligibility in noise and in quiet. *Journal of the Acoustical Society of America*, **130**, 2999–3012.

Rohdenburg, T. (2008) *Development and Objective Perceptual Quality Assessment of Monaural and Binaural Noise Reduction Schemes for Hearing Aids*, Ph.D. thesis, University of Oldenburg.

Roman, N., Wang, D., and Brown, G.J. (2003) Speech segregation based on sound localization. *Journal of the Acoustical Society of America*, **114** (4), 2236–2252.

Suzuki, Y., Tsukui, S., Asano, F., Nishimura, R., and Sone, T. (1999) New design method of a binaural microphone array using multiple constraints. *IEICE Transactions on Fundamentals of Electronics, Communications and Computer Sciences*, **E82-A** (4), 588–596.

Thiemann, J., Müller, M., Marquardt, D., Doclo, S., and van de Par, S. (2016) Speech enhancement for multimicrophone binaural hearing aids aiming to preserve the spatial auditory scene. *EURASIP Journal on Advances in Signal Processing*, **12**, 1–11.

Tohyama, M. and Suzuki, A. (1989) Interaural cross-correlation coefficients in stereo-reproduced sound fields. *Journal of the Acoustical Society of America*, **85** (2), 780–786.

van den Bogaert, T., Doclo, S., Wouters, J., and Moonen, M. (2008) The effect of multimicrophone noise reduction systems on sound source localization by users of binaural hearing aids. *Journal of the Acoustical Society of America*, **124** (1), 484–497.

van den Bogaert, T., Klasen, T.J., Moonen, M., Van Deun, L., and Wouters, J. (2006) Horizontal localisation with bilateral hearing aids: without is better than with. *Journal of the Acoustical Society of America*, **119** (1), 515–526.

Walther, A. and Faller, C. (2013) Interaural correlation discrimination from diffuse field reference correlations. *Journal of the Acoustical Society of America*, **133** (3), 1496–1502.

Wang, D. and Brown, G.J. (eds) (2006) *Computational Auditory Scene Analysis: Principles, Algorithms, and Applications*, Wiley-IEEE Press.

Welker, D., Greenberg, J., Desloge, J., and Zurek, P. (1997) Microphone-array hearing aids with binaural output — part II: A two-microphone adaptive system. *IEEE Transactions on Speech and Audio Processing*, **5** (6), 543–551.

Wightman, F.L. and Kistler, D.J. (1992) The dominant role of low-frequency interaural time differences in sound localization. *Journal of the Acoustical Society of America*, **91** (3), 1648–1661.

Wittkop, T. and Hohmann, V. (2003) Strategy-selective noise reduction for binaural digital hearing aids. *Speech Communication*, **39** (1-2), 111–138.

Woodruff, J. and Wang, D. (2013) Binaural detection, localization, and segregation in reverberant environments based on joint pitch and azimuth cues. *IEEE Transactions on Audio, Speech, and Language Processing*, **21** (4), 806–815.

Wouters, J., Doclo, S., Koning, R., and Francart, T. (2013) Sound processing for better coding of monaural and binaural cues in auditory prostheses. *Proceedings of the IEEE*, **101** (9), 1986–1997.

19

Perspectives

Emmanuel Vincent, Tuomas Virtanen, and Sharon Gannot

Source separation and speech enhancement research has made dramatic progress in the last 30 years. It is now a mainstream topic in speech and audio processing, with hundreds of papers published every year. Separation and enhancement performance have greatly improved and successful commercial applications are increasingly being deployed. This chapter provides an overview of research and development perspectives in the field. We do not attempt to cover all perspectives currently under discussion in the community. Instead, we focus on five directions in which we believe major progress is still possible: getting the most out of deep learning, exploiting phase relationships across time-frequency bins, improving the estimation accuracy of multichannel parameters, addressing scenarios involving multiple microphone arrays or other sensors, and accelerating industry transfer. These five directions are covered in Sections 19.1, 19.2, 19.3, 19.4, and 19.5, respectively.

19.1   Advancing Deep Learning

In just a few years, deep learning has emerged as a major paradigm for source separation and speech enhancement. Deep neural networks (DNNs) can model the complex characteristics of audio sources by making efficient use of large amounts (typically hours) of training data. They perform well on mixtures involving similar conditions to those in the training set and they are surprisingly robust to unseen conditions (Vincent *et al.*, 2017; Kolbæk *et al.*, 2017), provided that the training set is sufficiently large and diverse. Several research directions are currently under study to get the best out of this new paradigm.

19.1.1   DNN Design Choices

The first obvious direction is to tune the DNN architecture to the task at hand. Several architectures, namely multilayer perceptron, deep recurrent neural network (DRNN), long short-term memory (LSTM), bidirectional LSTM, convolutional neural

Audio Source Separation and Speech Enhancement, First Edition.
Edited by Emmanuel Vincent, Tuomas Virtanen and Sharon Gannot.
© 2018 John Wiley & Sons Ltd. Published 2018 by John Wiley & Sons Ltd.
Companion Website: https://project.inria.fr/ssse/

network, and nonnegative DNN have already been covered in Section 7.3.2. Recently, a new DRNN-like architecture known as the *deep stacking network* was successfully employed (Zhang *et al.*, 2016). This architecture concatenates the outputs in the previous time frame with the inputs in the current frame. It is motivated by the fact that iteratively applying a DNN to the outputs of the previous iteration improves performance (Nugraha *et al.*, 2016a), but it avoids multiple passes over the test data. Another architecture recently proposed by Chazan *et al.* (2016) combines a generative Gaussian mixture model (GMM) and a discriminative DNN in a hybrid approach. The DNN is used to estimate the posterior phoneme probability in each time frame, and a soft time-frequency mask is derived by modeling each phoneme as a single Gaussian. New architectures are invented every year in the fields of automatic speech and image recognition, e.g. (Zagoruyko and Komodakis 2016), and it is only a matter of time before they are adapted and applied to source separation and speech enhancement. Fusing the outputs of multiple architectures is also beneficial. The optimal fusion weights can be learned using a DNN, as shown by Jaureguiberry *et al.* (2016).

Another interesting research direction concerns the design of the training set. The most common approach today is to generate simulated training data by convolving target and interference signals with real or simulated acoustic impulse responses and mixing them together. It is generally believed that the larger the amount of training data, the more diverse, and the closer to the test conditions, the better the separation or enhancement performance. This has led to several data augmentation approaches to expand the size and coverage of the training set and reduce the mismatch with the test set. Yet, surprisingly, Heymann *et al.* (2016) obtained similar performance by training on time-frequency masks generated by thresholding target short-time Fourier transform (STFT) coefficients (without using any interference signal in training), while Vincent *et al.* (2017) found that training on mismatched noise conditions can outperform matched or multicondition training. This calls for a more principled approach to designing the training set. The algorithm of Sivasankaran *et al.* (2017), which weights the training samples so as to maximize performance on the validation set, is a first step in this direction.

The cost function used for training also has a significant impact. Various cost functions have been reviewed in Sections 7.3.3 and 7.3.4, namely cross-entropy, mean square error (MSE), phase-sensitive cost, Kullback–Leibler (KL) divergence, and Itakura–Saito (IS) divergence. The studies cited in these sections reported better performance for MSE and the phase-sensitive cost. More recently, however, Nugraha *et al.* (2016a) found KL to perform best in both single- and multichannel scenarios. This calls for more research on the choice of the cost function depending on the scenario and other DNN design choices. Taking psychoacoustics into account is also a promising direction, as recently explored by Shivakumar and Georgiou (2016).

Finally, the use of DNNs also impacts the signal processing steps involved in the overall separation or enhancement system. For instance, Nugraha *et al.* (2016b) found that, when the source power spectra are estimated by a DNN, the conventional expectation-maximization (EM) update rule for spatial covariance matrices (14.79) is outperformed by a temporally weighted rule. Deeper investigation of the interplay between DNN and signal processing is thus required in order to get the best out of hybrid systems involving both DNN and signal processing steps.

19.1.2   End-to-End Approaches

End-to-end DNN-based approaches attempt to address the interplay between DNN and signal processing by getting rid of signal processing entirely and developing purely DNN-based systems as opposed to hybrid systems. This makes it easier to jointly optimize all processing steps in a DNN framework (see Section 17.4.2). End-to-end DNNs operate in the time domain or in the complex-valued STFT domain (Li *et al.*, 2016), which enables them to exploit phase differences between neighboring time-frequency bins (see Section 19.2.3). Current end-to-end DNN architectures that integrate target localization and beamforming perform only marginally better than delay-and-sum (DS) beamforming, and lie significantly behind the conventional signal processing-based beamformers derived from DNN-based masks reviewed in Section 12.4 (Xiao *et al.*, 2016). Nevertheless, it is a widespread belief in the deep learning community that they will soon outperform other approaches. Progress might come from *generative DNNs* such as Wavenet (van den Oord *et al.*, 2016), which are now used to synthesize time-domain audio signals and could soon be used to generate the source signals that best match a given mixture signal. As a matter of fact, synthesis-based speech enhancement has recently started being investigated (Nickel *et al.*, 2013; Kato and Milner, 2016).

19.1.3   Unsupervised Separation

DNNs are typically trained in a supervised fashion in order to discriminate a certain class of sounds, e.g. speech vs. noise, foreground vs. background speech, male vs. female speech, or a specific speaker vs. others. To do this, the number of sources must be known and the order of the sources must be fixed, i.e. training is permutation-dependent. This implies that separating two foreground speakers of the same gender is infeasible unless their identity is known and training data are available for at least one of them, a situation which arises in certain scenarios only. *Deep clustering* algorithms inspired by the spectral clustering algorithms in Section 7.1.3 have recently overcome this limitation for single-channel mixtures.

Hershey *et al.* (2016) proposed to apply a DRNN g_z to the log-magnitude spectrogram $\log |\mathbf{X}| = [\log |x(n,f)|]_{fn}$ to extract a unit-norm *embedding*, i.e. a feature vector $\mathbf{y}(n,f)$ of arbitrary dimension K for each time-frequency bin. The output $g_z(\log |\mathbf{X}|)(n)$ of the DRNN in a given time frame n consists of the concatenation of the embeddings for all frequency bins in that time frame:

$$\begin{bmatrix} \mathbf{y}(n,0) \\ \vdots \\ \mathbf{y}(n,F-1) \end{bmatrix} = g_z(\log |\mathbf{X}|)(n). \tag{19.1}$$

The key is to train g_z such that each embedding characterizes the dominant source in the corresponding time-frequency bin. Separation can then be achieved by clustering these embeddings so that time-frequency bins dominated by the same source are clustered together. The optimal assignment of time-frequency bins to sources is given by the $FN \times J$ indicator matrix $\mathbf{O} = [o_j(n,f)]_{fnj}$ where

$$o_j(n,f) = \begin{cases} 1 & \text{if source } j \text{ dominates in time-frequency bin } (n,f) \\ 0 & \text{otherwise.} \end{cases} \tag{19.2}$$

All embeddings are stacked into an $FN \times K$ matrix $\mathbf{Y} = [\mathbf{y}^T(n,f)]_{fn}$ and training is achieved by minimizing

$$C^{\mathrm{PI}}(\mathcal{Z}) = \|\mathbf{Y}\mathbf{Y}^T - \mathbf{O}\mathbf{O}^T\|_2^2. \tag{19.3}$$

This cost is permutation-invariant: indeed, the binary affinity matrix $\mathbf{O}\mathbf{O}^T$ is such that $(\mathbf{O}\mathbf{O}^T)_{fn,f'n'} = 1$ if (n,f) and (n',f') belong to the same cluster and $(\mathbf{O}\mathbf{O}^T)_{fn,f'n'} = 0$ otherwise, and it is invariant to reordering of the sources.

Once the embeddings have been obtained, the sources are separated by soft time-frequency masking. The masks $w_j(n,f)$ are computed by a soft k-means algorithm, which alternately updates the masks and the centroid embedding $\bar{\mathbf{y}}_j$ of each source:

$$w_j(n,f) = \frac{e^{-\alpha\|\mathbf{y}(n,f)-\bar{\mathbf{y}}_j\|_2^2}}{\sum_{j'=1}^{J} e^{-\alpha\|\mathbf{y}(n,f)-\bar{\mathbf{y}}_{j'}\|_2^2}} \tag{19.4}$$

$$\bar{\mathbf{y}}_j = \frac{1}{\sum_{fn} w_j(n,f)} \sum_{fn} w_j(n,f)\mathbf{y}(n,f). \tag{19.5}$$

The parameter α controls the "hardness" of the clustering. In practice, only nonsilent time-frequency bins are taken into account in (19.3) and (19.5).

One limitation of this approach is that the training criterion (19.3) is not directly linked with the source separation performance. To address this, Isik *et al.* (2016) introduced a second DRNN $g_{z'}$ that takes as input the mixture amplitude spectrogram $|\mathbf{X}|$ and the amplitude spectrogram of a given source $|\mathbf{C}_j| = [w_j(n,f)|x(n,f)|]_{fn}$ estimated by the above clustering procedure and outputs an improved estimate of $|\mathbf{C}_j|$. An improved soft mask is then computed from these improved estimates and used to obtain the final source estimates. The iterations of the k-means algorithm are unfolded and trained along with this second DRNN according to the signal reconstruction cost (7.23).

The results in Table 19.1 show that this approach can separate mixtures of unknown speakers remarkably better than a DRNN trained to separate the foreground speaker. An alternative permutation-invariant DNN training approach which does not require intermediate embeddings was proposed by Yu *et al.* (2016). These approaches open up new perspectives for research in DNN-based source separation.

Table 19.1 Average signal-to-distortion ratio (SDR) achieved by the computational auditory scene analysis (CASA) method of Hu and Wang (2013), a DRNN trained to separate the foreground speaker, and two variants of deep clustering for the separation of mixtures of two speakers with all gender combinations at random signal-to-noise ratios (SNRs) between 0 and 10 dB (Hershey *et al.*, 2016; Isik *et al.*, 2016). The test speakers are not in the training set.

Method	SDR (dB)
Hu and Wang (2013)	3.1
DRNN foreground	1.2
Deep clustering g_z	10.3
Deep clustering $g_z + g_{z'}$	10.8

19.2   Exploiting Phase Relationships

The filters introduced in Chapters 5 and 10 for separation and enhancement exhibit two fundamental limitations. First, due to the narrowband approximation (2.11), they operate in each time-frequency bin independently and can generate cyclic convolution artifacts. Second, except for magnitude spectral subtraction, they rely on the assumption that phase is uniformly distributed, which translates into modeling the target and interference STFT coefficients as zero-mean. As a result, they exploit the magnitude spectrum of each source and the interchannel phase difference (IPD), but not the phase spectrum of each individual channel.[1] This assumption is motivated by the fact that the phase spectrum appears to be uniformly distributed when wrapped to its principal value in $[0, 2\pi)$. Yet, it does have some structure which can be exploited to estimate phase-aware filters or interframe/interband filters, as recently overviewed by Gerkmann *et al.* (2015) and Mowlaee *et al.* (2016).

19.2.1   Phase Reconstruction and Joint Phase-Magnitude Estimation

Griffin and Lim (1984) proposed one of the first algorithms to reconstruct the phase from a magnitude-only spectrum. They formulated this problem as finding the time-domain signal whose magnitude STFT is closest to this magnitude spectrum. Starting from a single-channel zero-phase spectrum $c(n,f)$, they iteratively updated it as follows:

$$c(n,f) \leftarrow |c(n,f)|\angle\text{STFT}(\text{iSTFT}(c))(n,f) \tag{19.6}$$

where iSTFT($\cdot$) denotes the inverse STFT. This update can be more efficiently implemented in the STFT domain (Le Roux *et al.*, 2010) and it can be shown to result in a local minimum of the above optimization problem. This algorithm exploits the so-called *consistency* of complex-valued spectra (Le Roux *et al.*, 2010), i.e. the fact that the STFTs of (real-valued) time-domain signals lie in a lower-dimensional linear subspace of the space of $F \times N$ complex-valued matrices, with F the number of frequency bins and N the number of time frames. In other words, magnitude and phase spectra are deterministically related to each other across frequency bins and, in the case when the STFT analysis windows overlap, also across frames.[2] In a source separation framework, while it can be employed to re-estimate the phase of each estimated source spatial image $\widehat{c}_j(n,f)$ independently of the others, it does not exploit the available complex-valued mixture spectrum $x(n,f)$. Gunawan and Sen (2010) and Sturmel and Daudet (2012) described alternative algorithms to jointly reconstruct the phase spectrum of all sources such that the sum of all $\widehat{c}_j(n,f)$ is closest to $x(n,f)$. Kameoka *et al.* (2009) combined this idea with a nonnegative matrix factorization (NMF) model for the magnitudes. The above algorithms only modify the phase spectra of the sources, therefore they can improve the

---

1 To be precise, the phase of the filtered signal depends on the phase of the individual channels, but the coefficients of the filter depend only on the IPD.
2 When the STFT analysis windows do not overlap, magnitude and phase spectra are not deterministically related to each other any more across frames since the subspace generated by the STFTs of time-domain signals become equal to the Cartesian product of the subspaces generated by individual frames. However, they remain statistically related to each other, as explained in Sections 19.2.2 and 19.2.3.

separation or enhancement performance only when the magnitude spectra have been accurately modeled or estimated.

To improve performance in practical scenarios where estimating the magnitude spectra is difficult, joint estimation of the magnitude and phase spectra is necessary. Le Roux and Vincent (2013) proposed a consistent Wiener filtering algorithm that iteratively estimates the magnitude and phase spectra of all sources. Assuming that the source STFT coefficients are zero-mean complex Gaussian, this is achieved by maximizing the logarithm of the posterior (5.35) under either a hard consistency constraint $c_j(n,f) = \text{STFT}(\text{iSTFT}(c_j))(n,f)$ for all n, f, or a soft penalty term $\sum_{nf} |c_j(n,f) - \text{STFT}(\text{iSTFT}(c_j))(n,f)|^2$. Mowlaee and Saeidi (2013) introduced an alternative algorithm based on a phase-aware magnitude estimator. Starting from an initial phase estimate, they estimated the magnitude via this estimator, updated the phase via (19.6), then re-estimated the magnitude from the updated phase, and so on. Both algorithms require several iterations to converge. More recently, Gerkmann (2014) designed a noniterative joint minimum mean square error (MMSE) estimator of magnitude and phase spectra. The resulting magnitude estimate achieves a tradeoff between phase-blind and phase-aware magnitude estimation, while the resulting phase achieves a tradeoff between the noisy phase and the initial phase estimate.

19.2.2 Interframe and Interband Filtering

So far, we have only addressed the structure of phases due to the consistency property. Yet, phases typically exhibit additional structure that is unveiled by considering phase differences between successive time frames or frequency bands. Figure 19.1 depicts the derivative of phase with respect to time, known as the *instantaneous frequency* (Stark and Paliwal, 2008), and its negative derivative with respect to frequency, known as the *group delay* (Yegnanarayana and Murthy, 1992). For robust computation of these quantities, see Mowlaee *et al.* (2016). Both representations reveal horizontal or vertical structures due to the periodic or transient nature of sounds.

Designing *interframe and/or interband filters* that exploit these phase differences can improve single-channel and multichannel enhancement performance. Interframe minimum variance distortionless response (MVDR) beamformers and multichannel Wiener filters (MWFs) can be designed by stacking N' successive frames

$$\bar{\mathbf{x}}(n,f) = \begin{bmatrix} \mathbf{x}(n,f) \\ \mathbf{x}(n-1,f) \\ \vdots \\ \mathbf{x}(n-N'+1,f) \end{bmatrix} \tag{19.7}$$

and applying the computations in Section 10.4 to $\bar{\mathbf{x}}(n,f)$ instead of $\mathbf{x}(n,f)$ (Avargel and Cohen, 2008; Talmon *et al.*, 2009; Huang and Benesty, 2012; Schasse and Martin, 2014; Fischer and Gerkmann, 2016), where each frame acts as an additional input channel. This implies estimating the interframe covariance matrix of target and interfering sources. Attias (2003) and Kameoka *et al.* (2010) estimated multichannel interframe filters using EM or variational Bayesian (VB) inference instead (see Table 14.1 for other examples). Interband filters can be obtained in a similar way by stacking successive frequency bins (Avargel and Cohen, 2008; Talmon *et al.*, 2009; Huang *et al.*, 2014). Linear prediction-based reverberation cancellation techniques (see Section 15.2)

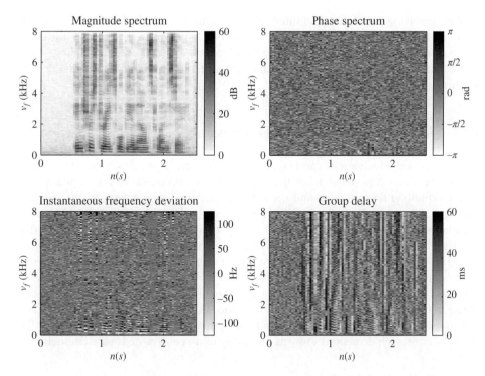

Figure 19.1 Short-term magnitude spectrum and various representations of the phase spectrum of a speech signal for an STFT analysis window size of 64 ms. For easier visualization, the deviation of the instantaneous frequency from the center frequency of each band is shown rather than the instantaneous frequency itself.

are also examples of single- or multichannel interframe filters. In addition to better exploiting the available phase information, another important attribute of these filters is that they can potentially overcome the circular convolution artifacts inherent to the narrowband approximation via subband filtering (2.10) or STFT-domain filtering (2.9).

19.2.3 Phase Models

While they can greatly improve performance in theory, interframe/interband filters are hard to estimate in practice due to the larger number of parameters involved. For instance, the number of entries of interframe covariance matrices grows quadratically with the number N' of stacked frames. To circumvent this issue, Fischer and Gerkmann (2016) considered fixed data-independent interframe coherence matrices trained on a wide range of interference signals. Yet, it is clear that data-dependent parameters are required to benefit from the full potential of such filters. This calls for prior models of phase along time and frequency based on the structure of sounds (see Section 2.2.2). Mowlaee and Saeidi (2013) exploited the fact that the group delay of harmonic sounds is minimum at the harmonics (Yegnanarayana and Murthy, 1992), while Krawczyk and Gerkmann (2014) and Bronson and Depalle (2014) reconstructed the phase of each harmonic assuming a sinusoidal model. Magron *et al.* (2015) used the repetition of musical notes to estimate the phases. Badeau (2011) proposed a probabilistic extension of NMF involving interframe and interband filters that can model both structured and random phases. End-to-end DNNs (see Section 19.1.2) provide a new take on this issue. These studies can be seen as first steps in the direction of better phase modeling. Their improvement and their combination with interframe and interband filtering hold great promise.

19.3 Advancing Multichannel Processing

With the advent of advanced spectral models such as NMF and DNN, accurately estimating the source power spectra or the source presence probabilities is now possible by jointly exploiting these models and the observed mixture. By contrast, accurately estimating their spatial parameters, e.g. relative acoustic transfer functions (RTFs) or spatial covariance matrices, still remains difficult today. Most methods do not rely on any prior constraint over the spatial parameters but on the observed mixture STFT coefficients only. They can provide accurate spatial parameter estimates only when the source power spectra or the source presence probabilities have been accurately estimated in the first place and the number of time frames is large enough. The difficulty is increased when the sources or the microphones are moving, since the spatial parameters vary for each time frame and statistics cannot be robustly computed from a single frame. We describe below two research directions towards designing more constrained spatial models.

19.3.1 Dealing with Moving Sources and Microphones

Methods designed for moving sources and microphones fall into three categories. The first category of methods simply track the spatial parameters over time using an online learning method (see Section 19.5.1 below). This approach was popular in the early days of sparsity-based separation (Rickard *et al.*, 2001; Lösch and Yang, 2009) and frequency domain independent component analysis (FD-ICA) (Mukai *et al.*, 2003; Wehr *et al.*, 2007), and it is still today for beamforming (Affes and Grenier, 1997; Markovich-Golan *et al.*, 2010). The activity pattern of the sources, i.e. sources appearing or disappearing, can also be tracked by looking at the stability of the parameter estimates over time (Markovich-Golan *et al.*, 2010). Although the estimated parameters are time-varying, the amount of variation over time cannot be easily controlled. Sliding block methods estimate the parameters in a given time frame from a small number of neighboring frames without any prior constraint, which intrinsically limits their accuracy. Exponential decay averaging methods can control the amount of variation over time by setting the decay factor, however the relationship between the actual temporal dynamics of the data and the optimal decay factor is not trivial.

A second category of methods track the spatial parameters over time by explicitly modeling their temporal dynamics. Duong *et al.* (2011) set a continuous Markov chain prior on the time-varying spatial covariance matrix of each source $\mathbf{R}_j(n,f)$

$$\mathbf{R}_j(n,f) \sim \mathcal{IW}(\mathbf{R}_j(n,f) \mid (m-I)\mathbf{R}_j(n-1,f), m) \tag{19.8}$$

where $\mathcal{IW}(. \mid \mathbf{\Psi}, m)$ is the *inverse Wishart* distribution over positive definite matrices with inverse scale matrix $\mathbf{\Psi}$ and m degrees of freedom and I is the number of channels. This prior is such that the mean of $\mathbf{R}_j(n,f)$ is equal to $\mathbf{R}_j(n-1,f)$ and the parameter *m* controls the deviation from the mean. The spatial covariance matrices can then be estimated in the maximum a posteriori (MAP) sense by EM, where the M-step involves solving a set of quadratic matrix equations. Kounades-Bastian *et al.* (2016) concatenated the columns of the $I \times J$ time-varying mixing matrix $\mathbf{A}(n,f)$ into an $IJ \times 1$ vector $\bar{\mathbf{a}}(n,f)$ and set a Gaussian continuity prior instead:

$$\bar{\mathbf{a}}(n,f) \sim \mathcal{N}_c(\bar{\mathbf{a}}(n,f) \mid \bar{\mathbf{a}}(n-1,f), \Sigma_{\bar{\mathbf{a}}}(f)). \tag{19.9}$$

They employed a Kalman smoother inside a VB algorithm to estimate the parameters. Such continuity priors help estimating time-varying parameters at low frequencies. Their effectiveness is limited at high frequencies due to fact that the spatial parameters vary much more quickly. Indeed, a small change in the direction of arrival (DOA) results in a large IPD at high frequencies. Also, they do not easily handle appearing or disappearing sources due to the ensuing change of dimension.

The last category of methods rely on time-varying DOAs and activity patterns estimated by a source localization method (see Chapter 4) in order, for example, to build a soft time-frequency mask (Pertilä, 2013) or to adapt a beamformer (Madhu and Martin, 2011; Thiergart *et al.*, 2014). These methods usually employ more sophisticated temporal dynamic models involving, for example, "birth and death" processes and tracking the speed and acceleration of the sources in addition to their position. They allow faster tracking at high frequencies, but they are sensitive to localization errors. Also, they typically exploit the estimated DOAs only, and do not attempt to estimate the deviations of the spatial parameters from their theoretical free-field values. The geometrical constraints reviewed in Chapter 3, which provide a prior model for the deviation of the RTF from the relative steering vector (3.15) and that of the spatial covariance matrix from its mean (3.21), could be used to improve estimation. The integration of DOA-based methods with advanced spectral models and continuity-based methods is also a promising research direction. A first attempt in this direction was made by Higuchi *et al.* (2014).

19.3.2 Manifold Learning

The models in Section 19.3.1 are valid for any recording room. A complementary research trend aims to learn a constrained model for the spatial parameters in a specific room. The problem can be formulated as follows: given a set of acoustic transfer functions sampled for a finite number of source and microphone positions, learn the acoustic properties of the room so as to predict the acoustic transfer function for any other source and microphone position. If the acoustic transfer function could be accurately predicted, source localization and spatial parameter estimation would be unified as a single problem, which would be much easier to solve.

The set of acoustic transfer functions in a given room forms a *manifold*. Although acoustic transfer functions are high-dimensional, they live on a small-dimensional nonlinear subspace parameterized by the positions and the orientations of the source and the microphone. This subspace is continuous: nearby source and microphone positions result in similar acoustic transfer functions. This property extends to RTFs, as illustrated in Figure 19.2. Once again, the continuity is stronger at low frequencies.

A series of studies have attempted to model this manifold. Koldovský *et al.* (2013) predicted the RTF in a given source position by sparse interpolation of RTFs recorded at nearby positions. Mignot *et al.* (2014) showed that accurate interpolation can be

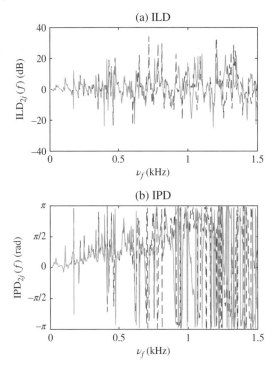

Figure 19.2 Interchannel level difference (ILD) and IPD for two different source positions $j = 1$ (plain curve) and $j = 2$ (dashed curve) 10 cm apart from each other at 1.70 m distance from the microphone pair. The source DOAs are 10° and 13°, respectively. The room size is $8.00 \times 5.00 \times 3.10$ m, the reverberation time is 230 ms, and the microphone distance is 15 cm.

achieved from few samples at low frequencies using a compressed sensing framework that exploits the modal properties of the room, i.e. the spatial frequencies related to the room dimensions. Deleforge *et al.* (2015) introduced a probabilistic piecewise affine mapping model that partitions the space of position and orientation coordinates into regions via a GMM and approximates the RTF within each region as a linear (affine) function of the coordinates, that is similar to the tangent of the manifold. They derived an EM algorithm for jointly learning the GMM and the linear functions from RTF samples. Wang *et al.* (2018) investigated the existence of global dependencies beyond the local tangent structure using a DNN with rectified linear unit activations, which results in a locally linear mapping whose parameters are tied across regions. This DNN outperformed probabilistic piecewise affine mapping and conventional linear interpolation for interpolation distances of 5 cm and beyond, especially at medium to high frequencies, while conventional linear interpolation performed best for shorter distances. Finally, Laufer-Goldshtein *et al.* (2016) demonstrated the limitations of linear approaches to infer physical adjacencies. They defined the diffusion distance related to the geodesic distance on the manifold and demonstrated its ability to arrange the samples according to their DOA and to achieve accurate source localization. This distance also combines local and global properties of the manifold.

While these studies have achieved some success in modeling the manifold from a theoretical point of view and in improving the source localization accuracy, their application to practical source separation and speech enhancement scenarios remains an open issue. Talmon and Gannot (2013) reported a preliminary study of applying these concepts to compute the blocking matrix of a generalized sidelobe canceler

(GSC). Deleforge *et al.* (2015) proposed a VB algorithm for joint source localization and separation by soft time-frequency masking, where both the source positions and the index of the dominant source in each time-frequency bin are considered as hidden data. In order to learn the manifold, a certain number of RTF samples must be recorded in the room. This is feasible in scenarios involving robots (Deleforge *et al.*, 2015), but less easily so in other scenarios. Asaei *et al.* (2014) made a first step towards circumventing this requirement by attempting to jointly estimate the room dimensions and the early part of the room impulse response in an unsupervised fashion from the mixture signal alone. Laufer-Goldshtein *et al.* (2016) and Wang *et al.* (2018) made another step by proposing semi-supervised approaches that require only a few RTF samples to be labeled with the source and microphone positions and orientations.

19.4 Addressing Multiple-Device Scenarios

Except for Chapter 18, the separation and enhancement algorithms reviewed in the previous chapters assume that signal acquisition and processing are concentrated in a single device. In many practical scenarios, however, several devices equipped with processing power, wireless communication capabilities, one or more microphones, and possibly other sensors (e.g., accelerometer, camera, laser rangefinder) are available. With current technology, these devices could be connected to form a *wireless acoustic sensor network*. Cellphones, laptops, and tablets, but also webcams, set-top-boxes, televisions, and assistive robots, are perfect candidates as nodes (or subarrays) of such networks. Compared with classical arrays, wireless acoustic sensor networks typically comprise more microphones and they cover a wider area. This increases the chance that each target source is close to at least one microphone, hence the potential enhancement performance. However, they raise new challenges regarding signal transmission and processing that should be addressed to fully exploit their potential.

19.4.1 Synchronization and Calibration

A first challenge is that the crystal clocks of different devices operate at slightly different frequencies. Therefore, even in the case when the analog-to-digital converters have the same nominal sampling rate, their effective sampling rates differ. The relative deviation ϵ from the nominal sampling rate can be up to $\pm 10^{-4}$. This deviation implies a linear drift ϵt over time that translates into a phase drift of the STFT coefficients $x_i(n,f)$ that is proportional to ϵ and to the signal frequency v_f. Figure 19.3 shows that, even for small ϵ, this quickly leads to large phase shifts at high frequencies which preclude the use of multichannel signal processing. The drift can be measured by exchanging time stamps (Schmalenstroeer *et al.*, 2015) or in a blind way by measuring phase shifts between the network nodes (Markovich-Golan *et al.*, 2012a; Miyabe *et al.*, 2015; Wang and Doclo, 2016; Cherkassky and Gannot, 2017). It can then be compensated by resampling the signals in the time domain or applying the opposite phase shift in the STFT domain.

Once it has been compensated, the signals recorded by different devices still have different temporal offsets, which vary slowly over time. In the case when the sources do not move, these offsets do not affect most source separation and speech enhancement methods that do not rely on a prior model of IPDs across nodes but estimate them

(a) Without sampling rate mismatch

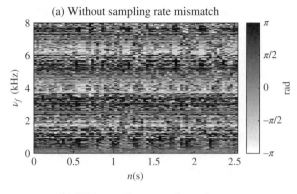

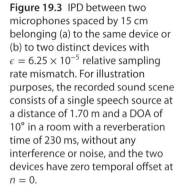

Figure 19.3 IPD between two microphones spaced by 15 cm belonging (a) to the same device or (b) to two distinct devices with $\epsilon = 6.25 \times 10^{-5}$ relative sampling rate mismatch. For illustration purposes, the recorded sound scene consists of a single speech source at a distance of 1.70 m and a DOA of 10° in a room with a reverberation time of 230 ms, without any interference or noise, and the two devices have zero temporal offset at $n = 0$.

(b) With sampling rate mismatch

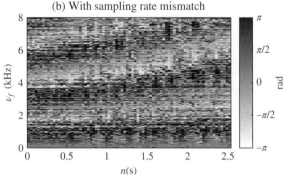

adaptively from the recorded signals. In the case where the sources are moving, tracking their location over time becomes desirable (see Section 19.3.1) and this requires estimating the offsets. The offsets can be estimated in a similar way as the sampling rate deviation from time stamps (Schmalenstroeer *et al.*, 2015) or phase shifts between the network nodes (Pertilä *et al.*, 2013). The latter approach assumes that the sound scene consists of diffuse noise or many sound sources surrounding the array, since the time difference of arrival (TDOA) due to a single localized source cannot be discriminated from the clock offset. Unfortunately, the above algorithms can estimate the temporal offset up to a standard deviation in the order of 0.1 sample to a few samples in the time domain, which is too large for source localization. Whenever precise localization is required, active self-localization algorithms based on emitting a calibration sound are required (Le and Ono, 2017).

Most of the above algorithms address the estimation of sampling rate mismatch and temporal offset between two nodes only. In the case where the number of nodes is larger than two, one arbitrary node is typically chosen as the master and all other nodes are synced to it. Schmalenstroeer *et al.* (2015) showed how to efficiently synchronize all nodes towards a virtual master node, which represents the average clock of all nodes, by exchanging local information between adjacent nodes using a *gossip* algorithm. This algorithm is robust to changes in the network topology, e.g. nodes entering or leaving the network.

Most of the above algorithms assume that the sensors do not move. Blind synchronization of moving sensors remains a challenging topic.

19.4.2   Distributed Algorithms

A second challenge raised by wireless acoustic sensor networks is that processing must be performed at each node without requiring the transmission of all signals to a master node. Markovich-Golan *et al.* (2015) reviewed three families of algorithms for distributed MVDR and linearly constrained minimum variance (LCMV) beamforming that rely on transmitting a compressed or fused version of the signals between neighboring nodes. One such family, called *distributed adaptive node-specific signal estimation* (Bertrand and Moonen, 2010), also allows for distributed implementation of the MWF (Doclo *et al.*, 2009). Several network topologies can be handled, e.g. fully-connected or tree-structured, and the algorithms are shown to deliver equivalent results to centralized processing where a single processor has access to all signals. Efficient adaptation mechanisms can be designed to adapt to changes in the number of available nodes and signals of interest (Markovich-Golan *et al.*, 2012b). Distributed implementations of MVDR and DS beamforming based on message passing, diffusion adaptation or randomized gossip algorithms have also been introduced (Heusdens *et al.*, 2012; O'Connor and Kleijn, 2014; Zeng and Hendriks, 2014). Gaubitch *et al.* (2014) described a practical setup with smartphones.

Aside from these advances on distributed beamforming, the distributed implementation of other families of speech enhancement and source separation methods remains an open problem. Souden *et al.* (2014) and Dorfan *et al.* (2015) made a step towards that goal by proposing distributed clustering schemes integrating intra- and internode location features for speech separation in wireless acoustic sensor networks.

19.4.3   Multimodal Source Separation and Enhancement

Although we have focused on audio modality, one must bear in mind that microphones are often embedded in devices equipped with other sensors. We have seen in Section 17.5 how video can be used to detect, localize, and extract relevant features for speech enhancement and separation. The application of these ideas to other sounds besides speech is a promising research direction: audiovisual processing is still in its infancy for music or environmental sound scenarios and it has been applied to related tasks only so far, e.g. (Joly *et al.* 2016; Dinesh *et al.* 2017). Multimodal separation and enhancement using other sensors, e.g. electroencephalogram (Das *et al.*, 2016), accelerometers (Zohourian and Martin, 2016), or laser rangefinders, remains largely untouched despite its great promise.

19.5   Towards Widespread Commercial Use

Today, source separation and speech enhancement technology can be found in smartphones, hearing aids, and voice command systems. Their extension to new application scenarios raises several research issues.

19.5.1   Practical Deployment Constraints

Generally speaking, commercial source separation and speech enhancement systems are expected to incur a small memory footprint and a low computational cost. Noniterative learning-free algorithms such as various forms of beamforming score well according to these two criteria, hence their popularity in hearing aids (see Chapter 18). With the advent of more powerful, energy-efficient storage and processors, learning-based algorithms are becoming feasible in an increasing number of scenarios. For instance, Virtanen *et al.* (2013) evaluated the complexity of several NMF algorithms with various dictionary sizes. Efforts are also underway to reduce the complexity of DNNs, which are appealing compared to NMF due to the noniterative nature of the forward pass. Kim and Smaragdis (2015) devised a bitwise neural network architecture, which showcases a comparable sound quality to a comprehensive real-valued DNN while spending significantly less memory and power.

In addition to the above constraints, several scenarios require processing the input signal in an online manner, i.e. without using future values of the mixture signal to estimate the source signals at a given time, and with low latency. Two strategies can be adopted to process the input signal as a stream: processing sliding blocks of data consisting of a few time frames (Mukai *et al.*, 2003; Joder *et al.*, 2012) or using exponential decay averaging (Gannot *et al.*, 1998; Rickard *et al.*, 2001; Lefèvre *et al.*, 2011; Schwartz *et al.*, 2015). Simon and Vincent (2012) combined both approaches into a single algorithm in the case of multichannel NMF. Whenever a very low latency is required, filtering the data in the time domain helps (Sunohara *et al.*, 2017).

19.5.2 Quality Assessment

In order to assess the suitability of a separation or enhancement algorithm for a given application scenario and to keep improving it by extensive testing, accurate quality metrics are required. The metrics used today, as reviewed in Section 1.2.6, correlate with sound quality and speech intelligibility only to a limited extent. The composite metric of Loizou (2007) and the perceptually-motivated metrics of Emiya *et al.* (2011) improved the correlation by training a linear regressor or a neural network on subjective quality scores collected from human subjects, but they suffer from the limited amount of subjective scores available for training today. Crowdsourcing is a promising way of collecting subjective scores for more audio data and from more subjects (Cartwright *et al.*, 2016) and increasing the accuracy of such machine learning-based quality metrics. Another issue is that sound quality and speech intelligibility are perceived differently by hearing-impaired vs. normal-hearing individuals, but also by different normal-hearing individuals (Emiya *et al.*, 2011). User-dependent quality metrics are an interesting research direction. Finally, most sound quality metrics are intrusive, in the sense that they require the true target signal in addition to the estimated signal. Developing nonintrusive quality metrics is essential to assess performance in many real scenarios where the true target signal is not available.

19.5.3 New Application Areas

Besides the applications to speech and music reviewed in Chapters 16, 17, and 18, source separation and speech enhancement are being applied to an increasing range of application scenarios, such as enhancing the intelligibility of spoken dialogue in television broadcasts (Geiger *et al.*, 2015), reducing the ego-noise due to fans and actuators in assistive robots (Ince *et al.*, 2009), rendering real sound scenes recorded via a microphone array in 3D over headphones in the context of virtual reality or augmented reality (Nikunen *et al.*, 2016), and encoding a recording into parametric source signals (Liutkus *et al.*, 2013) or sound objects (Vincent and Plumbley, 2007) for audio upmixing and remixing purposes. Some of these topics were studied many years ago for the first time but are still active today due to the lack of a fully satisfactory solution. Source separation is also useful every time one must analyze a sound scene consisting of multiple sources. Examples include recognizing overlapped environmental sound events (Heittola *et al.*, 2013), monitoring traffic (Toyoda *et al.*, 2016), and controlling the noise disturbance of wind turbines (Dumortier *et al.*, 2017). These and other emerging application areas will further increase the widespread commercial use of source separation and speech enhancement technology.

Acknowledgment

We thank T. Gerkmann and R. Badeau for their input to the second section of this chapter.

Bibliography

Affes, S. and Grenier, Y. (1997) A signal subspace tracking algorithm for microphone array processing of speech. *IEEE Transactions on Speech and Audio Processing*, **5** (5), 425–437.

Asaei, A., Golbabaee, M., Bourlard, H., and Cevher, V. (2014) Structured sparsity models for reverberant speech separation. *IEEE/ACM Transactions on Audio, Speech, and Language Processing*, **22** (3), 620–633.

Attias, H. (2003) New EM algorithms for source separation and deconvolution with a microphone array, in *Proceedings of IEEE International Conference on Audio, Speech and Signal Processing*, vol. V, pp. 297–300.

Avargel, Y. and Cohen, I. (2008) Adaptive system identification in the short-time Fourier transform domain using cross-multiplicative transfer function approximation. *IEEE Transactions on Audio, Speech, and Language Processing*, **16** (1), 162–173.

Badeau, R. (2011) Gaussian modeling of mixtures of non-stationary signals in the time-frequency domain (HR-NMF), in *Proceedings of IEEE Workshop on Applications of Signal Processing to Audio and Acoustics*, pp. 253–256.

Bertrand, A. and Moonen, M. (2010) Distributed adaptive node-specific signal estimation in fully connected sensor networks — part i: Sequential node updating. *IEEE Transactions on Signal Processing*, **58** (10), 5277–5291.

Bronson, J. and Depalle, P. (2014) Phase constrained complex NMF: Separating overlapping partials in mixtures of harmonic musical sources, in *Proceedings of IEEE International Conference on Audio, Speech and Signal Processing*, pp. 7475–7479.

Cartwright, M., Pardo, B., Mysore, G.J., and Hoffman, M. (2016) Fast and easy crowdsourced perceptual audio evaluation, in *Proceedings of IEEE International Conference on Audio, Speech and Signal Processing*, pp. 619–623.

Chazan, S.E., Goldberger, J., and Gannot, S. (2016) A hybrid approach for speech enhancement using MoG model and neural network phoneme classifier. *IEEE/ACM Transactions on Audio, Speech, and Language Processing*, **24** (12), 2516–2530.

Cherkassky, D. and Gannot, S. (2017) Blind synchronization in wireless acoustic sensor networks. *IEEE/ACM Transactions on Audio, Speech, and Language Processing*, **25** (3), 651–661.

Das, N., Van Eyndhoven, S., Francart, T., and Bertrand, A. (2016) Adaptive attention-driven speech enhancement for EEG-informed hearing prostheses, in *Proceedings of Annual International Conference of the IEEE Engineering in Medicine and Biology Society*, pp. 77–80.

Deleforge, A., Forbes, F., and Horaud, R. (2015) Acoustic space learning for sound-source separation and localization on binaural manifolds. *International Journal of Neural Systems*, **25** (1). 1440003.

Dinesh, K., Li, B., Liu, X., Duan, Z., and Sharma, G. (2017) Visually informed multi-pitch analysis of string ensembles, in *Proceedings of IEEE International Conference on Audio, Speech and Signal Processing*.

Doclo, S., Moonen, M., Van den Bogaert, T., and Wouters, J. (2009) Reduced-bandwidth and distributed MWF-based noise reduction algorithms for binaural hearing aids. *IEEE Transactions on Audio, Speech, and Language Processing*, **17** (1), 38–51.

Dorfan, Y., Cherkassky, D., and Gannot, S. (2015) Speaker localization and separation using distributed expectation-maximization, in *Proceedings of European Signal Processing Conference*, pp. 1256–1260.

Dumortier, B., Vincent, E., and Deaconu, M. (2017) Recursive Bayesian estimation of the acoustic noise emitted by wind farms, in *Proceedings of IEEE International Conference on Audio, Speech and Signal Processing*.

Duong, N.Q.K., Tachibana, H., Vincent, E., Ono, N., Gribonval, R., and Sagayama, S. (2011) Multichannel harmonic and percussive component separation by joint modeling of spatial and spectral continuity, in *Proceedings of IEEE International Conference on Audio, Speech and Signal Processing*, pp. 205–208.

Emiya, V., Vincent, E., Harlander, N., and Hohmann, V. (2011) Subjective and objective quality assessment of audio source separation. *IEEE Transactions on Audio, Speech, and Language Processing*, **19** (7), 2046–2057.

Fischer, D. and Gerkmann, T. (2016) Single-microphone speech enhancement using MVDR filtering and Wiener post-filtering, in *Proceedings of IEEE International Conference on Audio, Speech and Signal Processing*, pp. 201–205.

Gannot, S., Burshtein, D., and Weinstein, E. (1998) Iterative and sequential Kalman filter-based speech enhancement algorithms. *IEEE Transactions on Speech and Audio Processing*, **6** (4), 373–385.

Gaubitch, N.D., Martinez, J., Kleijn, W.B., and Heusdens, R. (2014) On near-field beamforming with smartphone-based ad-hoc microphone arrays, in *Proceedings of International Workshop on Acoustic Echo and Noise Control*, pp. 94–98.

Geiger, J.T., Grosche, P., and Lacouture Parodi, Y. (2015) Dialogue enhancement of stereo sound, in *Proceedings of European Signal Processing Conference*, pp. 874–878.

Gerkmann, T. (2014) Bayesian estimation of clean speech spectral coefficients given a priori knowledge of the phase. *IEEE Transactions on Signal Processing*, **62** (16), 4199–4208.

Gerkmann, T., Krawczyk-Becker, M., and Le Roux, J. (2015) Phase processing for single-channel speech enhancement: History and recent advances. *IEEE Signal Processing Magazine*, **32** (2), 55–66.

Griffin, D.W. and Lim, J.S. (1984) Signal estimation from modified short-time Fourier transform. *IEEE Transactions on Acoustics, Speech, and Signal Processing*, **32** (2), 236–243.

Gunawan, D. and Sen, D. (2010) Iterative phase estimation for the synthesis of separated sources from single-channel mixtures. *IEEE Signal Processing Letters*, **17** (5), 421–424.

Heittola, T., Mesaros, A., Virtanen, T., and Gabbouj, M. (2013) Supervised model training for overlapping sound events based on unsupervised source separation, in *Proceedings of IEEE International Conference on Audio, Speech and Signal Processing*, pp. 8677–8681.

Hershey, J.R., Chen, Z., Le Roux, J., and Watanabe, S. (2016) Deep clustering: Discriminative embeddings for segmentation and separation, in *Proceedings of IEEE International Conference on Audio, Speech and Signal Processing*, pp. 31–35.

Heusdens, R., Zhang, G., Hendriks, R.C., Zeng, Y., and Kleijn, W.B. (2012) Distributed MVDR beamforming for (wireless) microphone networks using message passing, in *Proceedings of International Workshop on Acoustic Echo and Noise Control*, pp. 1–4.

Heymann, J., Drude, L., and Haeb-Umbach, R. (2016) Neural network based spectral mask estimation for acoustic beamforming, in *Proceedings of IEEE International Conference on Audio, Speech and Signal Processing*, pp. 196–200.

Higuchi, T., Takamune, N., Nakamura, T., and Kameoka, H. (2014) Underdetermined blind separation and tracking of moving sources based on DOA-HMM, in *Proceedings of IEEE International Conference on Audio, Speech and Signal Processing*, pp. 3215–3219.

Hu, K. and Wang, D.L. (2013) An unsupervised approach to cochannel speech separation. *IEEE Transactions on Audio, Speech, and Language Processing*, **21** (1), 122–131.

Huang, H., Zhao, L., Chen, J., and Benesty, J. (2014) A minimum variance distortionless response filter based on the bifrequency spectrum for single-channel noise reduction. *Digital Signal Processing*, **33**, 169–179.

Huang, Y. and Benesty, J. (2012) A multi-frame approach to the frequency-domain single-channel noise reduction problem. *IEEE Transactions on Audio, Speech, and Language Processing*, **20** (4), 1256–1269.

Ince, G., Nakadai, K., Rodemann, T., Hasegawa, Y., Tsujino, H., and Imura, J. (2009) Ego noise suppression of a robot using template subtraction, in *Proceedings of IEEE/RSJ International Conference on Intelligent Robots and Systems*, pp. 199–204.

Isik, Y., Le Roux, J., Chen, Z., Watanabe, S., and Hershey, J.R. (2016) Single-channel multi-speaker separation using deep clustering, in *Proceedings of Interspeech*, pp. 545–549.

Jaureguiberry, X., Vincent, E., and Richard, G. (2016) Fusion methods for speech enhancement and audio source separation. *IEEE/ACM Transactions on Audio, Speech, and Language Processing*, **24** (7), 1266–1279.

Joder, C., Weninger, F., Eyben, F., Virette, D., and Schuller, B. (2012) Real-time speech separation by semi-supervised nonnegative matrix factorization, in *Proceedings of International Conference on Latent Variable Analysis and Signal Separation*, pp. 322–329.

Joly, A., Goëau, H., Glotin, H., Spampinato, C., Bonnet, P., Vellinga, W.P., Champ, J., Planqué, R., Palazzo, S., and Müller, H. (2016) LifeCLEF 2016: Multimedia life species identification challenges, in *Proceedings of International Conference of the CLEF Association*, pp. 286–310.

Kameoka, H., Ono, N., Kashino, K., and Sagayama, S. (2009) Complex NMF: A new sparse representation for acoustic signals, in *Proceedings of IEEE International Conference on Audio, Speech and Signal Processing*, pp. 3437–3440.

Kameoka, H., Yoshioka, T., Hamamura, M., Le Roux, J., and Kashino, K. (2010) Statistical model of speech signals based on composite autoregressive system with application to blind source separation, in *Proceedings of International Conference on Latent Variable Analysis and Signal Separation*, pp. 245–253.

Kato, A. and Milner, B. (2016) HMM-based speech enhancement using sub-word models and noise adaptation, in *Proceedings of Interspeech*, pp. 3748–3752.

Kim, M. and Smaragdis, P. (2015) Bitwise neural networks, in *Proceedings of International Conference on Machine Learning Workshop on Resource-Efficient Machine Learning*.

Kolbæk, M., Tan, Z.H., and Jensen, J. (2017) Speech intelligibility potential of general and specialized deep neural network based speech enhancement systems. *IEEE/ACM Transactions on Audio, Speech, and Language Processing*, **25** (1), 149–163.

Koldovský, Z., Málek, J., Tichavský, P., and Nesta, F. (2013) Semi-blind noise extraction using partially known position of the target source. *IEEE Transactions on Audio, Speech, and Language Processing*, **21** (10), 2029–2041.

Kounades-Bastian, D., Girin, L., Alameda-Pineda, X., Gannot, S., and Horaud, R. (2016) A variational EM algorithm for the separation of time-varying convolutive audio mixtures. *IEEE/ACM Transactions on Audio, Speech, and Language Processing*, **24** (8), 1408–1423.

Krawczyk, M. and Gerkmann, T. (2014) STFT phase reconstruction in voiced speech for an improved single-channel speech enhancement. *IEEE/ACM Transactions on Audio, Speech, and Language Processing*, **22** (12), 1931–1940.

Laufer-Goldshtein, B., Talmon, R., and Gannot, S. (2016) Semi-supervised sound source localization based on manifold regularization. *IEEE/ACM Transactions on Audio, Speech, and Language Processing*, **24** (8), 1393–1407.

Le, T.K. and Ono, N. (2017) Closed-form and near closed-form solutions for TDOA-based joint source and sensor localization. *IEEE/ACM Transactions on Audio, Speech, and Language Processing*, **65** (5), 1207–1221.

Le Roux, J., Kameoka, H., Ono, N., and Sagayama, S. (2010) Fast signal reconstruction from magnitude STFT spectrogram based on spectrogram consistency, in *Proceedings of International Conference on Digital Audio Effects*, pp. 1–7.

Le Roux, J. and Vincent, E. (2013) Consistent Wiener filtering for audio source separation. *IEEE Signal Processing Letters*, **20** (3), 217–220.

Lefèvre, A., Bach, F., and Févotte, C. (2011) Online algorithms for nonnegative matrix factorization with the Itakura-Saito divergence, in *Proceedings of IEEE Workshop on Applications of Signal Processing to Audio and Acoustics*, pp. 313–316.

Li, B., Sainath, T.N., Weiss, R.J., Wilson, K.W., and Bacchiani, M. (2016) Neural network adaptive beamforming for robust multichannel speech recognition, in *Proceedings of Interspeech*, pp. 1976–1979.

Liutkus, A., Durrieu, J.L., Daudet, L., and Richard, G. (2013) An overview of informed audio source separation, in *Proceedings of International Workshop on Image Analysis for Multimedia Interactive Services*, pp. 1–4.

Loizou, P.C. (2007) *Speech Enhancement: Theory and Practice*, CRC Press.

Lösch, B. and Yang, B. (2009) Online blind source separation based on time-frequency sparseness, in *Proceedings of IEEE International Conference on Audio, Speech and Signal Processing*, pp. 117–120.

Madhu, N. and Martin, R. (2011) A versatile framework for speaker separation using a model-based speaker localization approach. *IEEE Transactions on Audio, Speech, and Language Processing*, **19** (7), 1900–1912.

Magron, P., Badeau, R., and David, B. (2015) Phase reconstruction of spectrograms based on a model of repeated audio events, in *Proceedings of IEEE Workshop on Applications of Signal Processing to Audio and Acoustics*, pp. 1–5.

Markovich-Golan, S., Bertrand, A., Moonen, M., and Gannot, S. (2015) Optimal distributed minimum-variance beamforming approaches for speech enhancement in wireless acoustic sensor networks. *Signal Processing*, **107**, 4–20.

Markovich-Golan, S., Gannot, S., and Cohen, I. (2010) Subspace tracking of multiple sources and its application to speakers extraction, in *Proceedings of IEEE International Conference on Audio, Speech and Signal Processing*, pp. 201–204.

Markovich-Golan, S., Gannot, S., and Cohen, I. (2012a) Blind sampling rate offset estimation and compensation in wireless acoustic sensor networks with application to beamforming, in *Proceedings of International Workshop on Acoustic Echo and Noise Control*.

Markovich-Golan, S., Gannot, S., and Cohen, I. (2012b) Low-complexity addition or removal of sensors/constraints in LCMV beamformers. *IEEE Transactions on Signal Processing*, **60** (3), 1205–1214.

Mignot, R., Chardon, G., and Daudet, L. (2014) Low frequency interpolation of room impulse responses using compressed sensing. *IEEE/ACM Transactions on Audio, Speech, and Language Processing*, **22** (1), 205–216.

Miyabe, S., Ono, N., and Makino, S. (2015) Blind compensation of interchannel sampling frequency mismatch for ad hoc microphone array based on maximum likelihood estimation. *Signal Processing*, **107**, 185–196.

Mowlaee, P. and Saeidi, R. (2013) Iterative closed-loop phase-aware single-channel speech enhancement. *IEEE Signal Processing Letters*, **20** (12), 1235–1239.

Mowlaee, P., Saeidi, R., and Stylianou, Y. (2016) Advances in phase-aware signal processing in speech communication. *Speech Communication*, **81**, 1–29.

Mukai, R., Sawada, H., Araki, S., and Makino, S. (2003) Robust real-time blind source separation for moving speakers in a room, in *Proceedings of IEEE International Conference on Audio, Speech and Signal Processing*, pp. V–469–472.

Nickel, R.M., Astudillo, R.F., Kolossa, D., and Martin, R. (2013) Corpus-based speech enhancement with uncertainty modeling and cepstral smoothing. *IEEE Transactions on Audio, Speech, and Language Processing*, **21** (5), 983–997.

Nikunen, J., Diment, A., Virtanen, T., and Vilermo, M. (2016) Binaural rendering of microphone array captures based on source separation. *Speech Communication*, **76**, 157–169.

Nugraha, A.A., Liutkus, A., and Vincent, E. (2016a) Multichannel audio source separation with deep neural networks. *IEEE/ACM Transactions on Audio, Speech, and Language Processing*, **24** (10), 1652–1664.

Nugraha, A.A., Liutkus, A., and Vincent, E. (2016b) Multichannel music separation with deep neural networks, in *Proceedings of European Signal Processing Conference*, pp. 1748–1752.

O'Connor, M. and Kleijn, W.B. (2014) Diffusion-based distributed MVDR beamformer, in *Proceedings of IEEE International Conference on Audio, Speech and Signal Processing*, pp. 810–814.

Pertilä, P. (2013) Online blind speech separation using multiple acoustic speaker tracking and time-frequency masking. *Computer Speech and Language*, **27** (3), 683–702.

Pertilä, P., Hämäläinen, M.S., and Mieskolainen, M. (2013) Passive temporal offset estimation of multichannel recordings of an ad-hoc microphone array. *IEEE/ACM Transactions on Audio, Speech, and Language Processing*, **21** (11), 2393–2402.

Rickard, S.J., Balan, R.V., and Rosca, J.P. (2001) Real-time time-frequency based blind source separation, in *Proceedings of International Conference on Independent Component Analysis and Signal Separation*, pp. 421–426.

Schasse, A. and Martin, R. (2014) Estimation of subband speech correlations for noise reduction via MVDR processing. *IEEE/ACM Transactions on Audio, Speech, and Language Processing*, **22** (9), 1355–1365.

Schmalenstroeer, J., Jebramcik, P., and Haeb-Umbach, R. (2015) A combined hardware–software approach for acoustic sensor network synchronization. *Signal Processing*, **107**, 171–184.

Schwartz, B., Gannot, S., and Habets, E.A.P. (2015) On-line speech dereverberation using Kalman filter and EM algorithm. *IEEE/ACM Transactions on Audio, Speech, and Language Processing*, pp. 394–406.

Shivakumar, P.G. and Georgiou, P. (2016) Perception optimized deep denoising autoencoders for speech enhancement, in *Proceedings of Interspeech*, pp. 3743–3747.

Simon, L.S.R. and Vincent, E. (2012) A general framework for online audio source separation, in *Proceedings of International Conference on Latent Variable Analysis and Signal Separation*, pp. 397–404.

Sivasankaran, S., Vincent, E., and Illina, I. (2017) Discriminative importance weighting of augmented training data for acoustic model training, in *Proceedings of IEEE International Conference on Audio, Speech and Signal Processing*.

Souden, M., Kinoshita, K., Delcroix, M., and Nakatani, T. (2014) Location feature integration for clustering-based speech separation in distributed microphone arrays. *IEEE/ACM Transactions on Audio, Speech, and Language Processing*, **22** (2), 354–367.

Stark, A.P. and Paliwal, K.K. (2008) Speech analysis using instantaneous frequency deviation, in *Proceedings of Interspeech*, pp. 2602–2605.

Sturmel, N. and Daudet, L. (2012) Iterative phase reconstruction of Wiener filtered signals, in *Proceedings of IEEE International Conference on Audio, Speech and Signal Processing*, pp. 101–104.

Sunohara, M., Haruta, C., and Ono, N. (2017) Low-latency real-time blind source separation for hearing aids based on time-domain implementation of online independent vector analysis with truncation of non-causal components, in *Proceedings of IEEE International Conference on Audio, Speech and Signal Processing*.

Talmon, R., Cohen, I., and Gannot, S. (2009) Convolutive transfer function generalized sidelobe canceler. *IEEE Transactions on Audio, Speech, and Language Processing*, **17** (7), 1420–1434.

Talmon, R. and Gannot, S. (2013) Relative transfer function identification on manifolds for supervised GSC beamformers, in *Proceedings of European Signal Processing Conference*, pp. 1–5.

Thiergart, O., Taseska, M., and Habets, E.A.P. (2014) An informed parametric spatial filter based on instantaneous direction-of-arrival estimates. *IEEE/ACM Transactions on Audio, Speech, and Language Processing*, **22** (12), 2182–2196.

Toyoda, T., Ono, N., Miyabe, S., Yamada, T., and Makino, S. (2016) Vehicle counting and lane estimation with ad-hoc microphone array in real road environments, in *Proceedings of International Workshop on Nonlinear Circuits, Communications and Signal Processing*, pp. 622–625.

van den Oord, A., Dieleman, S., Zen, H., Simonyan, K., Vinyals, O., Graves, A., Kalchbrenner, N., Senior, A., and Kavukcuoglu, K. (2016) Wavenet: A generative model for raw audio, arXiv:1609.03499.

Vincent, E. and Plumbley, M.D. (2007) Low bit-rate object coding of musical audio using Bayesian harmonic models. *IEEE Transactions on Audio, Speech, and Language Processing*, **15** (4), 1273–1282.

Vincent, E., Watanabe, S., Nugraha, A.A., Barker, J., and Marxer, R. (2017) An analysis of environment, microphone and data simulation mismatches in robust speech recognition. *Computer Speech and Language*, **46**, 535–557.

Virtanen, T., Gemmeke, J.F., and Raj, B. (2013) Active-set Newton algorithm for overcomplete non-negative representations of audio. *IEEE Transactions on Audio, Speech, and Language Processing*, **21** (11), 2277–2289.

Wang, L. and Doclo, S. (2016) Correlation maximization based sampling rate offset estimation for distributed microphone arrays. *IEEE/ACM Transactions on Audio, Speech, and Language Processing*, **24** (3), 571–582.

Wang, Z., Li, J., Yan, Y., and Vincent, E. (2018) Semi-supervised learning with deep neural networks for relative transfer function inverse regression, in *Proceedings of IEEE International Conference on Audio, Speech and Signal Processing*.

Wehr, S., Lombard, A., Buchner, H., and Kellermann, W. (2007) "Shadow BSS" for blind source separation in rapidly time-varying acoustic scenes, in *Proceedings of International Conference on Independent Component Analysis and Signal Separation*, pp. 560–568.

Xiao, X., Xu, C., Zhang, Z., Zhao, S., Sun, S., Watanabe, S., Wang, L., Xie, L., Jones, D.L., Chng, E.S., and Li, H. (2016) A study of learning based beamforming methods for speech recognition, in *Proceedings of International Workshop on Speech Processing in Everyday Environments*, pp. 26–31.

Yegnanarayana, B. and Murthy, H. (1992) Significance of group delay functions in spectrum estimation. *IEEE Transactions on Signal Processing*, **40** (9), 2281–2289.

Yu, D., Kolbæk, M., Tan, Z.H., and Jensen, J. (2016) Permutation invariant training of deep models for speaker-independent multi-talker speech separation, arXiv:1607.00325.

Zagoruyko, S. and Komodakis, N. (2016) Wide residual networks, arXiv:1605.07146.

Zeng, Y. and Hendriks, R. (2014) Distributed delay and sum beamformer for speech enhancement via randomized gossip. *IEEE/ACM Transactions on Audio, Speech, and Language Processing*, **22** (1), 260–273.

Zhang, X., Zhang, H., Nie, S., Gao, G., and Liu, W. (2016) A pairwise algorithm using the deep stacking network for speech separation and pitch estimation. *IEEE/ACM Transactions on Audio, Speech, and Language Processing*, **24** (6), 1066–1078.

Zohourian, M. and Martin, R. (2016) Binaural speaker localization and separation based on a joint ITD/ILD model and head movement tracking, in *Proceedings of IEEE International Conference on Audio, Speech and Signal Processing*, pp. 430–434.

Index